Pharmaceutical Emulsions and Suspensions

DRUGS AND THE PHARMACEUTICAL SCIENCES

DRUGS AND THE PHARMACEUTICAL SCIENCES

A Series of Textbooks and Monographs

1. Pharmacokinetics, *Milo Gibaldi and Donald Perrier*
2. Good Manufacturing Practices for Pharmaceuticals: A Plan for Total Quality Control, *Sidney H. Willig, Murray M. Tuckerman, and William S. Hitchings IV*
3. Microencapsulation, *edited by J. R. Nixon*
4. Drug Metabolism: Chemical and Biochemical Aspects, *Bernard Testa and Peter Jenner*
5. New Drugs: Discovery and Development, *edited by Alan A. Rubin*
6. Sustained and Controlled Release Drug Delivery Systems, *edited by Joseph R. Robinson*
7. Modern Pharmaceutics, *edited by Gilbert S. Banker and Christopher T. Rhodes*
8. Prescription Drugs in Short Supply: Case Histories, *Michael A. Schwartz*
9. Activated Charcoal: Antidotal and Other Medical Uses, *David O. Cooney*
10. Concepts in Drug Metabolism (in two parts), *edited by Peter Jenner and Bernard Testa*
11. Pharmaceutical Analysis: Modern Methods (in two parts), *edited by James W. Munson*
12. Techniques of Solubilization of Drugs, *edited by Samuel H. Yalkowsky*
13. Orphan Drugs, *edited by Fred E. Karch*
14. Novel Drug Delivery Systems: Fundamentals, Developmental Concepts, Biomedical Assessments, *Yie W. Chien*
15. Pharmacokinetics: Second Edition, Revised and Expanded, *Milo Gibaldi and Donald Perrier*
16. Good Manufacturing Practices for Pharmaceuticals: A Plan for Total Quality Control, Second Edition, Revised and Expanded, *Sidney H. Willig, Murray M. Tuckerman, and William S. Hitchings IV*
17. Formulation of Veterinary Dosage Forms, *edited by Jack Blodinger*
18. Dermatological Formulations: Percutaneous Absorption, *Brian W. Barry*
19. The Clinical Research Process in the Pharmaceutical Industry, *edited by Gary M. Matoren*
20. Microencapsulation and Related Drug Processes, *Patrick B. Deasy*
21. Drugs and Nutrients: The Interactive Effects, *edited by Daphne A. Roe and T. Colin Campbell*

22. Biotechnology of Industrial Antibiotics, *Erick J. Vandamme*
23. Pharmaceutical Process Validation, *edited by Bernard T. Loftus and Robert A. Nash*
24. Anticancer and Interferon Agents: Synthesis and Properties, *edited by Raphael M. Ottenbrite and George B. Butler*
25. Pharmaceutical Statistics: Practical and Clinical Applications, *Sanford Bolton*
26. Drug Dynamics for Analytical, Clinical, and Biological Chemists, *Benjamin J. Gudzinowicz, Burrows T. Younkin, Jr., and Michael J. Gudzinowicz*
27. Modern Analysis of Antibiotics, *edited by Adjoran Aszalos*
28. Solubility and Related Properties, *Kenneth C. James*
29. Controlled Drug Delivery: Fundamentals and Applications, Second Edition, Revised and Expanded, *edited by Joseph R. Robinson and Vincent H. Lee*
30. New Drug Approval Process: Clinical and Regulatory Management, *edited by Richard A. Guarino*
31. Transdermal Controlled Systemic Medications, *edited by Yie W. Chien*
32. Drug Delivery Devices: Fundamentals and Applications, *edited by Praveen Tyle*
33. Pharmacokinetics: Regulatory • Industrial • Academic Perspectives, *edited by Peter G. Welling and Francis L. S. Tse*
34. Clinical Drug Trials and Tribulations, *edited by Allen E. Cato*
35. Transdermal Drug Delivery: Developmental Issues and Research Initiatives, *edited by Jonathan Hadgraft and Richard H. Guy*
36. Aqueous Polymeric Coatings for Pharmaceutical Dosage Forms, *edited by James W. McGinity*
37. Pharmaceutical Pelletization Technology, *edited by Isaac Ghebre-Sellassie*
38. Good Laboratory Practice Regulations, *edited by Allen F. Hirsch*
39. Nasal Systemic Drug Delivery, *Yie W. Chien, Kenneth S. E. Su, and Shyi-Feu Chang*
40. Modern Pharmaceutics: Second Edition, Revised and Expanded, *edited by Gilbert S. Banker and Christopher T. Rhodes*
41. Specialized Drug Delivery Systems: Manufacturing and Production Technology, *edited by Praveen Tyle*
42. Topical Drug Delivery Formulations, *edited by David W. Osborne and Anton H. Amann*
43. Drug Stability: Principles and Practices, *Jens T. Carstensen*
44. Pharmaceutical Statistics: Practical and Clinical Applications, Second Edition, Revised and Expanded, *Sanford Bolton*
45. Biodegradable Polymers as Drug Delivery Systems, *edited by Mark Chasin and Robert Langer*
46. Preclinical Drug Disposition: A Laboratory Handbook, *Francis L. S. Tse and James J. Jaffe*

47. HPLC in the Pharmaceutical Industry, *edited by Godwin W. Fong and Stanley K. Lam*
48. Pharmaceutical Bioequivalence, *edited by Peter G. Welling, Francis L. S. Tse, and Shrikant V. Dinghe*
49. Pharmaceutical Dissolution Testing, *Umesh V. Banakar*
50. Novel Drug Delivery Systems: Second Edition, Revised and Expanded, *Yie W. Chien*
51. Managing the Clinical Drug Development Process, *David M. Cocchetto and Ronald V. Nardi*
52. Good Manufacturing Practices for Pharmaceuticals: A Plan for Total Quality Control, Third Edition, *edited by Sidney H. Willig and James R. Stoker*
53. Prodrugs: Topical and Ocular Drug Delivery, *edited by Kenneth B. Sloan*
54. Pharmaceutical Inhalation Aerosol Technology, *edited by Anthony J. Hickey*
55. Radiopharmaceuticals: Chemistry and Pharmacology, *edited by Adrian D. Nunn*
56. New Drug Approval Process: Second Edition, Revised and Expanded, *edited by Richard A. Guarino*
57. Pharmaceutical Process Validation: Second Edition, Revised and Expanded, *edited by Ira R. Berry and Robert A. Nash*
58. Ophthalmic Drug Delivery Systems, *edited by Ashim K. Mitra*
59. Pharmaceutical Skin Penetration Enhancement, *edited by Kenneth A. Walters and Jonathan Hadgraft*
60. Colonic Drug Absorption and Metabolism, *edited by Peter R. Bieck*
61. Pharmaceutical Particulate Carriers: Therapeutic Applications, *edited by Alain Rolland*
62. Drug Permeation Enhancement: Theory and Applications, *edited by Dean S. Hsieh*
63. Glycopeptide Antibiotics, *edited by Ramakrishnan Nagarajan*
64. Achieving Sterility in Medical and Pharmaceutical Products, *Nigel A. Halls*
65. Multiparticulate Oral Drug Delivery, *edited by Isaac Ghebre-Sellassie*
66. Colloidal Drug Delivery Systems, *edited by Jörg Kreuter*
67. Pharmacokinetics: Regulatory • Industrial • Academic Perspectives, Second Edition, *edited by Peter G. Welling and Francis L. S. Tse*
68. Drug Stability: Principles and Practices, Second Edition, Revised and Expanded, *Jens T. Carstensen*
69. Good Laboratory Practice Regulations: Second Edition, Revised and Expanded, *edited by Sandy Weinberg*
70. Physical Characterization of Pharmaceutical Solids, *edited by Harry G. Brittain*
71. Pharmaceutical Powder Compaction Technology, *edited by Göran Alderborn and Christer Nyström*

72. Modern Pharmaceutics: Third Edition, Revised and Expanded, *edited by Gilbert S. Banker and Christopher T. Rhodes*
73. Microencapsulation: Methods and Industrial Applications, *edited by Simon Benita*
74. Oral Mucosal Drug Delivery, *edited by Michael J. Rathbone*
75. Clinical Research in Pharmaceutical Development, *edited by Barry Bleidt and Michael Montagne*
76. The Drug Development Process: Increasing Efficiency and Cost Effectiveness, *edited by Peter G. Welling, Louis Lasagna, and Umesh V. Banakar*
77. Microparticulate Systems for the Delivery of Proteins and Vaccines, *edited by Smadar Cohen and Howard Bernstein*
78. Good Manufacturing Practices for Pharmaceuticals: A Plan for Total Quality Control, Fourth Edition, Revised and Expanded, *Sidney H. Willig and James R. Stoker*
79. Aqueous Polymeric Coatings for Pharmaceutical Dosage Forms: Second Edition, Revised and Expanded, *edited by James W. McGinity*
80. Pharmaceutical Statistics: Practical and Clinical Applications, Third Edition, *Sanford Bolton*
81. Handbook of Pharmaceutical Granulation Technology, *edited by Dilip M. Parikh*
82. Biotechnology of Antibiotics: Second Edition, Revised and Expanded, *edited by William R. Strohl*
83. Mechanisms of Transdermal Drug Delivery, *edited by Russell O. Potts and Richard H. Guy*
84. Pharmaceutical Enzymes, *edited by Albert Lauwers and Simon Scharpé*
85. Development of Biopharmaceutical Parenteral Dosage Forms, *edited by John A. Bontempo*
86. Pharmaceutical Project Management, *edited by Tony Kennedy*
87. Drug Products for Clinical Trials: An International Guide to Formulation • Production • Quality Control, *edited by Donald C. Monkhouse and Christopher T. Rhodes*
88. Development and Formulation of Veterinary Dosage Forms: Second Edition, Revised and Expanded, *edited by Gregory E. Hardee and J. Desmond Baggot*
89. Receptor-Based Drug Design, *edited by Paul Leff*
90. Automation and Validation of Information in Pharmaceutical Processing, *edited by Joseph F. deSpautz*
91. Dermal Absorption and Toxicity Assessment, *edited by Michael S. Roberts and Kenneth A. Walters*
92. Pharmaceutical Experimental Design, *Gareth A. Lewis, Didier Mathieu, and Roger Phan-Tan-Luu*
93. Preparing for FDA Pre-Approval Inspections, *edited by Martin D. Hynes III*

94. Pharmaceutical Excipients: Characterization by IR, Raman, and NMR Spectroscopy, *David E. Bugay and W. Paul Findlay*
95. Polymorphism in Pharmaceutical Solids, *edited by Harry G. Brittain*
96. Freeze-Drying/Lyophilization of Pharmaceutical and Biological Products, *edited by Louis Rey and Joan C. May*
97. Percutaneous Absorption: Drugs–Cosmetics–Mechanisms–Methodology, Third Edition, Revised and Expanded, *edited by Robert L. Bronaugh and Howard I. Maibach*
98. Bioadhesive Drug Delivery Systems: Fundamentals, Novel Approaches, and Development, *edited by Edith Mathiowitz, Donald E. Chickering III, and Claus-Michael Lehr*
99. Protein Formulation and Delivery, *edited by Eugene J. McNally*
100. New Drug Approval Process: Third Edition, The Global Challenge, *edited by Richard A. Guarino*
101. Peptide and Protein Drug Analysis, *edited by Ronald E. Reid*
102. Transport Processes in Pharmaceutical Systems, *edited by Gordon L. Amidon, Ping I. Lee, and Elizabeth M. Topp*
103. Excipient Toxicity and Safety, *edited by Myra L. Weiner and Lois A. Kotkoskie*
104. The Clinical Audit in Pharmaceutical Development, *edited by Michael R. Hamrell*
105. Pharmaceutical Emulsions and Suspensions, *edited by Francoise Nielloud and Gilberte Marti-Mestres*

ADDITIONAL VOLUMES IN PREPARATION

Oral Drug Absorption: Prediction and Assessment, *edited by Jennifer B. Dressman and Hans Lenneras*

Drug Stability: Principles and Practices, Third Edition, Revised and Expanded, *edited by C. T. Rhodes and Jens T. Carstensen*

Containment in the Pharmaceutical Industry, *edited by James Wood*

Pharmaceutical Emulsions and Suspensions

edited by

Françoise Nielloud
Gilberte Marti-Mestres
Laboratoire de Technique Pharmaceutique Industrielle
Université Montpellier I
Montpellier, France

Marcel Dekker, Inc.   New York • Basel

ISBN: 0-8247-0304-9

This book is printed on acid-free paper.

Headquarters
Marcel Dekker, Inc.
270 Madison Avenue, New York, NY 10016
tel: 212-696-9000; fax: 212-685-4540

Eastern Hemisphere Distribution
Marcel Dekker AG
Hutgasse 4, Postfach 812, CH-4001 Basel, Switzerland
tel: 41-61-261-8482; fax: 41-61-261-8896

World Wide Web
http://www.dekker.com

The publisher offers discounts on this book when ordered in bulk quantities. For more information, write to Special Sales/Professional Marketing at the headquarters address above.

Current printing (last digit):
10 9 8 7 6 5 4 3

PRINTED IN THE UNITED STATES OF AMERICA

Preface

Numerous books dealing with emulsions and suspensions focus on surfactants and formulations, generally in such fields as agronomy, food sciences, and the paint and petroleum industries. Much less attention has been paid to pharmaceutical systems, which are unique and more specific in their applications, toxicity, health hazards, drug delivery, and legislation. In fact, there is no comprehensive and practical text available on this subject, notwithstanding the great demand. This book has been written to remedy this situation.

Pharmaceutical dispersed systems such as suspensions and emulsions are among the most used dosage forms. They are utilized for all routes of administration—oral, topical, parenteral, mucosal, and ophthalmic. These forms effectively present many significant advantages such as an easy dividing up of the forms for pediatric and geriatric patients. The reduction of drug particle size allowed by these formulations enhances the bioavailability of the active agents. Moreover, colloidal particles, such as microparticles, nanospheres, emulsions, and liposomes, have been developed as promising carrier systems for the delivery or the targeting of drugs. Emulsions are also particularly attractive as a vehicle for the administration of poorly soluble drugs.

Liquid-liquid and solid-liquid dispersed systems are complicated forms from a physicochemical point of view, because of the presence of two phases. Their formulation therefore necessitates comprehension of fundamental aspects controlling the behavior of these systems. With this end in view, we begin this volume with theoretical considerations concerning pharmaceutical surfactants, formulation concepts, and emulsion properties, and the related know-how to attain them. As the text progresses, each chapter becomes more advanced and specific. Thermodynamic and kinetic aspects of suspension formulations, as well as physicochemical properties leading to the stability of these systems, are described. An up-to-date analysis of emulsion applications as drug delivery systems is proposed. Most important are chapters describing the use of emulsions and suspensions with respect to the routes of administration. For example, the use of an emulsion or a nanosuspension may allow a poorly water-soluble drug to be administered parenterally or in an ophthalmic delivery system. Topical, transdermal, and gastrointestinal routes are also envisaged. The last part of this book is devoted to experimental designs. This methodology offers an excellent approach for the formulation of emulsions and suspensions. It reduces the expenditure of time and money by limiting the number of manipulations while retaining a very high quality of information. Finally, important aspects of dispersed systems stability are explored, such as rheology and determination of particle size.

As the editors of this volume—which represents the first complete coverage of pharmaceutical emulsions and suspensions—we chose the chapter authors on the basis of their experience and high degree of competence. The book should be useful to pharmacists, graduate students in the pharmaceutical sciences, professionals working in industrial research and development, and all those concerned with the health aspects of emulsions, suspensions, and dispersed systems.

In summary, this book covers fundamental and applied knowledge of emulsions and suspensions. It is hoped that it will provide much of the information and directions necessary to assist those working in all spheres of pharmacy in solving the difficult probelms posed by emulsion and suspension formulations and applications.

We are grateful to all the contributors for their invaluable work and Marcel Dekker, Inc., for its competent collaboration.

Françoise Nielloud
Gilberte Marti-Mestres

Contents

Contributors

Marianne Ashford, Ph.D. Pharmaceutical and Analytical Research Development, AstraZeneca, Macclesfield, Cheshire, United Kingdom

Joëlle Bardon, Ph.D. Laboratoire de Pharmacie Galénique Industrielle et Biogalénique, Université Claude Bernard Lyon I, Lyon, France

J. A. Bouwstra, Ph.D. Division of Pharmaceutical Technology, Leiden/Amsterdam Center for Drug Research, Einsteinweg, and Leiden University, Leiden, The Netherlands

María Isabel Briceño, M.Sc. Laboratory FIRP, University of Los Andes, Merida, Venezuela

Katrin Buszello, Ph.D. SBU Self-Medication, Boehringer Ingelheim GmbH, Ingelheim, Germany

Duncan Q. M. Craig, B.Pharm., Ph.D., M.R.Pharm.S. The School of Pharmacy, Queen's University of Belfast, Northern Ireland, United Kingdom

A. V. Delgado, Ph.D. Department of Applied Physics, University of Granada, Granada, Spain

V. Gallardo, Ph.D. Department of Pharmacy and Pharmaceutical Technology, University of Granada, Granada, Spain

Boris Giannaccini Department of Bioorganic Chemistry and Biopharmaceutics, University of Pisa, Pisa, Italy

Fernando González-Caballero, Ph.D. Department of Applied Physics, University of Granada, Granda, Spain

C. Jacobs Department of Pharmaceutics, Biopharmaceutics, and Biotechnology, Free University of Berlin, Berlin, Germany

O. Kayser, Ph.D. Department of Pharmaceutics, Biopharmaceutics, and Biotechnology, Free University of Berlin, Berlin, Germany

Karsten Klütsch, Ph.D. B. Braun Melsungen AG, Melsungen, Germany

Juan de Dios García López-Durán, Ph.D. Department of Applied Physics, University of Granada, Granada, Spain

Jörn-Stefan Lucks, Ph.D. Department of Pharmaceutics and Biopharmaceutics, Christian Albrecht University, Kiel, Germany

Howard I. Maibach, M.D. Department of Dermatology, University of California Medical School, San Francisco, California

Hélène Maillols, Ph.D. Laboratoire de Technique Pharmaceutique Industrielle, Université Montpellier I, Montpellier, France

Gilberte Marti-Mestres, Ph.D. Laboratoire de Technique Pharmaceutique Industrielle, Université Montpellier I, Montpellier, France

Didier Mathieu Laboratoire de Méthodologie de la Recherche Expérimentale—I.U.T. Département Informatique, Aix-en-Provence, France

Daniela Monti, Ph.D. Department of Bioorganic Chemistry and Biopharmaceutics, University of Pisa, Pisa, Italy

Bernd W. Müller, Ph.D. Department of Pharmaceutics and Biopharmaceutics, Christian Albrecht University, Kiel, Germany

R. H. Müller, M.D., Ph.D. Department of Pharmaceutics, Biopharmaceutics, and Biotechnology, Free University of Berlin, Berlin, Germany

Françoise Nielloud, Ph.D. Laboratoire de Technique Pharmaceutique Industrielle, Université Montpellier I, Montpellier, France

Manisha J. Patel, Ph.D. Center for Materials Science, School of Pharmacy, University of London, London, United Kingdom

Roger Phan-Tan-Luu Laboratoire de Méthodologie de la Recherche Expérimentale, Faculté de St. Jérôme, Marseille, France

E. R. Pike, Ph.D., F.R.S. Department of Physics, King's College, London, United Kingdom

Michel Rollet, Ph.D. Laboratoire de Pharmacie Galénique Industrielle et Biogalénique, Université Claude Bernard Lyon I, Lyon, France

M. A. Ruiz, Ph.D. Department of Pharmacy and Pharmaceutical Technology, University of Granada, Granada, Spain

M. Fabrizio Saettone, Ph.D. Department of Bioorganic Chemistry and Biopharmaceutics, University of Pisa, Pisa, Italy

Jean-Louis Salager, Ph.D. Lab. FIRP, Chemical Engineering School, University of the Andes, Mérida, Venezuela

Eric W. Smith, Ph.D. College of Pharmacy, Ohio Northern University, Ada, Ohio

Christian Surber, Ph.D. Department of Dermatology and Institute of Hospital Pharmacy, University Hospital, Basel, Switzerland

Pharmaceutical Emulsions and Suspensions

1

Main Surfactants Used in the Pharmaceutical Field

Gilberte Marti-Mestres and Françoise Nielloud
Université Montpellier I, Montpellier, France

I. INTRODUCTION

Surface-active agents, or surfactants, are molecules distinguished by the presence of both a polar and a nonpolar region. *Surface-active agent* is the general term that includes detergent, dispersing agents, emulsifying agents, foaming agents, penetrating agents, and wetting agents (1). In the pharmaceutical field surfactants are used especially as emulsifiers, solubilizers, and wetting agents. In pharmaceutics, the term *cream* is traditionally used to describe semisolids with a relatively fluid viscosity that are formulated as oil-in-water emulsions and that are aesthetically favorable and easy to spread. Creams may be capable of penetrating the different layers of the skin, particularly the horny layer (2). This form can be used for administrating drugs via the dermatological, ocular, and vaginal route. Conversely, the term *ointment* is used for semisolid formulations that help to keep drugs in lengthy contact and also confer an occlusive effect. Ointments are often water-in-oil emulsions. They are intended for application to skin or eye; in the latter case the product selected must be nonirritating to the eye and allow diffusion of the active ingredient. Surfactants are derived largely from petroleum, but may also be made from natural fats, oils, sugars, or other materials that are renewable resources characterized by their compatibility with the environment. The most important aspect of environmental protection is human protection. Classification of surfactants is quite arbitrary, and the most common is based on chemical structure. Some surfactants possess a hydrophilic region in the molecule which can carry a positive or negative charge. Respectively named cationic or anionic surfactants, their activity depends on the pH of the vehicle in which they are utilized. The nonionic group differs from these in the absence of charge on the molecule. This group is the largest and is the most widely utilized for pharmaceutical emulsions and suspensions; it will therefore be principally developed in the discussion that follows. Another group, the amphoteric or ampholytic surfactants, can have anionic or cationic properties depending on the pH. Anionic surfactants are broadly accepted as powerful irritants to human skin. Cationic surfactants are accepted as at least equally irritating, but more cytotoxic than the anionic, while the inherent irritation of nonionic surfactants is considered to be the lowest. This categorization does not permit exact determination of the potential toxicity of each product (3). The purpose of this chapter is to present a review of the wide diversity of surfactants that can be used or that have been included in the formulation of pharmaceutical systems. At the same time an indication of toxicity is given with reference to some examples of commercial products (pharmaceutical emulsions or suspensions) containing the surfactant (4,5). The main surfactants described in this chapter are those listed in pharmacopoeia sources. Trade names of these surfactants are not mentioned but for a complete list the reader should consult one of the treatises on this subject; trade names are extensively compiled by Ash et al. (6) for pharmaceutical excipients and by the

International Cosmetic Ingredient Dictionary (7) for cosmetic products. These books should be consulted for more detailed information. The total number of surfactants available worldwide is truly vast but the number used in the pharmaceutical field is very low. Inflated interest regarding toxicity and allergic adverse reactions to all inactive additives has contributed to their identification by pharmacopoeias or by other governmental legislation. The references are the following: Eur. Ph. means that a monograph for the surfactant exists in the European Pharmacopoeia (8). USP or NF indicates that a monograph for the product exists in either the United States Pharmacopoeia (USP 23 and updates) or the National Formulary (NF 18) (9). BP is used for the British Pharmacopoeia (10). If a surfactant is reported in Martindale, the Extra Pharmacopoeia (11), this will be also noted. Laws differ from country to country, and the measurement of the suitability of these products for the intended use is the responsibility of the formulator, who fully understands the route applications. Molecules on the market having interfacial activities are numerous but in the pharmaceutical field the available quantity is far more limited. Indeed, the toxicological aspect is the limiting factor for the utilization of these products. The description of surfactants in pharmacopoeia monographs makes the task of formulation and registration easier for the users. Therefore, an approved product may deliver the pharmaceutical group elaboration time and charge. With these circumstances, a new product will not be easily found within pharmaceutical formulations, and the choice remains very limited. A new surfactant in a pharmaceutical formulation will invariably oblige additional studies, adding both to development time and to the cost and labor of product development. For therapeutic use, the number of products is again more limited, particularly for parenteral and ocular routes. For a given drug, the risk of adverse events is higher if it is administered by injection as opposed to a nonparenteral route, and this limits the choice of excipients. A recent and very interesting review concerning excipients and their use in injectable products is given by Nema et al. (12). In practice, surfactants can sometimes be classified as inactive ingredients in pharmaceutical forms or in other cases used as an active constituent.

II. ANIONIC SURFACTANTS

Anionic surfactants are those that carry a negative charge on the hydrophilic part. The major classes of anionic surfactants used in this field are those containing carboxylate, sulfate, and sulfonate ions (13). The most frequently allied cations are sodium, calcium, magnesium, and zinc, multivalent ions producing marked water insolubility. The straight chain is a saturated or unsaturated C12–C18 aliphatic group. The degree of water solubility is greatly affected by the length of the alkyl chain and by the presence of double bonds (14). Depending on the chemical class and concentration, this group of surfactants may be irritating in certain conditions.

A. Anionic Surfactants with Carboxylate Ions

Monovalent and polyvalent alkyl carboxylates are respectively called "soap" and "metallic soap." The straight chain of the fatty acids varies from C12 to C20. Higher members are too hydrophobic to be used and lower members have little surfactant value. Stearic acid soaps are doubtless the most widely utilized emulsifiers in oil-in-water emulsions. Unsaturated fatty acid soaps such as oleate produce fluid emulsions (15). The most common associated cations are sodium, calcium, magnesium, potassium, ammonium, and triethanolamine. Monovalent salts of carboxylic acids are generally used to produce oil-in-water emulsions.

Aluminum monostearate or *stearate* are USP/NF- and Martindale-compliant, and is used as an emulsifier in topical preparations such as Neuriplege cream (Genevrier) or Efficort ointment (Galderma) and as a suspending agent in oily injectable suspensions (sterile penicillin G procaine with aluminum stearate suspension is described in USP23). It is also FDA-approved for implants, oral, rectal and topical use. *Calcium stearate* is USP/NF-, Eur.Ph., and Martindale-compliant and FDA-approved for oral, rectal, and topical use, and this product is chiefly used as a lubricant in manufacture of tablets, Solumedrol (Pharmacia-Upjohn) and Rowasa (Solvay Pharma). *Magnesium stearate* is also USP/NF-, Eur.Ph.-, BP-, and Martindale-compliant, and FDA-approved for buccals, parenterals, orals, and vaginals. This product can be used as an anticaking agent in suspensions, but is also found as a lubricant in tablets. Another product in this group is *sodium stearate*, not used in emulsion or suspension fields. *Zinc stearate* is USP/NF-, Eur. Ph.-, and Martindale-compliant and FDA-approved for orals. This surfactant is mentioned as an inactive ingredient in Madecassol neomycin hydrocortisone cream (Roche Nicholas S.A.) and has no known toxicity to skin. *Sodium*, *zinc*, and *potassium oleates* are presented in Martindale (11) as surfactants that may be used in skin preparations. In this group the last product is *sodium stearyl fumarate*, mentioned in Martindale and USP/NF but only used as inert tablet lubricant.

The primary disadvantage of soaps is that they can be used only over a limited pH range. In addition, soaps form insoluble salts with polyvalent ions.

B. Anionic Surfactants with Sulfate Group

A number of alkyl sulfates are available as surfactants. By far the most typical member of this group is *sodium lauryl sulfate* (SLS), which is utilized widely as a solubilizer, wetting agent, and emulsifier in pharmaceutical preparations (in Eur. Ph.-, USP/NF-, BP-, Martindale-compliant and FDA-approved for topicals, dentals, orals, and vaginals). The common synonym for sodium lauryl sulfate is sodium dodecyl sulfate. *Emulsifying wax* (in USP/NF, BP, Eur. Ph.) contains sodium lauryl sulfate not less than 8.7%, and cetostearyl alcohol. Unlike soaps, SLS is acid-stable. As an anionic surfactant, SLS reacts with cationic compounds. Sodium lauryl sulfate may be a human skin irritant and it has often been used in skin or mucosa irritancy investigations (16,17). This surfactant is the most thoroughly investigated for skin membrane irritations with concentration effects (18–21). Long contact under occlusion leads to cutaneous inflammation with increased transepidermal water loss (TEWL) (22), and the preferential affinity of sodium lauryl sulfate for skin lipids and proteins was confirmed by Patil et al. (23). Recently, Saija et al. have shown changes in the permeability of the blood–brain barrier following sodium lauryl sulfate administration in rat and they called attention to possible dangerous consequences of using SLS in injectable drug excipients (24). In vitro cytotoxic effects of *N*-alkyl sulfate with hydrocarbon chain lengths varying between C8 and C16 were studied and compared with the irritant response (measurement of erythema and TEWL (25). Sodium lauryl sulfate is found in Parfenac 5% cream (Wyeth Lederle), in Zovirax 5% cream (Glaxo Wellcome), and in the oral suspension Nizoral (Janssen Cilag). Other salts of lauryl sulfate may be used. One can find listed in Martindale (11) *mono-*, *di-*, and *triethanolamine* (*TEA*) *lauryl sulfate*, and *ammonium*, and *magnesium lauryl sulfate*. TEA lauryl sulfate is FDA-approved for topicals and was used as an emulsifier, foamer, or cleansing agent in skin preparations. Similar anionic surfactants include *sodium lauryl ether sulfate* found, for example, in Ketoderm 2% gel (Janssen-Cilag). All of these surfactants are listed only in Martindale (11). For similar purposes, *sodium cetostearyl sulfate* or *sodium cetearyl sulfate* (Eur. Ph.-, Martindale-compliant) is also found, for example, in Cutisan 2% cream (Boots Healthcare) and Synalar 0.025% cream (Cassenne Marion). *Sodium tetradecyl sulfate* (in BP and Martindale) is an anionic surfactant utilized exclusively as an active ingredient with sclerosing effects and found in Trombovar injectable (Promedica).

Sulfated oils are made by treating the oily fraction with sulfuric acid and neutralizing with a base (generally sodium hydroxide solution). *Sulfated castor oil* (listed in Martindale) consists primarily of the sodium salt of the sulfated triglyceride of castor oil. This surfactant is nonirritant and may be used as a detergent, wetting, and emulsifying agent for oil-in-water emulsions in the pharmaceutical field. Only Fisher et al. report a contact allergy to sulfated castor oil (26).

C. Anionic Surfactants with Sulfonate Group

Sulfonates are a group of compounds in which the sulfur atom is linked directly to the carbon atom. Various compounds are widely used in this surfactant field because of their excellent stability. This class shows a superior tolerance to calcium ions and does not hydrolyze as easily as the sulfate group.

The widely used surfactant of this group is *sodium docusate* (dioctyl sodium sulfosuccinate, or DSS) found in USP/NF, BP, and Martindale. In FDA, this compound is approved for injectable (IM), orals, topicals, vaginals. Considered as a severe eye irritant, sodium docusate can be used as a wetting agent in flocculated suspensions (27). DSS administered in the diet to three successive generations of rats at levels up to 1.0% had no effects on the reproductive function of either sex in any generation and produced no treatment-related antemortem or macroscopic observations (28). Sodium docusate is found in Efferalgan (Upsa), an effervescent powder for oral solutions. Calcium and potassium docusate are also mentioned in Martindale. Sodium alkyl sulfoacetates such as sodium lauryl sulfoacetate are Martindale-listed and FDA approved for topical administration.

III. CATIONIC SURFACTANTS

The cationic group has a positively charged cation. Negatively charged products strongly absorb cationic surfactants; some such products include hair, skin, and microorganisms. These type of surfactants are important pharmaceutically because of their bactericidal properties. Due to this activity, cationic surfactants are used on skin, especially in the cleansing of wounds and burns, or are used as preservatives. They may be very irritating to the eyes and skin. These compounds are applied infrequently as emulsifiers. Furthermore, the use of quaternary ammonium groups as emulsifying agents in creams is restricted because of their incompatibility with soap, many anionic compounds, and other inactive ingredients such as polymers (polyacrylate, carboxymethylcellulose) also used in these types of formulations. The principal cationic surfactants used in pharmaceutical preparations are quaternary ammonium salts, and the physicochemical properties and different applications of these surfactants will be described below.

Cetrimide consists of trimethyltetradecylammonium bromide, and may contain smaller amounts of dodecyl- and hexadecyltrimethylamonium bromides (Eur. Ph.-compliant and listed in Martindale). Hypersensitivity to topical application has been reported. Van Rij et al. reported corneal edema and toxic keratopathy due to accidental use of chlorhexidine and cetrimide during cataract surgery (29). Lee et al. (30) reported that 3% cetrimide has induced 18 cases of an ichthyosiform contact dermatitis. Lee (31) attributes this abnormal keratinization to dysfunction of lamellar bodies in the granular cells resulting from direct effects of cetrimide on the lipids and enzymes of lamellar bodies. Adverse effects following ingestion include nausea and esophageal damage (with stronger solutions). This surfactant can be found in the form of cetrimide cream (cetrimid, cetostearyl alcohol, liquid paraffin) and cetrimide emulsifying ointment (cetrimide 2.5–

3.3%, with soft paraffin and liquid paraffin and cetostearyl alcohol). Cetrimide is also used as an antiseptic detergent.

Cetrimonium bromide (cetyltrimethylammonium bromide, hexadecyltrimethylammonium bromide) was often formerly applied to cetrimide (BP- and Martindale-listed). This surfactant has similar properties to cetrimide.

Benzalkonium chloride is a mixture of alkylbenzyldimethylammonium chlorides where the alkyl groups have chain lengths of C8 to C18 (USP/NF-, Eur. Ph.-, BP-, and Martindale-compliant). Benzalkonium is used as wetting agent or solubilizer for ophthalmics, injectables, and topicals. Mixing benzalkonium chloride with soap or other anionic surfactants may decrease or destroy the bacteriostactic activity. Hatinen et al. (32) report that contact allergy was encountered with topical ophthalmological preparations containing benzalkonium chloride. *Cetalkonium chloride* (Martindale-compliant) is also used for its antimicrobial properties in pharmaceutical preparations and medical devices.

Cetylpyridinium chloride is the 1-hexadecylpyridinium chloride (USP/NF-, Eur.Ph.-, BP-, and Martindale-compliant). Lin et al. (33) compared the primary ocular and dermal irritations of several quaternary ammonium compounds. They concluded that the irritancy of these compounds is likely to be related to their solubility in addition to their cationic characteristics. Only cetylpyridinium chloride is very soluble in both lipids and water. It appears that not all of the quaternary ammonium compounds studied are irritant, but cetylpyridinium chloride was severely irritating to the skin of the test animal. In a recent work, Green et al. (34) confirmed that ophthalmic medications containing cetylpyridinium chloride are potentially hazardous to the corneal endothelium.

IV. NONIONIC SURFACTANTS

Nonionic surfactants differ from ionic surfactants in the absence of charge on the molecule. They are generally less irritant than anionic or cationic surfactants. They are compatible with other types of surfactants, but they may diminish the antimicrobial activity of some preservatives (11). The characteristics of nonionic surfactants are essentially dependent on the proportions of hydrophilic or hydrophobic groups in the molecule. The hydrophilic part contains the polyoxyethylene, polyoxypropylene, or polyol derivatives and the hydroxyl group. The hydrophobic part includes saturated or unsaturated fatty acids or fatty alcohols. By varying the number of hydrophilic groups or the length of the lipophilic chain, compounds are obtained with a wide range of hydrophilic–lipophilic balance (HLB) values. This is an empirical scale invented by Griffin that is useful to classify nonionic surfactants and to select surfactant mixtures for emulsification of particular oils (35) (see Chapter 2). Lipophilic surfactants ($0 < \text{HLB} < 10$) are known for their antifoaming, water-in-oil emulsifying or wetting properties.

Hydrophilic surfactants (10 < HLB < 20) have generally oil-in-water emulsifying or solubilizing properties. Due to the conditions of their fabrication, these surfactants are usually mixtures of associated substances, so there are sometimes variations in properties between different manufacturers.

Over many years, nonionic surfactants have become more and more important in the pharmaceutical field because of their ability to solubilize poorly soluble substances and their low toxicity (36). The principal groups used in this domain are polyol derivatives, polyoxyethylene esters and ethers, and poloxamers, but other surfactants are also included in this classification, such as nonylphenyl ethers, polyvinyl alcohol, propylene glycol diacetate, or alkanolamides.

A. Polyol Esters

Polyol surfactants are used in many pharmaceutical preparations as emulsifiers, solubilizers, dispersants, or stabilizers. They can be separated into two classes: glycol and glycerol esters and sorbitan derivatives.

Glycol and glycerol esters consist of fatty acid esters of glycols and glycerol. They are lipophilic compounds and are essentially used as stabilizers of both oil-in-water and water-in-oil emulsions because of their poor emulsifying properties.

The main product of this group is *glyceryl stearate* approved by the FDA for orals, ophthalmics, otics, rectals, and topicals and is USP/NF-, BP- and Eur. Ph.-compliant. Glyceryl stearate contains mainly stearate and palmitate monoglycerides and is used as a stabilizer or emollient in emulsions for internal and external use. This compound can be found in a large number of pharmaceutical emulsions, ointments, suspensions, or tablets (Dermoval 0.05% cream, Glaxo Wellcome; Alphosyl cream, Stafford Miller; Renutryl 500 oral suspension, Clintec Sopharga). Diverse other products with the same properties and uses as glyceryl monostearate are described in Martindale (11) and in BIAM (5). The *mono-* and *diglycerides* (USP/NF- and Eur-Ph-compliant), a mixture of glycerol mono- and diesters of fatty acid from edible oils (Cephaperos oral suspension, Bristol Myers Squibb), the *glyceryl monooleate* (USP/NF-approved and FDA-approved for oral use) (Alphacutanee, Leurquin Mediolatum), *glyceryl monolaurate* (Discotrine, 3M Sante), and *glyceryl ricinoleate* (FDA-approved for orals and topicals) (Selsun, Abbot France) are used as emulsifying or dispersing agents. *Glyceryl behenate* is used as a lubricant and binder in tablets.

Ethylene glycol and *diethylene glycol* or *propylene glycol monoesters* are the simplest polyol surfactants. *Stearate*, *oleate*, or *palmitostearate esters* are found in pharmaceutical preparations (Madecassol 1%, Hirucrème Roche Nicolas SA; Diprolene 0.05%, Shering Plough). They are very lipophilic and are used for imparting opacity to liquid preparations and consistency to emulsions, sticks, and pastes (37). *Glycol stearate* is FDA-approved for topicals; *propylene glycol stearate* is FDA-approved for rectals and vaginals, and is USP/NF- and Eur-Ph-compliant.

Sorbitan derivatives consist of esters of cyclic anhydrides of sorbitol with a fatty acid (C12–C18). They are named sorbitan esters. They are lipophilic surfactants used as water-in-oil emulsifiers. When these sorbitan esters are polyoxyethylened they are called polysorbates. A large range of properties may be obtained by varying the number of oxyethylene groups in the molecule. They have mainly oil-in-water emulsifying and solubilizing properties. Some sorbitan derivatives that are USP/NF-, Eur. Ph.-, BP-compliant and FDA-approved for oral administration are listed below:

Sorbitan monolaurate, *monooleate*, and *monostearate*. The acceptable daily intake of these is up to 25 mg, as total sorbitan esters, per kg body weight. These products may show adverse effects; some hypersensitive skin reactions after topical application in creams have been reported (11).

Polysorbate-20, *60*, and *80*. These have the same acceptable daily intake as sorbitan esters but are characterized by the property of enhancing the absorption of fat-soluble substances (38). This property can be an advantage or an adverse effect depending on the drug. Furthermore, polysorbates can also provoke hypersensitivity following topical application. They may disrupt normal membrane structures (39).

Other surfactants are only described in Martindale or in BIAM or USP/NF. They are *sorbitan monopalmitate* (USP/NF), *sesquioleate*, *trioleate*, or *tristearate*. Also found are *polysorbates-40* (USP/NF), *65*, and *85*. They have the same uses as the preceding surfactants. This class of sorbitan derivatives is widely used in pharmaceutical emulsions, oral or injectable suspensions, or solutions. It is also largely used in cosmetic preparations. Sorbitan esters can be associated with polysorbates to emulsify particular oils and obtain emulsions with a diversity of textures and viscosities. Many examples are found in the literature such as Alphosyl cream, Stafford Miller; Ketoderm 2%, Janssen Cilag; or Canesten 1%, Bayer Pharma (4,5). Hormones such as progesterone or corticoids and many other lipophilic drugs are solubilized or dispersed by polysorbates (Depo-prodasone, Upjohn Pharmacia; Hydrocortancyl, Roussel Diamant; Imukin, Boehringer Ingelheim). Both fat-soluble and water-soluble vitamins can be incorporated in the same formula by the use of polysorbates (Hydrosol polyvitamine BON, Doms Adrian). Suspensions of insoluble drugs in aqueous media often use sorbitan derivatives, either alone or in combination with other suspending agents (Povanyl oral suspension, Warner Wellcome).

Polyol derivatives can be associated with thickeners (free fatty acids or alcohols) that lead to products such as *self-emulsifying glyceryl stearate* (BP-compliant) or *emulsifying wax* (USP/NF-compliant), which are used as emulsifying agents in the preparation of self-emulsified bases.

B. Polyoxyethylene Esters and Ethers

Polyoxyl esters, also called **Macrogol esters**, are mixtures of mono- or difatty acids esters (from C12 to C18) of polyoxyethylene glycol (PEG). They differ in the number of PEG groups in the molecule and the nomenclature can use either the number of the oxyethylene group or the average polymer weight for qualifying the molecules. The main products found in the USP/NF and Mardindale Extra Pharmacopoeia are *stearate esters* (*PEG-40, PEG-50,* and *PEG-55*), but *laurate*, *oleate*, or *myristate esters* have also been used. (Only PEG-40 stearate is FDA-approved for dentals, ophthalmics, orals, otics, and topicals). These products have been reported to be used as emulsifying and dispersing agents in emulsions and suspensions (Pevaryl 1% cream, Janssen Cilag; Tegretol 100 mg oral suspension, Novartis Pharma). Polyoxyl stearates have been reported to diminish the antimicrobial activity of different preservatives with concentrations above 5%.

A compound can be added to this class, the *PEG-35 castor oil* (USP/NF- and Martindale-approved), which is a ricinoleate ester of ethoxylated glycerol. Polyethoxylated castor oils or hydrogenated castor oils (*PEG-40 hydrogenated castor oil*, USP/NF) have solubilizing properties and are used as injectable vehicles (40) (Daktarin, Janssen Cilab; Sandimmun 50, Novartis Pharma). Polyethoxylated castor oils in these vehicles have shown anaphylactoid reactions.

Polyoxyl ethers are ethers of magrocols and fatty alcohols. Virtually all fatty alcohols can react with ethylene oxide, and the level of oxidation can run from 1 to 100 or more. Therefore, a wide range of ethers is obtainable, and it is impossible here to describe all of them. When the chain length consists of 12 carbon atoms, the laureth compounds obtained may be used as surfactants or as active substances (spermicide). The other surfactants that are FDA-approved for topicals and USP/NF compliant are *steareth-20*, *ceteareth-20*, and *oleth-10*, which are respectively ethers of stearyl, cetostearyl, and oleyl alcohols. Polyoxyethers are not skin sensitizers and at dilutions of <0.1% are usually not skin or eye irritants (41). As in the foregoing examples, these surfactants are essentially emulsifiers (Diprolene 0.05% cream, Schering Plough).

C. Poloxamers

These are polyoxyethylene-polyoxypropylene derivatives, with polyoxyethylene groups forming the hydrophilic part and polypropylene groups the lipophilic part. The number of ethylene oxides may vary to give these products a wide range of properties. The surfactants that are USP/NF- and Martindale-compliant are *poloxamer-188* and *407*. They are water-soluble compounds used as emulsifiers, solubilizers in injectable forms, and wetting agents for antibiotics. They are also found in suppositories or ointments. (Erythrocine, oral suspension, Abbots France; Zovirax cream, Glaxo Wellcome).

D. Other Nonionic Surfactants

Nonylphenyl ethers, also called **nonoxinols**, are ethoxylated nonylphenols with a large range of ethylene oxide chain length. *Nonoxinol-9*, *10*, and *11* are described in Martindale and USP/NF. They have surface-active properties and may be used as solubilizing agents. Nonoxinol-9 may also be used as a spermicide and for this activity may present vaginal irritation.

Propylene glycol diacetate (USP/NF- and Martindale-compliant, FDA-approved for otics) is a nonionic water-soluble surfactant, which is used as emulsifier, solubilizer, or solvent in pharmaceuticals or as a food additive.

Polyvinyl alcohol is a nonionic surfactant used as stabilizer, lubricant, or viscosity builder in ophthalmic preparations and as a jelly builder for topical preparations. This product is FDA-approved for ophthalmics, orals, injectables, topicals, and vaginals, and is described in USP/NF and Martindale Extra Pharmacopoeia.

Alkanolamides are prepared from the reaction of fatty acids with mono- or diethanolamide. They are used as foam stabilizers and thickeners in cosmetic products intended for the cleansing of the skin and scalp. They have not found an important place in the pharmaceutical industry (37).

V. ADDITIONAL SURFACTANTS

A. Amphoteric Group

This class of surfactant includes those that carry both positive and negative charges. Depending on the pH of the preparation, this group behaves as a cationic, anionic, or nonionic species. In the cosmetics field, these surfactants are frequently applied in skin or hair formulations as relatively mild detergents. They are also used for their capacity for reducing the irritation of anionic surfactants. These surfactants are commonly accepted to be nonirritant to the eyes and have consequently been used in baby shampoos. Amphoteric surfactants are very much used in cosmetics, but they are not found in the different pharmacopoeia studied here.

B. Natural Emulsifiers

The natural biological surfactants of pharmaceutical interest contain bile salts. Bile salts influence intestinal drug absorption because they can form mixed micelles with a wide class of products or drugs (42). Dissolution rates of insoluble drugs can be increased by bile salts (43). *Sodium desoxycholate* or 7-desoxycholic acid sodium salt (BP-compliant and FDA-approved for parenterals) is used in parenterals as an emulsifying agent (Fungizone, Bristol Myers).

Phospholipids (called also phosphatides) are esters consisting of glycerol

Table 1 Phospholipid Composition of Commercial Lecithins

Phospholipid	Soy	Corn	Sunflower	Rapeseed	Egg	Bovin brain
PC	21	31	14	37	69	18
PE	22	3	24	29	24	36
PI	19	16	13	14	—	2
PA	10	9	7	—	—	2
PS	1	1	—	—	3	18
SM	—	—	—	—	1	15

PC, phosphatidylcholine; PE, phosphatidylethanolamine; PI, phosphatidylinositol; PA, phosphatidic acid; PS, phosphatididylserine; SM, sphingomyelin.

in combination with fatty acids, phosphoric acid, and certain nitrogenous compounds. The phospholipids are essential components of all vegetable and animal cells and the variety is impressive. The only phospholipids with pharmaceutical applications are the *lecithins.* Only vegetable oil lecithins are described in USP/NF, Martindale-compliant. Extractions from egg yolk, plant tissues or brain tissue are used to obtain the different lecithins. Commercially, one definition for lecithin is that it is a complex mixture of phospholipids (phosphatidylcholine, phosphatidylethanolamine, phosphatidylserine, and phosphatidylinositol), glycolipids, and some triglycerides (44). Table 1 indicates the phospholipid composition of some plant, egg, and bovine brain lecithins. Lecithins are used as emulsifiers especially for intravenous fat emulsions. Many formulations contain this type of surfactant and some of them are reported below. Soybean lecithins are found in Atrovent oral suspension (Boehringer Ingelheim), Combantrin oral suspension (Pfizer), and purified egg phospholipid are used in Diprivan injectable emulsion (Zeneca Pharma) and in Intralipide injectable emulsion (Pharmacia Upjohn).

C. Sucrose Esters and Alkylpolyglucosides

Sucrose esters are nonionic sugar-derived surfactants consisting of a lipophilic part, namely, a fatty acid (C12–C18) or amines associated with a hydrophilic sugar structure such as saccharose, dextrose, or glucose. They can be mono-, di-, or tri-esterified leading to a large range of HLB values. Therefore, monoesters give HLB values near 16 and di- or triesters can have very low HLB values (between 3 and 1). This very large range is not accessible with other polyol surfactants such as glycerol or sorbitol (45). They are used as emulsifiers (oil-in-water or water-in-oil), dispersing or wetting agents. They present many advantages, such as the smooth consistency they give to emulsions, their antiirritant effect, and their antibacterial action. Nevertheless, sucrose esters are poorly developed in Europe and the United

States, whereas they have found an extensive market in cosmetics and food application in Japan (46). There are still interested researchers (47) who studied microemulsion systems for transdermal delivery with sucrose esters (laurate and oleate) as surfactants and aliphatic alcohols as cosurfactant.

Alkylpolyglucosides (APGs) are nonionic surfactants prepared from natural raw materials, specifically starch or its glucose derivatives reacted with fat or fatty alcohol derivatives. They have been known for many years, but they are now available on an industrial scale (48). They are modern, multifunctional ingredients for the cosmetics industry and their main advantages are synergism with other surfactants, mildness and nonirritatability to skin and eyes, and the fact that they are derived from renewable sources and are readily biodegradable. They are free from dioxane, ethylene, and nitrosamine and so can be considered as ecological and non-toxical substitutes for those products with suspected adverse effects, such as polyethoxylated alkyl surfactants or alkyl sulfates. They are now used extensively as very efficient surfactants for their foaming, wetting, and solubilizing properties in shampoos and cosmetic detergents (46). The hydrophilic C8–C16 and C12–C16 APG surfactants form oil-in-water emulsions and, conversely to ethoxylated surfactants, their emulsifying properties are almost independent of temperature. Furthermore, when associated with hydrophobic coemulsifiers they can produce microemulsions (49). Independently of their well-known applications, in order to obtain comprehensive characterization of these surfactants, several authors have studied the physicochemical properties of APGs, namely, phase behavior, interfacial activity, or micelle structure (48,50–52). They concluded that APGs have unexpected and unusual properties that do not correspond exactly to other nonionics, i.e., polyethoxylated surfactants, but are sometimes closer to the behavior of anionic surfactants. The research also focused on the preparation of nontoxic microemulsions (53–55). Microemulsions involve a high percentage of surfactant which could induce side effects such as skin irritancy. Therefore the use of original biocompatible amphiphiles is widely investigated. On the same topic, Van Roekeghem et al. (56) recently studied the physicochemical properties of 12 alkyl xylitols in order to establish their amphiphilic properties and to compare them with available data on glucose derivatives. For all of these reasons, alkylpolyglucosides are attractive and promising compounds in the pharmaceutical field. Many innovations can be expected in the future in pharmaceutics and APGs will certainly make a major contribution.

VI. CONCLUSION

In this chapter, the authors have reviewed the variety of surfactants used in the pharmaceutical field. The anionic cationic and nonionic surfactants that appear in the British, European, and U.S. Pharmacopoeia or that are listed in Martindale Extra Pharmacopoeia are reported in Tables 2 and 3, respectively.

Table 2 Anionic or Cationic Surfactants Listed in Different Pharmacopoeia or Extra Pharmacopoeia

Surfactants	Class	Pharmacopoeia/extra pharmacopoeia			
Aluminium monostearate	Anionic	USP/NF			Martindale
Ammonium lauryl sulfate	Anionic				Martindale
Calcium stearate	Anionic	USP/NF	Eur. Ph.	BP	Martindale
Dioctyl calcium sulfosuccinate	Anionic				Martindale
Dioctyl potassium sulfosuccinate	Anionic				Martindale
Dioctyl sodium sulfosuccinate	Anionic	USP/NF		BP	Martindale
Emulsifying wax	Anionic		Eur. Ph.	BP	Martindale
Magnesium lauryl sulfate	Anionic				Martindale
Magnesium stearate	Anionic	USP/NF	Eur. Ph.	BP	Martindale
Mono-, di-, triethanolamine lauryl sulfate	Anionic				Martindale
Potassium oleate	Anionic				Martindale
Sodium castor oil	Anionic				Martindale
Sodium cetostearyl sulfate	Anionic		Eur. Ph.	BP	Martindale
Sodium lauryl ether sulfate	Anionic				Martindale
Sodium lauryl sulfate	Anionic	USP/NF	Eur. Ph.		Martindale
Sodium lauryl sulfoacetate	Anionic				Martindale
Sodium oleate	Anionic				Martindale
Sodium stearate	Anionic	USP/NF			Martindale
Sodium stearyl fumarate	Anionic	USP/NF			Martindale
Sodium tetradecyl sulfate	Anionic			BP	Martindale
Zinc oleate	Anionic				Martindale
Zinc stearate	Anionic	USP/NF	Eur. Ph.		Martindale
Benzalconium chloride	Cationic	USP/NF	Eur. Ph.		Martindale
Cetrimide	Cationic		Eur. Ph.	BP	Martindale
Cetrimonium bromide	Cationic			BP	Martindale
Cetylpyridinium chloride	Cationic	USP/NF	Eur. Ph.	BP	Martindale

Table 3 Nonionic Surfactants Listed in Different Pharmacopoeia or Extra Pharmacopoeia

Surfactants	Pharmacopoeia/extra pharmacopoeia			
Polyols esters				
Glyceryl monostearate	USP/NF	Eur. Ph.	BP	Martindale
Monodiglyceride	USP/NF	Eur. Ph.		Martindale
Glyceryl monooleate				Martindale
Glyceryl behenate	USP/NF			Martindale
Sorbitan monolaurate	USP/NF	Eur. Ph.	BP	Martindale
Sorbitan monopalmitate	USP/NF	Eur. Ph.		Martindale
Sorbitan monooleate	USP/NF	Eur. Ph.	BP	Martindale
Sorbitan monostearate	USP/NF	Eur. Ph.	BP	Martindale
Sorbitan sesquioleate	USP/NF			Martindale
Sorbitan trioleate	USP/NF	Eur. Ph.		Martindale
Sorbitan tristearate				Martindale
Polysorbate-20	USP/NF	Eur. Ph.	BP	Martindale
Polysorbate-40	USP/NF			Martindale
Polysorbate-60	USP/NF	Eur. Ph.	BP	Martindale
Polysorbate-65				Martindale
Polysorbate-80	USP/NF	Eur. Ph.	BP	Martindale
Polysorbate-85				Martindale
Diethylene glycol monostearate				Martindale
Ethylene glycol monostearate		Eur. Ph.		Martindale
Propylene glycol monostearate	USP/NF	Eur. Ph.		Martindale
Self-emulsifying glyceryl stearate			BP	
Emulsifying wax NF	USP/NF			
Polyoxyethylene esters and ethers				
PEG-40 stearate	USP/NF*	Eur. Ph.		Martindale
PEG-50 stearate	USP/NF*	Eur. Ph.		Martindale
PEG-8 stearate	USP/NF*	Eur. Ph.		Martindale
Polyoxyl-35 castor oil	USP/NF*	Eur. Ph.		Martindale
Polyoxyl-40 hydrogenated castor oil	USP/NF			Martindale
Laureth-2		Eur. Ph.		Martindale
Laureth-4		Eur. Ph.		Martindale
Laureth-9		Eur. Ph.		Martindale
Ceteareth-20		Eur. Ph.		Martindale
Steareth-20		Eur. Ph.		Martindale
Oleth-10	USP/NF*	Eur. Ph.		Martindale
Poloxamers				
Poloxamer-188	USP/NF		BP	Martindale
Poloxamer-407	USP/NF			Martindale
Other nonionic surfactants				
Nonoxinols-9	USP/NF			Martindale
Nonoxinols-10	USP/NF*			Martindale
Nonoxinols-11				Martindale
Propylene glycol diacetate	USP/NF*			Martindale
Polyvinyl alcohol	USP/NF			Martindale

USP/NF*: present in USP 23/NF 18 but not in USP 24/NF 19.

REFERENCES

1. ASTM Standards. American Society for Testing and Materials. Standard Terminology Relating to Soap and Other Detergents. D-459-95. Philadelphia, 1995.
2. H. E. Junginger, Ointments and Creams as Colloidal Drug Delivery Systems. Colloidal Drug Delivery Systems, Vol. 66. Marcel Dekker, New York, 1994, pp. 1–30.
3. I. Effendy, H. I. Maibach, Surfactants and experimental irritant contact dermatitis. Contact Dermatitis, 33(4): 217–225, 1995.
4. Vidal. Le dictionnaire. Edition du Vidal, Paris, 1998.
5. BIAM, Medical data base, http://www.cri.ensmp.fr/biam/acceuil.html, 1998.
6. M. Ash, I. Ash, Handbook of Pharmaceutical Additives. Gower, Hampshire, 1995.
7. International Cosmetic Ingredient Dictionary, Vols. 1 and 2. (J. A. Wennninger, ed.), Washington, Cosmetic, Toiletry and Fragrance Association, 1993.
8. Pharmacopée Européenne, 3rd ed. Strasbourg, Conseil de l'Europe. 1997, 1998, 1999.
9. USP23/NF18 and USP24/NF19. The United States Pharmacopeia/The National Formaulary. Rockville, MD, United States Pharmacopeial Convention, 1995, 2000.
10. British Pharmacopoeia, 1993, Addendum 1995, London, Her Majesty's Stationery Office. 1995.
11. J. E. F. Reynold, Martindale: The Extra Pharmacopoeia, 31th Ed. London, Royal Pharmaceutical Society, 1996.
12. S. Nema, R. J. Washkuhn, R. J. Brendel. Excipients and their use in injectable products. PDA J. Pharm. Sci. Tech. 51, (4): 166–171, 1997.
13. A. R. G. Remington, The Science and Practice of Pharmacy, 19th Ed., Vol. 1. Easton, PA, Mack Publishing, 1995, pp. 250–251.
14. G. Zagrafi, Interfacial Phenomena, in: The Science and Practice of Pharmacy, 19th Ed., Vol. 1. Easton, PA, Mack Publishing, 1995, pp. 241–251.
15. B. Idson, Surfactants for cosmetic macroemulsions: properties and application, in Surfactants in Cosmetics. In Surfactants Science Series, Vol 16. New York, Marcel Dekker, 1985, pp. 1–28.
16. A. B. Skaare, G. Rolla, P. Barkvoll. The influence of triclosan, zinc or propylene glycol on oral mucosa exposed to sodium lauryl sulfate. Eur. J. Oral. Sci. 105: 527–533, 1997.
17. A. Di Nardo, K. Sugino, P. Wertz, J. Ademola, H. I. Maibach. Sodium lauryl surfate induced irritant contact dermatitis: a correlation study between ceramides an in vivo parameters of irritation. Contact Derm. 35(2): 86–91, 1996.
18. Kästner W, Local tolerance mucous membranes and skin, in Anionic Surfactants: Biochemistry, Toxicology, Dermatology. Surfactants Science Series, Vol. 10. New York, Marcel Dekker, 1980, pp. 137–307.
19. M. M. Singer, R. S. Tjeerdema, Fate and effect of the surfactant sodium dodecyl sulfate. Rev. Environ. Contam. Toxicol. 133: 95–149, 1993.
20. E. K. Anderberg, C. Nystrom, P. Artursson, Epithelial transport of drug in cell culture. VII. Effects of pharamaceutical surfactants excipients and bile acids on transepithelial permeability in monolayers of human intestinal epithelial (Caco-2) cells. J. Pharm. Sci. 81(9): 879–887, 1992.
21. K. P. Wilheilm, M. Samblebe, C. P. Siegers, Quantitative in vitro assessment of N-alkyl sulphate-induced cytotoxicity in human keratinocytes. Comparison with in vivo human irritation tests. Br. J. Dermatol. 130(1): 18–23, 1994.

22. P. G. M. Van Der Walk, J. P. Nater, E. Bleumink, Skin irritancy of surfactants as assessed by water vapour loss measurements. J. Invest. Dematol. 82: 291–293, 1984.
23. S. Patil, P. Singh, K. Ṣarousour, Maibach H. Quantification of sodium lauryl sulfate penetration into the skin and underlying tissue after topical application, pharmacological and toxicological implications. J. Pharm. Sci. 84(10): 1240–1244, 1995.
24. A. Saija, P. Princi, D. Trombetta, M. Lanza, A. De Pasquale, Change in the permeability of the blood–brain barrier following sodium dodecyl sulphate administration in the rat. Exp. Brain Res. 115(3): 546–551, 1997.
25. K. P. Wilheilm, M. Samblebe, C. P. Siegers, Quantitative in vitro assessment of N-alkyl sulphate-induced cytotoxicity in human keratinocytes (HaCaT). Comparison with in vivo human irritations tests. Br. J. Dermatol 130(1): 18–23, 1994.
26. L. Fisher, B. Berman, Contact allergy to sulfonated castor oil. Contact Dermatitis. 7(6): 339–340, 1981.
27. J. L. Zatz, L. Schnitzer, P. Sarportar. Flocculation of sulfamerazine suspension by a cationic polymer. J. Pharm. Sci. 68(12): 1491–1494, 1979.
28. K. MacKenzie, S. Henwood, G. Foster, F. Akin, R. Davis, P. DeBaecke, G. Sisson, G. McKinney. Three-generation reproduction study with dioctyl sodium sulfosuccinate in rats. Fund. Appl. Toxicol. 15(1): 53–62, 1990.
29. G. Van Rij, W. H. Beekhuis, C. A. Eggink, A. J. Geerards, K. Remeijer, E. L. Pels, Documenta Ophthalmol. 90(1): 7–14, 1995.
30. J. Y. Lee, B. J. Wang, Contact dermatitis caused by cetrimide in antiseptics. Contact Dermatitis 33(3): 168–171, 1995.
31. J. Y. Lee, Pathogenesis of abnormal keratinization in ichthyosiform cetrimide dermatitis; an ultrastructural sudy. Am. J. Dermatopathol. 19(2): 162–167, 1997.
32. A. Hatinen, M. Terasvirta, J. E. Fraki, Contact allergy to components in topical ophthalmologic preparations. Acta Ophthalmol. 63(4): 424–426, 1985.
33. G. H. Lin, M. Hemming, Occular and dermal irritation of some quaternary ammonium compounds. Food Chem. Toxicol. 34(2): 177–182, 1996.
34. K. Green, D. S. Hull, E. D. Vaughn, A. A. Malizia, K. Bowman, Rabbit endothelial response to ophtalmic preservatives. Arch Ophthalmol. 98(12): 2218–2221, 1997.
35. D. Attwood, A. T. Florence, Surfactant Systems. Their Chemistry, Pharmacy and Biology. London, Chapman and Hall, 1983.
36. M. J. Schick, Nonionic surfactants, Vol. 1. New York, Marcel Dekker, 1966, pp 1–8.
37. H. A. Lieberman, M. M. Rieger, G. S. Banker, Pharmaceutical Dosage Forms: Disperse Systems, Vol. 1. New York, Marcel Dekker, 1988.
38. K. Taniguchi, K. Itakura, K. Morisaki, S. Hayashi. Effects of Tween 80 and liposomes on the corneal permeability of anti-inflammatory steroids. J.Pharmacobiodyn. 11(10): 685–693, 1988.
39. R. L. Oberle, T. J. Moore, D. A. Krummel, Evaluation of mucosal damage of surfactants in rat jejunum and colon. J. Pharmacol. Toxicol. Meth. 33(2): 75–81, 1995.
40. J. D. Jonkman-de Vries, B. J. Van den Bemt, W. W. Ten Bokkel-Huinink, et al. Pharmaceutical development of parenteral formulation of the investigational anticancer drug clanfenur. J. Pharm. Sci. Technol. 51(2): 89–95, 1997.
41. S. S. Talmage, Environmental and Human Safety of Major Surfactants. Alcohol Ethoxylates and Alkylphenol Ethoxylates. Soap and Detergents Association. London, Lewis, 1994.

42. X. Cai, D. J. Grant, T. S. Wiedmann, Analysis of the solubilization of steroids by bile salt micelles. J. Pharm. Sci. 86(3): 372–377, 1997.
43. M. M. Rieger, Surfactants, in Pharmaceutical Dosage Forms: Disperse Systems, Vol. 1. New York, Marcel Dekker, 1988, pp. 285–366.
44. J. R. Silvius, Structure and nomenclature, in Phospholipids Handbook. New York, Marcel Dekker, 1993, pp. 1–22.
45. X. V. Gauthier, Tensioactifs naturels: sucroesters et APG en forte croissance, Parfums Cosmetiques Actualités 139: 44–47, 1998.
46. B. Brancq, Development and trends of sugar derived surfactants. SOFW-Journal 118(15): 905–906, 1992.
47. M. A. Thevenin, J. L. Grossiord, M. C. Poelman, Sucrose esters/cosurfactant microemulsion systems for transdermal delivery: assessment of bicontinuous structures. Int. J. Pharm. 137: 177–186, 1996.
48. D. Balzer, Alkylpolyglucosides, their physico-chemical properties and their uses. Tenside Surf. Det. 28(6): 419–427, 1991.
49. T. Förster, H. Hensen, R. Hofmann, B. Salka, Using alkylpolyglucosides in personal-care products, Cosmet. Toiletries 110: 23–29, 1995.
50. D. Balzer, Cloud point phenomena in the phase behavior of alkylpolyglucosides in water. Langmuir, 9:3375–3382, 1993.
51. G. Platz, C. Thunig, J. Policke, W. Kirchhoff, D. Nickel, Phase behavior of alkylpolyglucosides in combination with fatty alcohols and alkylsulfates. Colloids Surfaces A 88: 113–122, 1994.
52. D. Nickel, C. Nitsch, P. Kurzenörfer, W. von Rybinski, Interfacial properties of surfactant mixture with alkylpolyglucoside. Progr. Colloid Polym. Sci. 89: 249–252, 1992.
53. M. Kahlweit, G. Busse, B. Faulhaber, Preparing microemulsions with alkylmonoglucoside and the role of n-alkanols. Langmuir, 11:3382–3387, 1995.
54. M. Kahlweit, G. Busse, B. Faulhaber, Preparing nontoxic microemulsions with alkylmonoglucosides and the role of alkanediol as cosolvents. Langmuir 12: 861–862, 1996.
55. K. Fukuda, O. Söderman, B. Lindman, K. Shinoda. Microemulsions formed by alkylpolyglucosides and on alkylglycerolether. Langmuir. 9: 2921–2925, 1993.
56. P. Van Roekeghem, M. P. Savelli-Boukhaled, O. Douillet, G. Cavé, P. Godet, G. Ronco, P. Villa, Physicochemical properties of a novel series of amphiphilic sugar-based molecules: 1-O-n-alkyl-D, L-xylitols. STP Pharma Sci. 7(5): 354–364, 1997.

2

Formulation Concepts for the Emulsion Maker

Jean-Louis Salager
Lab. FIRP, Chemical Engineering School,
University of the Andes, Mérida, Venezuela

I. INTRODUCTION AND DEFINITIONS

This chapter and the following one deal with the formulation and the properties of emulsions, i.e., dispersions of drops of some liquid in another immiscible one, which exhibit more or less stability depending on the application (1,2). The stability against drop coalescence is provided by the presence of a small amount of a third component, so-called emulsifier, which is in general a surface-active agent or surfactant that adsorbs at the drop interface and produces some interdrop repulsion according to a variety of static and dynamic phenomena (3).

The word *emulsion* is found in both *microemulsion* and *macroemulsion* (as well as the recently introduced terms *miniemulsion* and *nanoemulsion*), so that some confusion arises from this labeling. Macroemulsions (or simply emulsions) are liquid-in-liquid dispersions, with drop size ranging typically from 1 to 100 μm (that can be extended in special cases down to 0.5 μm or up to 500 μm). In this range, the drops are in general large enough to settle under gravity influence. Emulsions are thermodynamically unstable systems because their decay does result in a decrease in free energy. However, the kinetic mechanisms involved in emulsion breaking can be so slow that the corresponding emulsion may be considered as (meta)stable as far as the application is concerned. In what follows, the emulsion properties, with exception to stability, will be considered to remain constant during the time scale required for the experiment or for the practical application to be carried out.

Surfactant molecules contain both a polar and an apolar group. Hence, they won't be completely at ease either in an aqueous phase or in an organic phase, since one of their groups would not get the right match with the solvent, be it aqueous or organic. For such a reason, surfactant molecules exhibit a very peculiar behavior, that is inherent to their structure. First, they tend to adsorb at interfaces, where they can fulfill their dual affinity with the hydrophilic group located in the aqueous phase and the hydrophobic one in oil or air. Second, they reduce the mismatch with the solvent through a very specific kind of aggregation process known as micellization (4–6).

Micelles are colloidal aggregates that form above some concentration, the so-called critical micelle concentration (CMC), according to the pattern shown in Fig. 1. In a normal micellar aggregate (left), the surfactant hydrophilic groups are in contact with the aqueous solvent, while the hydrophobic "tails" are located in the micelle core away from the aqueous environment—a favorable situation from an energetic point of view. Inverse micelles (right) are similar in structure but this time the solvent is an oil phase and the surfactant hydrophilic groups are located inside (3).

The micellar aggregation is favorable but it is opposed by molecular random agitation. Thus, a certain micellar size is attained as the best dynamic com-

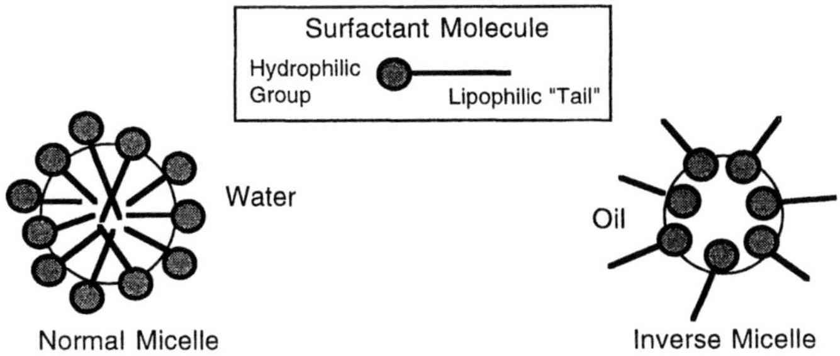

Figure 1 Micellar type aggregates.

promise between the two effects. It is worth remarking that since micelles exchange surfactant molecules with the bulk phase at a very fast rate, the shape is spherical only in average. According, micelles size and shape can change very quickly to satisfy new physicochemical conditions, such as those produced by a variation in formulation or temperature.

The core of a micelle is an exclusion region where substances that are incompatible with the solvent can enter spontaneously in a process called *solubilization* (4,7). Because of solubilization, micelles become swollen and may attain the size of a small droplet, i.e., 1000 Å or 0.1 μm. If the surfactant concentration increases well above the CMC, many micelles are formed. If another phase is present, e.g., oil if the solvent is water, and provided that the physicochemical formulation is appropriate, the micelles would solubilize large amounts of this phase and become swollen until they start interacting in a phenomenon called percolation. Such packed swollen micelle structures that could solubilize large amounts of both oil and water have been called microemulsions because they were first thought to be extremely small droplet emulsions (8–11). Actually this is a misnomer for at least two reasons. First, a microemulsion is before all a single-phase system, that is thermodynamically stable. Second, many microemulsions cannot be considered as dispersions of very small droplets, but rather as percolated or bicontinuous structures (12) in which there is no internal or external phase, and no possibility of dilution as in normal emulsions.

Conversely, miniemulsions (13) are macroemulsions, i.e., two-phase systems in which the drop size has been made extremely small by some tricky process (not only by stirring); miniemulsion drop size is actually of the same order of magnitude as the microemulsion structure, which is why they have been called nanoemulsions as well (14). However, and contrarily to microemulsions, they are not thermodynamically stable and they can be diluted with their external phase. Because of their extremely small drop size miniemulsions are often transparent and viscous, and have also been called gel emulsions (15–17).

There are several types of emulsions depending on how the oil and water phases are located in the dispersed system. Note that in all cases the terms oil and water are used in a broad sense i.e., the less polar and more polar of the two immiscible phases.

Figure 2 indicates the different types of emulsions. Simple emulsions are labeled as oil-in-water (O/W) when they exhibit oil drops dispersed in an aqueous phase, or water-in-oil (W/O) if the opposite occurs, while multiple or double emulsions are symbolized either by $W_1/O/W_2$ or $O_1/W/O_2$. W_1 (respectively O_1) and W2 (respectively O_2) indicate the most internal phase and the most external one. Note that phases with subscript 1 and 2 may be identical or different. If they are not the same a likely difference in chemical potential may drive a mass transfer process, a phenomenon that is advantageously harnessed for controlled-release applications. Biemulsions are emulsions containing two different internal phase droplets, either of the same nature (but different size) or of different nature (whatever the size). The first kind of biemulsion is used to control some property, as, for example, emulsion viscosity, whereas the second may be used to produce controlled chemical reaction or mass transfer between the two internal phases.

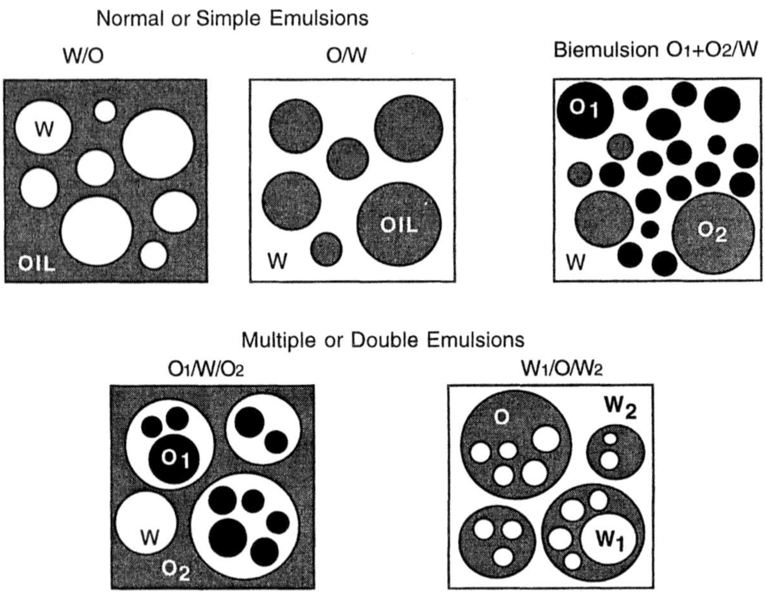

Figure 2 Different Types of Macroemulsions.

II. EMULSION MAKING: IMPORTANCE OF FORMULATION

In a broad sense, the making of an emulsified system involves several choices and activities, that may be classified in three categories according to the type of variables that are dealt with. The two first elections, concerning what to put into the emulsion and how much of it, are the subjects of this chapter. The third one, which addresses the question of how to make the emulsion, will be discussed in the next and several other chapters of this book.

The first choice to be made is the nature of the substances that will make the emulsion, will be the oil and aqueous phases, as well as the emulsifier. In most cases, none of them will be a pure substance but rather a mixture, sometimes a very complex one. For instance, the aqueous phase generally contains different kinds of electroylytes in variable concentration.

Naturally occurring oil phases, either hydrocarbons or glyceride esters, contain scores of chemical species. Commercial surfactants are rarely pure substances. Either for economic limitations or for chemical synthesis reasons, they are often mixtures. In certain cases a surfactant mixture is purposely made to attain some intermediate property or to produce some synergetic effect.

These so-called *field* or *formulation variables* are, with temperataure and pressure, the intensive physicochemical variables that define the thermodynamic equilibrium conditions of the system through the equality of chemical potentials. The typical expression of the chemical potential of a surfactant molecule in solution (either in oil or water) can be written as:

$$\mu = \mu^0 + RT \ln a \tag{1}$$

where a is the activity (or the concentration if an ideal solution behavior is assumed, as, for instance, below the CMC), and μ^0 is the standard chemical potential in some reference state. This standard chemical potential depends on the nature of both the solute and the solvent, as well as on temperature and pressure.

In the simplest practical case there are four variables: three formulation variables that define the surfactant, oil, and water nature, in addition to temperature, since the effect of pressure is generally neglected in liquid systems for the sake of simplicity. In a typical commercial application, as many as ten formulation variables may be involved. It is worth noting that such a high number of degrees of freedom is a challenge to any systematic study. For instance, if only 5 alternatives are to be picked out for each of the 10 formulation variables, this means that 100,000 different cases would have to be surveyed—an appalling number for any practical situation.

This is why in many instances the selection of formulation variables has been so far a matter of expertise or experience. Fortunately, the practical understanding of the concept of physicochemical formulation and its effects on phase behavior and other system properties has improved considerably in the past decade and a comprehensive approach is now available. This will be discussed in the following sections.

The second choice deals with the so-called *composition variables*, which define the relative amounts of the different substances involved in the system. In a surfactant–oil–water (SOW) ternary system, two independent composition variables are sufficient, since the third one is the complement to 100%. Most often, the surfactant concentration (or proportion) and the water-to-oil ratio (WOR) are the selected independent variables.

In many practical instances there are many more than three components, but the basic ternary (SOW) accounts for most of the system mass or volume, and minor additives such as alcohol or electrolyte may be neglected in the overall composition inventory, although they can completely change the phase behavior. It means that the representation of the composition effect can be carried out on a two-dimensional graph like a triangular ternary diagram, while taking the other composition variables as external conditions (i.e., not represented in the diagram), like temperature and pressure.

If for some reason there are two surfactants, or a surfactant and a cosurfactant, that are to be treated as independent components, a quaternary representation is required, as for instance in a tridimensional tetrahedron diagram. Although such a three-dimensional diagram is not going to be handled as straightforwardly as a ternary diagram, currently available tools like 3D software may provide the means to cope with the third dimension.

The last category involves several choices dealing with the *emulsification protocol* that can range from straightforward to whimsical, depending on the degree of nonequilibrium features that are introduced.

For instance, some emulsions are attained by simply stirring a biphasic system in a more or less violent fashion during a certain time. A variety of emulsifying devices are available from slow-motion ribbon mixers for very viscous materials, to highly turbulent stirrers as colloid mills or high-pressure homogenizors. It is often very difficult to precisely describe what exactly happens in this type of device, particularly because most of the research on two-phase stirring and mixing deals with oil/water systems in the absence of surfactants, i.e,. dispersions rather than emulsions (18,19). Nevertheless, when only fluid mechanical variables are involved, it seems that the process can be repeated and is relatively easy to master and manipulate.

Other emulsification protocols are based on a transient phenomenon, in which a dominant role is played by an unsteady mechanism, e.g., mass transfer through interface, which is not easy to ascertain or control. Often the nonequilibrium is driven by a continuously programmed change in a single variable such as temperature, amount or type of surfactant, water-to-oil ratio, etc., so that a phase behavior frontier is crossed and some event such as an emulsion inversion is triggered. In such a case a precise recipe has to be closely followed to avoid divergence from the sought-after result.

The making of an emulsion involves many nonequilibrium features, at least from the mechanical point of view. Actually the product of the interfacial tension by the produced surface area $\gamma\,\Delta A$, which is the interfacial energy, is always much smaller than the mechanical energy put into the system by the stirring device. A significant characteristic is the way and the efficiency in which the energy is provided to the drop so that breaking is favored over coalescence. This has to do not only with the device but with formulation and eventual transient events.

In effect, the two phases and the interface can be in physicochemical equilibrium or not prior to emulsification. If the SOW system is preequilibrated there will be no mass transfer or chemical potential driving force, only adsorption on newly created surface. If not, mass transfer through interface could produce spontaneous emulsification and other unusual phenomena, both during the emulsification and during the decay.

After emulsification is carried out, the emulsion is generally handled or stored for some length of time, which is long compared to the duration of emulsification. The changes during this period are related to emulsion stability, which deals with how the emulsion decay or breaking takes place. This involves many different mechanisms to be discussed later on. For now it is enough to point out that the direction of change is indicated by the free energy decrease produced by drop coalescence, since a $\Delta G = \gamma\,\Delta A$ decrease implies a reduction in surface area. It is worth remarking, however, that this decay can be delayed for so long, that it would not happen in practice. When the emulsion is poured in a vessel and left to rest, or when it is pumped into a pipeline or carried in a tanker, it is submitted to a variety of decay conditions. Emulsion decay, either slow or fast, involves different steps, such as settling and thin-film drainage. The first depends on physical variables, while the second is commanded by the phenomenology of two approaching interfaces, involving various phenomena, all transient in nature.

The role of the surfactant as the essential ingredient in the formation and persistence of an emulsion is associated with several nonequilibrium processes. When the drops are formed by effect of mechanical shearing the presence of surfactant reduces the interfacial tension, so that breaking is favored. On the other hand, the adsorption of surfactant onto the freshly produced interface results in a repulsion that prevents the immediate coalescence of neighboring droplets. Thus, the surfactant plays a role in both the breaking and coalescence steps, i.e., the dynamic balance that determines the drop size.

After all these comments on the presence of several transient events and unsteady-state phenomena involved in many instances of the emulsion life, how can it be useful to scrutinize the effect of physicochemical formulation on phase behavior, which is an equilibrium concept? The answer is because it is helpful to our understanding the equilibrium-related situation to substantiate the nonequilibrium phenomenology, which would otherwise be impossible to interpret because it involves changes that are kinetically slow, mechanistically complex, and often result in metastable configurations. As recently stated by Laughlin (20), ''*an independent determination of the equilibrium phase behavior is essential if one is to diagnose correctly the kinetics and colloidal phenomena that exist. Doing so is of considerable value industrially, because diagnosing problems correctly is the key to addressing them efficiently.*''

It is found that whenever a SOW system is at physicochemical equilibrium prior to emulsification, things become much easier to forecast. As a matter of fact, if an isotropic turbulent stirring is applied for emulsification, a preequilibrated SOW system (PES) would lead to predictable emulsion type and properties, according to the phenomenology to be described in the next chapter.

If tricky programming is carried out for some nonequilibrium purposes, things are not so easy to foresee, but the PES phenomenology could be helpful whether the changes are slow so that quasi-equilibrium is reached or, conversely, very rapid so that a memory feature generates a metastable state by quenching any possible change.

Before going into the topic of formulation, it is worth remembering some basic concepts about phase equilibria. Since emulsions are two-phase systems, their physicochemical formulation should be such that there are two phases in equilibrium, which are in due time stirred to make it a dispersion. Thus, formulation is to be studied according to its influence on the phase behavior of the SOW system, with emphasis on cases where two (or more) phases coexist in equilibrium. The well-known phase rule can be stated as:

$$P = C + 2 - F \tag{2}$$

where P is the number of phases at equilibrium, C the number of components, and F the degrees of freedom, i.e., the number of system variables that must be specified to define the state of the system (see, for instance, Ref. 20).

In the case of a ternary, the number of components is $C = 3$, and the field variables are up to four: two chemical potentials (linked with two independent composition variables), and temperature and pressure.

Single-phase behavior ($P = 1$) results in $F = 4$; temperature and pressure, as well as two composition variables, must be fixed to define the state of the system. Thus the single-phase region would take over a bidimensional zone in the ternary diagram.

When $P = 2$, and at a fixed temperature and pressure, then $F = 3$, and only one composition variable is free to be selected, at constant temperature and pressure. This means that the two-phase equilibrium occurs along a unidimensional line, i.e., the binodal boundary.

Three-phase occurrence ($P = 3$) requires the system to be invariant at fixed temperature and pressure. In other words, there is no degree of freedom in composition, and the three phases occur at fixed (composition) points in the diagram.

According to the above discussion, the ternary SOW systems are expected to exhibit up to three-phase behavior at equilibrium. Note that this statement implies that temperature and pressure, as well as nature of the three components, can be fixed.

SOW systems are special cases among ternary systems because the surfactant is compatible with both other (otherwise incompatible) components. In effect, surfactants are amphiphilic substances with a strong affinity toward both oil and water. Consequently, many surfactants are completely miscible both in water thanks to their hydrophilic group and in oil thanks to their lipophilic group. This situation may be taken as general, since it is often true only at low surfactant concentration, up to a few times the CMC. However, when the surfactant concentration is higher than a few percentage points, mesophases of the liquid crystal type often occur, although they might be destroyed by raising the temperature or by introducing a disorder factor such as a cosurfactant.

The SOW ternary is characterized by the fact that it has only one miscibility gap that takes place when oil and water are mixed. The surfactant may be viewed as an additive that reduces this miscibility gap. These concepts will be made clear in the following section.

III. FORMULATION INFLUENCE ON PHASE BEHAVIOR

A. Phase Separation and Phase Diagrams

The thermodynamic potential that is used to analyze the phase behavior of multicomponent systems at constant temperatrue and pressure is the Gibbs free energy, G. A spontaneous change will take place any time it is associated with a reduction in free energy ($\Delta G < 0$).

To make things as simple as possible, let us start with the case of a binary mixture containing components 1 and 2. Its composition depends on a single independent variable that is identified as some fraction of component 1, i.e., X_1. In this case the variation of its free energy vs. composition variable (X_1) readily indicates whether the mixture would be miscible in all proportions or would exhibit a miscibility gap, i.e., a region in which two phases would separate as in the case of oil and water (see Fig. 3).

The left diagram case indicates the typical aspect of the G graph for a completely miscible binary. The minimum indicates that the mixing results in a decrease of free energy, probably because of the associated disorder and entropy increase.

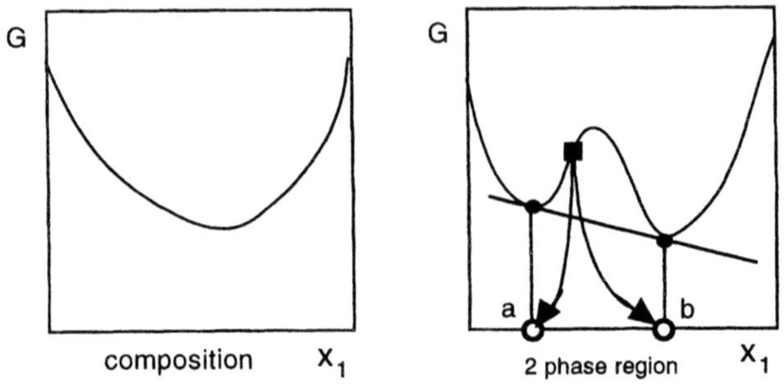

Figure 3 Shapes of free energy variation v. composition for a binary system.

In the other case that is illustrated in the right diagram, the free energy curve exhibits two minima. Actually the two minima are not required for what follows, only the fact that there is an inflexion point so that a common tangent can be drawn through two points. In the middle zone, the free energy of the single-phase system (square dot) is higher than the free energy of the split phases located at the points of contact of the common tangent (black dots). Hence, when the composition of the system is located between "a" and "b," then the system spontaneously separates into two phases, represented by composition "a" and "b" (white dots).

The system is thus biphasic in the a-b zone (called a miscibility gap), while it is monophasic elsewhere. In the two-phase region there is an equilibrium between phase "a" that contains component 2 saturated with 1, and phase "b" that contains mainly 1 saturated with 2. The relative amount of a and b are readily calculated according to the level rule.

A similar situation occurs with ternary systems, but this time a three-dimensional representation (with two independent variables) is required, and the free energy is a surface depending on two independent variables (called X_1 and X_2). As predicted by the phase rule, the free energy surface may exhibit up to three phases, so that the tangent plane at this surface may contact it in one, two, or three points, depending on the case.

Figure 4 is a tridimensional diagram in which the free energy *G* is plotted (vertically) against two independent composition variables X_1 and X_2, which are arranged to produce a equilateral triangular diagram at the bottom. Figure 4 indicates the case in which the tangent plane touches the surface at three points (black dots).

The composition of the phases represented by each of these black dots is indicated as a white dot on the bottom triangular phase diagram. Any monophasic system whose composition lies inside the triangle made by these three points has

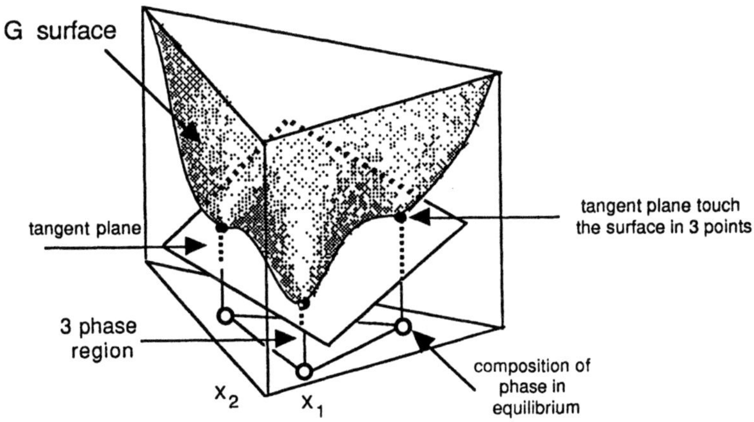

Figure 4 Shape of free energy surface v. composition for a ternary system.

a representative point on the *G* surface that is above the tangent plane, and thus a split into three phases (represented by the white dots, which have a fixed position because of the invariance of the system) results in a free energy decrease and thus takes place spontaneously. The phase behavior analysis is generally carried out in the bottom bidimensional phase diagram for the sake of simplicity.

The SOW ternary system may be essentially viewed as a nonmiscible OW binary, which is made more and more compatible by adding a surfactant. This is essentially similar to the way the temperature effect, as a miscibility enhancing factor, is studied on a binary diagram.

Figure 5 shows a series of binary diagrams that illustrate the variation of the free energy *G* vs. water–oil composition at increasing surfactant content. In the absence of surfactant (bottom graph), there is a large miscibility gap that ranges essentially from pure water to pure oil, with two extremely narrow miscible regions at both extremes, corresponding to the mutual solubilities of oil in water and water in oil.

The second diagram from bottom (same figure) characterizes a system that contains a small amount of surfactant. The black dot represents the same overall composition on the binary OW (plus S) graph and ternary SOW triangular phase diagram. The line that connects the two minima on the *G* curve is the same (tie) line that passes through the black dot on the ternary diagram and ends at the frontier of the two phase region (called binodal frontier). The compositions of the two phases in equilibrium are registered at the two extremes of this tie line on both diagrams. The ternary representation provides additional data on the surfactant partitioning through the tie line slope.

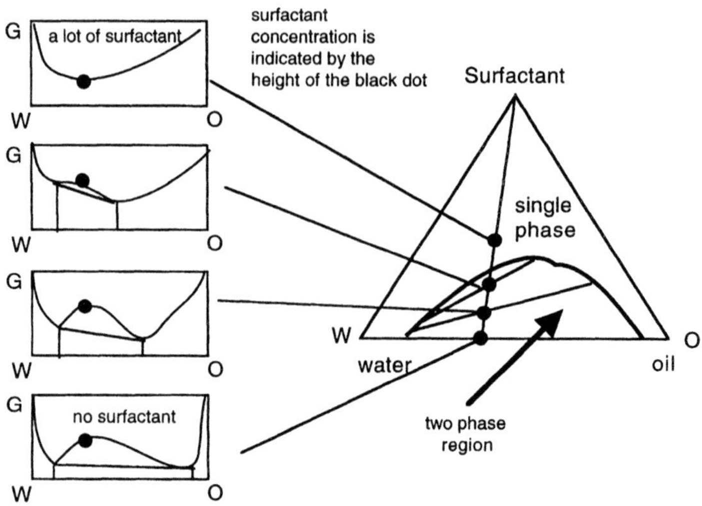

Figure 5 Shape of free energy curves v. composition for a W/O binary to which different amounts of surfactant are added, and equivalent ternary diagram.

In the next graph a narrower two-phase region indicates that the miscibility gap is shrinking. At some point (between the third and fourth *G* graphs) there is no longer a tangent common to two points, and a single phase is attained at all compositions.

If the two contact points merge together, the crossing of the binodal frontier takes place at a critical point.

When more and more surfactant is added, the miscibility gap shrinks and the compositions of the phases in equilibrium are shifted. In Fig. 5, the aqueous phase composition remains essentially the same, while the oil phase composition is displaced from right to left. The only way an oil phase can be made to contain more water is to solubilize it inside its (inverse) micelles. The right extreme of the tie line on the ternary diagram also moves not only to the left (because the oil phase is solubilizing water) but also vertically, an indication that the oil phase is taking up most of the surfactant. Such an oil phase that contains a lot of surfactant and that has solubilized a lot of water is probably a microemulsion or a liquid crystal, since no consolute phase would exist in these conditions. This particular case was predicted in Winsor's analysis half a century ago.

B. Winsor Phase Behavior Diagrams

In a series of papers (21) and a review book (22), Winsor reported the relationship between the phase behavior of amphiphile–oil–water and the nature of the different components of a ternary system, i.e., what we now call the formulation variables. By studying various ternary systems, he surveyed and rationalized the phase behavior situations into three typical cases of amphiphile–oil–water ternary system. In what follows the amphiphile is generally a pure surfactant, although it may also be a mixture of surfactant and cosurfactant (alcohol) with similar affinities for the oil and water phases.

Figure 6 indicates the three different phase behavior diagrams according to Winsor's classification. All three diagram types display a widely spread single-phase region from the A vertex down, which extends on both sides down to the W and O vertices. In practice the single-phase region does not always extend down to both the oil and water, but this is the case for an ''exemplary'' amphiphile that ought to be miscible in all proportions with both water and oil. In real cases this region may include mesophase zones that can, however, be ''made liquid'' by increasing temperature or disorder (e.g., by adding alcohol).

The unique miscibility gap is located near the OW side, where there is a polyphasic region in the three diagram cases. This region width tends to shrink as more and more amphiphile is added, as a confirmation of the capability of the amphiphile to diminish the miscibility gap. It may be said that the diagram type essentially relies on the configuration of its polyphasic region.

In type I diagram the polyphasic region phase behavior has been labeled 2̲, since it becomes visible as a two-phase separation with an amphiphile-rich aqueous phase being lower phase, from which the lower bar. As its surfactant

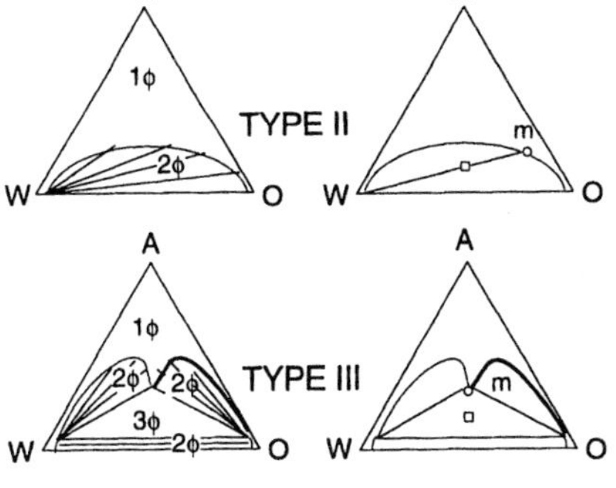

Figure 6 Winsor's three ternary diagram types.

content increases and the solubilized oil amount augments, the amphiphile-rich aqueous phase may contain S_1 type normal micelles, extremely swollen micelles, an O/W microemulsion, and then a percolated microemulsion in which the oil islands touch one another to end up forming some kind of bicontinuous structure. This surfactant-rich phase is located at the boundary of the single phase region in equilibrium with almost pure oil, as indicated by the tie line slope. Note that in this type I diagram, the critical point where the tie line becomes tangent to the binodal boundary is located on the extreme right of the binodal frontier.

Winsor's type II diagram and phase behavior, which is noted $\overline{2}$ for a similar reason as exposed previously, embodies the opposite situation, in which the polyphasic equilibrium consists of an inverse micellar oil solution S_2 that eventually solubilizes enough water to become a W/O microemulsion with separated or percolated water droplets or a bicontinuous one, in equilibrium with an essentially pure aqueous phase. In this case the tie line slope is slanted the other way and the critical point is located at the extreme left of the binodal curve.

Between type I and II cases, an intermediate diagram with a biphasic zone with horizontal tie lines could be expected. Such an occurrence does happen with some amphiphiles like alcohols, but not generally for surfactants, in which case a type III diagram and phase behavior is exhibited instead (23). Though it looks much more complex than the two previous ones, type III diagram is no exception and can be found in many systems. The polyphasic region possesses a three-phase zone surrounded by three two-phase zones according to Schreinemaker's rule of alternation (20,24).

Systems whose composition is located inside the three-phase triangle split into an amphiphile-rich phase (m), which is in the center of the diagram at the boundary of the single-phase region, and two excess phases, which are essentially pure aqueous phase and pure oil. The amphiphile-rich phase, which is found to be a bicontinuous microemulsion, has been called the middle phase because it appears in-between the oil and water phase in a test tube. Since the middle phase is at equilibrium with both excess phases, it cannot be diluted either by water or oil, and it is thus neither water- nor oil-continuous.

The two biphasic regions located above the three phase triangle are quite like type I (right) and type II (left) biphasic cases. However, in both cases the point representing the amphiphile-rich phase is very near the center of the diagram, a fact that reveals that this phase would contain similar amounts of oil and water. For instance, a system whose representative point is inside the left lobe would separate into an amphiphile-rich phase that contains almost as much water as oil, and an excess water phase that is essentially pure water. The amphiphile-rich phase is thus to be considered as the oil phase which has solubilized a lot of water, up to the point that such an oil phase could actually contain more water than oil. The composition of such a phase is similar to that of the middle phase, i.e., a bicontinuous microemulsion.

Since oil and water are not miscible with each other, there is necessarily a biphasic region near the OW side of the ternary diagram, so-called bottom biphasic region, which is a thin strip of very small extension in most diagrams. Experimental evidence indicates that the bottom tie line of the three-phase triangle is located at a height that corresponds to the CMCs so that no aggregation structure is found in the excess phases, a situation that precludes the existence of any solubilization.

Winsor reported that the single-phase region that appears in all types of diagrams could also exhibit liquid crystalline phases of different structure (lamellar, hexagonal, cubic) containing both oil and water in-between the surfactant bilayers. This circumstance appears above the multiphase region, i.e., at high amphiphile concentration. As in the case of a single liquid phase microemulsion, it does not bring any specific information about the type of diagram and is thus of no help as far as the type of diagram is concerned.

C. Phase Behavior Diagram Transition—Winsor's *R*

Experimental evidence gathered by Winsor and scores of other researchers afterward have shown that as a field variable (one of the formulation variables or temperature) is changed as in Fig. 7, the phase behavior diagram exhibits a continuous transition from type I to type II passing through type III, or vice versa, depending on the selected scan variable (11,22,25–27).

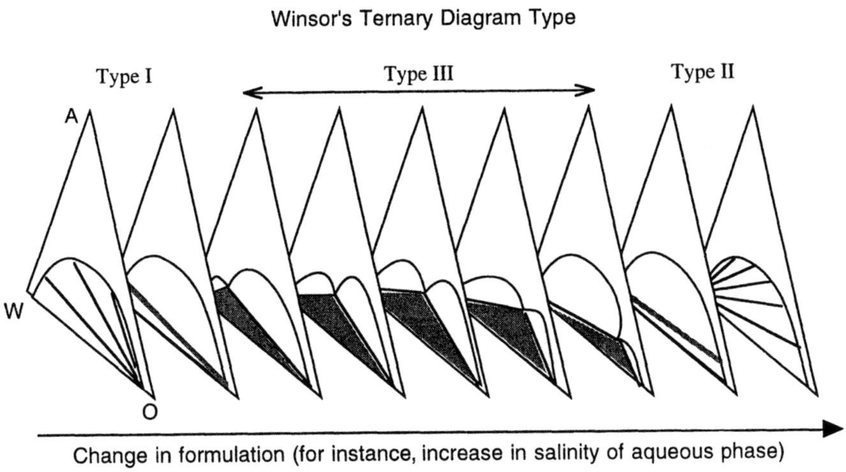

Figure 7 How Winsor's ternary diagram evolves when one formulation variable is changed.

The point that represents the microemulsion middle phase in central type III diagram moves from one side of the diagram to the other, as indicated in Fig. 7. It is seen that at both extremes of type III diagram occurrence, the three-phase tie triangle collapses when an upper lobe and bottom tie lines merge together (penultimate diagram on each side) (28,29).

Thanks to the experimental data he gathered on different systems, Winsor was able to associate the phase behavior with the physicochemical situation at interface. He proposed a pedagogical approach to interpret the results that would make use of the ratio of the interaction energies at interface.

Figure 8 represents the molecular population near the interfacial limit between the oil (O) and water (W) phases, where C represents the amphiphile or amphiphile mixture layer. The A's represent the molecular interaction energies per unit area. In Winsor notation:

$$A_{CW} = A_{HCW} + A_{LCW} \tag{3}$$

where A_{CW} represents the interaction between the amphiphile and the aqueous phase, A_{HCW} the interaction between the hydrophilic group of the amphiphile and the water, and A_{LCW} the interaction between the lipophilic group and the water, which is by the way probably much smaller than A_{HCW}. For a similar reason the interaction between the hydrophilic group of the surfactant and the oil, A_{HCO} is probably much smaller than the interaction of the lipophilic group and oil noted as A_{LCO}.

According to Winsor, A_{CO} and A_{CW} promote miscibility, while A_{OO}, A_{CC}, and A_{WW} promote the segregation of the components in different phases or in regions of higher local concentration of these components. The cosolubilizing effect of the amphiphile C will be greater, the higher both A_{CO} and A_{CW}.

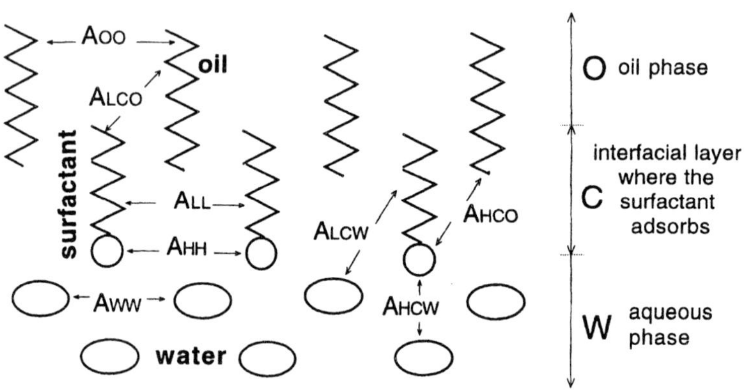

Figure 8 Molecular interactions in the interfacial region according to Winsor notation.

It was postulated by Winsor that the ratio of A_{CO} to A_{CW} would provide a rationale to determine the curvature of the C layer, which is directly linked with the phase behavior situation. The original R ratio was written as A_{CO}/A_{CW}, and it was suggested later that a better R ratio would be

$$R = \frac{A_{CO} - A_{OO} - A_{LL}}{A_{CW} - A_{WW} - A_{HH}} \tag{4}$$

because it takes into account all of the neat differences in interactions between the SOW mixed system and the separated component case.

For a given O-W component pair the C region would become concave toward W whenever the interactions between the amphiphile and the O phase are larger than the interaction of the amphiphile and the W phase or conversely (see Fig. 9). This is essentially equivalent to the so-called oriented wedge model (30,31) but with each amphiphile molecule associated with a certain amount of both O and W molecules, so that the resulting wedge units determine the bending when they are placed aside in the micellar or microemulsion structure.

Then, if $R < 1$ the interfacial layer has more interaction with the water phase and the interfacial layer would exhibit a concavity toward oil, i.e., it would form normal micelles. This situation corresponds to Winsor type I diagram occurrence. If there is enough surfactant, say a few percentage points, and if the micelles are solubilizing enough oil phase in their core, then this aqueous phase is a microemulsion above the $\underline{2}$ biphasic region. The tie line slope in the $\underline{2}$ biphasic region indicates that the surfactant-rich phase is the aqueous phase, in accordance with the fact that the surfactant has more interactions with that phase since $R < 1$.

If $R > 1$, the opposite occurs. Micellar systems are of the inverse type, and if there is enough surfactant to produce a microemulsion it will be the type encountered in Winsor type II diagram above the biphasic region.

According to this model, the $R = 1$ case would be associated with a zero curvature C layer that could be satisfied by a lamellar liquid crystal structure with alternating O and W flat layers, or a zero curvature surface of the Schwartz

Figure 9 Wedge theory model for micellar and interface curvature.

type or as a transient and fluctuating combination of micelles and inverse micelles (Fig. 10). It is now well accepted that middle phase microemulsions, which are in equilibrium with both oil and water excess phases (as in type III diagrams), exhibit some zero average curvature bicontinuous structures (12,32,33) that are not far from a fluctuating mixture of swollen normal and inverse micelles predicted by Winsor. This fluctuating model makes sense because micellar structures are not rigid. Actually, the micelle spherical shape is only an average, and the rate of exchange of surfactant molecules between the micelles' palisade region and the bulk phases is very fast, with relaxation times in the microsecond range.

By changing systematically one formulation variable (or temperature) at a time, a diagram transition type I ($R < 1$) $\rightarrow$ type III ($R = 1$) $\rightarrow$ type II ($R >$

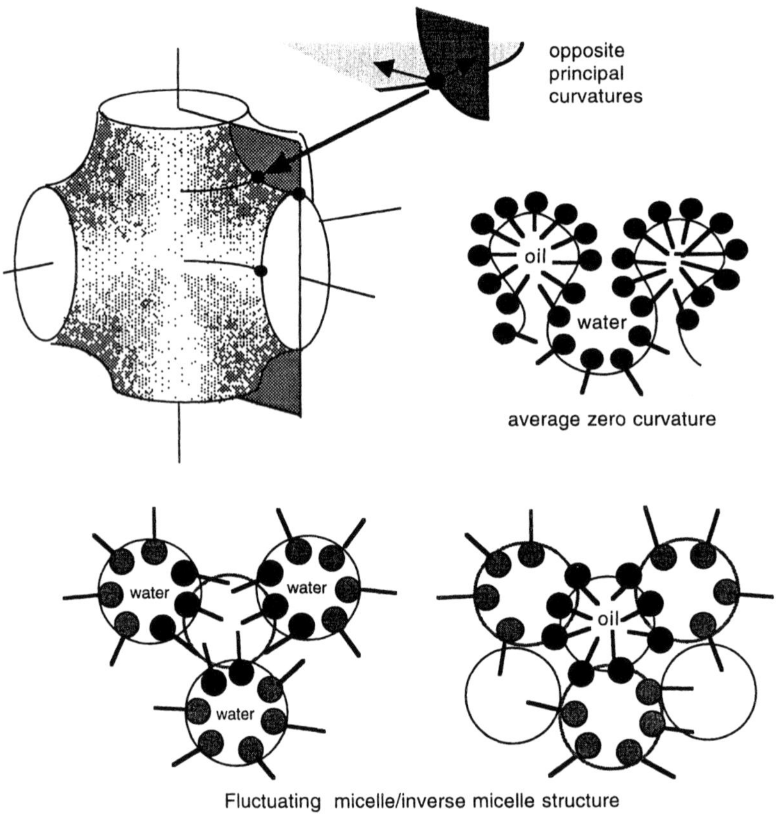

Figure 10 Schwartz surface of zero curvature and other models for bicontinuous microemulsion.

1) or inversely takes place. Winsor was able to corroborate the effect of each physicochemical formulation variable and temperature, and to correlate it to change in R value, as indicated in Table 1.

The first three effects are readily explained by the change in R numerator and denominator. As the surfactant lipophilic ''tail'' gets longer, the interaction with the oil phase A_{CO} is increased. In a similar way, a longer polyethylene oxide chain (hydrophilic group) of a nonionic surfactant results in an increase of the interaction with water A_{CW}.

The third effect is much more important with ionic surfactant systems than with nonionic ones, since it is linked with the shield effect produced by electrolyte ions with respect to the ionic group–water A_{CW} interaction.

The forth case is related to the fact that with most anionic surfactants, alkaline metal salts are more dissociated than their bivalent cation counterparts. Thus, the calcium or magnesium salts are less hydrophilic. This is also true for nonionic surfactants, although to a much lesser degree.

The effect of the oil chain length (alkane carbon number, ACN) is more complex to understand, since it modifies both A_{CO} and A_{OO} interactions with opposite effects. However, it is possible to solve this stalemate by remarking that London-style interactions may be taken as proportional to the number of methylene groups in each interacting molecules (34). Thus:

$$A_{CO} = \alpha\,\text{SACN ACN} \qquad \text{and} \qquad A_{OO} = \alpha\,\text{ACN}^2 \tag{5}$$

where SACN is the number of carbon atoms in the surfactant alkyl group. As oil ACN increases the second interaction increases faster (as the square) than the first one. As a consequence, the A_{CO}–A_{OO} summation decreases (provided that ACN is large enough). In other words, a higher ACN oil exhibits a lower neat

Table 1 Effect of Formulation Variables on Winsor R Ratio

When this variable is increased	Effect on interaction energies	and R
Amphiphile hydrophobe chain length	A_{CO} increases	Increases
Ethylene oxide polyether chain length (nonionic surfactant)	A_{CW} increases	Decreases
Aqueous phase salinity	A_{CW} decreases	Increases
Aqueous phase electrolyte valence ($Na^+ \rightarrow Ca^{2+}$)	A_{CW} decreases	Increases
Oil chain length (alkane hydrocarbon) increases	A_{CO} increases	
	A_{OO} increases more	
	A_{CO}-A_{OO} decreases	Decreases
Temperature increases (nonionic surfactant)	A_{CW} decreases	Increases
Temperature increases (ionic surfactant)	A_{CW} increases	Decreases

interaction energy with the C layer because of its more favorable self-association energy.

Nonionic surfactants of the polyethoxylated type exhibit a temperature dependency that is illustrated by the so-called cloud point phenomenon. As a matter of fact, the hydrophilicity of the polyether chain depends on a directional interaction between each -O- atom and surrounding water molecules. As temperature increases this interaction becomes looser because of the amplified randomness of the molecular motion. Hence, the hydrophilic character decreases and finally vanishes, as does the A_{CW} interaction.

With ionic surfactants, an increase in temperature tends to slightly increase the A_{CW} interaction, in accordance with the increase in solubility exhibited by most ionic species, and an opposite effect is attained, although of much lesser magnitude.

This is in agreement with the overall effect that is corroborated by experimental data. It is worth noting, however, that the actual temperature effect is probably much more complex to ascertain in Winsor's R ratio framework. In effect, temperature is likely to change not only A_{CW} but many or all other interactions as well, a situation that complicates the interpretation of experimental data. Moreover, the temperature range that can be studied without too much experimental trouble is not very extended, since it matches the liquid state of water. This is why a change in temperature is not the easiest to handle or to interpret experimental variation in order to produce a change in R value. As a matter of fact, the electrolyte concentration (salinity) and ethylene oxide number (EON) have often been preferred by experimentalists as scanning variables, for ionic and nonionic systems, since their effect is easier to manipulate, predict, and interpret.

The main lesson behind Winsor's pioneer work is that it reveals that there are essentially three-phase diagram elementary cases, although the number of formulation variables (including temperature and pressure) can be much higher than three. Winsor's R ratio allows one to bring together the effect of all formulation variables into a single one, i.e., the R ratio value, that commands the phase behavior of the SOW system. Consequently, the effects of the different formulation variables and temperature are competitive, and a certain R value may be attained in different ways—a favorable feature as far as practical formulation work is concerned.

It can be said that Winsor's R ratio has the best pedagogical value of all formulation concepts because it is simple to understand and at the same time allows quite good qualitative predictions. If it were amenable to numerical calculations there would be no need for alternate formulation concepts. This is not the case, however, because the molecular interactions A_{XX} cannot be calculated with sufficient accuracy yet. The R ratio would thus be the choice instrument to carry out qualitative inferences and to predicts trends. However, a quantitative instrument would be required when it comes down to serious formulation business.

D. Other Concepts Related to Formulation

The three cases of phase behavior related to Winsor R ratio have been disclosed through a different approach by other researchers. The oldest one is Bancroft's rule of thumb, i.e., the emulsion external phase is the one in which the surfactant preferentially partitions (35,36). Since it is obvious that the surfactant preferentially partitions into the phase for which it exhibits a greater affinity, a rather hydrophilic amphiphilic layer ($R < 1$) would produce O/W emulsions and vice versa, a result that is essentially the one resulting from the oriented wedge theory applied to an emulsion droplet. Although the wedge theory could be useful in interpreting the interactions whenever the colloid size is 10 or 20 times larger than the molecular size, i.e., in a micelle or up to a miniemulsion 200-Å drop, it is obvious that at a 10-μm drop interface the surfactant molecule cannot "feel the curvature." Nevertheless, the oriented wedge model has been recently brought back to life, when it was shown that the monolayer bending energy is related to the free energy of a transition state that affects coalescence rate and thus decides the emulsion type (37).

In presence of a solvent, e.g., water, it is known that mesophase can be lamellar (zero curvature), hexagonal, or cubic (20). In the presence of both oil and water, both phases may penetrate between the surfactant bilayers, and liquid crystals of the curved type could occur with the water or the oil toward concavity. In between, i.e., for the $R = 1$ case, a zero curvature, that may be attained either with a lamellar structure of the LC type, or a bicontinuous microemulsion with monkey saddle zero curvature geometry everywhere, is expected.

Thus the curvature may be considered as a state variable that can describe the situation of the system as Winsor's R ratio (38). The problem is that the substitution of the R ratio by the curvature does not produce any awesome advance in practice because curvature cannot be calculated directly in a quantitative way from the system formulation but rather through computations that are probably too complex for the practitioner (39).

An interesting feature is however that there is a natural or spontaneous curvature that indicates the preferred state of the interfacial layer, which is not necessarily the actual state. The preferred curvature could be the one related to the R concept, i.e., physicochemical formulation effects, while the actual one would result from the emulsification process. The difference between the two is some kind of bending energy frustration, which by the way could be related to the concept of stability of the interfacial structure, and of the resulting emulsion.

This could be linked to a model that connects the emulsion type to the relative coalescence rates of O/W and W/O emulsions (40).

IV. EMPIRICAL APPROACH TO FORMULATION QUANTIFICATION

While it is extremely important to develop fundamental concepts to understand and interpret the observed phenomena, it is tantamount important for the practitioner to have at hand effective tools to culminate his research path in a pragmatic solution to his problem. It has been said that even a simple formulation problem may exhibit many degrees of freedom, often too many for a systematic study to be carried out. The fundamental concepts enable the practitioner to know the expected trend and the direction in which he should start looking. They do not give him much information on the equivalence of different variables as far as the effect is concerned, critical information when economic matters or toxicity problems are dealt with. They do not give him any information either on the magnitude of the change he should apply to attain his purpose. Accordingly, the expertise or previous experience is the most quoted value. This experience is often translated into a rule of thumb or other intuitive or in house ''guesstimating'' method.

In the following review, it is seen that several alternate practical methods are paving the way to a full quantification of the formulation effect.

A. Hydrophile–Lipophile Balance

The first approach consisted of simplifying the problem by considering only one or two of the formulation variables, assuming that the selected variables were the most important for the particular problem and/or that the other formulation variables were either constant or had a negligible effect.

This was the stance taken in the first empirical yardstick, i.e., Griffin HLB introduced at the end of the 1940's (41,42). The HLB is a parameter characteristic of the surfactant and the required HLB (or HLB_{req}) is an attribute of the oil phase, which is determined as the HLB of the surfactant that provides the best match with a given oil to produce the maximum stability of an emulsion along a surfactant HLB scan.

All surfactants are assigned an HLB number according to a procedure that was first arbitrary, e.g., for oleic acid HLB = 1 and for potassium oleate HLB = 20, and was then deduced from consistency tests, particularly by comparing several kinds of properties. Table 2 indicates the HLB number of several well-known commercial surfactants. For polyethoxylated nonioic surfactants (with no other hydrophilic group) the HLB may be estimated by the following formula (43):

$$\mathrm{HLB} = \frac{100}{5} \quad \frac{\text{weight of polyethylene oxide chain}}{\text{total molecular weight}} \tag{6}$$

As an essential feature of the HLB scale, it was assumed that the HLB of a surfactant mixture could be calculated from a linear average mixing rule based on weight composition. Thus intermediate HLB could be attained simply by mixing two or more surfactants by using the relationship:

$$\text{HLB mixture} = \sum_{i=1}^{n} X_i \mathrm{HLB}_i \tag{7}$$

For anionic surfactants, Lin (44) proposed the empirical equations also listed in Table 2 as a first approximation. These relations indicate that the HLB varies in a linear fashion with the lipophilic ''tail'' length.

This follows the earlier suggested group contribution splitting proposed by Davies (40) to calculate HLB as an algebraic summation using group contributions as follows:

$$\begin{aligned} \mathrm{HLB} &= 7 + \Sigma\,(\text{hydrophilic group numbers}) \\ &\quad + n\,(\text{group number per C atom in tail}) \end{aligned} \tag{8}$$

where n is the number of carbon atoms in the tail. The group number values are also listed in Table 2. It is worth remarking that the ''neutral'' or balanced HLB is 7 instead of 10 in other calculations. Note also that the ethylene oxide group is hydrophilic (positive value) while the propylene oxide group is slightly lipophilic (negative), although less than a single carbon group (methyl or methylene). This group contribution feature is easy to handle and attractive for the practitioner who does not want to get into too much physicochemical trouble. However, the calculated values are not guaranteed to match other calculations better than ± 3 HLB units, a large inaccuracy that indicates that the decimal digits listed in the table could be rounded with no grief.

The physical or physicochemical signification of the HLB scale could be ascertained by associating the HLB value with some surfactant properties, e.g., water dispersibility. Surfactants with HLB <8 are not dispersible in water, whereas those with HLB >12 are fully soluble in water. Intermediate HLB surfactants are more or less dispersible depending on other factors. Scores of papers have related HLB to other surfactant properties, as reported in an exhaustive computer screening prepared by Becher (2).

The HLB scale essentially pretends to indicate that while HLB is <10, $=10$, or >10, then the conditions attained for Winsor's $R<1$, $=1$, or >1

Table 2 HLB Number Values from Different Sources

Surfactant type	Griffin's (42) HLB number
Sorbitan trioleate/tristearate (85:65)	1.8–2.1
Glycerol monostearate	3.8
Sorbitan monooleate (80)	4.3
Sorbitan monostearate (60)	4.7
Diethylenglycol monostearate/monooleate	4.3–4.7
Diethylenglycol monolarate	6.1
Sorbitan monopalmitate (40)	6.7
Tetraethylenglycol monostearate/monooleate	7.7
Sorbitan monolaurate (20)	8.6
Polyoxyethylene nonylphenol (5EO)	10.0
Polyoxyethylene (5EO) sorbitan monooleate (81)	10.0
Polyoxyethylene (20EO) sorbitan trioleate (81)	11.0
Polyoxyethylene (7EO) tridecanol	12.1
Polyoxyethylene (5EO) sorbitan monolaurate (21)	13.3
Polyoxyethylene (20EO) sorbitan monooleate (80)	15.0
Polyoxyethylene (20EO) sorbitan monolaurate (20)	16.7

Anionic surfactants (44)	HLB relationship
$RnOSO_3Na$	HLB = 47.5–0.475 n
RnCOOK	HLB = 28.1–0.475 n
RnCOONa	HLB = 26.1–0.475 n
$RnSO_3Na$	HLB = 18.0–0.475 n

Groups		(Davies' HLB Group numbers) (40)	
$-OSO_3$ Na	+38.7	-COO Na	+19.1
-COOK	+21.1	-N tertiary amine	+9.4
Ester (sorbitan ring)	+6.8	Ester (free)	+2.4
-COOH	+2.1	-O-	+1.3
-OH (free)	+1.9	-OH (sorbitan ring)	+0.5
$-CH_2-CH_2-O-$ (EO group)	+0.33	$-CH(CH_3)-CH_2-O-$ (PO group)	−.015
$-CH_3$, $-CH_2-$, $=CH-$	−0.475		

are met. Of course Winsor *R* takes into account not only the surfactant type but also several other variables such as the oil and aqueous phase nature, as well as temperature, while HLB number does not. This means that the HLB number could exhibit discrepancies whenever systems with different oil and aqueous phases are compared.

In order to improve the HLB number's practical interest, a second parameter was introduced to describe the oil phase nature, the so-called required HLB or HLB_{req}, so that a better description of the physicochemical environment would be attained. The oil HLB_{req} is the HLB of the surfactant that results in the most stable emulsion. In order to estimate the oil HLB_{req}, an HLB scan for emulsion stability has to be carried out. This is done by preparing several systems with the same oil and aqueous phases but different HLB surfactants or surfactant mixtures, so that a whole HLB scan is carried out, say, from definitely lipophilic ($HLB = 2$–4) to definitely hydrophilic ($HLB > 14$).

In many cases and in whatever way the stability is measured, two maxima appear on such a plot as in Fig. 11, one at low HLB (for W/O emulsions) and another one at high HLB (for O/W emulsions). There is thus an oil HLB_{req} for O/W emulsions and another one for W/O emulsions.

This experimental protocol proposed by Griffin is, however, tedious and lacks accuracy for several reasons. While the maximum at low HLB is essentially at the same position for most HLB scans, so that HLB_{req} for W/O is about 5–6 for most oil phases, the maximum at high HLB is often difficult to pinpoint with accuracy because the O/W emulsion stability does not change very much with HLB. On the other hand, the stability curve and thus minima position are not strictly dependent on the HLB scale but are found to change when the surfactant type is changed.

Although the HLB number is still used nowadays because it is based on extremely simple arithmetic, it should be stressed that it gives approximate infor-

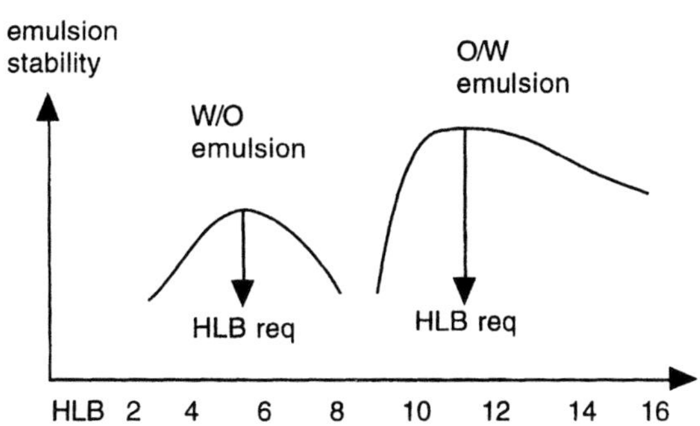

Figure 11 Determination of oil phase HLB required for W/O and O/W emulsions.

mation only. Also, it cannot take into account formulation effects that are known to be relevant such as the nature and structure of both the hydrophilic and lipophilic groups, the aqueous phase salinity, the presence of an alcohol, or the influence of the temperature, since none of these variables appear in the HLB rationale. On the other hand, it may be advantageously used as a yardstick to compare surfactants to one another, particularly of the same family, and provided they are going to be used in the same physicochemical environment (oil, brine, temperature, etc).

Since the previous relationships indicate that the HLB number would increase steadily as the hydrophilic group increases in size, it is no wonder that HLB numbers can be correlated with many properties that depend on the relative hydrophilicity (2,45–47).

B. Phase Inversion Temperature

The second intent to numerically characterize the formulation concept was the so-called phase inversion temperature (PIT), originally introduced by K. Shinoda in 1964 as the temperature at which a polyethoxylated nonionic surfactant switched its dominant affinity from the aqueous phase to the oil phase to produce a change in emulsion type. This was both easy to determine experimentally and simple to understand as far as the related phenomenology was concerned (48–50).

The PIT (with water and a given oil phase, i.e., hexadecane) was first correlated with the cloud point, i.e., the temperature at which a surfactant phase separates from a low concentration, e.g., 3%, surfactant solution (48). It is worth noting that the cloud point indicates a property of the surfactant, somehow the HLB, while the PIT takes into account the interaction with both the aqueous and oil phase. Early experimental evidence indicated that the PIT increased as the nonionic surfactant polyethylene chain length increased, increased as the hydrocarbon oil chain length increased, and decreased as the salt concentration increased in the aqueous phase. Since the PIT is actually the physicochemical situation in which Winsor $R = 1$, and since an increase in temperature tends to reduce the A_{CW} interaction, the interpretation of the previously mentioned trends is straightforward when the numerator and denominator of Winsor R are equated.

$$A_{CO} - A_{OO} - A_{LL} = A_{CW} - A_{WW} - A_{HH} \tag{9}$$

An increase in polyether chain length compels A_{CW} to increase, and an increase in temperature is required to decrease it, so that the equation still holds. When the hydrocarbon chain length increases, the $A_{CO} - A_{OO}$ term tends to decrease as discussed previously, and an increase in temperature is required to decrease A_{CW} in the same amount and keep the equality. Finally when electrolyte are added the A_{CW} term decreases, and a decrease of temperature is required to turn it back to its previous value. The PIT was thus an equivalent to the Winsor approach, but for the first time a formulation concept had a numerical attainable value, that could be related to the SOW system.

Recent work on the partitioning of ethoxylated alkylphenol surfactants (51) has shown that the energy of transfer of a surfactant molecule from water to oil, $\Delta G_{w \to o}$ can be written as:

$$\Delta G_{w \to o} = \Delta G^{*}(\text{alkylphenol}_{w \to o}) + \text{EON} \times \Delta G^{*}(\text{EO}_{\text{groupw} \to o}) \tag{10}$$

It is found that the ΔG for the transfer of an EO group from water to heptane decreases from 0.6 to 0.4 calorie as temperature increases from 25°C to 50°C. It means that it is easier to transfer an EO group from water to oil at higher temperature. Oil phase nature also seems to be quite important in this energetic barrier. Recent evidence indicates that with aromatic oil phases, the ΔG^{*} $(\text{EO}_{\text{groupw} \to o})$ may be much smaller than with aliphatic oils, may be zero with pure benzene. This corroborates the trends found on the PIT some 40 years ago.

It is worth noting that the PIT of a surfactant mixture can be calculated according to a linear mixing rule, where the X's are the mole fractions.

$$\text{PIT}_{\text{mixture}} = \sum_{i=1}^{n} X_i \, \text{PIT}_i \tag{11}$$

This is an odd formula at first sight since temperatures generally do not add up, but it turns out to be easy to embrace when it is understood that the PIT is not a temperature but a numerical indicator of the dual affinity of the surfactant for its physicochemical environment, i.e., an energy term that is usually summed up as in the thermodynamics first law.

Although it is now obvious that a tentative approach for assimilating or adapting the PIT as the experimental measurement of Winsor's R would have evolved into a quite general and practical formulation concept, Shinoda's first goal was rather to relate it to HLB (52), which was at this time the quite widely accepted reference, in spite of its drawbacks. As a consequence, the PIT did not evolve into the generalized formulation concept it could have been for any practitioner in the mid-1960s, and was only used by scholars working with nonionic surfactants who could sense its underlying significance.

Because of its basic phenomenon, the PIT is however limited to ethoxylated nonionic surfactant systems, and its experimentally attainable range (that fits the liquid water state at atmospheric pressure) is sometimes rather narrow, particularly in dealing with highly ethoxylated detergents and emulsifiers.

Nevertheless, it is important to stress again that the PIT procedure was the first method to make use of the technique of unidimensional formulation scan, in which a physicochemical formulation variable (temperature) is changed continuously with all other formulation variables held constant, until a very clear-cut situation arises that can be detected experimentally.

At the light of the current formulation state of the art, it can be said that the PIT was a clever and precursory technique. Today the PIT can be viewed as the optimum temperature in the multivariant surfactant affinity difference (SAD) concept to be discussed later on.

C. Cohesive Energy Ratio

In a tentative approach to attain a formulation concept with both the theoretical content of Winsor's *R* and the down-the-bench numerical data feature of the HLB, Beerbower and collaborators introduced the cohesive energy ratio (CER) approach in 1971 (53,54). From the conceptual point of view it was very similar to Winsor interaction energies ratio, but this time it was the ratio between the adhesion energy of the surfactant "layer" with the oil phase and the adhesion energy of the surfactant "layer" with the water phase. It must be recalled that the cohesion energy between molecules of a pure component system is calculated as:

$$\delta^2 = \frac{\Delta H_{\text{vap}}}{v_{\text{L}}} \tag{12}$$

where ΔH_{vap} is the enthalpy of vaporization per mole and v_L the molar volume in the liquid state, both measurable quantities. δ is the solubility parameter, a direct measurement of the intermolecular cohesion forces. When a mixed system is considered, the adhesion force between the two kinds of molecules is calculated according to the London's geometric mean relationship.

$$\text{if} \quad \delta^2_{\text{AA}} = \frac{\Delta H_{\text{A. vap}}}{v_{\text{A.L}}} \quad \text{and} \quad \delta^2_{\text{BB}} = \frac{\Delta H_{\text{B. vap}}}{v_{\text{B.L}}} \tag{13}$$

$$\text{then} \quad \delta^2_{\text{AB}} = \delta_{\text{AA}}\,\delta_{\text{BB}} \tag{14}$$

Solubility parameters have been measured and tabulated for hundreds of substances, and they are often used in relation to the regular solution model to estimate the activity coefficient in a mixture and the eventual separation into two phases (55). The adhesion between the surfactant and the oil phase was estimated by calculating a δ^2_{AB} term where A represented the oil phase and B stood for the lipophilic part of the surfactant that was assumed to be similar to that of a hydrocarbon with the same chain length. This assumption happened to be perfectly reasonable and satisfactory. Unfortunately, the corresponding adhesion energy on the water side of the interface was not easily calculated, in particular because of the lack of experimental data for the hydrophilic group of some surfactants, since a hydrophilic group cannot be analyzed separately from the rest of the molecule. It also seems that the theoretical basis of the regular solution theory was not strong enough to handle the extreme incompatibility between oil and water.

Due to these drawbacks, the final numerical value of the cohesive energy ratio was as inaccurate as the HLB number and even worse in some cases, and it had to be dropped.

D. Empirical Correlations for Optimum Formulation

The formulation concept was one of the aims of a large-scale research effort to develop enhanced oil recovery methods after the 1973 oil embargo. The goal was to inject a surfactant solution in the oil reservoir in order to overwhelm the capillary forces that trap the almost 75% of the original oil in place remaining in the reservoir after waterflooding.

It was soon found that the best conditions to attain an ultralow interfacial tension coincides with a Winsor type III physicochemical situation in which a bicontinuous microemulsion is in equilibrium with both oil and water excess phases. Such a case was called optimum formulation since it was associated with the successful mobilization of petroleum (56,57). The adjective optimum is still used to describe the $R = 1$ case, no matter whether it corresponds or not to the best case in the studied application. Since there are many ways to attain $R = 1$, particularly in real systems with many components, one of the goals was to find out a numerical relationship that conveys the fact that $R = 1$.

The experimental technique used to find an optimum formulation, known as unidimensional scan, goes on as follows. Series of surfactant–oil–water systems are prepared in test tubes, all with identical composition, and with the same formulation with the exception of the scanned variable, that is in general the aqueous phase salinity for ionic systems, and the average number of ethylene oxide groups per molecule (EON) if the systems contain an ethoxylated nonionic surfactant mixture.

The fixed composition is typically selected so that the representative point is in the polyphasic region of a ternary diagram; in this case Winsor-type, and thus R value can be deduced from the phase behavior observation. It is found that 1–3% surfactant and equal proportions of oil and water is a fairly good choice in most systems. In ionic systems a very low percentage of alcohol is often added to avoid the formation of mesophases. If the formulation should not be altered by the alcohol a good choice is *sec*-butanol.

In these conditions it is obvious that from one test tube to the next the only difference is the value of the scanned variable. The test tubes are closed with a screw cap or sealed, and are left to equilibrate at constant temperature until there is no apparent change, often no more than 2 or 3 days in practice. Then the phase behavior is deduced from a visual observation of the transition expected in a scan.

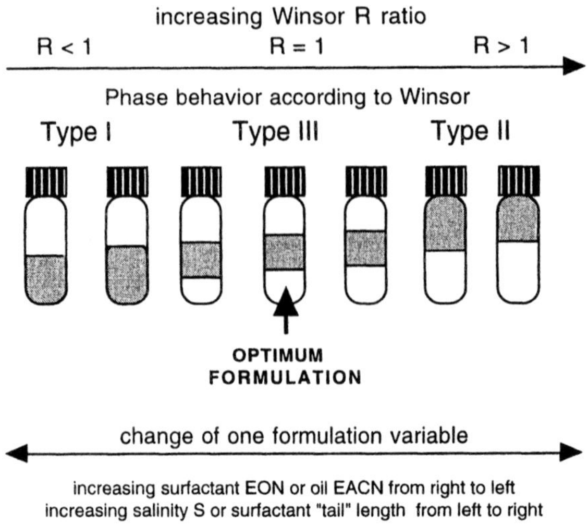

Figure 12 Aspect of test tube series in a unidimensional formulation scan at constant composition.

The optimum formulation of the scan is the value of the variable that corresponds to the center of the three-phase behavior range as indicated in Fig. 12. If the three-phase range is quite spread, an additional measurement could be required to pinpoint the optimum formulation, such as the solubilization or interfacial tension, as will be discussed later.

The quantification of formulation effects requires a method to find the equivalence between two changes in the physicochemical situation. The most used experimental technique has been the bidimensional scan, which is based on a double variation of competing effects, i.e., two counteracting changes as illustrated in Fig. 13.

Starting at an optimum formulation (where $R = 1$), a first formulation variable X_1 is changed by an amount ΔX_1, which results in a change in one of the

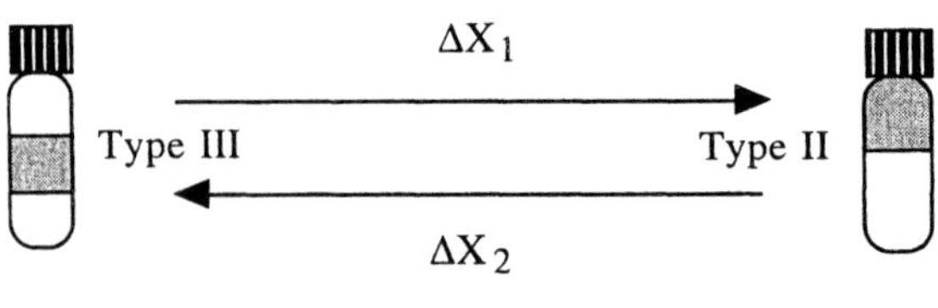

Figure 13 Double formulation change to produce compensating effects.

interaction energies, and the resulting change in the R ratio value. For instance, suppose that X_1 is the salinity of the aqueous phase. The increase in salinity ΔX_1 produces a decrease in A_{CW}, which means that the R ratio will increase from $R = 1$ to some $R > 1$ value, and that the phase behavior will exhibit a type III → type II change as indicated in Fig. 13.

Let us now select a second formulation variable X_2 and let us change it in the direction where the change in R would be opposite to the previous one, i.e., a decrease in R, until a type II → type III transition is attained again, as well as $R = 1$. For instance, an increase in the oil phase EACN will produce a reduction of the A_{CO}–A_{OO} term in the numerator of R. Such an increase ΔX_2 is pursued until R goes back to 1, as evidenced by the return to a type III three phase behavior situation.

At this point it can be said that the two formulation changes ΔX_1 and ΔX_2 have produced exactly opposite effects on R and are thus equivalent from the physicochemical point of view, but with opposite effects.

For instance, it is found that for an alkylbenzenesulfonate–alkane–NaCl brine system an increase in ACN of four units is compensated by an increase in salinity measured by an increase of 0.64 of the natural logarithm of the salinity, whatever the other variables, as seen in Fig. 14.

This means that a relationship such as

$$\Delta \ln S = K \Delta \text{ACN} \qquad \text{where } K = 0.16 \tag{15}$$

can be integrated as

$$\text{Ln } S - K \text{ ACN} = C \tag{16}$$

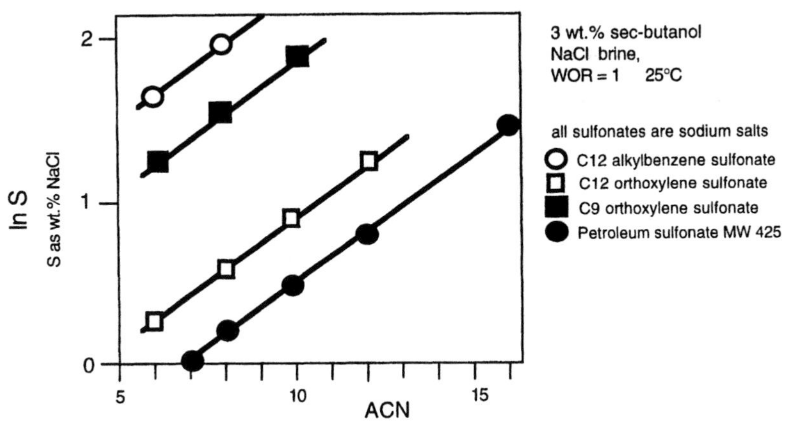

Figure 14 Relation between aqueous phase salinity (as a natural logarithm) and oil phase Alkane Carbon Number so that optimum formulation is attained.

where C does not depend on salinity or oil EACN, but on other formulation variables and temperature.

This technique of double change with compensating effects allows one to quantify the respective influence of the different formulation variables, and it has been widely used in performing a large amount of experimental work to quantify the Winsor concept. Extensive experimental work showed that a three-phase system of type III phase behavior was attained whenever a certain empirical relationship, the so-called correlation for optimum formulation, was satisfied. For anionic surfactants the correlation has been expressed as (11,58):

$$\ln S - K\,\mathrm{ACN} - f(A) + \sigma - a_{\mathrm{T}}\,\Delta\mathrm{T} = 0 \qquad (17)$$

Later on it was showed that the same correlation was valid as well for cationic surfactant systems with slightly different values of the parameters (59). For nonionic surfactants a similar correlation was found (60–62):

$$\alpha - \mathrm{EON} + bS - k\,\mathrm{ACN} - \phi(A) + c_{\mathrm{T}}\,\Delta\mathrm{T} = 0 \qquad (18)$$

In these expressions, S is the salinity, and $\ln S$ is the neperian logarithm of the salinity expressed in wt% NaCl with respect to the aqueous phase, ACN or alkane carbon number is a characteristic parameter of the oil phase, $f(A)$ and $\phi(A)$ are functions of the alcohol type and concentration, σ and α are parameters characteristic of the surfactant structure, and EON is the average number of ethylene oxide groups per molecule of nonionic surfactant. ΔT is the temperature deviation measured from a certain reference (25°C), b, k, K, a_{T}, and c_{T} are empirical constants that depend on the type of system.

K is 0.16 for alkylbenzenesulfonates, about 0.10 for alkyl sulfate and fatty acid sodium soaps, 0.17 for n-alkylammonium salts at low pH, and 0.19 for quaternary n-alkyltrimethyl ammonium salts, while k is found to be 0.16 for nonionic systems. The coefficient of the salinity effect for nonionic surfactants is small, i.e., 0.13 and 0.10 (in EON unit per wt% salt in aqueous phase) respectively for sodium and calcium chloride.

Not many investigations have been dedicated to the salinity effect when the electrolyte is not sodium chloride, maybe because the effect is more complex, not amenable to a simple expression in particular with divalent cations and ionic surfactants. However some trends are available in applied publications (27,63). Worth noting is a systematic study of the effect of the electrolyte anion on the equivalent salinity of different sodium salts, that showed that the correlation is followed for all sodium salts and that the effective or equivalent molar salinity only depends on the valency of the anion (64).

The nonionic surfactant characteristic is split into its ethylene oxide number (EON), i.e., the average number of ethylene oxide groups per molecule, and the hydrophobe contribution α. Only scarce data are available for α, which has a value near 6.6 and 6.3, respectively, for the branched nonyl and octyl phenol species, and 4.4 for the linear tridecanol base group. Much more information is available for ionic surfactants, particularly anionic ones, under the form of σ/K which is characteristic of the contribution of the surfactant:

$$\sigma/K = \sigma_o/K + 2.25 \text{ SACN} \tag{19}$$

where σ_o is characteristic of the hydrophilic group and SACN is the surfactant alkyl carbon number, i.e., the number of carbon atoms located in its single long hydrophobic tail. The data that are gathered in Table 3 fit a normal synthetic surfactant isomer mixture (58,65,66).

Note that the correlation may be written as:

$$\sigma/K = \text{ACN} + \frac{-\ln S + f(A) + a_T \, \Delta T}{K} \tag{20}$$

From the above relation, it is obvious that σ/K is the oil ACN for attaining an optimum formulation when the numerator of the right-hand fraction is equal to zero, a situation that happens at unit salinity ($S = 1$ and $\ln S = 0$), no alcohol so that $f(A) = 0$, and ambient temperature ($\Delta T = 0$).

For such a reason σ/K was labeled as EPACNUS, an acronym for extrapolated preferred ACN at unit salinity (58).

It can be seen from the table that the introduction of two methyl groups in ortho of the main alkyl on the benzene ring results in an increase in σ/K of 5 units, which is equivalent to the addition of two carbon atoms. The variation from alkanesulfonate to alkylbenzenesulfonate is about 18 units, i.e., the benzene group is contributing the equivalent of about 8 carbon atoms rather than 6. This discrepancy of 2 units may be attributed, according to the previous discussion, to the fact that the alkanesulfonate is probably much more linear in average than the alkyl group of the alkylbenzenesulfonate.

Table 3 Hydrophilic Group Contribution to the Surfactant Characteristic Parameter

Surfactant family	σ_o/K	
Alkyl sulfate Na salt	−55	±3
Fatty acid Na soap	−48	±5
Alkanesulfonate Na salt	−48	±3
Alkyltrimethylammonium salt (Br^- or Cl^-)	−45	±2
n-Alkylammonium salt (Br^- or Cl^-) at ph 3	−32	±3
Alkylbenzene sulfonate Na salt	−30	±1
Alkylorthoxylenesulfonate Na salt	−25	±2

The characteristic parameter of ionic surfactant mixture can be readily calculated from a linear mixing rule on its σ/K so that (65,67):

$$\sigma/K \text{ mixture} = \sum_{i=1}^{n} X_i(\sigma/K)_i \tag{21}$$

Let us now come back to the correlation and turn the attention to the effect of the oil phase nature. If the oil phase is not an alkane, its equivalent ACN, i.e., the ACN of the alkane that exhibits the same behavior, is used instead (11, 68). Since the EACN of an oil mixture can be deduced from its composition by a linear mixing rule based on a molar fraction, it has been possible to measure the EACN of many oils.

The lower the EACN of an oil, the higher its polarity. For instance, EACN is zero for benzene, and it is equal to the number of carbon atoms in the alkyl group of alkylbenzenes. Alkyl cyclohexanes have an EACN that is equal to the number of carbon atoms in their alkyl group augmented by 3. In other words, the saturated cyclohexane structure accounts for three units in the EACN, while the aromatic benzene ring does not count.

The ester polar group cuts the lipophilic contribution quite a lot, about 12 EACN units, since ethyl oleate is found to exhibit an EACN = 6, as well as oleic acid–based triglyceride oils such as Soja oil (69,70). The EACN of pinene (double-cycle terpene) and limonene (single-cycle) essential oils are found at about 7 and 8.5, respectively (71).

The EACN of chorinated solvents was recently reported as well in relation to soil remediation methods by microemulsion cleaning (72–74).

The alcohol functions $f(A)$ and $\phi(A)$ depend both on the type and concentration of the added alcohol as reported in the two correlation papers (58,61) that are the only ones to give a numerical dependency. Very few other papers report some data on alcohol effect both as a cosurfactant (23,75), and as an additive that tends to reduce the interfacial rigidity (76). In Fig. 15, the concentration is expressed in gram per deciliter (gpdl) with respect to the whole system, very close to wt.% unit. Both expressions are very much alike but different in numerical value, so that the following discussion applies for all systems.

Water-soluble alcohols present a slightly positive function, i.e., their addition increases the hydrophilicity of the surfactant/alcohol amphiphile, but very little indeed. As a consequence isopropanol effect in the correlation is negative, i.e., the same as decreasing salinity. *sec*-Butanol and *ter*-pentanol have practically no formulation effect since their functions are essentially nil whatever the concentration. These are the alcohols that are the more interface-seeking ones, and their main effect is to dilute the surfactant adsorption density without changing the formulation.

On the contrary, water-insoluble alcohols produce an increasingly negative function as the alcohol concentration increases, and the more lipophilic the alco-

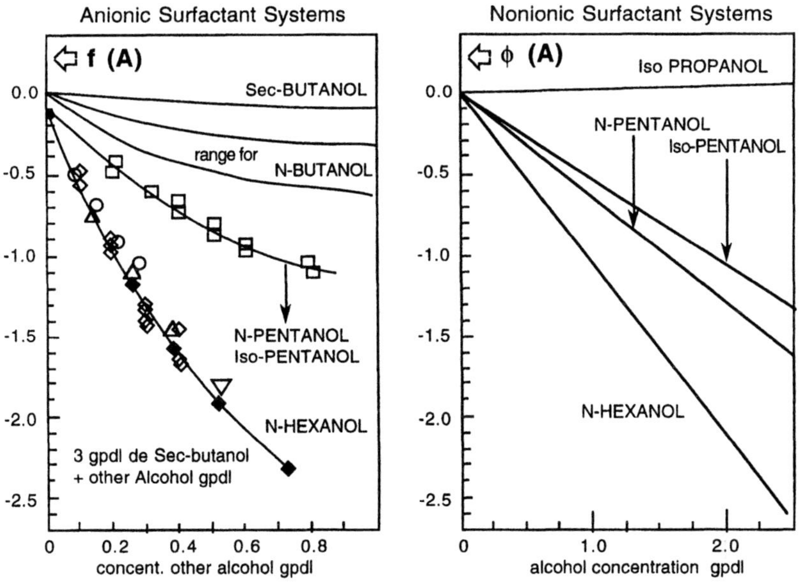

Figure 15 Effect of alcohols on optimum formulation as f(A) and ϕ(A). (Adapted from Ref. 11, 58, 61.)

hol is, the more negative the function is, with an almost linear variation. These functions are independent of the system starting with pentanol and above, which is a nice feature, since it is much easier to handle the alcohol this way than through a fourth component in a phase diagram. Thus they produce a positive contribution in the correlation, as an increase in salinity.

N-Butanol and *sec*-pentanol are intermediate affinity alcohol that are more or less water-soluble depending on the nature of the oil phase in equilibrium. For them the alcohol function is found to depend on the oil–water system. In any case the function does not follow a straight line variation as in other cases but tends toward an asymptote located between -0.5 and -1.0 depending on the system.

More information can be found in the original articles or in a review book on the subject (11). Relations conceptually similar to Eqs. (17) and (18), sometimes more accurate but limited to the effect of the temperature or another one or two variables, were proposed by several investigators (57,63,77–81).

The temperature effect (58–59,82–83) is mentioned in correlation [Eqs. (17) and (18)] as a linearized approximation over the narrow interval of experimental range, and may be considered as a trustworthy estimate for ionic surfactants, whereas is it rather approximate for nonionic ones.

The direction and intensity of change are indicated by the coefficient signs and values. The temperature coefficient a_T is 0.01°C^{-1} for anionic systems (58) and about twice as much for cationic ones (59). However, these numbers are not very precise since the temperature effect on ionic systems is rather small and difficult to estimate with accuracy. In any case, the sign before the coefficient is negative, which means that an increase in temperature is equivalent to a reduction in salinity.

On the contrary, the temperature coefficient for nonionic surfactant c_T is positive and much larger, typically 0.06, although it has been reported to change with the surfactant average EON, typically increasing from 0.04 to 0.08 when EON increases from 3 to 10 (84). The inaccuracy comes from the fact that most nonionic systems produce a three-phase behavior range that extends over a large zone of temperature, in which the optimum temperature must be selected (61,85). Because of the positive sign before the coefficient in the correlation, an increase in temperature would be equivalent to an increase in salinity or a decrease in EON.

The correlation for optimum formulation may be written with the temperature term on the left side, such as

$$T\text{–Tref} = \frac{\text{EON} - \alpha - bS + k\,\text{EACN} + \phi(A)}{c_T} \tag{22}$$

In this relation, T is the temperature at optimum formulation where $R = 1$, i.e., the PIT according to Shinoda's premise, an expression that deserves the latter label HLB-temperature. This relationship is very close to the one deduced by some researchers (78) who used the surfactant HLB instead EON-α to arrive to a similar result as far as the combined effects of temperature, salinity, and oil ACN are concerned. The above formula indicates how the PIT increases with the number of ethylene oxide groups in the surfactant molecule, increases with oil chain length, and decreases with electrolyte concentration and surfactant tail length (proportionally to α). It also predicts a variation with the alcohol type and concentration, a decrease with lipophilic species.

It is seen that the above relationship implies that the linear mixing rule on EON or EACN would compel a linear mixing rule on optimum temperature, which is an a posteriori justification of the PIT mixing rule.

E. Surfactant Affinity Difference: Generalized Formulation Concept

The correlations are the numerical translations of the $R = 1$ concept in Winsor terminology. The correlations are linear forms of all the formulation variables with no crossed term, so that the different variable effects are independent, which might seem to be too much of a coincidence. A few years after the correlations were empirically found, it was showed that they are actually a numerical expression proportional to the "surfactant affinity difference" (SAD), i.e., the difference between the negative of the standard chemical potential of the surfactant in the oil phase and the corresponding term for the water phase, in a three-phase optimum system (38,86–87):

$$\text{SAD} = -\mu_o^0 - (-\mu_w^0) = \mu_w^0 - \mu_o^0 \tag{23}$$

Now, it is clear that SAD is a form of the generalized formulation parameter, hopefully expressed in terms of measurable and manipulable variables. Again, it is worth noting that the correlations do not contain any cross-term, and are, on the contrary, linear forms of independent terms.

$$\frac{\text{SAD}}{RT} = \ln S - K\,\text{ACN} - f(A) + \sigma - a_T\,\Delta T \tag{24}$$

$$\frac{\text{SAD}}{RT} = \alpha - \text{EON} + bS - k\,\text{ACN} - \phi(A) + c_T\,\Delta T \tag{25}$$

Each term can be viewed as an energetic contribution to the overall interaction balance, which is expressed as an algebraic sum instead of a ratio as in Winsor R. The positive contributions (including the alcohol functions for lipophilic alcohol) tend to produce the transition in one way (from $R < 1$ to $R > 1$) while the negative term does it the other way.

Whenever SAD is respectively negative, zero, or positive, R is respectively inferior, equal, or superior to 1. An optimum formulation can be thus taken as a reference state that can be accurately pinpointed by experiment.

On the other hand, an off-optimum formulation can be defined as a deviation from optimum formulation by its SAD value. Thus, it is possible to compare systems in a same physicochemical state (same deviation from optimum formulation) although they have no single formulation variable with a common value. This concept has been used successfully to correlate and predict the influence of the formulation on the properties of macroemulsified systems in many applications (88–94).

Aside from Shinoda and collaborators who introduced the PIT, other investigators have presented similar conceptual approaches that are worth mentioning, although they did not achieve the detailed numerical quantification of the SAD expression.

Krugliakov (95–97) proposed a concept called hydrophile–lipophile ratio (HLR), which is the ratio of the energy of adsorption of the surfactant molecule from the water phase to its energy of adsorption from the oil phase. The HLR is a good alternative, but it suffers from the same drawback that Winsor R ratio does, i.e., it is difficult to evaluate numerically and it is a ratio of energy terms instead of an algebraic difference.

The early proposal by Davies (40) to use the partition coefficient of the surfactant between the oil and water phase as a formulation parameter was retaken by Marquez and collaborators (98) as a way of measuring the free energy of transfer from water to oil, a successful technique to deal with complex fractionating surfactant mixtures (99). The free energy of transfer of a molecule of surfactant from water to oil is:

$$\Delta G_{(w \to 0)} = RT \ln (C_w/C_0) \tag{26}$$

where the C_w and C_o are the surfactant concentrations in water and oil, respectively. This time the partitioning was measured between the excess phases of a three-phase Winsor type III system at optimum formulation, and there were plenty of degrees of freedom available to estimate the change of partitioning with the different formulation variables. Marquez data (51) showed that the partition coefficient of a surfactant specie is an excellent yardstick of the concept of generalized formulation. Based on the assumption that the activity coefficients are unity, which is legitimated by the low surfactant concentration found in the excess phases of Winsor type III systems, then the equilibrium between the water and oil phases can be written in terms of the chemical potentials of the surfactant:

$$\mu_w = \mu_o = \mu_w^* + RT \ln C_w/C_{w.ref} = \mu_o^* + RT \ln C_o/C_{o.ref} \tag{27}$$

where the standard chemical potentials are indicated with an asterisk, while the concentration references bear the subscript ref. If the partition coefficient is defined as $K = C_w/C_o$, then

$$RT \ln K + \text{constant} = \mu_o^* - \mu_w^* = \Delta G_{(w \to o)} = -\text{SAD} \tag{28}$$

Thus $RT \ln K$ and SAD should exhibit similar (and opposite) variations with respect to the formulation variables. By measuring the partition coefficient with different systems exhibiting a variety of oils, surfactant characteristics, alcohol content, brine salinity, and temperature, Marquez was able to correlate the term $RT \ln K$ with the formulation variables in a linear relationship very similar to the correlations. For instance, it was found that (98):

$$\begin{aligned} RT \ln K(\text{EON, SACN}) = {} & \text{EON} \times \Delta\mu^*(\text{EO group}) + \text{SACN} \\ & \times \Delta\mu^*(\text{CH2 group}) + \Delta\mu^*(\text{other factors}) \end{aligned} \tag{29}$$

where $\Delta\mu^*(\text{EO group}) = 0.62$ cal. by transferred EO group from the aqueous phase to the oil phase, and $\Delta\mu^*(\text{CH2 group}) = -0.06$ cal. by transferred CH2 group from the aqueous phase to the oil phase.

In the past 10 years several studies on the effect of different oils, electrolytes, and other surfactants have brought additional data for using the numerical correlations to help the practititoner make use of the SAD approach as a generalized formulation yardstick (27,64,72,74,80,83,100–105).

There is not yet a comprehensive treatise to gather these data and to explain how to handle it beyond the book of Bourrel and Schechter (11), but only some attempts like this chapter and other reviews with different approaches in recent books (106,107). Accordingly, the formulator should start from the basics that are explained here and find his way to his own case, keeping in mind that a similar SAD value does mean a similar physicochemical environment. The correlations will provide both the direction of change to follow and the equivalence between the different possible changes, a valuable feature when cost and toxicity constraints are present.

The previous section dealt with formulation variables, i.e., those that are able to change the type of diagram according to Winsor rationale. In order to study the formulation effect, a specific composition was selected (typically 1–3% surfactant and equal amounts of oil and water) which lies inside the polyphasic region so that the diagram type could be deduced from the observed phase behavior at this point. This is satisfactory only if the diagram is of the Winsor type, i.e., in the presence of an ideal ternary behavior. In other cases, particularly in the presence of a dual amphiphile like a mixture of surfactant and alcohol, it is necessary to analyze the whole diagram and to locate the different phase behavior regions as a function of the composition.

V. EXPERIMENTAL METHODS APPLIED TO REAL CASES

A. Some Experimental Diagram Cases

In many truly diagrams where the Winsor phase behavior regions are found, additional regions where different types of mesophases, specially liquid crystals of various types depending on the formulation and the relative amount of the components, are present. Often these mesophase regions tend to disappear as temperature is increased, although it could be at the expense of a change in formulation, particularly with nonionic surfactants.

In many instances the SOW ideal ternary case is not sufficient to describe the behavior of an actual system. It has been noted that the addition of alcohol as a disordering agent is often required to avoid viscous or solid-like mesophases, particularly to produce microemulsions with ionic surfactant systems (108). According to the correlations for optimum formulation, the alcohol effect can also be that of a cosurfactant that modifies the overall balance of affinity through the $f(A)$ and $\phi(A)$ terms. The use of two surfactants is also often recommended to attain a better emulsion stability, a statement that should not be taken for granted in all cases, although it could prove correct in some cases. Hence, it is often necessary to carry out a study of a system whose amphiphile is composed of two substances, two surfactants or a surfactant and an alcohol, which exhibit different degrees of hydrophilicity and thus cannot be included in a single pseudocomponent.

Quaternary systems, such as surfactant–alcohol–oil–water (SAOW), are analyzed in a quaternary diagram made in a regular tetrahedron as indicated in Fig. 16. The brute force method that consists of selecting hundreds of composition points in the diagram (located at some grid pattern) and in analyzing the phase behavior in all these points is of course too tedious to be carried out.

The typical way to handle this problem is to use pseudoternary cuts, i.e., plane cuts through the tetrahedron. Since the two amphiphiles are often the principal source of unknown asymmetry, the usual tactic is to make a cut as indicated in Fig. 16, which is a triangle passing through the oil and water vertices, as well as one point of the surfactant–alcohol side located by the ratio of surfactant to alcohol. All of the points located in the gray triangle (Fig. 16, right) will contain a same ratio S/A, which is called a pseudocomponent.

The phase behavior is then studied in the S/A-O-W pseudo ternary diagram (Fig. 16, left), according to usual techniques, which may be grid scanning or others whenever some extra information is available to guide the selection of the trial points.

When the S/A pseudocomponent point is moved along the SA side, the gray triangle sweeps the whole tetrahedron volume. A systematic way to map the phase behavior in the tetrahedron is thus to map it in several bidimensional cuts made at S/A constant. Let us further discuss this point on a first example, in which the amphiphile contains a rather hydrophilic surfactant (octylphenol with 7.5 EO groups in average) and a lipophilic alcohol (*n*-pentanol)

Figure 17, upper left is a tetrahedron representation of a quaternary diagram containing a hydrophilic surfactant and a quite lipophilic alcohol (109). The other diagrams in Fig. 17 are different cuts at constant S/A ranging from a very hydro-

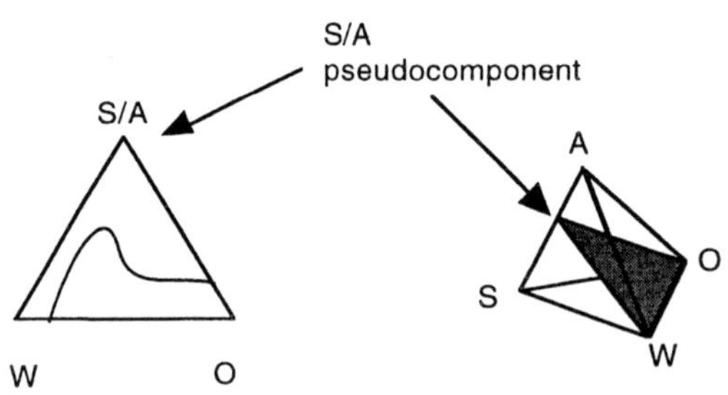

Figure 16 Quaternary diagram and pseudoternary cut.

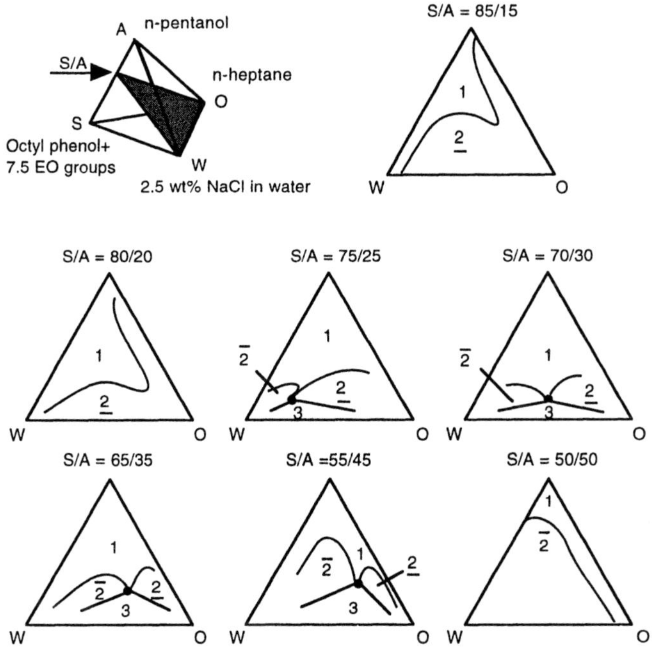

Figure 17 Quaternary diagram and different pseudoternary cuts at S/A constant. (Adapted from Ref. 109.)

philic mixture (85% surfactant and 15% alcohol) to a rather lipophilic one (50% surfactant and alcohol) in which the lipophilic character of the alcohol prevails. It is worth noting that the proportions are expressed in weight percentage, so that 50% alcohol represents a composition with a large excess of alcohol on a molar basis.

As the S/A ratio decreases a sequence of phase behavior takes place, so that the amphiphilic mixture becomes more and more lipophilic since the content of the more lipophilic component increases. The first diagram at S/A = 85:15 shows that the amphiphilic mixture is completely miscible with water but not with the oil phase, which is essentially the case for the surfactant alone.

It is essentially a Winsor type I diagram with this modification near the (S+A) − O side. At S/A = 80:20 the situation is basically unchanged, although the single-phase region has augmented in size, i.e., the miscibility gap has diminished.

From S/A= 75:25 to 55:45 the diagram exhibits a three-phase behavior region which is not very different from the Winsor type III diagram, with the middle phase composition moving from left to right, i.e., taking up more and more oil phase.

It is worth noting that the tie lines are not fully indicated in the pseudoternary diagrams because they are not actually laying in the plane of the cut. In effect, the amphiphile mixture content of the phases in equilibrium does not match the S/A ratio because of the so-called preferential partitioning. For instance, in a two-phase split, the oil phase would contain more alcohol than indicated by the S/A ratio, while the aqueous phase would contain more surfactant than indicated by the S/A ratio. Since the binodal curve is determined by trial and error by diluting a single-phase system located just above the polyphasic region with an O+W mixture, the middle phase microemulsion (black dot) is in general extremely near the plane of cut if not exactly in it (110). Between S/A = 55:45 and S/A = 50:50, the phase behavior changes to mostly lipophilic, and the Winsor type II region invades the whole diagram. In this case the surfactant is not hydrophilic enough to compensate for the alcohol effect. It is also probable that such a high amount of alcohol results in a consolute system in which most of the alcohol is miscible in the oil phase at equilibrium with an almost pure water phase. Note that a nonionic surfactant of the alkylphenol type with even 7.5 EO groups in average is certainly at least 10 times more soluble in a mixture of heptane and pentanol than in water (98).

The changes of the S/A ratio result in a transition not unlike the one seen in a formulation change $R < 1 \rightarrow R = 1 \rightarrow R > 1$. This is due to the fact that an increase in alcohol content in the amphiphilic mixture is essentially equivalent to an increase in lipophilicity in this combined amphiphile. The change of S/A, i.e., a composition change, is thus also a formulation change, just like the one in oil nature or in salinity could produce. The same sequence of Winsor diagrams when the surfactant/alcohol ratio is swept has been found for different systems (78,111–115). The three-phase volume body in a quaternary diagram has a very peculiar shape that has been studied both from the experimental and thermodynamic point of view (116,117).

Not all the cases are so simple. For instance, let us examine a case in which there is a very hydrophilic surfactant and a slightly lipophilic alcohol. Figure 18 indicates the phase behavior of such a case, with a pseudoternary cut at an S/A near unity. The phase behavior that is represented in Fig. 18, right is hypothetical and even simplified with respect to real cases, but it is reasonably representative of a system containing sodium dodecyl sulfate and *n*-butanol. In this system, the excessive hydrophilicity of the dodecyl sulfate is compensated by factors that are known to reduce the interaction of the surfactant for the aqueous phase (a high-salinity brine) and to favor its interaction for the oil phase (short alkane or aromatic oil) (118,119). It is seen that the polyphasic region map found in such a cut does not match any of the Winsor diagrams. However, the different phase

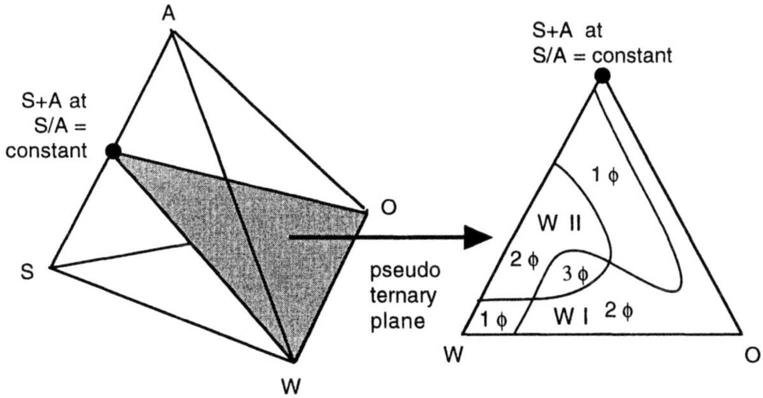

Figure 18 Hypothetical non Winsor type diagram in a ternary cut, in a quaternary like dodecylsulfate, high salinity brine, aromatic oil, and n-butanol.

behaviors do, and there are regions in which one-phase (a microemulsion), two-phase (Winsor type I or II) , and even three-phase behavior are exhibited.

The present phase behavior is typical of the case in which the surfactant is very hydrophilic while the alcohol is only slightly lipophilic. In the present case it is obvious that a change in total amphiphile concentration, i.e., a vertical displacement on the left side of the diagram, would produce a type I → type III → type II transition. This is because the increase in S + A really produces a change due to the increase in alcohol effect, whereas the surfactant effect is roughly independent of its concentration. This is because the surfactant is very hydrophilic from the start but then does not change anymore, while the alcohol is water-soluble at low concentration, reaching its saturation in water and becoming more and more oil-soluble with a resulting lipophilic effect. This behavior of intermediate alcohols such as *n*-butanol has been typified to be difficult to handle since it changes from one system to another.

This is a typical case with an alcohol that is slightly water-soluble but much less so than the surfactant. On the right part of the diagram (more oil than water) there is quite an extended W/O type of microemulsion region.

In the previous case the phase behavior transition was produced by a change in the ratio S/A, i.e., in the amphiphilic mixture hydrophilicity. In the present one the phase behavior change is mainly due to the total concentration of amphiphile mixture, which drives a considerable hydrophilicity variation. These are two cases of quaternary systems in which a composition change produces a phase behavior change typical of the formulation effect in Winsor's diagrams. Such a coupling of formulation and composition is not an exception but rather a rule when the amphiphile is composed of various substances. Many other diagrams exhibit a lot of similarities with these two cases (102,111,120).

B. Solubilization-Related Phase Diagrams

The coupling of the two kinds of variables has been advantageously examined in the so-called solubilization vs. formulation diagrams that have been used by several industry researchers, particularly Bourrel and co-workers (63,121–125). Such diagrams are concerned with a very practical problem, i.e., the amount of surfactant and alcohol that is required to reach the binodal curve frontier with the single-phase microemulsion region, since this is obviously related to the cost of cosolubilizing oil and water.

Figure 19 indicates the way to construct such plots. In each diagram (see upper left) an amphiphile concentration scan allows one to detect the height of the polyphasic region. This is done by adding surfactant and alcohol (S+A) at an fixed S/A ratio to an W+O system (generally at unit water-to-oil ratio for the sake of simplicity) until a monophasic region appears. In reality the contrary

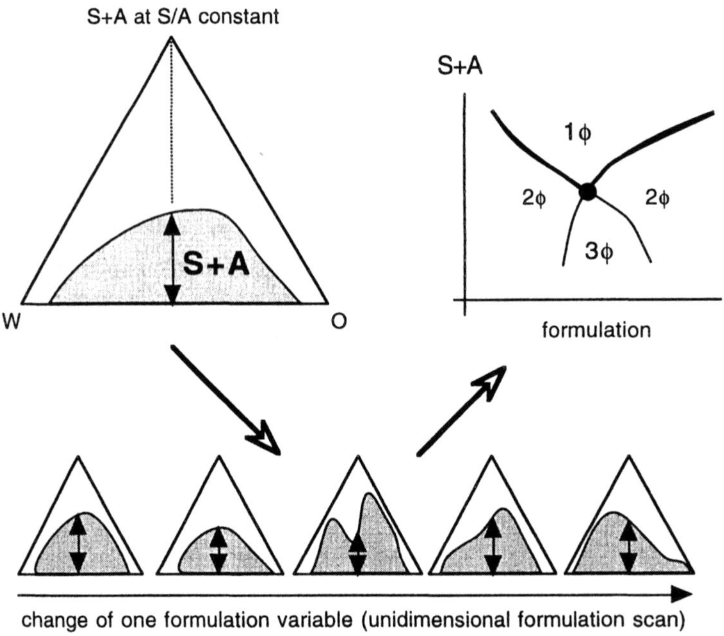

Figure 19 How to construct a S+A v. formulation plot.

is easier to do, i.e., a single phase SAOW system containing the S/A fixed ratio as well as equal amounts of oil and water is diluted with equal aliquots of oil and water until a turbidity appears.

The procedure is repeated with different diagrams along a unidimensional formulation scan and the S+A data are recorded. Then they are plotted as S+A composition to reach the single phase vs. formulation. As a matter of fact, the whole phase behavior is reported, i.e., the frontier between two- and three-phase behavior as well. The typical S+A vs. formulation plot is indicated in Fig. 19 (above right) where it is seen that the S+A amount passes through a minimum just at optimum formulation for three-phase behavior. In effect, below the lowest point of the curve (black dot) occurs a three-phase region, not a biphasic one. Note that at this formulation the required amount of surfactant+alcohol to cosolubilize equal amounts of oil and water undergoes a minimum. It is said that the microemulsion solubilization, which is the amount of oil or water that is solubilized per gram of amphiphile (the inverse of the plotted data), passes through a maximum—a now well-recognized property of optimum formulation.

Extensive studies to separate the effects of the different formulation variables on solubilization were carried out by Bourrel and collaborators in the past decade (11,121–125). The first diagram to be dealt with is a formulation scan with a perfect ternary that strictly follows Winsor phase behavior. This may be, for instance, reasonably represented by a system containing an alkylbenzenesulfonate and *sec*-butanol amphiphile mixture, an alkane, and NaCl brine, in which the relation surfactant/alcohol is kept constant.

Figure 20 shows the general aspect of such a diagram in which the phase transition frontier looks like a lowercase Greek gamma. Note that if rotated 90° as done in studies that use temperature instead of formulation as the field variable,

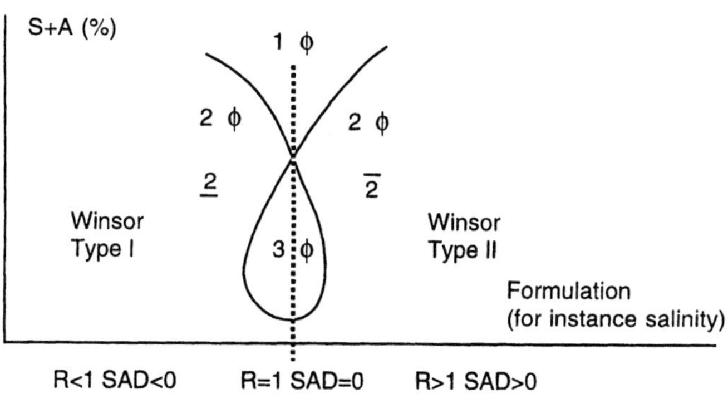

Figure 20 Normal S+A v. formulation phase behavior diagram with the total S+A concentration.

it looks like a fish (126). Coming back to Fig. 20, it is seen that the maximum solubilization is attained at the minimum of the single-phase region, i.e., at the crossing from one-phase to three-phase regions, which is the exact position of the middle phase microemulsion that contains equal amounts of oil and water.

The optimum formulation corresponds to the center line of the single- and three-phase regions that here is a vertical (dashed) line passing through the crossing point. It is indicated as SAD = 0 or $R = 1$ on the abscissa axis.

In the second so-called twisted diagram, illustrated in Fig. 21, the gamma shape is not in the vertical direction, but this time it is inclined. At each S+A level, the optimum formulation is also the center line of the single- and three-phase regions since it represents the value of the formulation variable for which the center of the three- or single-phase range is attained. In other words, it may be said that with this type of diagram the optimum formulation (dashed line) changes.

Note that if a vertical path is drawn from bottom to top (arrow in Fig. 21) along an increasing amount of S+A, the phase behavior changes from Winsor type I to type III and to type II, and finally to a single phase, as was the case in the S/A = 1 ternary cut through a quaternary diagram with sodium dodecyl sulfate and butanol seen in a previous section (Fig. 18). It is worth remarking that what seemed an extremely complex phase behavior is simply the consequence of the twisting of the diagram. This twisting is due to the fact that the surfactant and the alcohol do not exhibit the same hydrophilicity; thus one of them dominates. Here the lipophilic amphiphile dominates because at high S+A the formulation (salinity) is lower; thus everything occurs as if the amphiphile were less hydrophilic according to the relation between the salinity and the surfactant char-

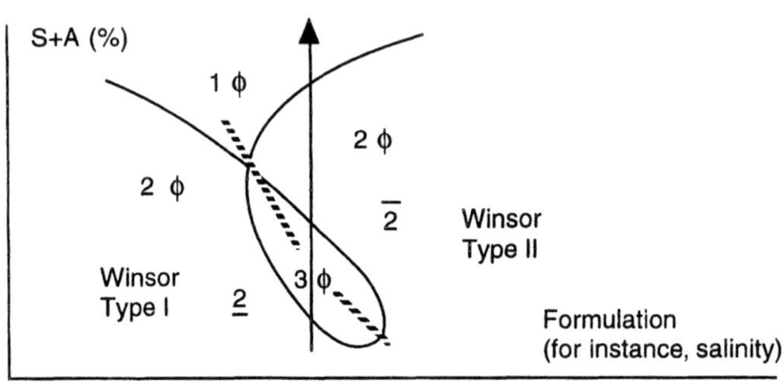

Figure 21 Twisted S+A versus formulation phase behavior diagram.

acteristic parameter discussed in the previous section. The opposite has also been found to occur with hydrophilic alcohols (63).

C. Formulation in Practice

It has been shown that the generalized formulation corresponds to the departure from an optimum formulation, and this is the way it is found useful when correlating the emulsion properties. It is thus important to find out an experimental method to know the formulation state of a given system.

If the system contains a single phase, it is sufficient to dilute it with oil and water to fall in the polyphasic region, and if the system exhibits three phases, it is at or very near optimum formulation, so that these cases do not result in any interpretation problem. In most cases, however, the system would exhibit two phases at equilibrium and the determination of the phase behavior type is of paramount importance. According to all reviewed concepts, the partitioning of the surfactant preferentially into the water (respectively oil) phase indicates a Winsor type I (respectively type II) phase behavior. In many cases the phase that contains surfactant in an amount above the CMC can be detected by the light diffusion known as the Tyndall effect, which turns it bluish or slightly opaque. However, if the micelles are small enough the phase is completely transparent to visible light, and it is recommended to check the light scattering with a low-power laser (something like a laser pointer is sufficient). Whenever there are some micelles, the laser beam path through the phase will turn visible.

On the other hand, the aspect of the surfactant-rich phase may serve as a tell-tale information about the formulation departure from optimum. It has been seen that near optimum formulation the solubilization is high, i.e., a lot of oil and water is enclosed in the surfactant structures, whatever they are. The enclosure is like a surfactant layer, and the amount of surfactant is thus proportional to the surface area of the structure. On the other hand, the solubilized phases are inside the structure islands and thus their amount is proportional to the structure unit volume. If it is assumed that the structure is composed of spherical or pseudo-spherical units of radius R, the surface area is $4\pi R^2$ while the volume is $4/3\pi R^3$, so that the volume per unit surface is proportional to R. In other words, a high solubilization should be associated with large structural units, which are likely to produce a lot of light scattering. This is corroborated by experience, and it may be said that the micellar phase or microemulsion tends to become more opaque whenever optimum formulation is approached from both sides. A turbid microemulsion phase is thus a good hint that the system is very near optimum formulation.

Once the system is found to be one of the basic Winsor types, it is important to know how near or how far from optimum formulation it is located. With the exception of the light scattering discussed in the previous paragraph, there is no other way but to carry a formulation scan to find out how far the optimum formulation is.

If the system is found to be of Winsor type I, then the affinity of the surfactant toward the aqueous phase dominates and some formulation changes have to be accomplished to produce a shift toward a lesser hydrophilic interaction or a larger lipophilic interaction, until optimum formulation is reached.

With an ionic system, such displacement results from an increase in salinity of the aqueous phase, a reduction of the oil EACN either by using a shorter alkane or by adding polar or aromatic components, the addition of some lipophilic alcohol starting with *n*-pentanol up to a few vol %, less with hexanol or higher alcohols. Finally the surfactant may be mixed with a less hydrophilic surfactant of the same family (e.g., with a longer tail). If the system contains nonionic surfactants an increase in temperature would decrease the hydrophilic interaction quite rapidly while the electrolyte effect would be almost negligible, with the exception of polyvalent cation salts.

If the system is of Winsor type II, either an increase in hydrophilic interaction or a reduction in lipophilic interaction is required to approach optimum formulation. This may be done by increasing the oil EACN, particularly by reducing its polar or aromatic content, by reducing the salinity of the aqueous phase, particularly if divalent ions are presents and/or if the system contains ionic surfactants, by reducing the lipophilic alcohol concentration and/or chain length, by reducing the temperature if the system contains nonionic surfactants. Again the surfactant can be mixed, this time with a more hydrophilic member of the same family.

It is worth remarking that the temperature is used for nonionic systems, while the salinity is mainly used to alter ionic ones.

In all experimental cases, the optimum formulation can be detected by any of the phenomena that happen there, such as (a) three-phase behavior, (b) single-phase behavior in-between two biphasic behaviors at high surfactant concentration, (c) interfacial tension minimum, which is particularly suitable when a very small amount of surfactant is present so that no microemulsion is formed, (d) change in preferred partitioning, and (e) several emulsion properties such as emulsion inversion and minimum stability that will be discussed in detail in the next chapter.

Once the change or combination of several changes has shifted the formulation to optimum, some accounting is required to measure the distance from optimum. The "intensity" of each change is calculated according to the correlation expression, and in cases of various concurrent changes they are summed up. Since both correlations contain the term ACN with a coefficient that is nearly the same, a comparison in terms of deviation in ACN unit is preferred. Moreover, ACN units are also used to estimate EPACNUS, the ionic surfactant characteristic parameter.

If the actual formulation change was not due to a variation in oil ACN or EACN, the equivalent change in ACN units is readily found by applying the correlation. For instance, a ΔACN $= +10$ units is equivalent to a fivefold decrease in salinity so that $\Delta \ln S = -0.16 \times 10 = -1.6$, a change in surfactant EPACNUS $= \Delta\sigma/K = -10$, which is equivalent to remove 4.4 carbon atoms from the ionic surfactant tail, or to increase ΔEON $= +1.6$ EO group in the polyether chain. It also corresponds to an increase of about 160°C for ionic surfactants and a ΔT change of about -20 to -40°C (decrease) for nonionic systems, or to the removal of a lipophilic alcohol effect $f(A)$ or $\phi(A)$ = -1.6 equivalent to about 1.5 (respectively 2.5) % of *n*-pentanol or 0.4 (respectively 1.4) % *n*-hexanol, for ionic (respectively nonionic) systems.

The point is now to evaluate how many ACN units is a small or a large deviation from optimum formulation. But first it is worth finding out the accuracy of the numerical calculations that are carried out. If the formula to calculate the HLB of a nonionic surfactant is applied to ethoxylated nonylphenols containing five and six ethylene oxide groups symbolized as NP5 and NP6, it turns out that

$$\mathrm{HLB}_{(\mathrm{NP5})} = \frac{100}{5}\,\frac{5 \times 44}{5 \times 44 + 220} = 10.0$$

while

$$\mathrm{HLB}_{(\mathrm{NP6})} = \frac{100}{5}\,\frac{6 \times 44}{6 \times 44 + 220} = 10.9$$

Thus, near HLB = 10, a change of one unit in HLB is roughly equivalent to a change in one EO group in the polyether chain, which is equivalent to a change in 6–7 ACN units according to the previous paragraph.

The accuracy of the detection of the position of the optimum formulation is about half an ACN unit or a 10% change in relative salinity in good conditions and one ACN unit as routine. This is equivalent to about 0.2 units of HLB, a fairly good accuracy compared with the ± 2 unit uncertainty generally associated with HLB.

Experience indicates that the amount of change that can be viewed as significant, as far as a modification in emulsion property is concerned, is about 1 ACN unit near optimum formulation and 2 ACN units at some distance from it. It will be seen that the change from a very stable O/W emulsion to an unstable one can occur typically over an 8- to 10-ACN-unit change in formulation, and a complete shift from the maximum instability of a O/W emulsion to the maximum stability of the W/O (see Fig. 11) could be from 20 to 40 ACN units depending on the system.

As a rule of thumb it can be said that a 1- or 2-ACN unit change is the minimum to try to get some variation, while a 10-ACN-unit change could produce a complete alteration of the situation and properties. As a matter of fact, the phenomenology associated with the vicinity of the optimum formulation is noted at about ±5 ACN units from it.

This chapter was dedicated to set up the formulation and composition variables framework, which is going to be used extensively in the next chapter to describe the relationship between the formulation and the emulsion properties.

ACKNOWLEDGMENTS

The author thanks the local and national research councils, CDCHT-ULA and CONICIT, for supporting the lab. FIRP research program at Universidad de Los Andes.

REFERENCES

1. P Becher. Emulsions: Theory and Practice. Reprint, Robert Krieger, Huntington, NY (1977).
2. P Becher, ed. Encyclopedia of Emulsion Technology. Four volumes. Marcel Dekker, New York (1985–96).
3. M Rosen. Surfactants and Interfacial Phenomena. John Wiley, New York, 1978.
4. K Shinoda, ed. Solvent Properties of Surfactant Solutions, Marcel Dekker, New York (1967).
5. C Tanford. The Hydrophobic Effect: Formation of Micelles and Biological Membranes, John Wiley, New York (1973).
6. K Mittal, ed. Micellization, Solubilization and Microemulsions, 2 vols. Plenum Press, New York (1977).
7. ME McBain, E Hutchinson. Solubilization and Related Phenomena, Academic Press, New York (1955).
8. JH Schulman, W Stoeckenius, LM Prince. J. Phys. Chem. 63: 1677 (1959).
9. LM Prince. Microemulsions: Theory and Practice, Academic Press, New York (1977).
10. ID Robb, ed. Microemulsions, Plenum Press, New York (1982).
11. M Bourrel, RS Schechter. Microemulsions and Related Systems, Marcel Dekker, New York (1988).
12. S Scriven. In KL Mittal, ed. Micellization, Solubilization and Microemulsions, Plenum Press, New York (1977) p. 887.
13. M Miñana-Pérez, C Gutron, C Zundell, JM Andérez, JL Salager. J. Dispersion Sci. Technol. 20, 893 (1999).
14. H Nakajima, S Tomomasa, M Okabe. Proceedings First World Congress on Emulsion, Paris, October 19–22 (1993). Paper 1-11-162.

15. R Pons, I Carrera, P Erra, H Kunieda, C Solans. Proceedings First World Congress on Emulsion, Paris, October 19–22 (1993). Paper 1-12-159.
16. C Solans, J Esquena, N Azemar. Tenth Int. Symposium on Surfactants in Solution, Caracas, Venezuela (1994).
17. H Kunieda, Y Fukui, H Uchiyama. C Solans, Langmuir 12: 2136 (1996).
18. S Nagata. Mixing: Principles and Applications, Wiley, New York (1975).
19. GB Tatterson. Fluid Mixing and Gas Dispersion in Agitated Tanks, MacGraw-Hill New York (1991).
20. R Laughlin. The Aqueous Phase Behavior of Surfactants, Academic Press, London (1994).
21. PA Winsor. Trans. Faraday Soc. 43: 376, 382, 387, 390, 451, 455, 459, and 463 (1947).
22. PA Winsor. Solvent Properties of Amphiphilic Compounds, Butterworth, London (1954).
23. M Bavière, RS Schechter, WH Wade. J. Colloid Interf. Sci. 81: 266 (1981).
24. A Prince. Alloy Phase Equilibria, Elsevier, Amsterdam (1966).
25. DO Shah, RS Schechter, eds. Improved Oil Recovery by Surfactant and Polymer Flooding, Academic Press, New York (1977).
26. D Anderson, H Bidner, HT Davis, C Manning, LE Scriven. Paper SPE 5811, 1rst SPE Improved Oil Recovery Symposium, Tulsa (1976).
27. K Bennet, C Phelps, HT Davis, LE Scriven. Paper SPE 9351, 55th Annual Fall Techn. Conf., Dallas (1980).
28. M Kahlweit, R Strey, G Busse. Phys. Rev. E47: 4197 (1993).
29. M Kahlweit, R Strey, G Busse. J. Phys. Chem., 94: 3881 (1990).
30. WD Harkins, EC Davies, EC Clark. J. Am. Chem. Soc., 39: 541 (1917).
31. I Langmuir. J. Am. Chem. Soc. 39: 1849 (1917).
32. S Scriven. Nature 263:123 (1976).
33. S Friberg. J. Colloid Interf. Sci. 56: 19 (1976).
34. M Bourrel, J Biais, P Bothorel, B Clin, P Lalanne. J. Dispers. Sci. Technol. 12: 531 (1991).
35. WD Bancroft. J. Phys. Chem., 17: 501 (1913).
36. WD Bancroft. J. Phys. Chem.,19: 275 (1915).
37. A Kalbanov, H Wennerström. Langmuir, 12: 276 (1996).
38. JL Salager. In P Becher, ed. Encyclopedia of Emulsion Technology, Vol. 3, Marcel Dekker, New York (1988) Chap. 3.
39. A Kalbanov, J Weers, Langmuir 12: 1931 (1996).
40. JT Davies. Proceedings 2nd Int. Congress Surface Activity, Vol. 1, Butterworth, London (1957) p. 426.
41. WC Griffin. J. Soc. Cosm. Chem. 1: 311 (1949).
42. WC Griffin. J. Soc. Cosm. Chem. 5: 249 (1954).
43. WC Griffin. Off. Digest Fed. Paint Varnish Prod. Clubs 28: 446 (1956).
44. I Lin. Tenside Det. 17: 119 (1980).
45. P Becher, WC Griffin. HLB: A Bibliography 1949–1969, Atlas Chemical Industries (1970).
46. P Becher, WC Griffin. In McCutcheon's Detergents and Emulsifiers, North American Edition, Allure Publishing, Glen Rock, (1974).

47. P Becher. International Symposium on Surfactants in Solution, Lund, Sweden, July 1982.
48. K Shinoda, H Arai. J. Phys. Chem. 68: 3485 (1964).
49. H Arai, K Shinoda. J. Colloid Interf. Sci. 25: 396 (1967).
50. K Shinoda, H Kunieda. In P Becher, ed. Encyclopedia of Emulsion Technology, Vol. 1, Marcel Dekker, New York (1985) Chapter 5.
51. N Marquéz. Influence de la formulation physico-chimique sur le partage des surfactifs nonioniques polyéthoxylés entre l'eau et l'huile de systèmes Winsor III. Doctoral dissertation, Université de Pau P. A., Pau France (1994).
52. K Shinoda. Proceedings 2nd Int. Congress Surface Activity, Vol. 2, Butterworth, London (1957) p. 275.
53. A Beerbower, M Hill. In MacCutcheon's Detergents and Emulsifiers, Allure Publishing, Glen Rock, IL, (1971) p. 223.
54. A Beerbower, M Hill. Am. Cosmetics Perfum. 87: 85 (1972).
55. AF Barton. Handbook of Solubility Parameters and Other Parameters, CRC Press, Boca Raton (1983).
56. RL Reed, RN Healy. In DO Shah, RS Schechter, eds. Improved Oil Recovery by Polymer and Surfactant Flooding, Academic Press, New York (1977) p. 383.
57. MC Puerto, WW Gale. Soc. Petrol. Eng. J. 17: 193 (1977).
58. JL Salager, J Morgan, RS Schechter, WH Wade, E Vasquez. Soc. Petrol. Eng. J. 19: 107 (1979).
59. RE Antón, N Garcés, A Yajure. J. Dispers. Sci. Technol. 18: 539 (1997).
60. M Hayes, M Bourrel, M El-Emary, RS Schechter, WH Wade. Soc. Petrol. Eng. J. 19: 349 (1979).
61. M Bourrel, JL Salager, RS Schechter, WH Wade. J. Colloid Interf. Sci. 75: 451 (1980).
62. M Bourrel, C Koukounis, RS Schechter, WH Wade. J. Dispers. Sci. Technol. 1: 13 (1980).
63. CJ Glover, MC Puerto. JM Maerker, E Sandvik. Paper SPE 7053, 5th Improved Methods for Oil Recovery Symposium, Tulsa OK, April 16–19, 1978.
64. RE Antón, JL Salager. J. Colloid Interf. Sci., 140, 75 (1990).
65. WH Wade, J Morgan, J Jacobson, RS Schechter. Soc. Petrol. Eng. J. 17: 122 (1977).
66. J Morgan, RS Schechter, WH Wade. In DO Shah, RS Schechter, eds. Improved Oil Recovery by Polymer and Surfactant Flooding, Academic Press, New York (1977).
67. JL Salager, M Bourrel, RS Schechter, WH Wade. Soc. Petrol. Eng. J. 19: 271 (1979).
68. R Cash, JL Cayias, G Fournier, D MacAllister, T Shares, RS Schechter, WH Wade. J. Colloid Interf. Sci. 59: 39 (1977).
69. M Miñana-Pérez, A Graciaa, J Lachaise, JL Salager. Progr. Colloid Polym. Sci., 98: 177 (1995).
70. M Miñana-Pérez, A Graciaa, J Lachaise, JL Salager. Colloids Surfaces A, 100: 217 (1995).
71. S Lettreux, MScT thesis, Université de Montpellier 2, Montpellier, France (1997).
72. JR Baran , G Pope, WH Wade, V Weerasooriya, A Yapa, J. Colloid Interf. Sci., 168: 67 (1994).

73. JR Baran, G Pope, WH Wade, V Weerasooriya. Langmuir 10: 1146 (1994).
74. KH Oh, JR Baran, WH Wade, V.Weerasooriya. J. Dispers. Sci. Technol. 16: 165 (1995).
75. SC Jones, KD Dreher. Soc. Petrol. Eng. J., June 1976: 161.
76. JM DiMeglio, M Dvolaitzky, C Taupin. J. Phys. Chem. 89: 871 (1985).
77. M Puerto, R Reed. Soc. Petrol. Eng. J., 23: 669 (1983).
78. M Buzier, JC Ravey. J. Colloid Interf. Sci. 91: 20 (1983).
79. H Kunieda, N Ishikawa. J. Colloid Interf. Sci., 107: 122 (1985).
80. H Kunieda, K Hanno, S Yamaguchi, K Shinoda, J. Colloid Interf. Sci., 107: 129 (1985).
81. MC Puerto, R Reed. SPE Reservoir Eng., May 1990: 198.
82. P Fotland, A Skange. J. Dispers. Sci Technol. 7: 563 (1986).
83. A Skange, P Fotland. SPE Reservoir Eng., Nov. 1990: 601.
84. RE Antón, F Mosquera, M Oduber. Progr. Colloid Polym. Sci. 98: 85–88 (1995).
85. M Bourrel, JL Salager, RS Schechter, W Wade, Colloques Nat. CNRS. 938: 337–343 (1978).
86. WH Wade, J Morgan, J Jacobson, JL Salager, RS Schechter. Soc. Petrol. Eng. J. 18: 242 (1978).
87. JL Salager. Rev. Inst. Mex. Petrol. 11: 59 (1979).
88. E Dickinson. Annual Reports C, Roy. Soc. Chem. (1986) p. 31.
89. P Jarry, M Miñana-Pérez, JL Salager. In K Mittal, P. Bothorel, eds. Surfactants in Solution, Vol. 6. Plenum Press, New York, (1987) p. 1689.
90. BW Brooks, HN Richmond. Colloids Surfaces 58: 131 (1991).
91. HT Davis. Colloids Surfaces A 91: 9 (1994).
92. BW Brooks, HN Richmond. Chemical Eng. Sci., 49: 1053 (1994).
93. BW Brooks, HN Richmond. J. Colloid Interf. Sci. 162: 67 (1994).
94. JL Salager. In: A Chattopadhay, K Mittal, eds., Surfactants in Solution. Surfactant Science Series Vol. 64. Marcel Dekker, New York (1996) p. 261.
95. P Kruglyakov. Dokladi Akademia Nauk CCCP, 1971, T. 197:1106.
96. P Kruglyakov. Isveztia Sib. Otd. Akademia Nauk CCCP, 9(4): 11 (1971).
97. P Kruglyakov. Tenth Int. Symposium on Surfactants in Solution, Caracas, Venezuela (1994) 57.
98. N Marquéz, RE Antón, A Graciaa. J Lachaise, JL Salager. Colloids Surfaces A: 100: 225 (1995).
99. A Graciaa, J Lachaise, JG Sayous, P Grenier, S Yiv, RS Schechter, WH Wade. J. Colloid Interf. Sci. 93: 474 (1983).
100. W Hsieh, DO Shah. Paper SPE 6594, Symp. on Oilfield and Geothermal Chemistry, La Jolla, CA (1977).
101. VK Banzal, DO Shah. J. Colloid Interf. Sci. 65: 451 (1978).
102. P Kilpatrick, LE Scriven, HT Davis. Soc. Petrol. Eng. J. 25: 330 (1985).
103. H Katayama, T Tagawa, H Kunieda. J. Colloid Interf. Sci. 153: 429 (1992).
104. G Nonaka, M Harada, A Shioi, M Goto, F. Nakashio. J. Colloid Interf. Sci. 176: 1 (1995).
105. XF Li, GH Zhao, EH Lin, SQ Qin. J. Dispers. Sci. Technol. 17: 111 (1996).
106. C Solans, H Kunieda, eds. Industrial Applications of Microemulsions. Surfactant Science Series Vol. 67. Marcel Dekker, New York (1996).

107. P Kumar, K Mittal, eds., Handbook of Microemulsion Science and Technology. Marcel Dekker, New York (1999).
108. M Kahlweit, R Strey, G Busse. J. Phys. Chem. 95: 272 (1995).
109. RE Antón, Ciencia e Ingeniería 17: 79 (1985).
110. RE Antón, MSc thesis ChE, Universidad de Oriente, Puerto La Cruz, Venezuela (1981).
111. H Kunieda, K Shinoda. J. Colloid Interf. Sci. 75: 601 (1980).
112. H Kunieda, S Friberg. Bull. Soc. Jpn., 54: 1010 (1981).
113. M Kahlweit. J. Colloid Interf. Sci. 90: 197 (1982).
114. H. Kunieda, K Shinoda. J. Dispers. Sci. Technol., 3: 233 (1982).
115. M Kahlweit, E.Lessner, R Strey. J. Phys. Chem. 87: 5032 (1983).
116. JC Lang, B Widom. Physica 81A: 190 (1975).
117. B Widom. J. Chem. Phys., 62: 1332 (1975).
118. AM Bellocq, J Biais, B Clin, A Gelot, P Lalanne, B Clemanceau. J. Colloid Interf. Sci. 74: 311 (1980).
119. C Blom, J Mellema. J. Dispers. Sci. Technol. 5: 193 (1984).
120. AM Bellocq, D Bourdon, B Lemanceau. J. Colloid Interf. Sci. 79: 419 (1981).
121. M Bourrel, C Chambu, RS Schechter, WH Wade. Soc. Petrol. Eng. J. 22: 28 (1981).
122. M Bourrel, C Chambu. Paper SPE 10676, 3rd Joint Symp. SPE/DOE on Enhanced Oil Recovery, Tulsa (1982).
123. M Bourrel, C Chambu. Soc. Petrol. Eng. J. 22: 28 (1982).
124. M Bourrel, C. Chambu, RS Schechter, WH Wade. Soc. Petrol. Eng. J. 23: 327 (1983).
125. M Bourrel, F Verzaro. C Chambu. SPE Reservoir Eng. 2: 41 (1987).
126. M Kahlweit, R Strey. J. Phys. Chem. 91: 1553 (1987).

3

Emulsion Properties and Related Know-how to Attain Them

Jean-Louis Salager
Lab. FIRP, Chemical Engineering School,
University of the Andes, Mérida, Venezuela

This chapter deals with emulsion properties, i.e., type, drop size, stability, and viscosity. It shows how to estimate or measure them in practice and how they are related to formulation, composition, and other variables. The current know-how is presented in a comprehensive way so that the emulsion maker can use it to attain some suitable property. The case of emulsion made from preequilibrated surfactant–oil–water systems is treated first. Then follows a discussion on how to modify emulsions to seek specific properties, which leads to the introduction of the dynamic inversion phenomena, discussed from the practitioner's point of view.

I. EMULSION CHARACTERISTICS AND PROPERTIES

A. Conductivity and Type

Maybe the first and most important emulsion property to be determined is type, i.e., either oil-in-water (O/W) or water-in-oil (W/O), and eventually the occurrence of a multiple emulsion. This is essentially equivalent to determine which is the emulsion external phase.

There are different methods to do so. Since the external phase is the one that first touches whatever is in contact with the emulsion, the external phase and the emulsion exhibit similar wettability and dispersion properties. For instance, a dash of O/W emulsion would extend on a hydrophilic substrate such as a clean glass slab or a filter paper, while a W/O would not. If a small amount of an O/W emulsion is poured in an aqueous medium its external phase would dissolve in the aqueous phase and its oil droplets would disperse. On the contrary, a W/O emulsion globule would not do so (1).

These are simple qualitative methods that require only the withdrawal of a small sample, but they are not amenable to in line or in situ measurement, such as to continuously monitor the emulsion type and to detect inversion occurrence.

In most cases the aqueous phase contains some electrolyte while the oil phase does not, so that the electrolytic conductivity is the property that is measured to detect the emulsion type.

As a first approximation it may be said that the electrolytic conductivity of an emulsion κ_{em} that is proportional to the conductivity of its external phase κ_{ext} and to its volumetric proportion ϕ_{ext}.

$$\kappa_{em} = \kappa_{ext}\, \phi_{ext} \tag{1}$$

Actually the variation depends on the drop size average and distribution, but it is not very sensitive unless there are large changes in these characteristics. This means that the conductivity of an emulsion typically varies vs. the external phase proportion as indicated in Fig. 1 (left). The variation is proportional while it is an O/W type and it is essentially nil (at the same scale) when it is a W/O emulsion because the typical electrolytic conductivity of the oil phase is 100 or 1000 times smaller than the conductivity of the aqueous brine. Consequently, the point where the emulsion switches type corresponds to a very large change in conductivity that may be detected even with a very rugged apparatus. The only requirement is to make sure that there is enough mixing so that the emulsion does not separate inside the conductivity cell. This is why conductivity cells that have a bell-type antishock protector are inappropriate for this purpose since the dispersed phase drops are likely to settle in the unstirred zone and to fill it. The best suitable shape is an open geometry, eventually with a nonconductor bridge between the electrodes as a reinforcement.

If an accurate conductimeter apparatus is used, a deviation from the straight line law may indicate the occurrence of a multiple emulsion. Consider, for instance, a $W_1/O/W_2$ type multiple emulsion with the proportions ϕ_1 and ϕ_2 of most internal (W_1) and most external (W_2) water phases, respectively. If the total aqueous phase (proportion ϕ_{TW}) were located in the external phase the conductivity would be:

$$\kappa_{emT} = \kappa_2(\phi_1 + \phi_2) = \kappa_2\phi_{TW} \tag{2}$$

However the real external phase that conduces the electrical current is W_2, so that:

$$\kappa_{emR} = \kappa_2\phi_2 \tag{3}$$

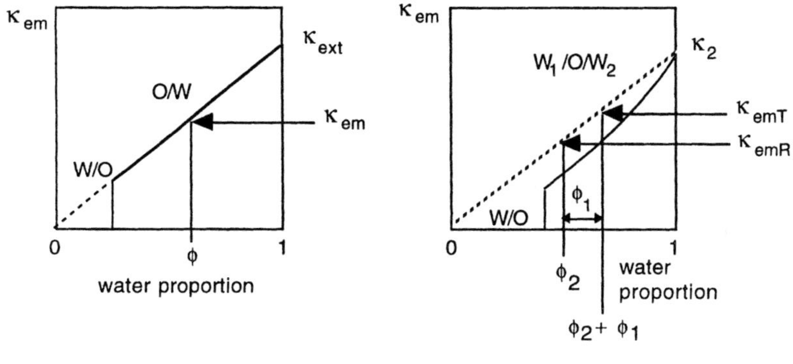

Figure 1 Variation of emulsion conductivity as a function of the water fraction, for a SAD < 0 (respectively SAD > 0) formulation at left (respectively right). (After Ref. 2).

which is the same expression as the previous one, but shifted a smaller value of the water phase proportion (see Fig. 1 right), the shift being the proportion of water ϕ_1 which is inside the oil drops (2).

$$\phi_1 = \phi_{TW} - \frac{\kappa_{emR}}{\kappa_2} \tag{4}$$

The presence of a multiple emulsion can be detected by conductimetry when ϕ_1 is about 2%, whereas it can be measured when it is more than 5%.

The application of multiple emulsion to controlled release or controlled transfer involves nonequilibrium effects in which the most internal and most external phases, e.g. W_1 and W_2 in the previous case, have different compositions, so that a water transfer would take place under the osmotic pressure gradient.

In such a situation special care is advisable since the water transfer would change the external phase proportion ϕ_2, as well as its electrolyte concentration κ_2. If the most inside droplets (W_1) contain a higher electrolyte concentration, water would be transferred from outside water (W_2) to inside water (W_1), i.e., ϕ_2 would decrease, while the electrolyte concentration in W_2 would increase, as would κ_2. These opposite effects on the emulsion conductivity would make it quite insensitive to the mass transfer occurrence, and another detection method would have to be found.

B. Drop Size

After characterizing the emulsion type, the second most crucial information is the size of the drops. Since emulsification is generally a more or less random stirring process in which the breaking and coalescence steps are in dynamic equilibrium, the resulting emulsion is a polydispersed system in which small and big drops coexist. The outcome of this equilibrium depends on a large number of factors that can influence the two antagonistic steps. Generally speaking, a lower tension or an increase in stirring energy and duration are expected to increase the break up, while a higher fluid viscosity would slow it down, and temperature would increase the coalescence rate (3–7). These statements may be true or false, however, because in many cases a single variable can have opposite effects, as for instance the case of formulation and temperature, which will be discussed in detail later on. It is thus safe not to make too straightforward guesses.

The best description of an emulsion is through its drop size distribution, which gives a statistical inventory of the dispersed phase fragmentation. This information is extremely valuable in practice because both the stability and the viscosity depend on the drop size distribution.

The drop size can be measured by several techniques from rugged to sophisticated and from cheap to expensive (8,9). The first type of method deals with the determination of an average property linked to the dispersion size such as turbidity or reflectance. The problem is that they also depend on the proportion of dispersed phase and other factors like the wavelength used and the refractive indexes. As a consequence, turbidity measurements in the so-called nephelometry technique are no longer used in practice but in visual guessed diagnostics, which by the way could turn out to be pretty accurate if the observer is experienced. In effect, it is known that the maximum turbidity is attained when the drop size is in the 2- to 5-μm range, and that it decreases whether the drops become smaller or larger.

The interaction between light and matter is a complex phenomenology that has been theoretically solved a long time ago, but has been applied to wide range instrumentation techniques only very recently because of computational problems that could not be solved quickly prior to the arrival of digital computers.

The individual drop-by-drop analysis started with microscopic observation several decades ago. The method is tedious and sometimes erroneous, such as when big drops are flattened in-between glass slides, or when the field depth is so small that the adjustment of the image plane cuts the drop above or below its equator, resulting in an apparent smaller size. These problems were not amenable to ready solution with the arrival of shape recognition software in the 1980s.

Orifice obscuration methods have been used for about two decades as the first automated counting techniques. In these methods the diluted emulsion is forced to flow through an orifice or a short capillary, by a pressure gradient. When a drop passes through the orifice, a conductivity property is altered by the obstruction produced by the drop. It may be electrical conductivity between two electrodes placed on the two sides of the orifice, or light transmission through the orifice from a source to a photodetector. In both cases the obscuration of the property signal is translated to a drop size information by an appropriate data-processing device. If the emulsion is diluted enough, drops cross the orifice one by one so that an effective counting can be carried out and registered. After counting about 500 drops a statistical analysis can be carried out. These methods were, however, limited by the accuracy in detecting small drops, i.e., smaller than 10% of the orifice diameter, and by the eventual plugging of the orifice by big drops that could not pass through. As a matter of fact, the range is typically 0.1 to 0.8 time the orifice diameter and a change of orifice is often necessary to accommodate different ranges. If the emulsion is very polydispersed, measurements with two or more different orifices are required to be matched to cover the whole statistical distribution—a risky exercise as far as accuracy and reliability are concerned.

In the past 10 years, the laser light scattering analyzer has displaced every other equipment, whenever the high price tag of this apparatus is no limitation. The method is based on the matter–light interaction according to Mie's law. In the usual macroemulsion drop size range, the diffraction angle increases as the drop size decreases, and the relation, although complex, is well known. Laser light pulses are sent at a quick pace through a cell in which a diluted emulsion quickly flows to get a rapid renewal of the sample. With an expanded laser beam several drops can be intercepted simultaneously, with essentially no limit in size. On the other side of the cell an optical system measures the diffracted light at various angles at the same quick pace. Under these conditions only a few seconds is necessary to capture the signal from thousands of drops, a number that ensures a high statistical significance. Thanks to improved optics and with the help of the computational muscle attained by personal computers, the light scattering signals can be translated into wide range drop size statistics in a very few seconds.

In the near future acoustic wave absorption might provide a way to estimate the drop size in opaque high-internal-phase ratio emulsions without dilution. However, the technique is still on probation.

Figure 2 indicates different types of distribution found in emulsions: unimodal of the log normal type produced by turbulent homogeneous stirring; narrowly shaped or highly polydispersed; bimodal emulsions resulting from a mixture of two emulsions, which by the way might be intentionally made to attain a low viscosity.

The distribution frequency coming from the analyzer is generally in volume, i.e., the proportion of drops included in a class is the volume proportion of these drops with respect to the whole internal phase volume. In some cases the distribution in number or in surface is also required. It is worth remarking that if the distribution follows a log normal statistic, as is often the case, the Hatch-Choate relationships may be used to readily translate volume distribution frequencies into equivalent number or surface distribution frequencies.

The notion of average size or distribution central tendency is a tricky one since there are many ways to calculate the mean, which are not at all equivalent, particularly with asymmetrical distributions. The mode is the most common size,

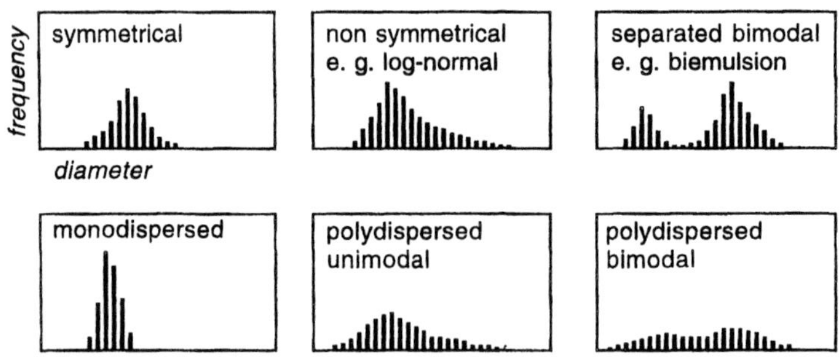

Figure 2 Different shapes of the drop size distribution.

while the median is the size so that half the internal phase volume is fragmented in smaller drops and half in larger ones. It is often symbolized as $D(1/2)$ or $D(\mathrm{v},0.5)$.

Other used mean values are the mean of the distribution in volume or mass, so-called $D(4,3)$ because it is the ratio of the fourth moment of the distribution in number to the third one, or the mean of the distribution in surface, so-called Sauter mean diameter SMD or D(3,2), i.e., the ratio of the third moment to the second one. This last mean diameter is the diameter of the sphere that has the same surface/volume ratio as the whole population, the most significant value as far as the surfactant adsorption is concerned.

$$D(4,3) = \frac{\sum_{i=1}^{n} f_i\, x_i^4}{\sum_{i=1}^{n} f_i\, x_i^3} \qquad D(3,2) = \frac{\sum_{i=1}^{n} f_i\, x_i^3}{\sum_{i=1}^{n} f_i\, x_i^2} \tag{5}$$

where f_i is the frequency in number in class i, characterized by its central diameter x_i. The arithmetic mean m or first moment is calculated as:

$$m = M_1 = \sum_{i=1}^{n} f_i x_i \tag{6}$$

The degree of polydispersity is given by the standard deviation, symbolized by σ, which is the square root of the second moment around the first moment value.

$$\sigma = \sqrt{\sum_{i=1}^{n} f_i\,(x_i - m)^2} = \sqrt{M_2 - M_1^2} \tag{7}$$

where m and σ are the two parameters that appear in the theoretical expression of the normal distribution density:

$$f(x) = k \exp\left(\frac{(x - m)^2}{2\sigma^2}\right) \tag{8}$$

where k is a numerical coefficient so that the integral below the frequency curve is the unity. If the distribution is log normal, then x, m, and σ are replaced by their natural logarithms.

Drop size distributions are often plotted as frequency vs. logarithm of diameter to test the symmetry around the first arithmetic mean value or first moment m of an assumed log normal distribution, which is a quite common case.

If the distribution is symmetrical (in the selected scale) the mode corresponds to the median, i.e., 50% of the cumulated drop volume or number whatever the distribution is. Another way to estimate the asymmetry is to calculate the third moment of the distribution around the mean, which is zero for perfectly symmetrical distributions.

$$M_3 = \sum_{i=1}^{n} f_i (x_i - m)^3 \tag{9}$$

The distribution shape and its changes are handy information on the emulsion formation process or its evolution. For instance, the presence of two modes indicates two separate stirring processes, and could be related to the mixing of two emulsions or to improper stirring. An asymmetrical distribution with a long tail on the higher drop size, particularly in logarithmic scale, is indicative of incomplete mixing. The increase of the frequency in the same region as time elapses is a precursor signal of future stability problems. The disappearance of the left part of the main peak (smaller drop) is indicative of a slow ripening process. It is recommended that all practitioners practice extracting information from the drop size distribution.

Another way to handle this statistical data is to plot the accumulative or integral distribution on a Gaussian or log-Gaussian graph, sometimes called probability plot. In such plots the normal or log normal distributions become straight lines. It is thus very easy to test the goodness of fit and to calculate the two parameters that define the distribution. The splitting into several straight line segments is also indicative of multimodal distributions that can be easily separated. It is worth noting that such a way to represent a normal or log normal distribution by a straight line is readily amenable to very simple and transparent regression analysis on any personal computer.

C. Stability

Although emulsion stability is not a concept with a well-agreed-on definition, it is always linked either to the persistence or to the decay of the dispersed system under certain circumstances. As a matter of fact, it is a fundamental emulsion property and a lot of attention has been dedicated to its study (10,11).

1. Involved Phenomena

The first question to answer is, stability against what? In fact, an emulsion can remain unchanged under certain circumstances while it breaks readily under others. Of particular importance are the eventual changes in temperature that have a formulation effect as seen in the previous chapter. Other emulsion breaking processes could include physical effects like artificial gravity, Brownian motion, or stirring. However, the most common case is when an emulsion is kept in a container to rest at constant temperature under normal gravity conditions.

This is the one to be dealt with here and later to be considered as a reference based on which other cases could be discussed. This breaking process comprises several steps: (a) long-distance approach between drops or between drop and flat interface, (b) interdrop film drainage; and, finally, (c) coalescence (10–13).

Since the continuous and dispersed phase have generally different densities, there is a neat Archimedes pull on the dispersed phase drops that drives a separation process called *settling*. This separation tends to gather the drops in a region that becomes a high internal phase ratio emulsion, sometimes referred to as a *cream*.

The settling process is essentially similar in nature to the Stokes falling sphere sedimentation problem, eventually modified according to Hadamard's work (14) to account for the fluid motion inside the drop. The Stokes calculation of limiting falling velocity is solved by equating the gravity force to the friction force, which is taken as proportional to velocity (creeping flow) and proportional to the radius of the sphere R. Thus, the limiting velocity turns out to be:

$$v = \alpha \frac{R^3\, \Delta\rho}{\eta R} = \alpha \frac{R^2\, \Delta\rho}{\eta} \tag{10}$$

where α is a constant ($2\pi/9$ in the case of Stokes's law), $\Delta\rho$ is the density difference, and η the continuous phase viscosity.

Since Stokes's problem addresses the case of a single rigid sphere sedimentation in an infinite medium, the actual problem with plenty of falling spheres (with different diameters, thus different velocities) could be depicted only qualitatively by such a law. For instance, the simultaneous falling of many drops results in a countercurrent in the external phase that would reduce the actual falling velocity. The interaction of falling drops with one another is also a factor that would reduce the falling velocity. Even more serious objections rise from the possible retardation effect produced by tension gradients along the drop surface (15).

This means that the value of the constant is of course to be determined experimentally in each case, and that Stokes's law is only indicative, and may turn out to be wrong when the internal phase proportion is high. In any case, Stokes's law indicates that the approach of the drops would be facilitated and accelerated whenever the drop size or density difference increases, while it would be slowed down by an increase in external phase viscosity.

In the macroemulsion range (1–100 μm) the settling may be rather quick unless the external phase is viscous or the density difference very small. In the miniemulsion range (100–500 Å) the settling is generally very slow because of the importance of the squared radius factor. Thus, it may be said that an extremely small drop size would slow down the first step of the breaking process and as such would slow down the whole process. This statement is true in the event that the second step (drainage) is rather quick, i.e., when the surfactant does not provide the appropriate stabilization conditions, so that the first step controls the decay.

The maximum internal phase ratio that can be attained without deforming the drop spherical shape depends on the drop size distribution. For monodispersed rigid spheres, the most dense tessellation is the hexagonal packing at about 74% of internal phase. As an example, for randomly settled monodispersed spheres it could be 65%. For very polydispersed emulsions it might be higher than 90% (16).

The final arrangement of the drops in the cream depends on the segregation process between drops. In effect, if the density difference is large, the R^2 factor results in a considerably increased force on the large drops that would settle more quickly and end up accumulating at the far end of the cream. The shape of the settling vessel is also important in the segregation process because the segregation could be large if the settling path is long and vice versa.

Of course, if the emulsion contains more than 60–70% of internal phase, particularly with small drops, the settling would not take place significantly. In any case the Brownian motion or other unusual effects may provide some driving force for the drop–drop interaction to be dealt with in the next paragraph (17, 18).

When the drops are close together, whatever the driving force that makes them approach, the film that is located between neighboring drops exhibits a complex drainage process that involves several different mechanisms, and this controls the second step of the emulsion decay. Some of them depend on the drop volume like the van der Waals attractive forces, or the Archimedes pull, while others depend on the interdrop film physical properties such as viscosity, or on the interfacial phenomena that occur whenever two interfaces approach at sub micrometer range. The first class of interfacial phenomena deals with static attractive or repulsive forces, like electrical, steric, or entropic repulsions, while the second one has to do with dynamic processes like the steaming potential and interfacial viscosity effects, as well as the more classical hydrodynamic considerations (19–23).

In some cases the interdrop film becomes very thin, i.e., only a few micellar or molecular sizes across. It decreases not continuously but stepwise, layer by layer, and could end up in a surfactant bilayer with no solvent content (sometimes referred to as black film because of its color). Such extremely thin films could exhibit a high resistance to rupture as it occurs in so-called foam emulsions (23,24).

In most cases, there exists an electrical double layer near the interface, even in the absence of charged surfactant, in which case the differential adsorption of OH^- and H_3O^+ ions results in a net charge per unit area. As a matter of fact, nonionic surfactants often produce a negatively charged interface at neutral pH because the smaller OH^- is more abundant in the vicinity of interface than the hydrated proton. In any case it can be considered that the double layer is a very general feature at liquid interfaces. Of course, the charge separation occurs in a very thin region near the interface, whereas the whole electrical balance per unit area is neutral. As indicated in Fig. 3, the region is split into an adsorbed or fixed layer and the ''diffuse'' layer in which the ions are free to move with the fluid, hence the name ''double layer.'' In the diffuse layer the ions are submitted to an attractive force that pulls them back toward the opposite charge interfacial layer, and the random molecular motion, i.e., their escaping tendency. Several models have been proposed that are discussed in classic textbooks on surface science (25–27) or specialized publications (28,29).

The overall result is typically an exponential distribution of ionic charges in the diffuse layer that results in an exponentially decreasing potential according to the so-called Debye approximation:

$$\psi = \psi_0 \exp(-x/\lambda) \tag{11}$$

where x is the distance from interface and λ a characteristic parameter known as Debye length. ψ is the potential at distance x from interface while ψ_0 is its value at interface.

Debye length is somehow a measurement of the diffuse layer thickness, which is directly linked to the distance at which electrical repulsion between two approaching interfaces begins. In effect, the pressure (force per unit interfacial area) produced by two diffuse layers overlapping may be expressed as follows in the case of approaching flat surfaces separated by a distance x:

$$P = \frac{2\varepsilon\psi_0^2}{\lambda^2 \cosh^2(x/\lambda)} \tag{12}$$

It is seen that the repulsion is a function of both λ and ψ_0. It may be shown that as ψ_0 and λ increase the repulsion takes place at a larger distance, a favorable

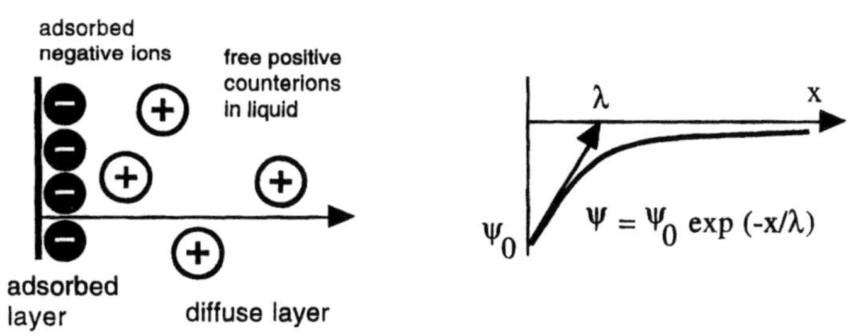

Figure 3 Electrical double layer and potential variation vs. distance from interface.

feature for emulsion stability. While ψ_0 is directly associated with the amount of adsorbed charges, e.g., ionic surfactant adsorption, λ depends on several factors as seen in the following equation

$$\lambda = \sqrt{\frac{\varepsilon RT}{F \sum C_{i\infty} Z^2_i}} \tag{13}$$

where F is the Faraday constant, ε the permittivity of the medium, $C_{i\infty}$ the concentration of ion specie ''i'' in the bulk (out of the double layer), and Z_i its valence. This relationship indicates that an increase in electrolyte concentration, particularly with polyvalent species, would decrease the double-layer thickness and thus the distance at which the repulsion is effective. Since attractive van der Waals forces are generally dominating at short distance, a repulsion effective double layer should be thick, so to speak, since this could mean 100 Å only. This antagonism of electrical repulsive forces and van der Waals attractive forces was the subject matter of the DLVO theory [name given after the people who proposed it simultaneously in Russia (Derjaguin and Landeau) and Netherlands (Vervew and Overbeek)] to explain lyophobic colloid stability in the presence of electrolyte (30–32). This theory was a landmark in colloid chemistry history as the first successful attempt to explain complex experimentally known phenomena.

Today it contributes as a model for emulsion stability on a qualitative basis, since the drop size is much bigger than a colloidal particle, so that DLVO interaction calculations do not apply straighforwardly. Moreover, it is now recognized that the repulsion between approaching drops may come from other (non-DLVO) interactions such as the steric repulsion between adsorbed molecules, which is the most common case with nonionic surfactant and polymer emulsifiers both in water and in oil (33,34). There is first an enthalpic contribution that has been referred to as osmotic repulsion as well, because it has to do with the departure of the chemical potential of the solvent in the region where adsorbed surfactant layers overlap (35). On the other hand, there exits another repulsion mechanism, often called entropic, that is related to the molecular organization or degree of freedom with respect to interactions of surfactant or polymer molecules with the solvent that changes as the two interfaces get closer (36). Figure 4 illustrates different cases of repulsion: electrical repulsion in the presence of ionic surfactants and steric repulsions with nonioic surfactants, such as the polyethoxylated ones. The third type, i.e., entropic repulsion, occurs, for instance, when polymeric amphiphilic molecules are forced to rearrange and sometimes to lose solvatation as they are compressed.

As in DLVO theory, the overall balance of repulsive and attractive forces may result in different cases since their variations are not necessarily at the same scale or manner. For instance, molecular attractive forces act at short range (x^{-6}), while electrical ones are rather long range forces (x^{-2}). Steric and entropic forces

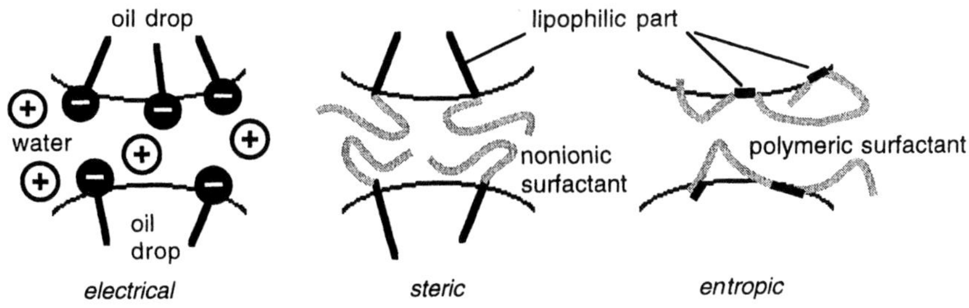

Figure 4 Three cases of interdrop repulsion.

do not produce any effect until some kind of contact takes place between adsorbed antagonist layers, at which point they can increase slowly or rapidly depending on the compressibility or elasticity of the adsorbed layer. In many cases several effects can contribute.

There are essentially three cases, as in DLVO theory, that are characterized by the shape of the repulsive–attractive potential vs. distance (see Fig. 5). In the first case (left), the attractive forces dominate at all distances and thus the two approaching interfaces will get into contact sooner or later depending on kinetic phenomena, as seen later on. The minimum of the curve (maximum attractive potential) corresponds to the notion of contact and occurs at an essentially zero distance (exaggerated in the figure). A further reduction in distance would be opposed by the compressibility of matter and would result in a strong repulsion. At the maximum attraction distance, solid coagulation and liquid drop coalescence would take place at once. The second case (center) is typical of a long-

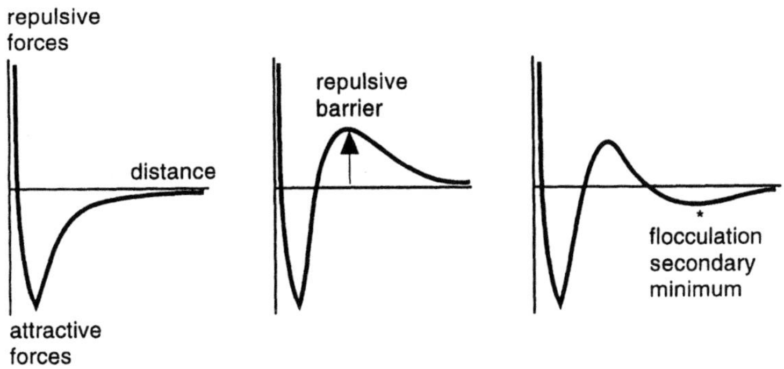

Figure 5 Three cases of variation of the interdrop forces vs. interdrop distance according to the extension of the DLVO theory.

range repulsion and a short-range attraction as in ionic surfactant systems with low electrolyte content. As the distance between approaching interfaces decreases, the electrical repulsion dominates and reaches a maximum value. If this ''barrier'' is high enough, no drop would possess the sufficient kinetic energy to surpass it, and the repulsion would end up dominating, with a resulting stable emulsion. If the barrier is not high enough, a fraction of the approaching drops, those with enough kinetic energy, would reach the distance where the forces become attractive and would end up getting into contact and coalescing.

The third and most complex case exhibits the so-called secondary minimum. As interdrop distance decreases the force is first attractive, then it becomes repulsive, them attractive again. The long-range attractive forces are weak forces, so that the secondary minimum, which is located at a ''large'' distance, is not very deep. Hence, a small amount of kinetic energy, like mechanical mixing, can offset it to redisperse the flocculated drops. It is worth noting that the distance at which the secondary minimum occurs is too long for the drops to coalesce. They are just linked by some weak attractive force and can be separated again, so that this is not to be considered as an instability phenomenon but rather as a reversible metastable association. This phenomenon often occurs when the repulsion forces of the second case are weakened, for instance, by adding electrolytes. Too much weakening could, however, lower the repulsive barrier so that the primary deep minimum is attained. There the distance is so short that coalescence will take place, and the energy hole is so large that redispersion would require an energy input similar to the one that allowed to make the emulsion in the first place.

Interdrop flocculation is not an uncommon phenomenon in high internal phase or creamed emulsion, and it generally results in a temporary increase in viscosity, yet this effect may usually be reversed by applying a low-energy shear.

The previous discussion dealt with forces and potentials, i.e., equilibrium concepts, so that it indicated the possibility of occurrence of some events, not the velocity at which they could happen. These potential considerations may be brought into a kinetic model similar to a bimolecular reaction rate, in which the probability of coalescence is taken as a function of the repulsive potential. This leads to a second-order kinetic model first proposed by Smoluchovski in 1916. It is usefull, to describe an emulsion ripening process in which the reciprocate of the number of drops in a given system increases linearly as time elapses. This model has been improved on to calculate the coalescence dynamics of nondeformable drops (37,38) as well as deformable drops associated with a thin-film hydrodynamic drainage (39,40).

It is worth mentioning that a special issue of the *Journal of Dispersion Science and Technology* (41) was recently dedicated to the research work of Bulgarian investigators on thin films and related topics, as a tribute to their outstanding contribution to the field.

There are also other dynamic phenomena to be somehow accounted for (20,42). For instance, the approaching of two droplets occurs simultaneously with the drainage of the interdrop liquid, in which fluid dynamic phenomena may bring in several other kinetic features. For example, in presence of an ionic surfactant such as sodium dodecyl sulfate, the amphiphilic dodecyl anion would adsorb at interface while the sodium counterions would stay in the aqueous bulk phase. In such a case the negative charges are somehow ''attached'' to the interface, while the positive ones are linked to the liquid. When the two drops get closer, the liquid film withdrawal strips away electrical charges from the electrical double layer, which creates a charge separation. This results in the steaming potential effect, which tends to return the charges to their initial position. As a consequence, the film drainage is slowed down and everything happens as if the film liquid were much more viscous than it actually is. This is why this phenomenon is also referred to as electroviscosity.

The presence of adsorbed molecules that bear interactions with the film liquid phase, e.g., by solvatation, may result in the interfacial viscosity effect, in which the motion of the liquid is transmitted to the adsorbed molecules that cannot move freely because of lateral interactions with other adsorbed molecules (43,44). This may happen particularly with macromolecular species, often known as colloid protectors, that also exhibit a viscosity enhancing effect in the bulk phase. All of these effects can reduce the interdrop film drainage rate and thus determine the long-term persistence of the emulsion.

It is considered that unless the final instantaneous step, i.e., drop coalescence, has occurred, the emulsion is not broken since it can be redispersed with an amount of energy that is often much smaller than that required to make the emulsion in the first place.

These were the typical situations dealt with whenever an emulsion was left to decay in the gravity field at constant temperature. If this is not the case, the actual occurrence might depend quite a lot on the circumstances of the emulsion exposure to decay, like changes in formulation, temperature, or WOR (drying), as well as stirring. Only trends are known as far as these effects are concerned, and it can be said that even the commonplace emulsion decay at rest under normal gravity is too complex to be fully interpreted in a quantitative way.

2. How to Measure Stability

It is said that an emulsion is stable when it does not change its aspect in 3 years or so, and that it is unstable if it has completely separated after a few minutes. Anything in-between these extremes requires a quantitative measurement of the emulsion evolution with time.

The unique absolute measurement would be to count the number of drops in a given container, information that cannot be collected without altering the system, so that only a single measurement may be done, which is not very sagacious anyway. The second choice would be to analyze a sample of the emulsion from time to time to determine whether the drop size average and/or distribution is changing. This could be done with some accuracy, but a lot of inaccuracy may come from the sampling process. In effect, if the emulsion is left to rest in the gravity field, the settling would result in a segregation by size according to a vertical distance. Thus, the sample drop size is likely to depend on the location where the sample is withdrawn. The answer to this problem is to not take the sample at the same location, because such a tactic could backfire. For instance, if the sampling is taken exactly at the middle of a test tube that contains an emulsion containing small and big oil drops in water, the first change to be observed could be an increase in drop size because of the transit of the big drops that were initially in the lower part of the test tube, then a return to the initial average size, then finally a decrease in drop size after all big drops have settled into the upper part. It is obvious that such a segregation could completely hide the way the drop size changes because of coalescence.

The experimenter could be rid of this obstacle by completely remixing the whole system before taking a sample. However, such a procedure would destroy the rest state of the emulsion by interrupting any segregation or flocculation process. The final solution of this problem is to prepare several small samples in separated vials to be kept exactly in the same conditions until each of them is opened and fully mixed before carrying out a drop size measurement. This may not be satisfactory for some unconventional studies that would require large amounts of emulsion to be kept or turned upside down in a jar or submitted to heating–cooling cycles as in real life occurrence.

In many instances the concept of emulsion stability is linked with a visual change in appearance or tactility, i.e., because some portion of one or two phases separates from the emulsion, or because some other crucial property such as viscosity is significantly altered. If the emulsion is not too stable this may be a nice way to follow up the decay. If it takes too long a time to get visible measurable changes, then some time acceleration tactics would be advised.

Figure 6 indicates the typical variation of separated oil or water volume from the emulsion as times elapses. The separated volume V_c could come from the internal phase drops that have coalesced or could be the external phase which has been cleared from settling drops. V_∞ is the same phase volume value at infinite time, i.e., after complete settling, so that the V_c/V_∞ ratio varies from zero to unity.

The second occurrence (external phase clearing) is not necessarily a measurement of instability since the gathering of drops in a cream without coalescence could be reversed by an appropriate slow mixing.

This clearing and creaming is common in low-internal-phase-ratio emulsions, in which there is a large density difference and large drops. If the experimenter is looking for a solid basis to study formulation effects, for instance, a

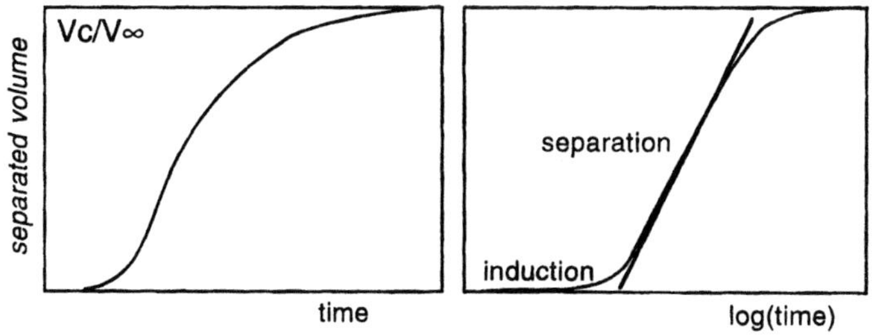

Figure 6 Variation of the separated phase volume vs. time.

unit WOR is advised as well as the making of rather small drops to eliminate this eventuality. If this is not possible, the monitored phase separation should refer to the internal phase only.

Coming back to Fig. 6, it is seen that the apparent decay occurs in three periods. First, there is a period of time in which no separation takes place. This induction period may be related to the drop approach and film drainage, with no coalescence occurrence, so that no drop big enough to settle quickly is formed.

In the second period the separated volume increases steadily with time according to an S-shaped curve as indicated in Fig. 6. This is when most of the coalescence takes place at a steady rate.

The third period is some kind of asymptotic behavior during which the last and smaller droplets complete their separation. This could trail for quite a long time since the scarcity of droplets shrinks the probability of encounter and coalescence. Often the central part of the separation curve is found to be linear vs. a logarithmic time scale, and the use of such a scale is advised to compare emulsified systems with very different stability.

3. Acceleration Tactics to Measure Stability

Some emulsions are sometimes so stable that they require an extremely long time to start separating, which is quite inconvenient when experimental time is money. It is thus worthwhile to try to artificially accelerate the emulsion breakup in order to compare formulation effects at a shorter time scale.

The first thing that comes to mind is to use a centrifuge to speed up both early sedimentation and ensuing creaming, and to augment the force that pushes the drops against one another and thus helps in the film drainage. This will in general accelerate the separation process, but it may also convey false or misleading information. In effect, some emulsions would never coalesce in normal gravity, while they would do so under artificial gravity, because of different relative strengths between gravity and repulsive forces. In other cases, the creamed emulsion often will not coalesce with natural or artificial gravity because the gravity pull has very little or no effect on the surfactant role in slowing down the interdrop film drainage that commands the second and often crucial step in the separation process. The separation under centrifugation is thus a useful technique but not necessarily successful even for comparison purposes.

Another method is to make an emulsion with a larger drop size so that the settling process is accelerated. A larger drop size also means a smaller surface-to-volume ratio, i.e., a lower repulsive effect (that depends on the adsorbed surfactant at interface) vs. a higher attractive force (that depends on the drop volume). This will lead to a reduction in separation time, although probably no more than one order of magnitude, which might be too little in many instances. This method influences the separation second step and is thus more reliable than centrifugation to predict trends.

A clever method is to reduce the stabilizing agent concentration at interface so that all of the stabilizing phenomena taking place in the second step (that dominates the decay of stable emulsions) would be weakened. A reduction in surfactant concentration in the whole system is not appropriate to produce such a reduction in surfactant adsorption. In effect, it is known that the surfactant adsorption starts occurring at very low concentration, much lower than its CMC, while the surfactant concentration in stable emulsions is often much higher in order to ensure a quick drop coverage during emulsification. Accordingly, a reduction in surfactant concentration is likely to first produce alterations in the emulsification process, and thus the emulsion would be quite different from the one that must be studied.

It is preferred to reduce the interfacial surfactant concentration without reducing the surfactant concentration in the bulk phase. This may be done by introducing a cosurfactant that would compete with the surfactant for interfacial occupancy but would not provide the stability features the surfactant molecules do. The most likely candidates for such a purpose are short chain alcohols in the three- to five-carbon range. In order to occupy a larger interfacial area, the alcohol should not be too hydrophilic or lipophilic. The best intermediate choice seems to be in-between *n*-propanol and *n*-butanol, the first one being rather hydrophilic and the second more lipophilic. A mixture of these two *n*-alcohols, as well as secondary butanol, and even tertiary pentanol, would do the job fairly well at a 2–3 wt% concentration level in the system. It is worth noting that these alcohols will exhibit a neutral hydrophilic–lipophilic tendency that will not significantly affect the formulation at interface. In the previous chapter, *sec*-butanol was attributed an $f(A)$ value of -0.16 at a concentration above 3%, which is equivalent to a 15% increase in salinity or a reduction in one ACN unit in oil nature—a change that is smaller than the typical increment in a formulation scan, and thus is within formulation accuracy limits unless special fine tuning is sought.

Methanol and ethanol are mostly water-soluble, but if they are added at the 10% level in water they will produce a significant lowering in surfactant interfacial effect by reducing the polarity difference between oil and water. Higher alcohols (*n*-butanol to *n*-hexanol) would dilute the surfactant at interface, but would shift the formulation toward more lipophilicity, and thus alter the physicochemical formulation of the system. Even more lipophilic alcohols (above hexanol) are found to be mostly miscible with oil with little adsorption at interface, so that their diluting capability vanishes. However, they are known to produce an interesting effect in the oil side of the interface either as a lipophilic linker or as a polar oil. If some alcohol is already present in the emulsion, the addition of 1% or 2% of *sec*-butanol might be effective or not depending on the effect of the alcohol added in the first place.

The recommended addition of up to 3% *sec*-butanol could be realized before or after emulsification, but preferably before as an admixture to the aqueous phase, since it guarantees the best homogeneization. In many cases the addition of *sec*-butanol is found to considerably reduce (100-fold or 1000-fold) the time scale of the emulsion decay without altering its characteristics. However, the actual ''accelerating'' factor cannot be forecast by any means.

4. Quantifying Stability Measurements

There are two ways to quantify the notion of stability from the separated volume data as indicated in Fig. 7, where the coalesced volume V_c (referred to as a fraction of its maximum value V_∞) is plotted versus time (as logarithmic scale) for two emulsions labeled A and B.

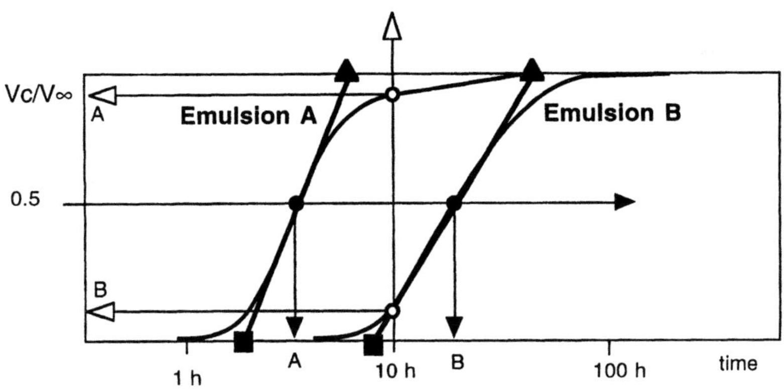

Figure 7 Comparison of the stability of two emulsions (A and B) according to the criterion defined in Fig. 6.

From the first glance it is obvious that the same trends apply to both emulsions, but earlier in the case of emulsion A which may be qualified as less stable. The way to quantify this concept is to intercept the two curves by a single line. For instance, the (vertical) white-headed arrow located at a fixed time (here 10 h) intercepts both curves at a point (white dot) whose ordinate indicates the coalesced volume after 10 h. It is larger for emulsion A, which is thus less stable. In the present case there is quite a difference in coalescence fraction V_c/V_∞ between the two emulsions. However, this is not the general case, since the emulsion-coalesced volume fraction does not vary but over a sometimes narrow period of time. For instance, if the interception is made at 100 h instead of 10 h, then both emulsions are completely coalesced, while after 1 h none of them has started to coalesce. Thus at 1 h and 100 h the diagnostic is the same for both emulsions, in discrepancy with the actual difference.

This is why the preferred method is to intercept the curve with a horizontal line, e.g., the (horizontal) black-headed arrow located at $V_c/V_\infty = 0.5$. The corresponding (black circle) intercepts indicate the times required for both emulsions to exhibit one-half settling. It is worth noting that since the curves are very similar in shape (with a logarithmic time scale), the result is essentially the same whatever the position of the intercept line. In some cases the incipient (respectively final) separation is the critical criterion and a 10% (respectively 90%) separation may be taken instead. Since in most cases the lower and upper regions of the curve do not exhibit abundant experimental points, an extrapolation of the central linear trend to zero and total separation (squares and triangles) could lead to a good comparison criterion.

D. Emulsion Viscosity

1. What is Really the Emulsion Viscosity

At a sufficiently large scale, say 100 or 1000 times the drop size, the emulsion is generally considered to be a homogeneous substance, with constant rheological properties, provided that no change in emulsion characteristics takes place during the motion. This assumption means that there is no segregation phenomena, either by gravity or centrifugal acceleration, that would rule out the homogeneity assumption, or increase in drop size (e.g., by coalescence) due to the development of instability, or decrease in drop size (e.g., by breakup) due to shearing effects. In other words, emulsion rheology deals with stable nonsettling emulsions submitted to mechanical stresses that do not produce any characteristic change.

It is worth noting that these conditions might be too restrictive for many practical applications, so that only an approximation to real-world effects could be expected. The situation is made more complicated by the fact that in the presence of surfactant interfacial effects are likely to produce a pseudodiscontinuity at the macroscopic scale, e.g., a slipping velocity at the wall or a tension gradient at an interface. If the wall is not completely wetted by the external phase, as it could happen with rough surfaces, then boundary conditions often become intractable. Moreover, emulsions are often non Newtonian time-variable fluids with complex rheological behavior, so that it is a clever practical stance to keep expectation at the level of an apparent viscosity concept that would indicate the resistance to flow in some specific conditions and would allow comparisons.

This is the posture taken in this chapter since rheological intricacies are discussed in details in another chapter of this book. The next section deals with the effects of physical variables on viscosity (1,46). Later in this chapter the crossed effects between physical properties, formulation, composition, and stirring will be analyzed.

2. Influence of Physical Variables on Emulsion Viscosity

The dependency of emulsion viscosity on different physical variables has been known for quite a while. It is generally written as proportional to the external phase viscosity:

$$\eta_{em} = \eta_{ext} f(\text{other variables}) \qquad (14)$$

where the subscripts "em" and "ext" refer, respectively, to the emulsion and to its external phase, and where $f()$ stands for the other-effects contribution, i.e., those that do not depend on the external phase. Obviously $f()$ is required to tend to unity as the internal phase content (habitually as volume fraction ϕ_{int}) tends to zero. This relationship is assumed to be valid in all cases, even with high-internal-phase-ratio emulsions, and other real-world situations for which it cannot be experimentally verified. If the external phase exhibits a special behavior such as non-Newtonian or viscoelastic, this behavior is directly transferred to the emulsion through the external phase viscosity term.

The ratio of the emulsion viscosity to the external phase viscosity is often called relative viscosity, and it is noted η_r instead of $f()$.

$$\eta_r = \frac{\eta_{em}}{\eta_{ext}} = f(-) \rightarrow 1 \quad \text{as} \quad \phi_{int} \rightarrow 0 \qquad (15)$$

The second and often most important influence is due to the internal phase presence. If drops are scarce, they are too far away to interact between them and the only interaction beyond the homogeneous fluid case is that of each drop with its surrounding fluid. The increase in viscosity due to this interaction in the case of rigid spherical drops was calculated by Albert Einstein, who has been generally known for other scientific achievements, as:

$$\eta_r = 1 + 2.5\phi_{int} \qquad (16)$$

Equation [16] is essentially valid up to $\phi_{int} = 0.02$, and is useful only to indicate the trend at origin. When the number of drops increases, the drop–drop interactions become predominant and the resulting frictional effects drive the viscosity increase.

If all of the drops are rigid spheres with identical size, the densest tridimensional arrangement is the so-called compact hexagonal packing that fills 74% of the space. Thus, in the most favorable case, the maximum value for ϕ_{int} is 0.74, at which viscosity is infinite. In practice, such an arrangement cannot be attained and essentially infinite viscosity is reached with a monodispersed system at 65–70% internal phase content. However, emulsions are not generally monodispersed systems but contain a distribution of drop sizes, so that small drops can enter the void space between larger drops. Moreover, liquid drops are not rigid spheres either and some deformation is likely to take place, up to the superlative case of polyhedral foam like structures. As a consequence, it may be said that there is essentially no limit to ϕ_{int}. Nevertheless, it is worth remarking that emulsions with more than 60–65% internal phase generally behave as complex fluids for which the concept of viscosity is no longer sufficient to describe the flow behavior.

Figure 8 indicates a typical variation of viscosity with internal phase content. It is seen that the increase in viscosity with the internal phase proportion starts slowly then turns faster and faster, until an almost vertical variation is registered near the emulsion inversion ϕ_{inv} value, here at about 82% of internal

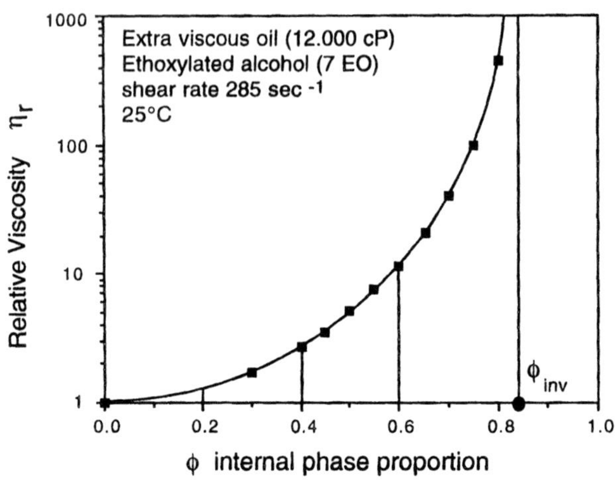

Figure 8 Typical variation of an O/W emulsion viscosity vs. the amount of internal phase (49).

phase. In most cases, the concavity points up, even when the viscosity scale is logarithmic. Many studies have reported empirical relationships to describe this behavior, but none are valid in the general case because many other effects, particularly the non-Newtonian behavior, that are dissimulated in the η_r relative viscosity are to be considered as well (47,48).

Hence, such a plot should be prepared for the particular application and emulsifying device to be dealt with. As far as some advice could be helpful, it may be said that above 60–70% internal phase, most emulsions become pseudoplastic fluids and their viscosity depends on applied shear rate, often according to a power law model. Above 85–90% internal phase, the emulsion no longer behaves as a simple fluid and a viscoelastic description is often useful.

In many cases Pal's empirical equation (48) has been found to provide a fair approximation up to 70–80% internal phase, maybe because it contains experimental data ϕ_{100} as the ϕ_{int} value at which $\eta_r = 100$. This experimental value has to be attained under the same conditions, particularly stirring characteristics, which may be why it significantly embodies the global effects of all remaining variables beyond η_{ext} and ϕ_{int}.

$$\eta_r = \left[1 + \frac{\phi_{int}/\phi_{100}}{1.187 - \phi_{int}/\phi_{100}}\right]^{2.49} \tag{17}$$

As a matter of fact, the internal phase content is the most important variable as far as the viscosity of high internal phase ratio emulsion is concerned, and provided that the formulation ensures a proper stability.

The third kind of variable to be dealt with is the drop size average value and distribution. Since the overall emulsion surface area depends on the square of the drop size, while the internal phase volume depends upon the cube of the drop size, the surface-to-volume ratio changes as the inverse of the size. In other words, small-drop emulsions exhibit a higher surface area per unit volume than big-drop emulsions. Since the interdrop friction effect is related to the surface area of the drops, an increase in viscosity is expected to be associated to a decrease in drop size.

Most experimental studies on monodispersed systems were carried out on solid dispersions rather than emulsions, since it is quite tedious to get a monodispersed emulsion. Experimental evidence from narrowly dispersed emulsions shows that the viscosity increases as the drop size (average) decreases, often according to an inverse power law such as:

$$\log \frac{\eta}{\eta_0} = -B \log \frac{d}{d_0} \tag{18}$$

where the "0" subscript indicates a reference state. In some cases of not-too-severe stirring, B is found to be near unity, although it is less if an energetic stirring is provided. Also in this case, an experimental approach is advised since many other variables are interfering, as well as the very selection of the formula to compute the drop size average.

The drop size distribution is found to play an important role as well. In effect, a monodispersed system contains at least a 26% void volume, that may be filled with much smaller droplets, and which could occupy up to 74% of this volume, thus leaving 7% (= 26% × 26%) void and so on. This means that a polydispersed emulsion is generally less viscous than a monodispersed one with the same drop size average, no matter which way the average is calculated. There is no empirical relationship to render this effect, but it may be said that it is significant when the ratio between the extreme diameters at 90% and 10% of the distribution in volume, respectively D(v,0.9) and D(v,0.1), exceeds 5.

The most spectacular viscosity reduction effect is with bimodal emulsions, which exhibit a distribution curve with two maxima, in most cases as the result of mixing two emulsions. These biemulsions are frequently found to be less viscous than their base emulsions whenever the difference in average size or mode separation is large enough. Most of the bimodal dispersion studies have been carried out on solid suspensions instead of emulsions (50–52); however, the results seems to be directly applicable to emulsions. Figure 9 indicates such a case attained by mixing emulsions with identical internal phase ratio but different sizes, with a mode separation measured as the corresponding diameter ratio is 3 (53).

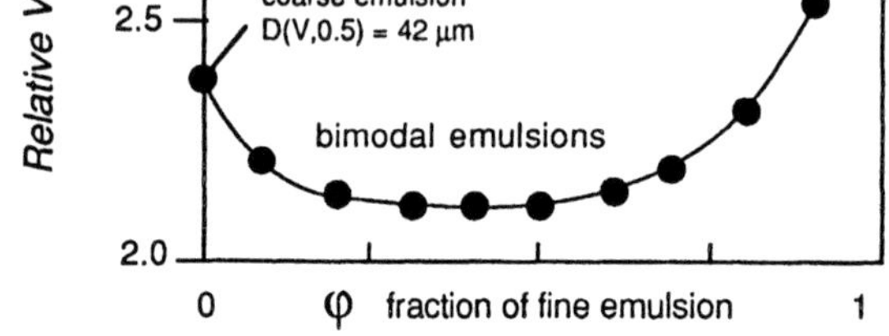

Figure 9 Variation of the viscosity of biemulsions made with two emulsions with the same water/oil ratio but different drop size, vs. the proportion of fine emulsion in the mixture (54).

It is seen that a considerable viscosity reduction is attained by mixing a fine emulsion with a coarse one, over a wide range of composition (54). However, the effect depends on the relative polydispersity of each emulsion, and it is most spectacular with a narrowly dispersed coarse emulsion and a polydispersed fine emulsion with no overlapping. This technique has been proposed for reducing viscosity in crude oil emulsified transport (55), or coal slurry flow (56), but it may also be readily applied for producing less viscous cosmetic milks or lotions and pharmaceutical emulsions with high internal phase ratio.

Some literature reports confirm the intuitive feeling that the internal phase viscosity should have an effect on the emulsion viscosity. This is misleading for the following reasons. First of all, most emulsion droplets are so small that capillarity makes them perfectly spherical. Second, the shear imposed to get an emulsion flowing is generally much less severe than the one used to make the emulsion in the first place. Note that if this is not true, one of the basic assumptions stated in the first paragraph of this section is not complied with. With a low shear stress level, the momentum is not transferred inside the drops that may be regarded as rigid spheres. If there is no motion in the internal phase, there is no way to relate any effect to the viscosity of this phase since viscosity is a meaningless concept at rest.

Therefore, the internal phase viscosity must not alter the emulsion viscosity. The often reported trend that a higher internal phase viscosity is associated with a reduction in emulsion viscosity is indirectly due to a change in drop size. In effect, under identical stirring conditions, a viscous internal phase might not break up as easily as a less viscous one, resulting in larger drops and lower emulsion viscosity. These phenomena are linked with the rupture and coalescence mechanisms during emulsification. They have been discussed in detail in the literature, although not often in the presence of a surfactant (57–65).

Other effects are likely to influence emulsion viscosity, particularly high-internal-phase-ratio emulsions in which the drops are separated by thin films. Both dynamic phenomena such as streaming potential or interfacial viscosity retardation could take place at the drainage rate of these films. Electrical and steric repulsion, as well as surface hydration, are probably also important in these kinds of emulsions. These phenomena, that are due to the presence of adsorbed surfactant onto approaching interfaces, have been studied in relation to stability problems but not for their effect on viscosity. However, it can be said as a rule of thumb that any effect that would tend to reduce the flow of interdrop film would result in increased viscosity or, rather, in impaired flowing ability.

On the other hand, it was discussed in the previous chapter that formulation can greatly influence interfacial tension, a crucial factor in the determination of the breakup–coalescence dynamic equilibrium and thus the drop size average and distribution. Consequently, it is no wonder that—last but not least—the physicochemical formulation influences emulsion viscosity. The approach to be presented here is based on experimental facts because theoretical considerations are not proficient enough to give a whole comprehensive picture.

For the sake of simplicity, the influence of formulation is to be presented on all emulsion properties simultaneously, which is the most logical way of analyzing the results since the properties are not independent, e.g., the drop size influences viscosity and stability.

II. INFLUENCE OF FORMULATION ON EMULSION PROPERTIES

Experimental studies that were carried out to analyze the influence of formulation on the emulsion conductivity, stability, viscosity, and drop size indicate that there exists a general phenomenological pattern of variation for all tested systems.

The following results are quite general for systems containing roughly equal amounts of oil and water, e.g., from 30 to 70% of any of them, and a very few percentages of surfactant. This limitation on the water-to-oil ratio is stated to exclude high-internal-phase-ratio emulsions, for which the composition is found to have a strong influence, as will be discussed in the third section. The small amount of surfactant ensures that the system's representative point falls inside the multiphase region of a ternary diagram. In order to avoid any special effect coming from the mixing protocol, a turbine blender is used to provide isotropic turbulent stirring.

The investigation technique is the unidimensional formulation scan as in the phase behavior studies discussed in the previous chapter. For the sake of simplicity, the scanned variable is often taken as the salinity for ionic surfactant systems, and as surfactant EON or temperature for nonionic systems, but it should be well understood that other formulation variables would produce exactly the same effects. In the reasoning, the formulation will be referred to as SAD, the deviation from optimum formulation, whatever the variable used to produce the scan.

A. Conductivity–Emulsion Type Change

Figure 10 indicates the change in conductivity along two formulation scans (66). For each formulation a test tube is left to equilibrate during several hours or days, and it is then emulsified according to a standard stirring protocol. The electrolytic conductivity is then measured. The left plot indicates the conductivity change along a salinity scan with an anionic surfactant system. At low salinity SAD $<$ 0, and the surfactant dominant affinity toward the aqueous phase results in a

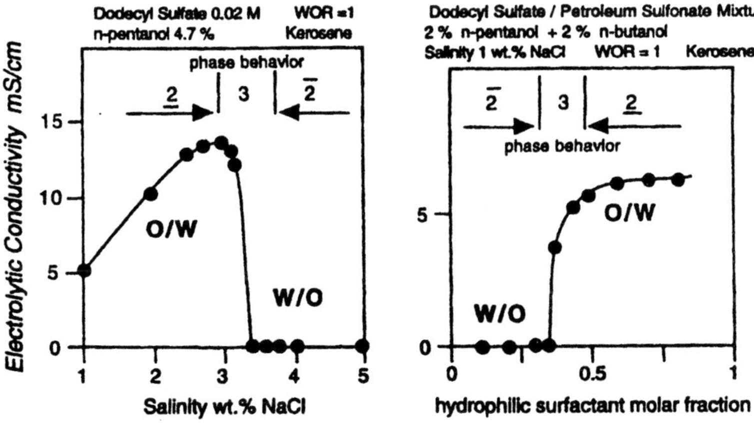

Figure 10 Variation of the emulsion conductivity along two formulation scans. Salinity scan (left) and surfactant hydrophilicity scan (right). (After Ref. 66).

curvature that favors the O/W type, as corroborated by the high conductivity value. As the aqueous phase salinity is increased, the conductivity increases, since the κ_{ext} proportionality term increases (see previous section on conductivity). In the vicinity of optimum formulation (SAD = 0), the conductivity decreases rapidly as an indication that the emulsion is now of the opposite W/O type. The formulation at which the emulsion changes from one type to the other is thus near optimum formulation, inside the three-phase behavior zone.

The right plot indicates the variation of conductivity as a function of surfactant hydrophilicity, by mixing a low hydrophilicity petroleum sulfonate with sodium dodecyl sulfate. As the mole fraction of hydrophilic dodecyl sulfate increases, SAD decreases, as opposite to the previous scan. It is also seen that the conductivity increases near SAD = 0 in the three-phase behavior region. This time the conductivity stays at a constant value for SAD < 0, since the salinity of the aqueous phase is unchanged during the scan.

Many experimental results show that when the formulation is scanned through optimum formulation, the emulsion conductivity changes from high to low (or vice versa depending on the direction of the effect of the scanned variable).

This means that the emulsion type switches at optimum formulation, with O/W emulsion on the negative SAD side and W/O on the positive one, according to Banckroft's rule. Fine-tuning experimentation shows that the change from O/W to W/O, or vice versa, takes place not at a precise point but rather over a short range of formulation. As illustrated in figure 10 right plot, it is sometimes possible to pinpoint some intermediate conductivity value in the three-phase behavior region, which is probably related to the existence of bicontinuity either in the emulsion or in its external phase which could be the microemulsion.

B. Emulsion Stability

In what follows, the emulsion stability is measured after the settling curve, as indicated in the previous section on stability. However, the results of this section are quite general and independent of the method used to estimate stability.

Figure 11 indicates the separated volume of internal phase (as $V_c/V_{c\infty}$) vs. time (log scale) for several emulsified systems corresponding to the formulation scan already studied in Figure 10 (left). The formulation is indicated by the salinity of the aqueous phase, which changes from 1.8 to 7.6 (wt% NaCl). Three-phase behavior ranges from 2.8% to 3.8% and optimum formulation is located at about 3.4% salinity. Figure 11 indicates the relative amount of separated aqueous phase ($V_c/V_{c\infty}$).

It is worth noting that in this kind of plot, all settling curves exhibit the same shape, a quite remarkable fact since the time scale extends over four orders of magnitude. This allows one to classify the emulsion stability according to the position of the curves in respect to one another.

As the salinity of the aqueous phase increases, the settling curve moves from the extreme right (high time equivalent to high stability) to the left (low time equivalent to low stability), and then moves back to the right (high stability again). The minimum stability is attained for a salinity (3.4%) corresponding to the center of the three-phase range, i.e., at or very near optimum formulation.

A better diagnostic could be apprehended if the stability is taken as the

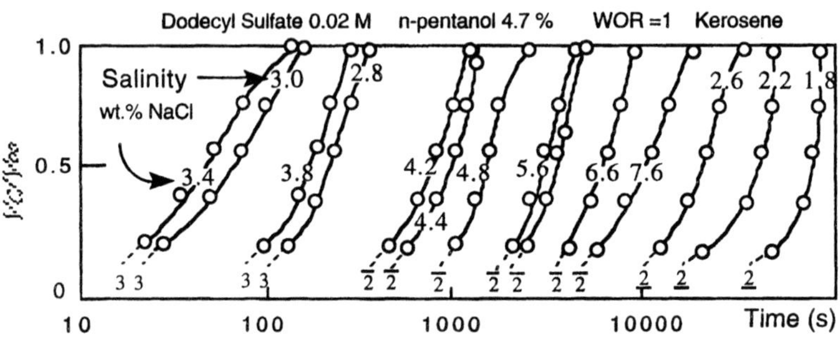

Figure 11 Relative separated volume of aqueous phase vs. time for a series of emulsified systems along a salinity scan (66).

time t_c required for a certain settling to take place, e.g., $V_c/V_{c\infty} = 1:3$, $1:2$, or $2:3$ as plotted vs. formulation in Fig. 12. On such a plot the minimum stability is even more evident, and it is seen to be essentially independent from the selected $V_c/V_{c\infty}$ value. The occurrence of a stability minimum at optimum formulation is a very general feature that has been found in many different systems with no exception known (42,66–72). Various explanations have been advanced by different researchers to interpret this occurrence, such as the trapping of the surfactant in the microemulsion phase (73), the formation of liquid crystalline bridges (74,75), or the instability of a hole in the interdrop film (76). Whatever the right justification, the result is consistent with the wedge theory prediction that at optimum formulation the natural interface curvature is zero, and thus the surfactant cannot stabilize any type of emulsion that would imply the existence of a natural nonzero curvature.

Note that the variation of the emulsion stability is extremely rapid near optimum formulation, while it is much less at some "distance" from it. Also note that in Fig. 12 the variation of emulsion stability is not exactly symmetrical with respect to optimum formulation. In fact it could be more or less symmetrical, depending on the case, and there is no general rule on that since there are other reasons for O/W and W/O emulsions to exhibit different decay rates. In particular, the formulation variable selected to carry out the scan could affect the two types of emulsion in very different ways.

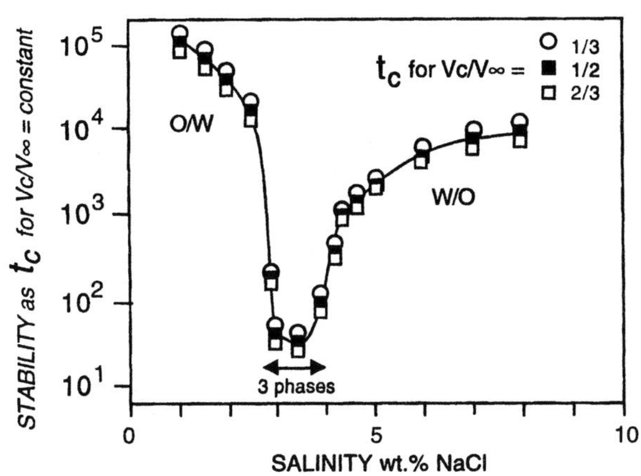

Figure 12 Variation of the stability of the emulsified systems, as the time required to separate a given relative volume of aqueous phase, for a salinity scan (66). Same systems as Fig. 11.

C. Viscosity

As for the study of conductivity, a series of emulsions are prepared along a formulation scan, and their viscosity is measured by any standard method. Figure 13, left indicates the variation of the emulsion viscosity in the case of a near unity water-to-oil ratio, and oil and aqueous phases with similar viscosities. In such a case the variation is roughly symmetrical with respect to optimum formulation (77). The viscosity minimum located at optimum formulation is easy to detect but difficult to measure since the corresponding emulsions are extremely unstable (see previous section). It is believed that this low viscosity is linked with the occurrence of extremely low interfacial tension, which allows the drops to elongate very easily along the streamlines.

Figure 13 right indicates a similar scan but this time with a paraffinic oil, which is much more viscous than water. As expected from the proportionality relationship between the emulsion viscosity and its external phase viscosity, W/O emulsions are much more viscous than their O/W counterparts. Nevertheless, a slight minimum is found at optimum formulation in spite of a strongly asymmetrical shape.

The viscosity minimum is found in all cases in the vicinity of optimum formulation, and it is generally more discernible with nonviscous fluids and extremely low tension systems (77–79).

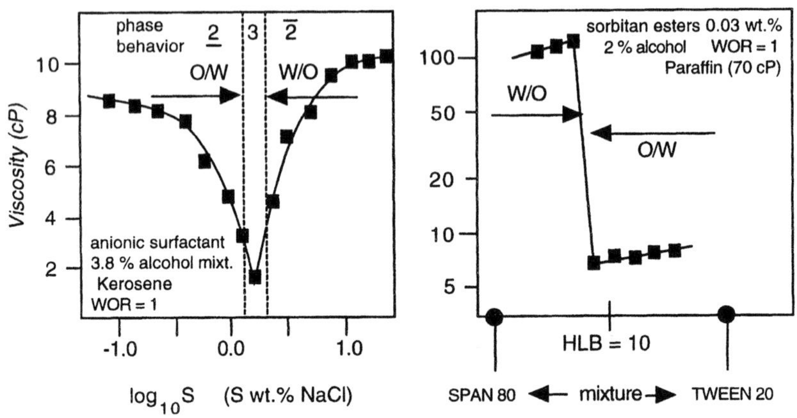

Figure 13 Variation of the emulsion viscosity along a formulation scan. Low-viscosity oil (left) and high-viscosity oil (right). (After Ref. 77.)

D. Drop Size

The drop size variation is similarly studied along a formulation scan in which all emulsions are made according to a standard protocol using the same stirring device and conditions (rotational speed and duration). In the following figures the average drop size is the median, but it could be calculated by any of the classical methods with no effect on the generality of the conclusions.

Figure 14 (right) indicates the variation of the emulsion drop size along a formulation scan, which here is a temperature scan with a nonionic surfactant. Optimum formulation, here optimum temperature T^*, essentially corresponds to Shinoda's phase inversion temperature (PIT) (80) where the surfactant affinity switches from hydrophilic to lipophilic.

Figure 14 (right) indicates that there are two minima in drop size, one at some distance from optimum temperature but below it (O/W), and another slightly above it (W/O).

The occurrence of these two minima may be interpreted by analyzing the variation of the interfacial tension (Fig. 14, left) and emulsion stability (Fig. 14, center) along the same formulation scan (81,82). The difference between the two variations is essentially due to the fact that the tension variation takes place over a much wider temperature interval than the stability change. The combination of these two factors that have opposite effects on the drop size generates the shown variation. When optimum formulation is approached (from any side), the first effect to be felt is the tension reduction, which makes breakup easier with a resulting drop size decrease. Then, when the rapid reduction in emulsion stability takes place, the coalescence rate increases very quickly and the trend is reversed to produce larger drops. Consequently, the minimum size drop is not attained at optimum formulation, where the tension exhibits its lowest value, but at some ''distance'' from it, where the best compromise between low tension and not too

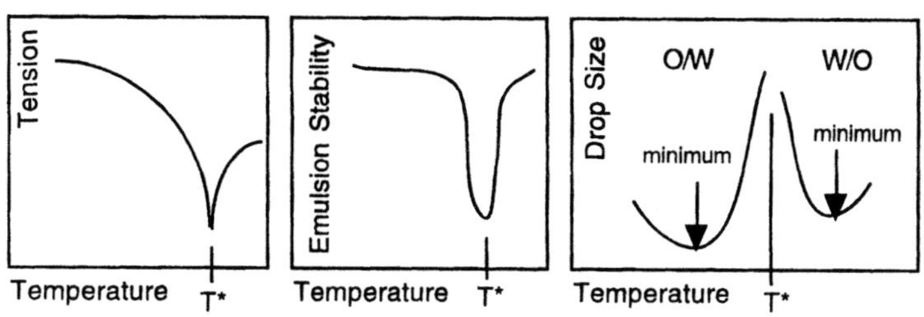

Figure 14 Typical variations of the interfacial tension (left), emulsion stability (center), and emulsion drop size (right), along a formulation scan. Case of nonionic system and temperature scan.

unstable emulsions is reached. Since the phenomenon takes place when approaching optimum formulation from both sides, there are two drop size minima, on the O/W and W/O sides, respectively. Since the relative strength of antagonist effects may vary from one side of optimum formulation to the other, the drop size variations are not necessarily symmetrical with respect to optimum formulation.

III. FORMULATION/COMPOSITION MAP

A. Crossed Effects on Bidimensional Map

The previously reported dependency of different properties on formulation is quite general for systems containing similar amounts of oil and water, but it is found to be no longer valid when the internal phase amount increases beyond some value, e.g., 70% or 75%. In such a case the emulsion type and properties are found to also depend on composition. In the simplest ternary case there are two independent composition variables, which are generally taken as the surfactant concentration and the water-to-oil ratio. The crossed effect of formulation and composition would be perfectly visible in diagrams exhibiting a property variation vs. the corresponding variables. For a bidimensional map one variable would be the generalized formulation while the other would be either the surfactant concentration or the water-to-oil ratio.

Although the surfactant concentration plays an important role, it is not the most critical variable, essentially because it is almost fixed by practical conditions. Actually a minimum surfactant concentration, often in the 0.3 wt% range, is required to attain a satisfactory stability level. On the other hand, a too high surfactant concentration, say for instance 15%, would often result in a monophasic system that would not produce an emulsion. In most practical cases the surfactant concentration is limited to a very small percentage just for cost considerations and toxicity constraints.

Accordingly, the bidimensional studies are carried out on a bidimensional map (formulation–water/oil ratio) at constant surfactant concentration and constant stirring with an emulsification protocol, called ''standard'' emulsification from preequilibrated system, which proceeds as follows. First the surfactant–oil–water system, typically 100 mL, is left to equilibrate in a capped beaker or a similar closed vessel at constant temperature for the duration, which is sufficient to attain physicochemical equilibrium, e.g., 2 or 3 days. The system is then stirred in a turbulent fashion with a turbine blender, for a short time, e.g., 20 or 30 s. The turbine blender impeller is almost the size of the vessel, and the rotating speed is high (e.g., 5000 rpm), so that instant mixing is attained. Under such conditions it is found that an extension of the stirring duration would not change the drop size very significantly. The electrolytic conductivity is measured immediately after emulsification to determine the emulsion type. Then, and provided that the emulsion is stable over several minutes, properties such as drop size distribution and viscosity are evaluated. Stability is estimated by pouring an emulsion sample in a graduated test tube, which is then capped and placed in a vertical rack kept in a constant-temperature enclosure. The separated phase volumes are monitored from time to time and the stability is expressed as the time required for some settling percentage to take place, e.g., 50% internal phase separation.

Such experiments are carried out on different points of the bidimensional formulation–WOR composition map to cover the region of interest. Constant grid size coverage is generally not the smartest experimental design policy since it does not minimize the overall task. Several judiciously placed one-dimensional scans, both formulation changes and WOR composition variations, would often produce a much better result, in particular because the variations along a scan may be interpolated, a feature of first importance to buildup the contour map. The isoproperty contours are drawn by joining the points with the same value of the property, using interpolation between points as needed. The availability of well-fitted continuous variations of the property along some directions is extremely useful to decide how to fill the gaps. The selection of the proper scans to do so will be discussed after the analysis of the typical features of such a map.

B. Conductivity Map and Emulsion Type

Figure 15 indicates the variations of the emulsion electrolytic conductivity vs. formulation (aqueous phase salinity) at different constant water-to-oil ratios, indicated by the water volume fraction f_w (83). When f_w ranges from 0.4 to 0.7, the conductivity exhibits the typical pattern, already mentioned in the previous section. Suboptimum salinity (SAD < 0) is associated to O/W emulsions, while overoptimum salinity results in W/O ones. There is, however, a slight change in

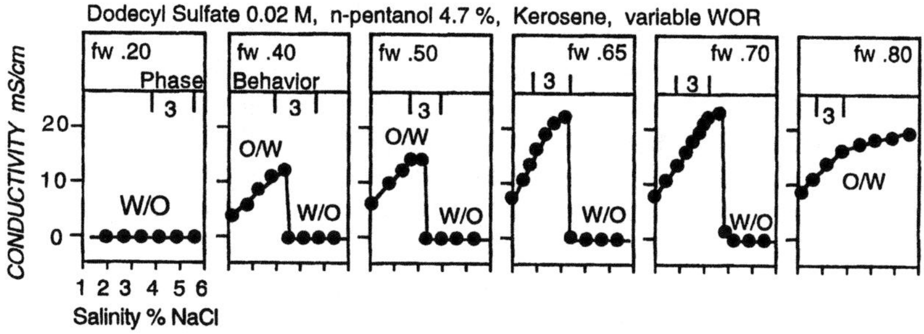

Figure 15 Variation of emulsion conductivity along a formulation (salinity) scan at different water proportions. (After Ref. (83.)

the conductivity curves as the water fraction increases, and it is worth remarking the weak shift in the conductivity drop (inversion point) from the extreme left (case $f_w = 0.4$) to the extreme right (case $f_w = 0.65$ and 0.7) of the three-phase behavior zone.

Whenever f_w is lower than some critical value (here 0.3) a low conductivity everywhere indicates a W/O type emulsion, even when the salinity is below optimum (case $f_w = 0.2$). On the other hand, at high water content (case $f_w \geq 0.8$) the opposite occurs, i.e., the conductivity remains at a high value, indicating a O/W type emulsion, even when the salinity is above optimum. It is worth noting, however, that an unambiguous break in conductivity variation is exhibited at optimum formulation (case $f_w = 0.8$). Indeed, the conductivity increases with a weaker slope after optimum formulation, an indication that the emulsion is actually not a simple O/W type.

By combining the data from all previous (and other) curves, an isoconductivity map is constructed as indicated in Fig. 16 (left). The bold line is the inversion geometrical locus, i.e., the frontier between high (O/W) and low (W/O) conductivity emulsions, which is quite easy to determine according to the shape of the curves.

This line exhibits a stair shape that has been found to be typical of many systems. In the central region (e.g., $0.3 < f_w < 0.7$) the inversion line is essentially horizontal, while it exhibits two vertical branches on both extremes. It is fair remarking that this kind of inversion map was mentioned in early work on the phase inversion temperature with nonionic surfactant systems (84), although the vertical branches were not entirely studied.

This is not surprising since the temperature is known to produce a formulation effect on these systems, as fully confirmed recently in terms of the bidimensional map (85).

Isoconductivity contours at 2, 4, . . . , 24 mS/cm are plotted in the O/W zone of Fig. 16 (left). In the central region, these contours are straight slanted lines because both an increase in salinity and an increase in brine content would roughly result in a proportional conductivity augmentation. In the O/W zone that is located above optimum salinity, i.e., on the upper right of the plot, the isoconductivity contours appear to be much more slanted. This means that at a given water content, the emulsion conductivity is lower than expected from the typical phenomenology. This is an indication of the existence of a W/O/W type multiple emulsion, which is easily corroborated by microscopic observation. Similarly, the W/O region located in the lower left part of the plot often exhibits a O/W/O type multiple emulsion.

The map has been divided into six regions indicated in Fig. 16 (right) symbolized by a letter and a sign (83), a setup that has been adopted by other investigators (86–89). The A region corresponds to water-to-oil ratio midrange, while the B and C regions are associated with low and high water contents, respectively.

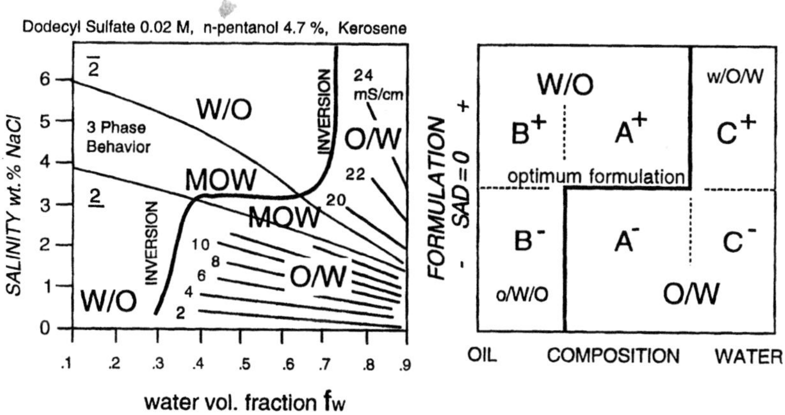

Figure 16 Mixed bidimensional (formulation-composition) map showing the phase behavior at equilibrium and the emulsion iso conductivity contours (left). Simplified map patterns with inversion locus and region labels (right). (After Reference 83.)

The frontiers are typically 30% and 70% water, with emulsions containing low-viscosity fluids, though they could be quite different in other cases. The formulation effect is rendered through the SAD value sign, i.e., (+) for above optimum situation and (−) for the opposite case.

The A^- C^- and A^+ B^+ are called normal O/W and W/O regions, respectively, because the emulsion type matches the interface curvature expected from phase behavior according to the wedge theory and similar rules of thumb.

B^- and C^+ regions exhibit multiple emulsions, of the O/W/O and W/O/W type, respectively. They are the so-called abnormal regions because the "principal" or "outside" emulsion is not of the type expected from the formulation. Nevertheless, it is worth noting that the most "inside" emulsion, i.e., the small droplets that are located inside the drop of the multiple emulsion, do comply with the curvature requirements set by formulation.

The stair shape of the inversion line may be determined easily by a relatively small number of unidimensional scans. First, one or two formulation scans located from 40% to 60% water content allow one to locate the (formulation position of) horizontal branch of the inversion line. Then, a few composition (f_w) scans located both below and above the horizontal branch allow one to locate the position of the vertical branches of the inversion line. Whenever there is no previous information on the system, the first composition scans to be carried out may be located at two or three "formulation distance" units from optimum formulation. These distances are evaluated from the correlation for optimum formulation, as discussed at the end of previous chapter. This dual-scan technique ensures that the inversion line is found by crossing it perpendicularly, which is of course the most accurate method.

The horizontal branch of the inversion line has been associated with the so-called ''transitional'' inversion, while the vertical branches correspond to the ''catastrophic'' inversion, a labeling whose origin will become evident later on.

C. Stability Map

The emulsion stability map is constructed in a similar way. Emulsions made from preequilibrated systems according to the standard procedure are left to settle and their stability is measured as the time for some settling, e.g., 50%, to take place. Sufficient measurements should be carried out so that isostability contours might be drawn. As with conductivity measurements a clever location of tested systems according to the expected pattern could greatly improve the experimental yield.

Figure 17 indicates a typical isostability map representative of different reported results (82–90). The number typifying each isostability contour is the decimal logarithm of the time (in seconds) to produce a $V_c/V_\infty = 2:3$ settling according to the previously discussed method. It is seen that high-stability regions are located in the center of the A^- (O/W) and A^+ (W/O) zones, with some extension on the C^- and B^+ normal zones, respectively, with low-internal-phase-ratio emulsions, which are however likely to exhibit creaming-type separation. The stability minimum found near optimum formulation occurs at all compositions, and it appears as an elongated ''valley'' that closely follows the three-phase behavior band, whatever its slanting.

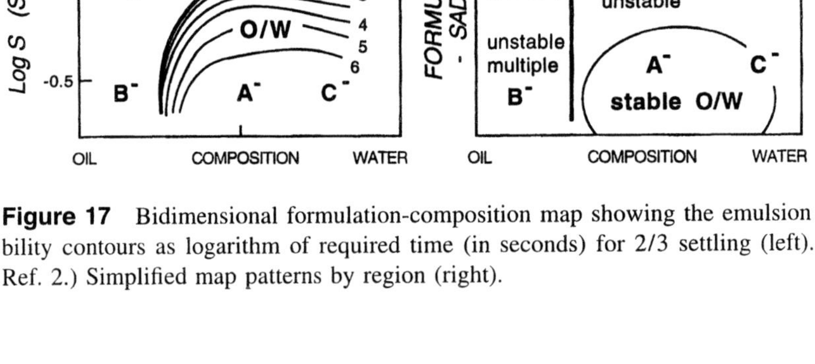

Figure 17 Bidimensional formulation-composition map showing the emulsion isostability contours as logarithm of required time (in seconds) for 2/3 settling (left). (After Ref. 2.) Simplified map patterns by region (right).

Abnormal regions C^+ and B^- also exhibit low-stability emulsions, a feature that is consistent with the fact that the emulsion type, and thus the interface curvature, is opposite to the one favored by the formulation effect. Since multiple emulsions are often made in these regions, a closer look is warranted. For instance, a multiple $W_1/O/W_2$ emulsion is found in the C^+ region. W_1 represents the most internal phase, i.e., the water droplets that are located inside the oil drops. It may be considered that the ''principal'' or ''outside'' O/W2 emulsion has the W_1/O ''inside'' emulsion as internal phase. Since the W_1/O inside emulsion matches the expected type from formulation effects, it is certainly stable, whereas the outside emulsion is not. Thus, such a $W_1/O/W_2$ emulsion would quickly decay in a two-layer system, consisting of a W_2 phase and an oil layer that would actually be a W_1/O emulsion, which is expected to be quite stable if the formulation is sufficiently away from optimum. This means that such ''unstable'' multiple emulsions do not necessarily yield a quick and complete phase separation unless the formulation is appropriate, e.g., near-optimum. This feature could be useful for applications dealing with controlled release or capture through mass transfer.

It is obvious that a multiple emulsion would be fully stable only if both inside and outside emulsions are stable at the same time. From the previous results it is clear that only one of the inside or outside emulsions may be stabilized by a surfactant at equilibrium according to any repulsion mechanism. Hence, the other emulsion has to be stabilized by some other effect, generally a dynamic one. For instance, in the stabilization of a $W_1/O/W_2$ multiple emulsion, the A^+ region may be selected to formulate first the W_1/O emulsion, e.g., with a lipophilic surfactant called LS. Then the W_1/O emulsion is poured into a vessel containing W_2 phase under constant stirring, so that an O/W emulsion may be made. If a hydrophilic surfactant (called HS) is dissolved in W_2 to stabilize the O/W_2 emulsion, in which O is really the inside W_1/O emulsion, then the system ends up containing two surfactants, LS and HS. These two surfactants will conceivably migrate to both W_1/O and O/W_2 interfaces, and equilibrium will be reached after some time, so that the same mixture of LS and HS will be present at both interfaces. Since this mixture cannot be simultaneously lipophilic and hydrophilic, it will not stabilize both emulsion types but only one of them. The trick (91) is to avoid the migration of the second (HS) surfactant from the W_2 phase to the W_1–O interface, and as a second priority, the migration of the LS surfactant from W_1–O interface to an O–W_2 interface. This is attained by using an HS surfactant with a very slow desorption rate, e.g., a polymeric substance that possesses several lipophilic and hydrophilic groups. Such amphiphilic polymers are attached to the interface by several ''plugs,'' so that desorption requires the unplugging of all attachment points at once, which is extremely unlikely. Moreover, the HS polymeric surfactant may be selected to be quite hydrophilic with a very low solubility in the oil phase, so that transfer through it is improbable.

The close examination of typical maps indicates that the most stable emulsions are not always found at the center of the corresponding A region, but sometimes rather on the side of the vertical branch of the inversion line, i.e., toward the high-internal-phase-ratio region. However this trend may be misleading as far as its interpretation is concerned. In effect, an increase in internal phase content often results in a change in viscosity and drop size for a given stirring procedure, which could be the actual justification for the stability change.

Depending on the case, the high-stability region is very limited in area or extends away from optimum formulation. In any case, the stability is often found to decrease far away from optimum formulation because the surfactant becomes too hydrophilic or too hydrophobic.

D. Viscosity Map

The viscosity map is in general simpler than the stability one, probably because the emulsion viscosity is directly related to formulation and composition. In effect, in both A regions the viscosity decreases as optimum formulation is approached and augments as internal phase content increases. As a consequence, the isoviscosity curves follow the inversion line as indicated in Fig. 18. The highest (relative) viscosity zones are located near the vertical branches of the inversion line, inside the A regions, where the effect of the internal phase ratio increase is enhanced by the occurrence of the smallest drop size for a given stirring process (see next section). The arrows in Fig. 18 (right) indicate the direc-

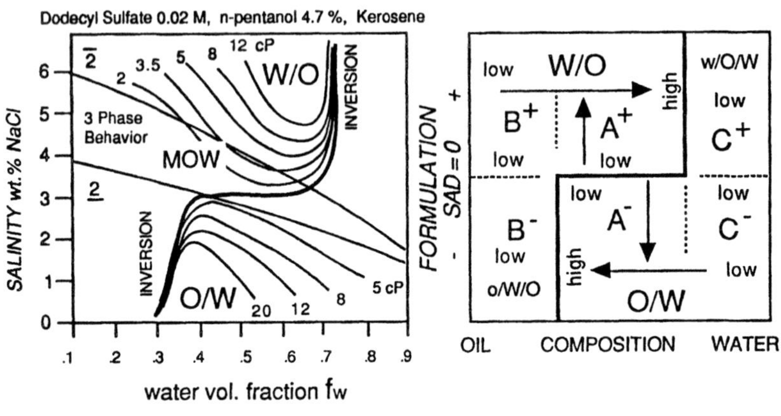

Figure 18 Bidimensional formulation-composition map showing the emulsion isoviscosity contours (left). Simplified map patterns indicating viscosity increase in the direction of arrows (right). (After Ref. 2.)

tions of increasing viscosity at constant formulation (horizontal) and constant composition (vertical).

It is worth remarking that when the inversion line is trespassed across the horizontal branch, the viscosity goes through a minimum, whereas it proceeds through a maximum when any of the vertical branches is crossed.

Low-viscosity emulsions are found in the vicinity of optimum formulation or in the low-internal-phase-ratio regions B^+ and C^-. Besides these general trends, it is of course important to remember that the viscosity may be modified as well, by changing the continuous phase viscosity and the drop size average and distribution.

E. Drop Size Map

The emulsion drop size depends on such a high number of different variables that it is difficult to separate and impute all effects. In what follows a standard emulsification protocol is applied to a preequilibrated system, so that the change in drop size is only due to formulation or composition effects, or their direct consequences on other properties such as tension, stability, and viscosity. In fact, very few maps have been experimentally determined and only trends are available (81,92).

The complex formulation effect previously discussed is found in all cases studied, whatever the formulation variable. Hence, the minimum drop size is attained at some ''formulation distance'' from optimum in both A regions, as schematically indicated in Fig. 19. However, the exact position of this minimum strip depends on stirring conditions and physicochemical properties.

The composition effect is not simpler. It has been found with low-viscosity oil/water emulsions that the drop size tends first to increase and then to decrease as the internal phase ratio increases (93). This results in isodrop size contours as indicated in Fig. 19. On the other hand, it seems that with high-viscosity oil the drop size decreases steadily when the internal phase ratio increases.

It is important to remark that in both cases the drop size decreases considerably when approaching the inversion line by augmentation of the internal phase content in any A region (a path indicated as a black arrow). This effect seems to be due to a considerable improvement in stirring efficiency in the viscous high-internal-phase-ratio emulsions located in these zones. So far it is not known whether this fact is absolutely general, but it may be said that it is a quite common circumstance, and this is why Fig. 19 indicates the presence of a small drop size strip in the vicinity of the vertical branches of the inversion line.

F. Effect of Other Variables on the Inversion Line

In many maps, the property contours nearly follow the inversion line which exhibits a horizontal branch and two vertical branches. Actually the ''horizontal''

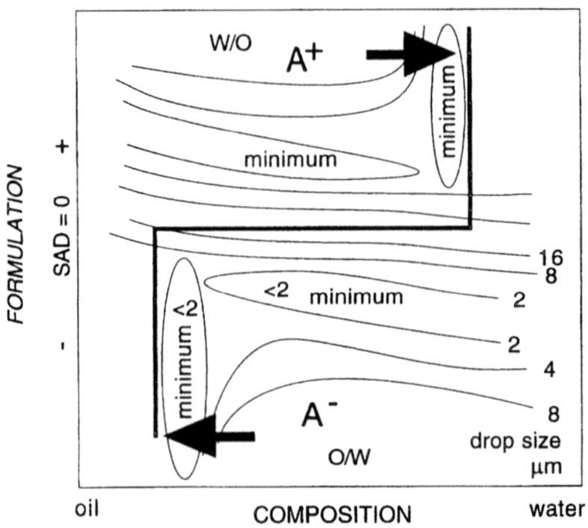

Figure 19 Scheme of a bidimensional formulation-composition map showing the most probable emulsion isodrop size contours. (After Ref. 93.)

branch is not necessarily horizontal, but it always follows closely the three-phase or optimum formulation strip in the middle of the diagram. In some cases the optimum formulation strip is found to be quite slanted, i.e., optimum formulation changes when the water/oil composition changes. This slanting seems to occur only with surfactant mixtures or surfactant–alcohol mixtures that exhibit the so-called fractionation phenomenon. This has been studied thoroughly (94–97) but is too complex to be exposed here in detail, and it will only be treated qualitatively.

The basic rationale is that when a surfactant–oil–water system contains two (or more) surfactant species with quite different hydrophilicities, the species partition into the two phases and at interface (or in the middle phase microemulsion) in different ways. For instance, for nonionic surfactant (commercial) mixtures, many of the most lipophilic oligomers partition into the oil phase, while the migration into the aqueous phase is severely limited by the CMC and is hence essentially negligible. Consequently, the interface (or microemulsion) contains the remaining oligomers which are more hydrophilic than the mixture average. Thus, it may be said that when fractionation occurs with nonionic surfactant mixtures, the interface composition is more hydrophilic than the overall or global composition. Since it is known that at optimum formulation the amphiphile hydrophilicity at interface is fixed whenever the oil, water, and temperature are fixed, it is actually equivalent but preferable to state that the overall composition is more lipophilic than the interface composition. It has been shown that when the composition variables (surfactant concentration and water/oil composition) change, the partitioning changes, so that a constant interfacial hydrophilicity mixture requires a changing overall hydrophilicity. This is why the apparent or global formulation (e.g., mixture EON) changes with the water/oil composition as indicated by the slanting.

The importance of this phenomenon increases with the difference in hydrophilicity of the surfactant and alcohol species present in the amphiphilic mixture. It is found to be almost negligible with pure ionic surfactants, even in the presence of intermediate alcohols (*sec*-butanol). It increases slightly when a surfactant and an alcohol of very different hydrophilicity are used, e.g., sodium dodecyl sulfate and pentanol. It is most significant and essentially unavoidable with commercial ethoxylated nonionic surfactants, particularly the low ethoxylation ones (EON < 10) that contain oligomers with extremely different hydrophilicities. Figure 20 shows such a case in which the optimum formulation region (shaded) is extremely slanted. It is worth remarking that in this case the strong slanting which could be normally attributed to the nonionic surfactant mixture partitioning, is exacerbated by the presence of a lipophilic alcohol such as *n*-pentanol (2).

The fractionation and the resulting slanting of the horizontal branch is enhanced when the surfactant concentration decreases. As a rule of thumb, it may be said that the effect is moderate at or above 3% surfactant, while it becomes dominant below 0.5% surfactant and quite worrisome below 0.1%.

In any case the slanting of the horizontal branch may be easily detected by carrying out two formulation scans near the map center, e.g., at 30% and 60% water. As far as the property mapping is concerned, it is only necessary to distort

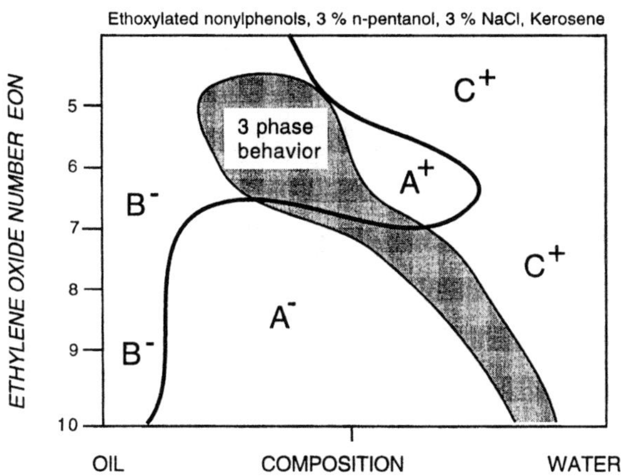

Figure 20 Bidimensional formulation-composition map, showing the emulsion inversion locus, in case of strongly slanted optimum formulation band. (After Ref. 2.)

it according to the slanting. This may of course produce some unusual situations as seen in Fig. 20, where a rather small A^+ region is squeezed in between the very slanted optimum formulation zone and a A^+/C^+ vertical branch normally at 50% water. In such a case it is clear that the occurrence of stable W/O emulsion (associated with A+) is restricted to a very small zone on the map.

The vertical branches of the inversion line have been found to depend on most of the other variables that are susceptible to participate in the balance of breaking and coalescence rates occurring during emulsification. No exhaustive information is available at the date but some trends have been found (83,98).

First to be detected was the effect of the phase viscosity at constant stirring energy. When the oil phase viscosity increases, it is found that the A^+/C^+ branch is shifted to the left, thus reducing the extension of the A^+ zone, where the oil phase is the external phase. Meanwhile the location of the A^-/B^- branch is essentially unchanged. The effect may be so drastic than the A^+/C^+ (right) branch can be shifted to 20% or 30% water, i.e., eventually to the left of the (left) A^-/B^- branch, with extremely viscous oil phases.

Exploratory investigations indicate that when the water phase viscosity is increased, e.g., by adding a polymer, then it is the A^-/B^- branch that is shifted, this time to the right (93). It may be said as a general statement that when the viscosity of one of the phases increases, the extension of the region where this phase is the external phase of a normal stable emulsion is reduced. In other words, when the viscosity of a phase increases it is more difficult to make emulsions in which this more viscous phase is the external phase. This is consistent with the well-known experimental fact that when two liquids are stirred (in absence of surfactant), the most viscous liquid turns out to be dispersed into the less viscous one.

As far as the effect of stirring is concerned, it is found that not only the energy but also the kind of stirrer could be important, keeping in mind that the most vigorous stirring is not the one that would necessarily lead to the smaller drop size. When working with a rotational stirrer and changing the speed, it has been found that increased stirring energy tends to shift both vertical branches toward the center of the map (99), and hence to reduce the extension of both A^- and A^+ zones, as if increased stirring were offsetting the formulation effect (central zone), and favoring the composition influence, i.e., the formation of an emulsion in which the most abundant phase is the external phase and vice versa (extreme zones). Note that this rectifies a misleading statement on the subject in a recent review (100), which otherwise sums up the state of the art on the topic.

An increase of surfactant concentration is found to produce the opposite effect, i.e., both vertical branches are shifted to the extremes and the A zone range extends (101). This is actually not surprising since an increase in surfactant concentration should logically favor the effect of formulation. It is worth noting that an increase in surfactant concentration is also associated with a decrease in drop size in most cases (82) because of an improved coalescence inhibition, at least below the CMC (3).

Judicious logical reasoning may be carried out by combining these effects. For instance, if a W/O high-internal-phase-ratio emulsion is to be made with a viscous oil, then it is favorable to decrease the stirring energy and to increase the surfactant concentration. An alternative would be to increase the temperature to reduce the oil viscosity, provided that the corresponding formulation effect is not detrimental (as could happen with ionic surfactants). Actually such situations can be solved also by using the memory feature discussed in the next section.

IV. DYNAMIC PHENOMENA IN EMULSION MODIFICATION

In the preceding sections, emulsions were prepared from preequilibrated surfactant–oil–water systems, and their properties were sufficiently persistent that they did not change significantly during the time scale of the measurement or the application. This is not always the case, and it is not uncommon practice to modify the emulsion formulation or composition according to some procedures involving a temperature variation, a continuous or lump addition of one or two phases with or without surfactant, and/or a changing stirring method, sometimes according to some whimsical protocol.

Since these changes are likely to modify the location of the representative point of the system in the bidimensional map, with a corresponding change in property, the question is to know whether the maps are still useful to predict the emulsion property evolution along the path, and when and how this can be done.

The problem may be split into two cases. The first one deals with a modifying path that does not trespass through the map inversion line, in which case it is often possible to make a good forecast concerning the property alterations. This is not the case whenever the inversion line is crossed over. In some cases, the dynamic crossing of the inversion line defined previously does not trigger the inversion, which is somehow delayed, and some "guesstimate" may be advanced. However, if the change proceeds, sooner or later the inversion would take place and a completely new emulsion would be formed through a chaotic transition. There are two types of emulsion inversions and in spite of some clever fundamental hints (89,102), in none of them is it known for sure what happens at the moment of inversion. Nevertheless, the transient phenomena taking place during the inversion have been advantageously harnessed to produce handy systems such as miniemulsions.

A. Modifying Emulsions Without Inversion

In this section it is assumed that a so-called initial emulsion is made from a preequilibrated surfactant–oil–water system located at some "initial" point in the bidimensional map, with properties corresponding to its position in the map. Then a change that may affect its formulation and/or composition is applied to the emulsion, so that the representative point is shifted to another place in the map located on the same side of the inversion line. The final ''shifted'' emulsion corresponds to a point on the map, in which the corresponding standard emulsion exhibits other properties. The question is whether the ''shifted'' emulsion keeps its ''initial'' properties or embraces the ''new'' ones that correspond to its final position in the map.

Since it may be generally assumed that the change duration is quite short, the answer is often elemental and utterly rational, simply because some properties are maintained during the formulation and/or composition change whereas others are not.

Provided that the emulsion remains stable over the change, the drop size stays constant, as do other related properties. If the internal phase content is increased by adding some amount of this phase (under constant stirring), as along path 1 in Fig. 21, emulsion viscosity is expected to augment from the increase in internal phase content, although other effects would have to be taken into account, such as the efficiency of the stirring required to proceed with emulsification of the added internal phase, or the eventual production of a bimodal distribution that could result in a viscosity reduction.

If the formulation or temperature is changed so that the emulsion representative point is shifted from a high-stability region to a low-stability one, as for instance by approaching optimum formulation as along path 2 in Fig. 21, the emulsion-accelerated decay becomes the main concern and drives other property changes like drop size growth and viscosity abatement. Path 2 is thus advised when emulsion breaking is sought.

Some composite path could take advantage of different helpful features, as in case 3, where the initial emulsification is carried out ''at some distance''

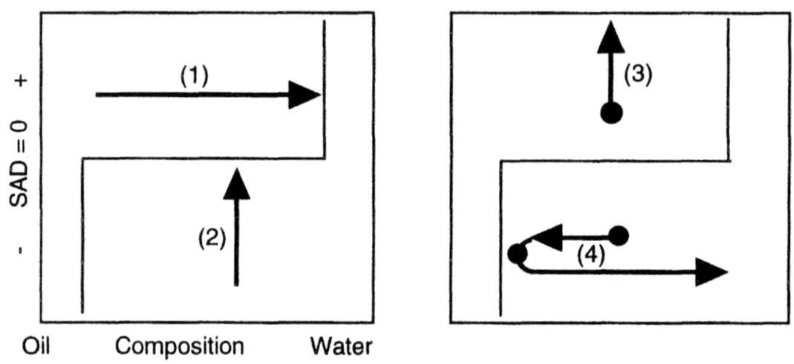

Figure 21 Emulsion dynamic changes that do not trespass the inversion locus.

from optimum formulation where the smallest drop size is expected (black circle at the start of path 3 in Fig. 21), although it is too near optimum formulation to exhibit a satisfactory stability. In order to counter this dilemma, the formulation or temperature is changed quickly after emulsification, so that the representative point moves away from optimum formulation until a high stability is secured. This method was suggested a long time ago under the term PIT emulsification method (103), where the selected formulation variable is the temperature. Note that it could be carried out too with any formulation variable that may be changed quickly through a lump addition of matter, e.g., surfactant hydrophilicity, alcohol contribution, salinity increase.

Path 4 illustrates an even more complex way to produce a very fine drop emulsion with a low internal phase ratio, e.g., 1:5, as in a make-up removal milk. At such low internal phase ratio, the emulsification is very inefficient and energetically expensive, and it is quite painstaking to produce a monodispersed fine emulsion even with a colloid mill. This obstacle is circumvented by the following procedure. A starting medium drop size emulsion with approximately 50–60% internal phase proportion is shifted to high internal phase content until the vicinity of the inversion line is attained, say, at 80%. Proceeding this way, there is generally no danger to trigger the inversion for the reasons discussed in the next section. The viscosity becomes quite high in this region, but a slow-motion whip-style stirring turns out to be very efficient and extremely small droplets can be produced with little expense of energy. When the fine high-internal-phase-ratio emulsion is finally made, it is then diluted with a large amount of external phase to match the required 20% internal phase composition. Often a slight formulation change is additionally applied to enhance the stability of the final emulsion, although the extremely small drop size could be sufficient to slow down the creaming kinetics considerably.

B. Emulsion Dynamic Inversions and Applications

Up to now the inversion line has been the limit between emulsion types when emulsification is carried out from a preequilibrated system according to the so-called standard procedure. In practice the emulsion inversion could also be the situation in which a change in formulation or composition triggers a switch in emulsion type. This kind of inversion is generally called dynamic inversion since it takes place as a consequence of the change. Depending on the circumstances it may be favorable or quite detrimental, and should be either harnessed or avoided.

Dynamic inversion is studied by producing a change that moves the point that represents the formulation and composition of an emulsion on the map from one side of the inversion line to the other side. In practice the system is first equilibrated and then emulsified, to produce the initial emulsion. Then its formulation or composition is altered continuously or by small increments, while a low-energy stirring is maintained to keep the emulsion from settling, until inversion is detected, usually by conductimetry. It is worth noting that the stirring should be much less energetic than that one applied to make the emulsion in the first place.

Figure 22 indicates the typical patterns found in dynamic inversion. The arrows indicate the direction and path of change while their respective heads indicate the inversion point along a formulation (white) or composition (black) shift. Depending on which branch is crossed, one of two possible inversion types is encountered.

When a vertical crossing of the inversion line is carried out (white arrows in Fig. 22, left), it is seen that the inversion takes place exactly at the same location, i.e., at optimum formulation, whatever the direction of change. Because of these reversibility characteristics, this inversion has been termed "transitional."

The dynamic inversion through any of the vertical branches of the inversion line is quite different. At both crossing boundaries (A^-/B^- and A^+/C^+) the composition is changed so that the emulsion internal phase ratio increases in the direction in which inversion takes place (black arrows). From A^- to B^- and from A^+ to C^+, it is from a normal emulsion to an abnormal one (left), while from B^- to A^- and to C^+ to A^+, it is the opposite case (center). Figure 22 indicates that the inversion location depends on the direction of change. Actually there is a region (shaded) in which the two types of emulsions can be found depending on the direction of change, i.e., depending on the previous history of the emulsion, which is why the term "emulsion memory" has been proposed. Everything happens as if the inversion were delayed along the path of change, with the resulting displacement of the inversion line over the shaded region, a phenomenon first noticed long ago (104).

These regions, known as hysteresis regions, exhibit a triangular shape, so that they vanish at optimum formulation whereas they become wider as formulation departs from optimum. The standard inversion line discussed in the previous

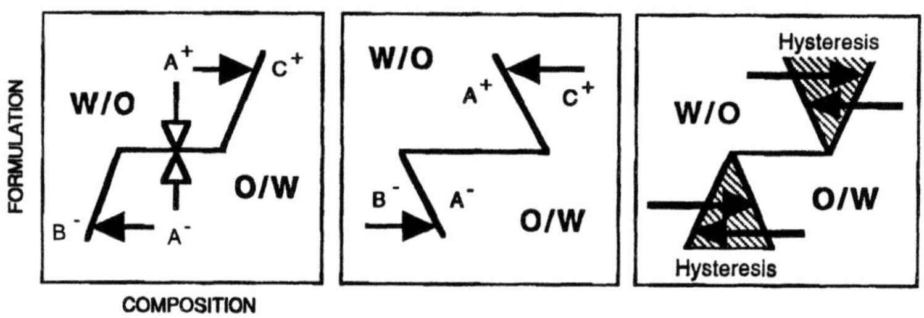

Figure 22 Dynamic changes that produce the emulsion inversion. (After Ref. 100.)

section is located somewhere inside the hysteresis region, not necessarily in the middle.

This second type of inversion has been called "catastrophic" due to its characteristic phenomenology, similar to the cusp as well as to the higher degree catastrophe (105–109) that has been used to interpret it in a quite satisfactory way, as discussed in a review (107).

In such an inversion, everything happens as if the internal phase ratio could be increased beyond the value corresponding to the standard inversion, with this extra increase being greater as formulation departs from optimum. This feature is of course very useful in practice and was used long ago, e.g., in mayonnaise making, in which oil is added little by little to an aqueous phase containing a surfactant (egg yolk). In this case an A^- emulsion is displaced to the left far beyond the point where an oil–egg yolk mixture results in an egg yolk dispersion in oil by standard emulsification. It is worth remembering that when the internal phase content reaches 70% or 80% the emulsion becomes so viscous that conventional turbulent stirring does not work anymore to incorporate more internal phase. Instead, a slow-motion blender is preferred, like a whipping device or a ribbon moving gear, which are often much more efficient. In such a high-internal-phase-ratio inversion scheme, the external phase is stretched as interdrop films that finally break into small droplets by capillary instability.

It was mentioned earlier that the B^- and C^+ regions often exhibit multiple emulsions, which is actually the simultaneous occurrence of both emulsion types. There is some evidence that multiple emulsions start occurring before the catastrophic inversion takes place, and some researchers have proposed a competitive kinetic model in which one of the emulsions could be more stable and thus would prevail (87,89,110). This is consistent with the fact that the variables susceptible to influence the breaking-coalescence mechanisms do influence the location of the inversion line and hysteresis region.

The concomitant occurrence of both types of emulsions near the catastrophic inversion has been used to make extremely fine emulsions that are often wrongly labeled microemulsions by manufacturers of antifoaming suspensions, floor wax, cosmetics, and pharmaceuticals. This occurs when an (unstable) abnormal emulsion is inverted into a (stable) normal one as illustrated in Fig. 22 (center). For instance, an abnormal W/O emulsion of the B^- type may be inverted by adding water (containing a hydrophilic surfactant) up to some moderate value, e.g., 40% or 50%, sometimes less if the oil phase is viscous and the surfactant concentration is high. Assume that an $O_1/W/O_2$ multiple emulsion occurs in the B^- region as the B^-/A^- boundary is approached. If an increasing number of extremely fine O_1 drops are formed inside the water drops of a W/O_2 emulsion just before the inversion, the apparent water drop volume (O_1 + W) may become much higher than the real water content (W), and a high-internal-phase-ratio inversion scheme may take place. In such a case the O_1 oil droplets that were trapped in the water drops are brought together to those coming from the (O_2) external oil film capillary instability. As a matter of fact, the drop size attained by this process could be considerably smaller than that to be reached by intense turbulent stirring.

Enhanced emulsification efficiency along a $B^- \rightarrow A^-$ path can be reached by dissolving the hydrophilic surfactant inside the original oil phase. As water is added the surfactant transfers from oil to water as soon as the fluids are in contact. This transient phenomenon often triggers a spontaneous emulsification that results in extremely small oil droplets. This is, however, too complex an occurrence to be properly interpreted with the formulation-composition map features.

On the other hand, transitional inversion is used as well to attain extremely small drop emulsions, sometimes referred to as miniemulsions or gel emulsions, because of the high viscosity resulting from the exceedingly small drop size, even at a low internal phase ratio (111–115).

The formulation transition is in most known cases carried out by changing the temperature with a nonionic surfactant system. According to the previous chapter, an increase in temperature can modify the phase behavior from Winsor type I to type II for nonionic surfactant systems (or vice versa for ionic ones) undergoing type III (three-phase) or type IV (single-phase) case at optimum formulation as indicated in Fig. 23.

Each phase behavior type is associated with an emulsion type, but near optimum formulation either a monophasic (microemulsion) or triphasic (microemulsion at equilibrium with excess oil and water) is exhibited, depending on the amphiphile surfactant/alcohol mixture (S + A) concentration. When temperature

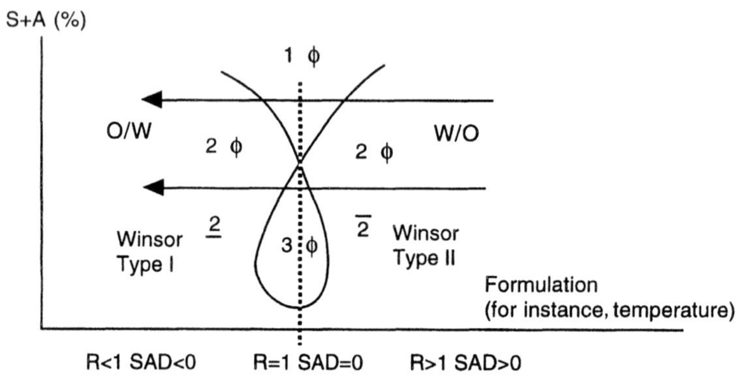

Figure 23 Phase behavior chart as amphiphile amount vs. formulation. Black arrows indicate the paths undergone by emulsified systems physicochemical state as temperature is decreased (nonionic surfactant).

is changed near optimum formulation, the microemulsion composition changes from mostly oil (type II side) to mostly water (type I side) or vice versa. When this change takes place, the microemulsion tends to exude oil or water droplets as in a nucleation process. If the change in temperature is slow, the droplets have the time to gather and to instantly coalesce, since the emulsion stability is at its minimum. However, if the temperature is very quickly changed after the nucleation takes place, the droplets may be prevented from coalescing and a quenched miniemulsion is formed. The V-shaped single-phase region (1ϕ) in Fig. 23 often displays liquid crystalline phases that may perform a determinant role in stabilizing the formed droplets, at least by preventing an early coalescence before quenching takes place (74,116,117).

This mechanism implies that all the oil and water are solubilized into a single-phase microemulsion at optimum formulation. This requires either high solubilization or a high surfactant concentration, and sometimes both, which might not be practical.

If there is not enough surfactant to produce a single-phase system at optimum formulation, some of the emulsion drops will not come from desorption but will be the result of the stirring process, and it would be important to take advantage of the presence of a most favorable compromise in drop size at some distance from optimum formulation (see Fig. 14). If the formulation or temperature is adjusted at the minimum drop size compromise and then a quench is applied to quickly shift the formulation from the unstable region, a small drop emulsion may be attained with extreme facility. Nevertheless, it is worth pointing out that no miniemulsion can be made this way or by any other means involving stirring emulsification, except when a dynamic inversion process, either transitional or dynamic, takes place.

Recent research has unveiled that the location and shape of the catastrophic inversion hysteresis zones depend on other variables, such as stirring, phase viscosity, or surfactant concentration, in a way that is consistent with the shift of the standard inversion line vertical branches with these variables (99,118). For instance, it has been found that an increase in surfactant concentration tends to shift the hysteresis region toward the extreme and to increase their extension (101). Increased stirring energy has just the opposite effect (99), and it could be conjectured that infinitely energetic stirring would result in the vanishing of the hysteresis zone into the standard inversion line.

ACKNOWLEDGMENTS

The author thanks the local and national research councils, CDCHT-ULA and CONICIT, for supporting the Lab. FIRP research program at Universidad de Los Andes.

REFERENCES

1. P Becher, Emulsion: Theory and Practice, Reprint, R. Krieger, Huntington (1977).
2. M Miñana-Pérez, P Jarry, M Pérez-Sanchez, M Rámirez-Gouveia, JL Salager, J. Dispersion Sci. Technol., 7: 331 (1986).
3. S Rehfeld, J. Colloid Interface Sci., 24: 358 (1967).
4. S Nagata, Mixing: Principles and Applications, John Wiley and Sons, New York (1975).
5. AM Ali, HHS Yuan, DS Dickey, GG Tatterson, Chem. Eng. Commun., 10: 205 (1981).
6. TP Chang, YH Sheu, GB Tatterson, DS Dickey, Chem. Eng. Commun., 10: 215 (1981).
7. GB Tatterson, Fluid Mixing and Gas Dispersion in Agitated Tanks, MacGraw-Hill, New York (1991).
8. T Allen, Particle Size Measurement, Chapman and Hall, London (1975).
9. MJ Groves, Particle Size Analysis, Heyden, London (1978).
10. Th Tadros, B Vincent, in Encyclopedia of Emulsion Technology, P Becher, ed., Vol. 1. Marcel Dekker, New York (1986) Chapter 3.
11. J Sjöblom, ed., Emulsions and Emulsion Stability, Marcel Dekker, New York (1996).
12. GV Jeffreys, GA Davies, in Recent Advances in Liquid/Liquid Extraction, C. Hanson. ed. Pergamon Press, New York (1971) p. 495.
13. BJ Carrol, in Surface and Colloid Science, E Matijevic, ed., Vol. 9:1. Wiley Interscience, New York (1976), p. 1.
14. H Hadamard, Comptes Rendu Acad. Sci. Paris, 152: 1975 (1911).
15. V Levich, Physicochemical Hydrodynamics, Prentice Hall, Englewood Cliffs (1962).
16. KJ Lissant, ed., Emulsions and Emulsion Technology, 2 vols., Marcel Dekker, New York (1974).
17. D Melik, HS Fogler, in Encyclopedia of Emulsion Technology, P Becher, ed., Vol. 3:1. Marcel Dekker, New York (1988), p. 1.
18. H Wang, R Davis, J. Colloid Interf. Sci., 159, 108 (1993).
19. XB Reed, E Riolo, S Hartland, Int. J. Multiphase Flow, 1: 411 and 437 (1974).
20. AD Barber, S Hartland, Can. J. Chem. Eng., 54: 279 (1976).
21. IB Ivanov, TT Traykov, Int. J. Multiphase Flow, 2: 397 (1976).
22. TT Traykov, ED Manev, IB Ivanov, Int. J. Multiphase Flow, 3:485 (1977).
23. DT Wasan, A Nikolov, First World Congress on Emulsion, Paris, Oct. 1993. Proceedings 4: 93 (1993).
24. IB Ivanov, PA Kralchevsky, Colloids Surfaces A, 128: 155 (1997).
25. AW Adamson, Physical Chemistry of Surfaces, 3rd Ed., Interscience, New York (1976).
26. M Rosen, Surfactants and Interfacial Phenomena, John Wiley and Sons, New York (1978).
27. P Hiemienz, Principles of Colloid and Surface Chemistry, Marcel Dekker, New York (1977).

28. D Schuhmann, ed., Propriétés électriques des interfaces chargées, Masson, Paris (1978).
29. A Kitahara, A Watanabe, eds., Electrical Phenomena at Interfaces: Fundamentals, Measurements, and Applications, Marcel Dekker, New York (1980).
30. EJ Verwey, JThG Overbeek, Theory and Stability of Lyophobic Colloids, Elsevier, Amsterdam (1948).
31. BV Derjaguin, Theory of Stability of Colloids and Thin Films, Plenum Press, New York (1989).
32. JN Israelachvili, Intermolecular and Surface Forces, 2nd Ed., Academic Press, London (1991).
33. B Vincent, Adv. Colloid Interf. Sci., 4: 193 (1974).
34. DW Osmond, B Vincent, FA Wait, Colloid Polym. Sci., 253: 676 (1975).
35. RH Ottewill, T Walker, Kolloid-Z. Polym., 227: 108 (1968).
36. EJ Clayfield, EC Lumb, J. Colloid Interf. Sci., 22: 269 and 285 (1966).
37. G Hadziioannou, S Patel, S Granick, M Tirrel, J. Am. Chem. Soc., 108: 2869 (1986).
38. X Zhang, RH Davis, Fluid Mech., 230: 479 (1991).
39. IB Ivanov, ed., Thin Liquid Films, Marcel Dekker, New York (1988).
40. KD Danov, ND Denkov, DN Petsev, IB Ivanov, R Borwankar, Langmuir, 9: 1731 (1993).
41. D Platikanov, ed., special issue Nos. 6 & 7, J. Disper. Sci. Technol. 18 (1997).
42. RW Flumerfelt, AB Catalano, CH Tong, in Solution Chemistry of Surfactants, Vol. 2. Plenum Press, New York (1979) p. 571.
43. LA Lobo, DT Wasan, in Food Emulsions and Foams: Theory and Practice, P J Wan, ed. A.I.Ch.E Symposium Series Vol. 86, No. 277 (1990).
44. DT Wasan, Chem. Eng. Ed., 27: 104 (1992).
45. A Graciaa, J Lachaise, C Cucuphat, M Bourrel, JL Salager, Langmuir, 9: 669 and 3371 (1993).
46. P Sherman, ed., Rheology of Emulsions, Macmillan, New York (1963).
47. P Sherman, in Encyclopedia of Emulsion Technology, P Becher, ed., Vol. 1, Marcel Dekker, New York (1983).
48. R Pal, E Rhodes, J. Rheology, 33: 1021 (1989).
49. C Bracho, Tech. Rep. FIRP No. 8302, Universidad de Los Andes, Mérida, Venezuela (1983).
50. RJ Farris, Trans. Soc. Rheol., 12: 281 (1968).
51. C Parkinson, S Matsumoto, P Sherman, J. Colloid Interf. Sci., 33: 150 (1970).
52. R Hoffman, J. Rheol., 36: 947 (1992).
53. JL Salager, M Ramirez-Gouveia, J Bullon, Progr. Colloid Polym. Sci., 98: 173 (1995).
54. M Ramirez-Gouveia, Tech. Rep. FIRP No. 9210, Universidad de Los Andes, Mérida, Venezuela (1992).
55. H Rivas, G Nuñez, C Dalas, Vision Tecnológica 1: 18 (1993).
56. MZ Sengun, RF Probstein, Rheol. Acta, 28: 382 and 394 (1989).
57. JO Hinze, A.I.Ch.E. J., 1: 289 (1955).
58. R Shinnar, JM Church, Ind. Eng. Chem., 52: 253 (1960).
59. FD Rumscheidt, SG Masson, J. Colloid Interf. Sci., 16: 210 and 238 (1961).

60. HT Chen, S Middleman, A.I.Ch.E. J., 13: 989 (1967).
61. FB Sprow, Chem. Eng. Sci., 22: 435 (1967).
62. HJ Karam, JC Bellinger, Ind. Eng. Chem. Fund., 7: 576 (1968).
63. P Walstraa, Dechema Monograph. 77: 87 (1974).
64. CA Coulaloglou, LL Tavlarides, A.I.Ch.E. J., 22: 289 (1976).
65. K Arai, M Konno, Y Matunaga, S Saito, J. Chem. Eng. Jpn., 10: 325 (1977).
66. JL Salager, I Loaiza-Madonado, M Miñana-Pérez, F Silva, J. Disper. Sci. Technol., 3: 279 (1982).
67. J Boyd, C Parkinson, P Sherman, J. Colloid Interf. Sci., 41: 359 (1972).
68. K Mandani, S Friberg, Progr. Colloid Polym. Sci. 65: 164 (1978).
69. M Bourrel, A Graciaa, RS Schechter, WH Wade, J. Colloid Interf. Sci., 72: 161 (1979).
70. JL Salager, L Quintero, E Ramos, JM Andérez, J. Colloid Interf. Sci., 77: 287 (1980).
71. JE Vinatieri, Soc. Petrol. Eng. J., 20: 402 (1980).
72. FS Milos, DT Wasan, Colloids Surfaces, 4: 91 (1982).
73. RE Antón, JL Salager, J. Colloid Interf. Sci., 111: 54 (1986).
74. S Friberg, L Mandell, J. Pharm. Sci., 59: 1001 (1970).
75. R Hazzlett, RS Schechter, Colloids Surfaces, 29: 53 (1988).
76. A Kalbanov, H Wennerström. Langmuir, 12: 276 (1996).
77. JL Salager, M Miñana-Pérez, JM Andérez, JL Grosso, CI Rojas, J. Disper. Sci. Technol., 4: 161 (1983).
78. S Vijayan, C Ramachandran, H Doshi, DO Shah, Third Int. Conf. Surface and Colloid Science, Stockholm, Sweden, August 1979.
79. JL Salager, JL Grosso, M Eslava, Revista Técnica INTEVEP, 2: 149 (1982).
80. K Shinoda, Proceedings 5th Int. Congress Surface Activity, Barcelona, Spain, vol. 2 (1969) p. 275.
81. JL Salager, M Pérez-Sánchez, Y Garcia, Colloid Polym. Sci., 274: 81 (1996).
82. JL Salager, M Pérez-Sanchez, M Rámirez-Gouveia, JM Andérez, M I Briceño, Récents Progrès en Génie des Procédés, Vol. 11, No. 52, Lavoisier, Paris (1997) p. 123.
83. JL Salager, M Miñana-Pérez, M Rámirez-Gouveia, CI Rojas, J Disper. Sci. Technol., 4: 313 (1983).
84. K Shinoda, H Arai, J. Colloid Interface Sci., 25: 429 (1967).
85. RE Antón, P Castillo, JL Salager, J. Disper. Sci. Technol., 7: 319 (1986).
86. E Dickinson. Annual Reports C, Royal Society of Chemistry, London (1986) p. 31.
87. BW Brooks, HN Richmond, Colloids Surfaces A, 58: 131 (1991).
88. HT Davis, Colloid Surfaces A, 91: 9 (1994).
89. GJ Vaessen, PhD Dissertation, Eindhoven University of Technology, Netherlands (1996).
90. P Jarry, M Miñana-Pérez, JL Salager, in Surfactants in Solution, Vol. 6, K Mittal, P Bothorel, eds., Plenum Press New York (1987) p. 1689.
91. C Py, J Rouvière, Th F Tadros, MC Taelman, P Loll, Colloids Surfaces A, 91: 215 (1994).
92. M Pérez-Sanchez, Y Garcia, Tenth Int. Symposium on Surfactants in Solution, Caracas, Venezuela, June 1994.

93. M Pérez-Sanchez, Tech. Rep. No. 9406, Lab. FIRP Universidad de Los Andes, Mérida, Venezuela (1994).
94. A Graciaa, J Lachaise, JG Sayous, P Grenier, S Yiv, RS Schechter, WH Wade, J. Colloid Interf. Sci., 93: 474 (1983).
95. A Graciaa, J Lachaise, M Bourrel, I Osborne-Lee, RS Schechter, WH Wade, SPE Reservoir Eng., 305 (1984).
96. N Márquez, R Antón, A Graciaa, J Lachaise, JL Salager, Colloids Surfaces A, 100: 225 (1995).
97. N Márquez, R Antón, A Graciaa, J Lachaise, JL Salager, Colloids Surfaces A, 131: 45 (1998).
98. JL Salager, G López-Castellanos, M Miñana-Pérez, J. Disper. Sci. Technol., 11: 397 (1990).
99. A Peña, Ch.E. thesis, Universidad de Los Andes, Mérida, Venezuela (1996).
100. JL Salager, in Surfactants in Solution, A. Chattopadhay, K. Mittal, eds., Surfactant Science Series Vol. 64. Marcel Dekker, New York (1996) p. 261.
101. F Silva, A Peña, M Miñana-Pérez, JL Salager, Colloids Surfaces A, 132: 221 (1998).
102. E Dickinson, J. Colloid Interf. Sci., 87: 416 (1982).
103. K Shinoda, H Saito, J. Colloid Interf. Sci., 30: 258 (1969).
104. P Becher, J. Soc. Cosmetic Chem., 9: 141 (1958).
105. E Dickinson, J. Colloid Interf. Sci., 84: 284 (1981).
106. JL Salager, in Surfactants in Solution, K. Mittal, P. Bothorel, eds., vol. 4. Plenum Press, New York (1987) p. 439.
107. JL Salager, in Encyclopedia of Emulsion Technology, P Becher, ed., Vol. 3: Plenum Press, New York, (1988) p. 79.
108. DH Smith, KH Lim, Langmuir, 6: 1071 (1990).
109. KH Lim, DH Smith, J. Colloid Interf. Sci., 142: 278 (1991).
110. GJ Vaessen, HN Stein, First World Congress on Emulsion, Paris, Oct. 19–22, 1993. Proceedings Vol. 1, paper 1-30-102.
111. H Sagitani, J. Am. Oil Chem. Soc., 58: 738 (1981).
112. R Pons, I Carrera, P Erra, H Kunieda, C Solans, First World Congress on Emulsion, Paris, October 19–22, 1993. Proceedings, vol. 1, paper 1-12-159.
113. B Brooks, H Richmond, Chem. Eng. Sci., 49: 1053 (1994).
114. C Solans, J Esquena, N Azemar, Tenth Int. Symposium on Surfactants in Solution, Caracas, Venezuela (1994).
115. M Miñana-Pérez, C Gutron, C Zundel, JM Andérez, JL Salager, J. Disper. Sci. Technol., 20: 893 (1999).
116. J Rouviere, JL Razakarison, J Marignan, B Brun, J. Colloid Interf. Sci., 133: 293 (1989).
117. J Rouviere, Information Chimie, 325: 158 (1991).
118. M Rondón, Ch.E. thesis, Universidad de Los Andes, Mérida, Venezuela (1997).

4

Suspension Formulation

Fernando González-Caballero and Juan de Dios García López-Durán
University of Granada, Granada, Spain

I. INTRODUCTION

Among the usual forms of dosage of pharmaceutical drugs such as solution, suspension, capsules, compressed tablets, or coated tablets, the most commonly used route of administration is through solid dosage forms (1,2). However, the bioavailability of a drug is greatly increased if the size of the particles is reduced, and for this reason other forms of dosage, mainly solutions or suspensions of the active agents, are most effective. In particular, the rate of absorption of a drug is known to decrease in the order solution > suspension > capsules > compressed tablets > coated tablets (3). According to Ofner et al. (4), delayed absorption of the drug can result in a prolonged clinically useful plasma concentration of the drug. Many practical advantages of suspensions for a better dosage of a drug in pediatric and geriatric populations are described in the literature (5). However, liquid disperse systems are undoubtedly among the various pharmaceutical dosage forms, the most complex from a physicochemical point of view, due to the coexistence of two different phases in the formulation. Being energetically unstable systems, the formulator must keep in mind that the suspension should have a half-life long enough to facilitate its transportation and storage in different environmental conditions (temperature, humidity, etc.). Thus, the adequate formulation of a drug suspension requires knowledge of the fundamental aspects, mainly thermodynamic and kinetic, regulating the behavior of the dispersed systems.

Pharmaceutical suspensions consist of solid particles of variable size dispersed in a liquid medium, generally an aqueous solution. Usually the size of the dispersed particles ranges in the colloidal domain (0.1–10 μm) and this makes pharmaceutical suspensions typical systems where the principles of colloid and surface science have to be used to deal with their properties. According to Hunter (6), the usual criterion to classify colloidal dispersions concerns mainly the nature of phases forming the system. Table 1 shows some typical denominations.

The small size of colloidal particles implies that the area/volume relation is very high for these systems, and the same can be said about their surface energy. The spontaneous tendency of such systems should be the aggregation of single particles to minimize the total energy, unless any energetic barrier would hinder it. From a thermodynamic point of view, the formation of an interface

Table 1 Classification of Dispersions

Dispersed phase	Dispersion medium	Technical name	Common name
Solid	Gas	Aerosol	Smoke, dust
Liquid	Gas	Aerosol	Fog
Solid	Liquid	Sol or colloidal sol	Suspension, slurry, jelly
Liquid	Liquid	Emulsion	Emulsion
Gas	Liquid	Foam	Foam, froth
Solid	Solid	Solid dispersion	Some alloys and glasses
Liquid	Solid	Solid emulsion	
Gas	Solid	Solid foam	

Source: From Ref. 6.

needs to provide a positive work to increase the surface area. The Gibbs free energy change, ΔG, associated with the formation of a dispersed phase is positive and for this reason a colloidal dispersion is thermodynamically unstable. Only the existence of an activation energy for the process of aggregation can delay the destruction of the disperse system by purely kinetic reasons. In short, it is only a matter of time before the dispersion becomes unstabilized. Any attempt to confer stability to a solid dispersion must look at the possible mechanisms that can influence the extent of that activation energy.

As in many other applications of disperse systems, the preparation of a drug suspension is carried out in steps: (a) milling of a solid until the required particle size is reached; (b) wetting of the solid by the liquid medium; (c) addition of suitable components to ensure proper stability and rheological requirements. If the suspension is designed to be orally administered (other possibilities are suspensions for topical or parenteral application; see Chapter 13 of this book), it is usually necessary to incorporate other additives like colorants, fragrances, tastes, and so forth to the final formulation. Points b and c are the steps where the recent advances in surface science can help in improving the formulation of pharmaceutical suspensions. This is so because, firstly, most pharmaceutical drugs are ''hydrophobic'' substances, and the wetting processes involved in the preparation of the suspension can be modulated through the adsorption of properly selected complex molecules. The usual wetting procedures require the adsorption of surfactants and/or polymers that improve the wettability of the solid material. This is another field where progress has been made in recent years and will be treated in this work. Second, coagulation or irreversible aggregation of the particles must be avoided and, at most, the suspension should stay in a flocculated state thus making it possible to reversibly change it to a homogeneous dispersion before dispensation. For this reason, a deep knowledge of the different mechanisms of stabilization/flocculation/coagulation is also required. Their description and usefulness for the formulation of drug suspensions is another aim of this contribution.

One further feature must be mentioned about pharmaceutical suspensions, namely, their desirable rheological properties (7). In practice, a ''Bingham plastic'' behavior is most used: a minimum shear stress (*yield stress*) is needed for the suspension to begin to flow. For lower stresses—and, of course, when the system is left undisturbed—the viscosity is so high that the particles will likely remain homogeneously dispersed. According to Falkiewicz (7), thixotropy is another flow characteristic that can be useful, since in thixotropic fluids a finite time is needed to rebuild the structure after, for instance, shaking it for administration. For this reason, most formulations contain thixotropic flow regulators intended to confer optimal viscous flow properties to the suspensions. The reader is referred to Chapter 5 of this book for details.

Let us finally remark that the formulation of a suspension implies the use of a number of components (the active principle plus wetting agents, dispersants, flocculating agents, thickeners, flavoring additives, preservatives, etc.) which, taken together, determine the global properties of the suspension. Wetting, stability, and rheology are in fact the macroscopic result of the interfacial properties of the dispersed individual particles. Throughout the next sections, the characteristics of the solid–liquid interface will be described, emphasizing in particular the fundamental thermodynamic and electrical quantities governing the overall macroscopic behavior. The methods for their experimental determination will be described, and we will finally be concerned with the modification of these interfacial properties by using proper components of the formulated suspension.

II. THERMODYNAMIC DESCRIPTION OF AN INTERFACE

When two homogeneous phases, e.g., a solid and a liquid, are placed in contact, a new nonhomogeneous phase appears at the interphase. The properties of this region, mainly density and composition, differ from those of the bulk phases. Clearly, the limits of the interphase region are not well defined because the transition from one phase to the other is progressive, and any quantity changes also more or less abruptly between the two homogeneous regions. The thermodynamic description of any system must include a precise definition of its physical limits, which play the role of the walls through which the exchange of matter and energy takes place. The difficulty of this definition in the interphase region was solved by Gibbs, who introduced the convention that the real interphase can be substituted by an ideal two-dimensional region that possesses the thermodynamic properties characterizing the real interphase. The separation between the two phases thus becomes an "interface" and its thermodynamic description is carried out in terms of the so-called surface excess quantities, X^σ, defined as $X^\sigma = X - X^s - X^l$, where the superscripts σ, s, and l correspond to the interface, and the solid and liquid phases, respectively. X represents any extensive thermodynamic quantity such as volume, number of moles of a component, Gibbs free energy, or entropy. An excellent comprehensive treatment of this field as applied to colloidal systems can be found in the recent monography by Lyklema (8).

The condition of thermodynamic equilibrium between phases in contact requires that the values of the intensive quantities, i.e., temperature, pressure, chemical potential of each species, etc., be constant throughout the phases. In disperse systems an important quantity is the surface tension, γ, defined by:

$$\gamma = \left(\frac{\partial G^\sigma}{\partial A}\right)_{T,n} \qquad [1]$$

where G^σ is the surface excess Gibbs free energy and A, T, and n are the interfacial area, temperature, and composition (number of moles) of the system, respectively. This quantity is usually named "surface tension" if the phases in contact are liquid and air, but in the applications where solids are dispersed in liquids it is commonly denominated "surface free energy." Comments on the use of this terminology can be found in the work by Ip and Toguri (9).

If two cylinders of unit cross-sectional area of a given material i are reversibly brought together to form a single continuous entity (see Fig. 1), the variation of free energy associated with this process is known as "free energy of cohesion," ΔG_{ii} ($= -W_i^c$, work of cohesion), defined by (10):

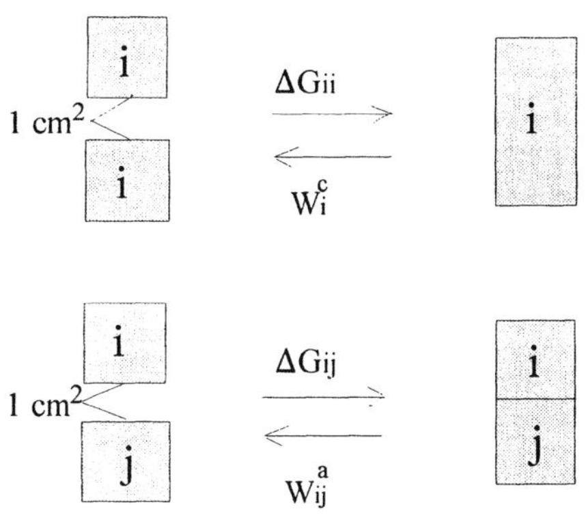

Figure 1 Diagrams illustrating: (a) the thermodynamic "cohesion" process, and (b) the thermodynamic "adhesion" process. See Eqs. (2) and (3). (Redrawn from Ref. 10, with permission.)

$$\Delta G_{ii} = -2\gamma_i = -W_i^c \qquad [2]$$

If the two cylinders are of different materials, *i* and *j*, the change of free energy of the process of adhesion, $\Delta G_{ij}(= -W_{ij}^a$, work of adhesion), is given by the Dupré equation:

$$\Delta G_{ij} = \gamma_{ij} - \gamma_i - \gamma_j = -W_{ij}^a \qquad [3]$$

where γ_{ij} is the interfacial tension between phases *i* and *j* and γ_i and γ_j the surface tension of materials *i*, *j*, respectively.

Both the cohesion and adhesion between phases are the result of the different types of intermolecular interactions in the condensed phases. Fowkes, and later van Oss, Good, and Chaudhury (11–15), proposed the separation of the free energy of interaction between condensed phases in two additive terms: (a) Lifshitz-van der Waals (LW) interactions, including dipole–dipole (Keesom), dipole–induced dipole (Debye), and induced dipole–induced dipole (dispersion, London) interactions, and (b) acid–base (AB) interactions, which, according to van Oss et al. (11–13), are of electron acceptor–electron donor nature and include, among others, the hydrogen bond interactions. Many solid materials have in their structure surface groups of a polar nature with either electron donicity properties (e.g., hydroxyl groups in hydrated oxides, sulfide groups in metal sulfides, aromatic rings, amine, carbonyl groups, etc., in organic materials, mostly polymers), or electron accepticity (ammonium, nitro groups, etc.). Thus:

$$\Delta G = \Delta G^{LW} + \Delta G^{AB} \qquad [4]$$

Taking into account Eqs. [2] and [4], the surface tension of a material *i* can be expressed as the sum of two components, LW and AB, respectively:

$$\gamma_i = \gamma_i^{LW} + \gamma_i^{AB} \qquad [5]$$

The original contribution of van Oss et al. (11) to this field is the determination of the acid–base component of the surface free energy in terms of two parameters: $\gamma^+ \equiv$ (Lewis) acid parameter of surface free energy; $\gamma^- \equiv$ (Lewis) base parameter of surface free energy. According to these authors, one can write:

$$\gamma_i^{AB} = 2\sqrt{\gamma_i^+ \gamma_i^-} \qquad [6]$$

The determination of a set of γ^+ and γ^- values for a series of different liquids is straightforward and can be found in the literature (13,16). The important point, however, is that the values are relative to those of water. The usual convention is that $\gamma_w^+ \equiv \gamma_w^-$, which together with $\gamma_w^{LW} = 21.8$ mJ/m^2 and $\gamma_w = 72.8$ mJ/m^2 at 20°C, gives [using Eqs. [5]–[6]]: $\gamma_w^+ = \gamma_w^- = 25$ mJ/m^2.

The interfacial tension between species *i* and *j* is expressed in terms of the components of individual surface tension as (13):

$$\gamma_{ij} = \gamma_{ij}^{LW} + \gamma_{ij}^{AB} = (\sqrt{\gamma_i^{LW}} - \sqrt{\gamma_j^{LW}})^2 + 2\,(\sqrt{\gamma_i^+\gamma_i^-} + \sqrt{\gamma_j^+\gamma_j^-} - \sqrt{\gamma_i^+\gamma_j^-} - \sqrt{\gamma_i^-\gamma_j^+}) \quad [7]$$

Using Eq. [4] for the free energy of adhesion between phases *i* and *j* gives:

$$\Delta G_{ij} = \Delta G_{ij}^{LW} + \Delta G_{ij}^{AB} \quad [8]$$

Now, ΔG_{ij}^{LW} and ΔG_{ij}^{AB} can be expressed in terms of the LW and AB components of surface tension of each phase using Eq. [3] and [5]–[7]:

$$\Delta G_{ij}^{LW} = -2\sqrt{\gamma_i^{LW}\gamma_j^{LW}},\ \Delta G_{ij}^{AB} = -2\sqrt{\gamma_i^+\gamma_j^-} - 2\sqrt{\gamma_i^-\gamma_j^+} \quad [9]$$

The free energy of interaction (note that the electrostatic contribution is not considered here) between two solid phases (1 and 2) immersed in a liquid phase (3) can be obtained from Eq. [3], written in the form:

$$\Delta G_{132} = \gamma_{12} - \gamma_{13} - \gamma_{23} \quad [10a]$$

where the interfacial free energies, γ_{ij}, can be obtained using Eq. [7]. The free energy of interaction between identical solid particles (1) in a liquid (3) is obtained from Eq. [10a] with $1 \equiv 2$:

$$\Delta G_{131} = -2\gamma_{13} \quad [10b]$$

Expressions [7]–[10] will allow, as it will be explained in Sec. III, establishment in a quantitative way of the hydrophilicity or hydrophobicity of a dispersed material use of it for the prediction of the stability of a suspension (see Sec. VIII).

III. WETTING PHENOMENA: HYDROPHOBICITY AND HYDROPHILICITY OF A DISPERSED SOLID

As mentioned above, the first requirement to produce a pharmaceutical suspension is to achieve adequate wetting of the solid particles by the liquid vehicle. Furthermore, the stability of the suspension demands that the energy of interaction between the dispersed particles be repulsive or, in thermodynamic terms, that the total free energy of interaction between the particles (ΔG_{131}) be positive. Wetting of particles dispersed in water is closely related to their hydrophilicity or hydrophobicity. These concepts are frequently employed in a qualitative way, and not always scientifically correct, as "water loving" or not. In this section the terms "hydrophilic colloids" and "hydrophobic colloids" will be described on a more quantitative basis.

The term *wetting*, in general sense, means the displacement from a surface of a fluid by another [17]. Applied to solid particles dispersed in a liquid, the substitution of a solid–air interface by a solid–liquid one is related to the so-called immersion wetting, which can be thermodynamically interpreted in terms of the work of immersion, defined by $W^i = \gamma_S - \gamma_{SL}$. The difference between the surface tension of a solid–air and a solid–liquid interface is thus the driving force determining the ease of complete immersion of solid particles in a liquid. This process, however, is most frequently described in terms of another wetting process, called *spreading wetting*, in which a thick liquid layer spreads on a solid surface. In this process, a solid–air interface is substituted by a solid–liquid and a liquid–air interface. The thermodynamic definition can also be made in terms of the *work of spreading*, W^s, sometimes called *spreading coefficient*, S_C, given by:

$$W^S = S_c = \gamma_S - \gamma_L - \gamma_{SL} \quad [11]$$

or, taking into account Eqs. [2] and [3]:

$$W^S = W^a_{SL} - W^c_L \quad [12]$$

Equation [12] means that the work of adhesion between the solid and the liquid must be higher than the work of liquid cohesion for spontaneous spreading to occur (18). It is clear from the definition of work of immersion and Eq. [11] that a positive value of S_C, i.e., $\gamma_S - \gamma_{SL} \geq \gamma_L$, implies that immersion of the solid particle is spontaneous. If the spreading coefficient is negative, work should be done to completely immerse the solid. In this case, the only way to get the particles fully immersed in the liquid is through modification of the values of surface tension of the several surfaces involved, until a positive value of the spreading coefficient is reached (17) (see below, Sec. VII). From the perspective of the preparation of a solid dispersion, the important point is that once the particles are conveniently wetted, they should remain separated. In this case the colloidal dispersion would be "hydrophilic"; on the contrary, i.e., if aggregation spontaneously appears, the dispersion would be "hydrophobic." Following Lyklema (8), it can be said that a colloidal system is hydrophilic when the affinity of the particles toward the solvent is higher than the mutual affinity between the particles; it will be hydrophobic in the opposite case.

It must be noted that "hydrophobicity" of a material does not mean that it has no affinity for water. In fact, all materials interact favorably with water. The change of free energy related to such interaction, i.e., the free energy of hydration, ΔG_{iw}, is negative even for the most hydrophobic materials. The enthalpies of wetting (immersion) per unit area shown in Table 2 give information on the relative affinities of the molecules of water for the surface. These data can serve to some extent as a measure of the hydrophobicity or hydrophilicity of the surface (19).

It can be observed in Table 2 that even for a material as hydrophobic as Teflon, the hydration is a favorable process from the enthalpic point of view. According to van Oss et al. (12), aliphatic hydrocarbons also attract water with a free energy of the order of -40 to -50 mJ/m^2. The criteria of the above authors

Table 2 Enthalpies of Wetting for Some Solid–Water Pairs[a]

Solid	-ΔH
$BaSO_4$	490–760
TiO_2 (anatase)	510–650
TiO_2 (rutile)	480–550
SiO_2 (quartz)	610–850
SiO_2 (amorphous)	210
SiO_2 (pyrogenic)	150
γ-FeOOH (goethite)	465
α-Fe_2O_3 (hematite)	280
β-Cu phthalocyanine (a pigment)	50
Graphon	32
Teflon	6
Smectites (clay minerals)	55–150
Vermiculites (clay minerals)	165–300
Kaolinites (clay minerals)	320–500

[a] ΔH in mJ/m^2 at 25° C.
Source: From Ref. 18, p. 1.34.

Table 3 Free Energy of Hydration, ΔG_{iw}, and Free Energy of Interfacial Interaction in Water, ΔG_{iwi}

Compound	ΔG_{iwi}	ΔG_{iw}
Hydrophobic:		
Benzene	−69.9	−66.7
Fibrin	−34.2	−99.7
Gelatin	−19.4	−100.7
Octanol	−15.8	−91.2
Agarose	−3.3	−112.2
Hydrophilic:		
Polyvinylpyrrolidone	+1.0	−114.8
Dextran	+41.2	−135.4
Fibrinofen	+41.9	−134.3
Polyethylene oxide (PEO)	+52.2	−142.0

[a] In mJ/m^2 of various compounds, in order of increasing ΔG_{iwi}.
Source: From Ref. 13, p. 209.

for the determination of free energy changes, from the components and parameters of the surface free energy of solid and water, can be used for the determination of free energy of hydration, ΔG_{iw}, (Eqs. [8] and [9]), and also for ΔG_{iwi}, the free energy of interaction between particles dispersed in water (or any other liquid) (Eqs. [7] and [10]). Table 3 shows some of these data for different types of materials.The use of the set of Eqs. [7] and [10] is the basis of the quantitative estimation of the hydrophobicity or hydrophilicity in a suspension. If ΔG_{iwi} is negative, attraction exists between the dispersed particles and then the hydrophobic effect appears, favoring coagulation in the system. If the sign is positive, the repulsive contribution will enhance the dispersion of the particles (hydrophilic suspension). Data in Table 3 illustrate these comments.

IV. SURFACE FREE ENERGY OF SOLIDS: MEASUREMENT AND APPLICATIONS

Surface tension of liquids is an experimentally accessible quantity. There are several well-known standard methods that will not be described here. The reader is referred to the monograph by Adamsom (20) and to Ref. 21. The concept ''surface tension of a solid'' has been subject of discussion for years, given the indeformability of the solid surface. However, this quantity exists and it is better understandable in terms of Eq. [2]: it accounts for the extent of cohesion energy in the material. Usually named ''solid surface free energy,'' γ_S, it obviously must take part in the determination of the interaction of solid particles with liquids (wettability), as well as between the dispersed particles themselves (stability). Although it has been frequently used in pharmaceutical research, mainly through correlations between γ_S and the wetting (7) or adhesion properties (22), in practice the knowledge of γ_S alone is by itself of little use for solids (13, p. 33); in order to make predictions on the wettability and stability of solid dispersions, it is, according to the recent models of interpretation of interfacial energetics (13), essential to know the surface free energy components (of both the liquid and the solid phases). Thus the interactions between the interfaces involved, as described by Eqs. [7]–[10], can be computed. However, some authors maintain the usefulness of an equation of state approach to interfacial tension to predict the interfacial properties in solid–liquid systems, despite arguments about its theoretical validity (22,23).

Concerning solids in general, and powdered forms in particular, there are not direct methods of determination of surface free energy, and one must use indirect methods that depend on the morphology of the solid material. In what follows some of the most useful techniques will be described.

A. Experimental Methods: Contact Angle and Thin-Layer Wicking Measurements

If the starting material is available as a solid with smooth and homogeneous (at the macroscopic level) surface, a method widely used for surface free energy determination is based on the measurement of the contact angle of a drop of selected liquid, deposited on the surface. The discussion of the results is made in terms of Young's equation:

$$\gamma_L \cos\theta = \gamma_S - \gamma_{SL} \quad [13]$$

where θ is the contact angle (Fig. 2) and the other symbols have the usual meaning. The application of this equation requires thermodynamic equilibrium between the three phases involved (see Ref. 21, chapters 1 and 2). Making use of Dupre's equation (Eq. [3]), the following relation is obtained:

$$-\Delta G_{SL} = \gamma_L(1 + \cos\theta) \quad [14]$$

which, substituted into Eqs. [8] and [9], yields:

$$(1 + \cos\theta)\gamma_L = 2\left(\sqrt{\gamma_S^{LW}\gamma_L^{LW}} + \sqrt{\gamma_S^{+}\gamma_L^{-}} + \sqrt{\gamma_S^{-}\gamma_L^{+}}\right) \quad [15]$$

If the values of γ_L and its Lifshitiz-van der Waals component γ_L^{LW}, as well as the acid–base parameters γ_L^{+} and γ_L^{-} are known (see Ref 13, p. 106, for the most commonly used liquids), the measurement of θ allows the determination of the free energy components of the solid. In any particular problem three different probe liquids are needed, and three equations of the Eq. [15] type have to be solved to obtain the unknowns, γ_S^{LW}, γ_S^{+}, and γ_S^{-}. Then, using Eqs. [5] and [6],

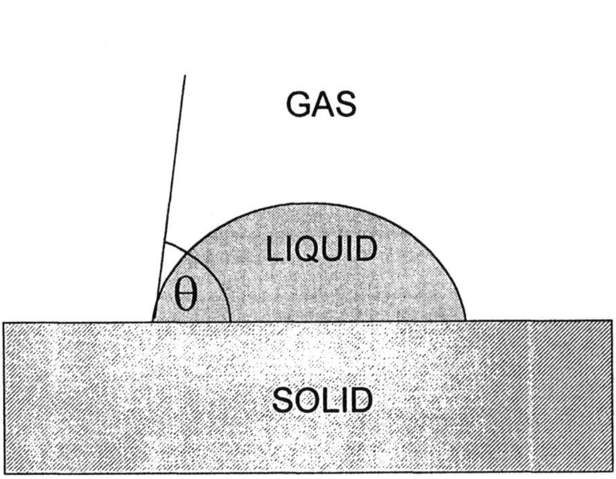

Figure 2 Contact angle of a liquid drop on a solid surface.

the total surface free energy of the solid can be obtained. Usually an apolar liquid ($\gamma_L^+ = \gamma_L^- = 0$), such as diiodomethane or α-bromonaphthalene, and two polar liquids (γ_L^+ and/or $\gamma_L^- \neq 0$), such as water and formamide, are used. Depending on the nature of the system to be studied, these liquids can be substituted by others (see Ref. 13) that also have a high surface tension.

Contact angle measurements can only by made when large single crystals or compact tablets are available. Even in the latter case the measurements cannot be made with confidence because of the porosity, and other methods are more convenient. This will be discussed below. Another problem is related to the reproducibility of contact angle measurements. Depending on the angle being measured (advancing or receding contact angles; see Ref. 21), as well as on the internal conditions in the measurement chamber (mainly vapor pressure of liquid forming the drop, see Refs. 24 and 25), hysteresis can be an important source of complications. Briefly, hysteresis is a phenomenon in which the contact angle formed by a liquid advancing across an unwetted surface is generally larger than the contact angle of the same liquid as it recedes across the previously wetted surface (10,22,24,25). The origin of this phenomenon is not presently well known, although it is generally accepted that it is related to the surface chemical heterogeneities (21). Some authors (24,25), however, note the role played by the liquid film behind the drop when it is receding, which would result in a change of the interfacial free energy at the solid–vapor interface. In any case, in most pharmaceutical applications only advancing contact angles are usually measured.

Frequently, solid surfaces are not available in homogeneous, flat-plate form. Especially in pharmaceutical applications, solids are usually powdered and in this case the estimation of surface free energy is better made by other methods, different from contact angle measurement.The two main techniques used with these solids are the sedimentation volume experiments and the so-called thin-layer wicking method. The former method (22,26) is based on the measurement of the volume of solid that is deposited on a tube previously filled with a liquid of known surface tension. Usually, the liquid is formed of a mixture of miscible liquids of different surface tension in order to ''build'' a liquid of a desired surface tension. When the sedimented volume of solid is maximum, then its surface tension (surface free energy) equals that of the liquid used in the experiment. Obviously, the partial solubility of the solid in the liquid mixtures might make it necessary to measure the surface tension of the saturated ''solutions.'' Furthermore, the need of using many liquid mixtures with a wide range of values of surface tension makes this procedure tedious and not always successful.

In this section only the thin-layer wicking method will be described in some detail. The physical basis of this technique has long been known, although its application to the determination of free energy of solids is very recent. The first investigations in this field were made by van Oss et al. (27), who later developed it in a series of papers (12,28; see also Ref. 13). The method was further modified by Chibowski et al., giving a detailed procedure to obtain the several components of the solid surface free energy (29–32). These authors successfully applied the method to several pharmaceutical drugs, as will be shown later. In the methodology used by Chibowski et al. (32), a thin layer of porous solid is formed by deposition and further drying of a concentrated dispersion of the solid on a glass slide placed horizontally. Then, a probe liquid is conducted from a small reservoir, and wicking occurs through the solid layer. The method is based on the measurement of the time needed for the liquid to penetrate a given distance through the porous solid layer (Fig. 3).

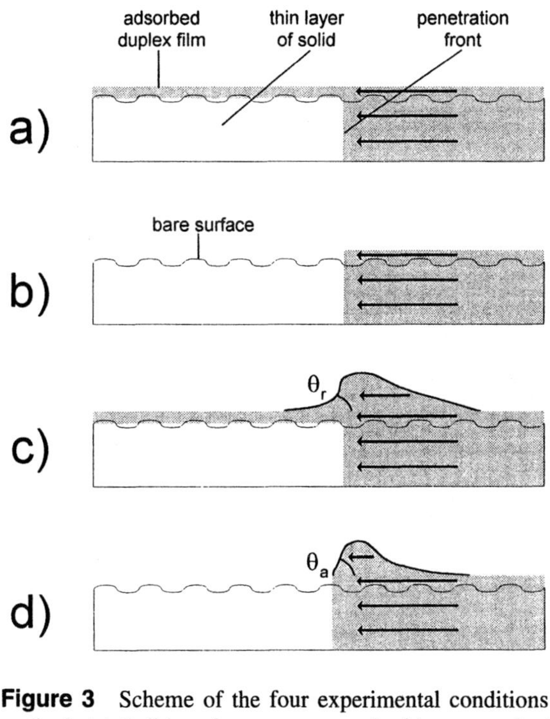

Figure 3 Scheme of the four experimental conditions used in the thin-layer wicking method. (a) Solid surface precontacted with vapor of a liquid that completely wets the surface. (b) Bare solid surface, penetration of a liquid that completely wets the surface. (c) Solid surface precontacted with vapor of a liquid that only partially wets the surface (it does not spread completely over the solid surface). (d) Same as in (c) but the surface is bare.

According to the original Washburn equation (20, p. 496), the square of the penetrated distance, x^2, should be a linear function of time t for a particular liquid of viscosity η and surface tension γ_L, at a constant temperature:

$$x^2 = \frac{Rt}{2\eta}\gamma_L \qquad [16]$$

Where the parameter R is an averaged value of the apparent capillary radius of the thin porous layer (29). However, as shown in (33–35), Eq. [16] is valid only if a precursor duplex liquid film is present ahead of the penetrating front of a liquid completely wetting the solid surface. Thus, from this equation it is clearly seen that the surface free energy of the substrate is not related to the rate of penetration of the liquid. Nevertheless, Eq. [16] is very useful for the determination of the R parameter of glass plates covered with the powder of the solid tested. Liquids most suitable for this purpose are the *n*-alkanes, like *n*-octane or *n*-decane.

Chibowski et al. (31,32) have considered four experimental procedures and thus four different forms of Washburn equation for the determination of the surface free energy components of a given solid. The starting equation is a generalized version of eq. [16] (34,35):

$$x^2 = \frac{Rt}{2\eta}\Delta G \qquad [17]$$

The important point is the appropriate description of the energy change accompanying the penetration process, ΔG. For this purpose, the following cases can be distinguished (the detailed justification can be found in Ref. 32):

1. $\Delta G = \gamma_L$. This is the case when the solid surface is precovered with the liquid duplex film and the liquid completely wets the solid surface (Fig. 3a). The original form of Washburn's equation, Eq. (16), is used.
2. $\Delta G = W^a - W^c$, where W^a is the work of adhesion of the liquid to the solid surface and W^c is the work of cohesion of the liquid, equal to $2\gamma_L$. This is the value of the free energy change in the experimental method where the solid surface is bare and the penetrating liquid completely wets the solid surface (Fig. 3b). Using van Oss et al.'s formulation of the interfacial interaction (13), it reads (32):

$$\Delta G = 2\sqrt{\gamma_S^{LW}\gamma_L^{LW}} + 2\sqrt{\gamma_S^+\gamma_L^-} + 2\sqrt{\gamma_S^-\gamma_L^+} - 2\gamma L \qquad [18]$$

3. $\Delta G = \gamma_L \cos\theta$. The solid surface is contacted with vapor of the liquid prior to the penetration experiment, but the liquid does not spread completely over the solid surface; a dynamic contact angle θ, which is not equal to the equilibrium contact angle of the Young equation, appears at the penetration front (Fig. 3c).

4. $\Delta G = \gamma_L \cos\theta + W^a - W^c$. The solid surface is bare and the liquid forms a dynamic contact angle, θ, during the penetration. Again, using van Oss et al.'s approach (13) we obtain:

$$\Delta G = \gamma_L \cos\theta + 2\sqrt{\gamma_S^{LW}\gamma_L^{LW}} + 2\sqrt{\gamma_S^{+}\gamma_L^{-}} + 2\sqrt{\gamma_S^{-}\gamma_L^{+}} - 2\gamma_L \qquad [19]$$

The value of $\Delta G(= \gamma_L \cos\theta)$ from case (3) can be introduced into Eq. [19]. Now, using three liquids, e.g., n-alkane, water, and formamide, for which the surface tension parameters and viscosities are known, it is possible to determine γ_S^{LW}, γ_S^{+}, and γ_S^{-} for the tested solid under the proper experimental conditions 1–4.

B. Some Examples of Surface Free Energy Determination

In pharmaceutical applications, the method of contact angle measurement has long been used for interpretation of the wettability of solids by the coating liquids, in terms of the critical surface tension of the solid, γ_C: by plotting $\cos\theta$ against γ_L , one can obtain γ_C by extrapolating to $\cos\theta = 1$. The higher γ_C is, the better is the wettability of the solid, interpreted as the ability of the liquid to spread onto the solid. This has been used, for example, for the characterization of compact tablets of acetylsalicyilic acid (36) and other drugs of a different nature (37–40) . This procedure, however, is limited to sufficiently compactable tablets and is insufficient for discussion of the several steps involved in the preparation of a solid dispersion (not only spreading, but also adhesion and immersion wetting processes) as well as for prediction of its stability once the solid has been wetted. In fact, as previously mentioned, it is the knowledge of surface free energy components that gives the possibility of calculating the energy of interaction between suspended particles and thus the prediction of their stability. For this purpose, either the contact angle method or the penetration of liquids into powdered solids can be used.

As an illustration of the procedure, some results will be shown, obtained with substances of biological and pharmaceutical interest. For this purpose, cholesterol and nitrofurantoin were selected. Cholesterol is an unsaturated alcohol with chemical formula $C_{27}H_{45}OH$. The molecule consists of four 6-carbon rings and a side-branched carbon chain. The hydroxyl group of the molecule occupies an equatorial position in the relatively rigid ring system. Moreover, the molecule also contains one unsaturated C═C bond. Nitrofurantoin is an antibiotic that is widely used for the treatment of infections of the urinary tract. Chemically it is *N*-(5-nitro-2-furfurylidene)-1-aminohydantoin and is a rather complicated molecule possessing several polar groups, such as nitro or imide.

In Fig. 4 are plotted the results of wicking experiments of n-decane, water, and formamide on powdered nitrofurantoin. As observed, a linear relationship is found between time and the square of the penetrated distance, as predicted by

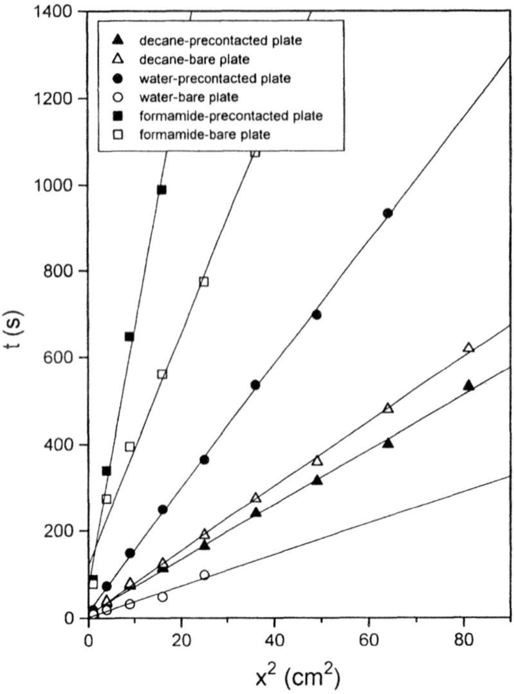

Figure 4 Time of penetration of *n*-decane, water, and formamide through thin layer of nitrofurantoin, vs. the squared distance. (Drawn from plots in Ref. 41, with permission.)

the Washburn equation, Eq. [17], under two different experimental conditions, i.e., precontacted and bare plates. Making use of Eqs. [17]–[19], the solid surface free energy components can be determined. Table 4 shows the results for nitrofurantoin (41). In this table are also shown the results obtained for cholesterol (42), using the contact angle method and Eq. [15]. Besides the agreement between the

Table 4 Surface Free Energy (mJ/m^2) and Components of Nitrofurantoin and Cholesterol

Material	Technique	γ	γ^{LW}	γ^{+}	γ^{-}	Ref.
Nitrofurantoin	Contact angle	74.5	48.4	7.22	23.5	42
	Wicking	74.5	47.7	6.7	26.8	
Cholesterol	Contact angle	37.9	35.7	0.53	2.21	43

results obtained using the two techniques, within the experimental error, it is worth mentioning that both materials, nitrofurantoin and cholesterol, are polar, as γ^+ and γ^- differ from zero, i.e., there is an acid–base contribution to the solid surface free energy. To justify the values of this contribution, it is worthwhile to have an insight into the structure of the molecules, which in the two cases consists of several polar groups. In particular, the surface energetic properties of nitrofurantoin could be explained taking into account that at least the secondary amine group -NH and the -C═N- group of its molecule are electron donor groups, thus supporting the relatively large value of the γ_S^- component. The electron acceptor component can result from the presence of a nitro group. On the other hand, cholesterol is practically monopolar due to the relevance of the electron donor character of hydroxyl surface groups. In this case, the acid–base component of the cholesterol surface free energy is negligible, although it contributes to the free energy of adhesion (Eqs. [8] and [9]), or the free energy of interaction between the particles immersed in water (hydrophobic or hydrophilic effect), (Eqs. [7]–[10b]). It is worth mentioning that while cholesterol is a material of low surface energy ($\gamma \approx 38$ mJ/m^2), the surface free energy of nitrofurantoin attains values of the order of 75 mJ/m^2. Most pharmaceutical drugs, however, have relatively low surface free energy, typically ranging from 30 to 50 mJ/m^2

Table 5 Surface Free Energy Components (mJ/m^2) of Nitrofurantoin with Preadsorbed Amino Acids

Amino acid	Concentration (mol/L)	γ_S^{LW}	γ_S^-	γ_S^+
	0	46.3	26.8	6.7
Lysine	10^{-5}	50.8	28.8	1.6
	10^{-4}	53.1	26.2	0
	5×10^{-4}	61.4	40.2	0.17
	10^{-3}	52.3	47.4	0
	5×10^{-3}	53.4	46.8	1.43
	10^{-2}	58.4	47.4	0.26
Alanine	10^{-4}	51.8	26.8	3.82
	10^{-3}	52.2	77.0	0
	10^{-2}	57.6	83.3	0
Glutamic acid	10^{-5}	51.8	37.1	0.43
	10^{-4}	54.4	45.5	0.26
	5×10^{-4}	54.0	80.7	0
	10^{-3}	50.9	85.9	0
	10^{-2}	52.1	89.9	0

Source: From Ref. 45.

[e.g., acetaminophen, γ = 40–46 mJ/m^2 (24), adipic acid, γ = 30–34 mJ/m^2 (22)]. The two methods considered above for the study of the surface free energy of solids and its components (thin-layer wicking and contact angle measurements) can also be used for the estimation of the effect of the adsorption of ions or molecules on the surface properties of solid particles dispersed in complex media. In the following paragraphs some examples will be given.

Besides the importance of amino acids in the living organisms, their use in controlling the stability of suspensions or emulsions of pharmaceutical interest has been largely reported in the literature (43,44). For a better understanding of the mechanisms involved in the stabilization of pharmaceutical suspensions, an example will now be discussed that deal with nitrofurantoin. Table 5 and Fig. 5 show, respectively, the surface free energy components of nitrofurantoin with preadsorbed amino acids (lysine, alanine, and glutamic acid) and the values γ_S^- of nitrofurantoin as a function of the equilibrium concentration of the adsorbed amino acid (45). It can be seen that the Lifshitz-van der Waals component,

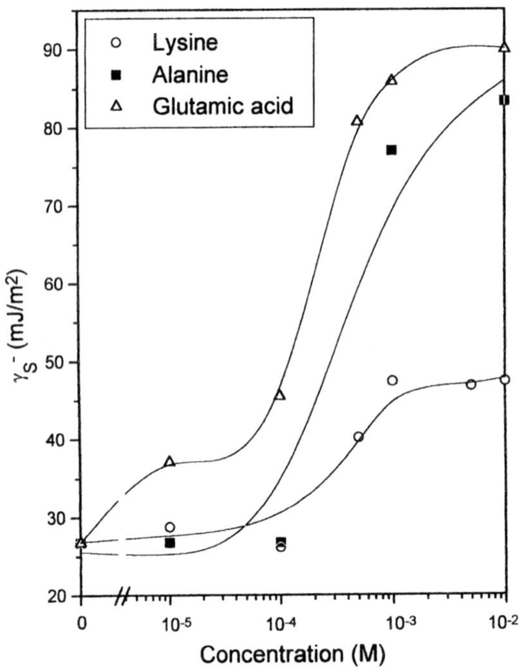

Figure 5 Electron donor component, γ_S^- (mJ/m^2), of nitrofurantoin with preadsorbed amino acid against the concentration of the solution from which the amino acid was adsorbed (After Ref. 45, with permission.)

γ_S^{LW}, increases slightly as a result of the adsorption. The γ_S^+ component (6.7 mJ/m^2 for the bare nitrofurantoin surface) is reduced to zero in virtually all cases studied. On the contrary, a sharp increase of γ_S^- takes place when the concentration of amino acid exceeds 10^{-4} M. The results can be interpreted taking into account the experimental conditions (pH and concentration of amino acid in the suspension) and the isoelectric point of the amino acids (9.47, 6.11, and 3.08, respectively, for lysine, alanine, and glutamic acid). The natural pH of 10^{-2} M solutions of the amino acids were 6.2, 4.8, and 3.0 for the same set of amino acids, respectively. Thus the glutamic acid molecules are in dipolar form, with no net charge, whereas those of alanine and lysine are to some extent positively charged. The ζ potential of nitrofurantoin suspensions at the pH of the experiments is negative in the whole range of concentrations (46). Thus, the adsorption of amino acids mainly neutralizes the positively charged sites of the molecules, and the remaining carboxyl groups confer their electron donor characteristics to the surface, provoking the changes of the γ_S^- component of surface free energy by the adsorp-

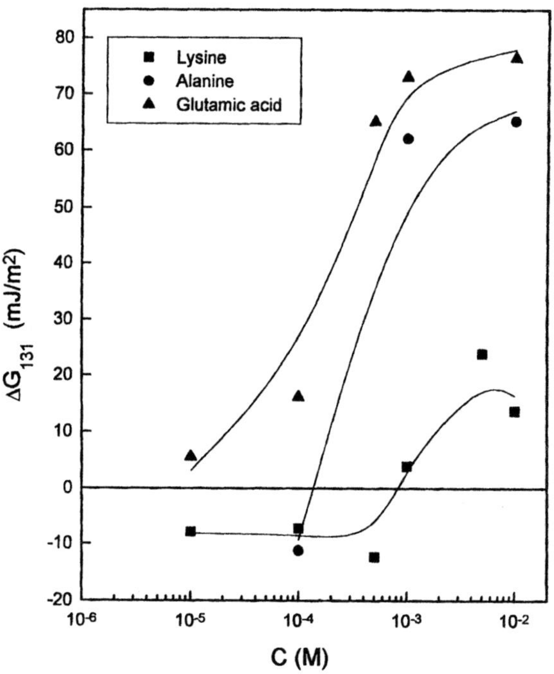

Figure 6 Free energy of interaction between particles of nitrofurantoin immersed in solutions of amino acids (lysine, alanine, and glutamic acid) of different concentration. (Plotted from data in Ref. 45.)

tion of the amino acids. This effect is more noticeable for glutamic acid due to the existence of two carboxyl groups in its molecule.

The important increase of γ_S^- as the amino acid is adsorbed on nitrofurantoin should in turn influence the energy of interaction between the suspended particles, ΔG_{131}, as given by Eqs. [7] and [10b], and thus the stability of the suspension. In Fig. 6 are represented the values of ΔG_{131} as a function of amino acid concentration. It can be seen that positive free energy of interaction between the particles (and thus repulsion between them) exists in the higher range of amino acid concentration. Thus, from the thermodynamic point of view, interfacial interactions between nitrofurantoin particles dispersed in amino acid solution favor their stability when the concentration is higher than 10^{-5} M, 10^{-4} M, and 10^{-3} M, respectively, for glutamic acid, alanine, and lysine. Of course, electrostatic interactions must also be considered for the complete prediction of the stability conditions, as it will be discussed in Sec. 8.1.

Another interesting example is cholesterol in the presence of bile salts. Due to the relationship of cholesterol to many diseases, such as atherosclerosis or

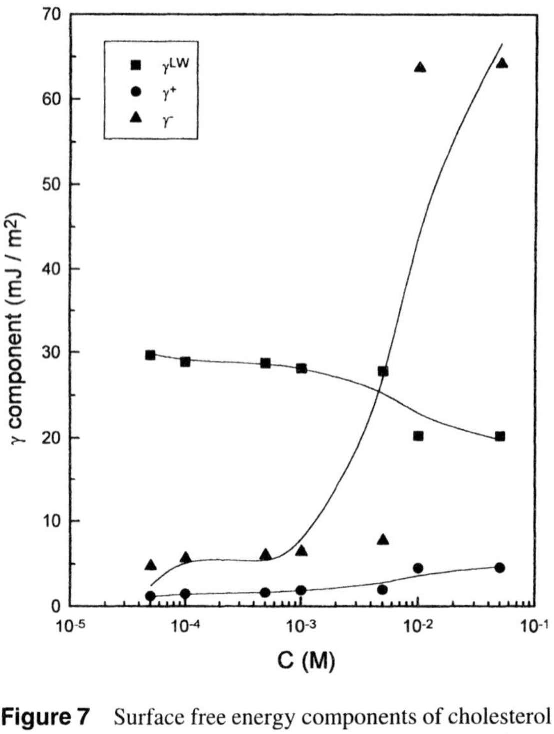

Figure 7 Surface free energy components of cholesterol with adsorbed bile salt (sodium cholate, NaC) from solutions of different concentrations. (Plotted from data in Ref. 47.)

gallstone formation, most investigations on this substance have focused on biological or clinical aspects. However, the first step in the formation of a cholesterol calculus is the existence of an unstable bile with a high cholesterol content. Then, the growing of particles will depend primarily on the characteristics of the solid–solution interface. For the purpose of pharmaceutical applications, a better knowledge of the interfacial interactions in these systems of such high complexity would be of interest to determine the dosage and particular composition of a drug used to preserve the stability of the suspension. In this study, some results will be presented on the effect of a bile salt, sodium cholate, on the surface free energy of cholesterol and the free energy of interaction between cholesterol particles with preadsorbed bile salt. Figure 7 shows the surface free energy components of cholesterol as a function of sodium cholate (NaC) concentration in solution (47). Similar to the results obtained with nitrofurantoin, γ_S^{LW} is only slightly dependent on the concentration of the adsorbed molecules. Also, a great increase manifests in the electron donor component, γ_S^- at sodium cholate concentrations

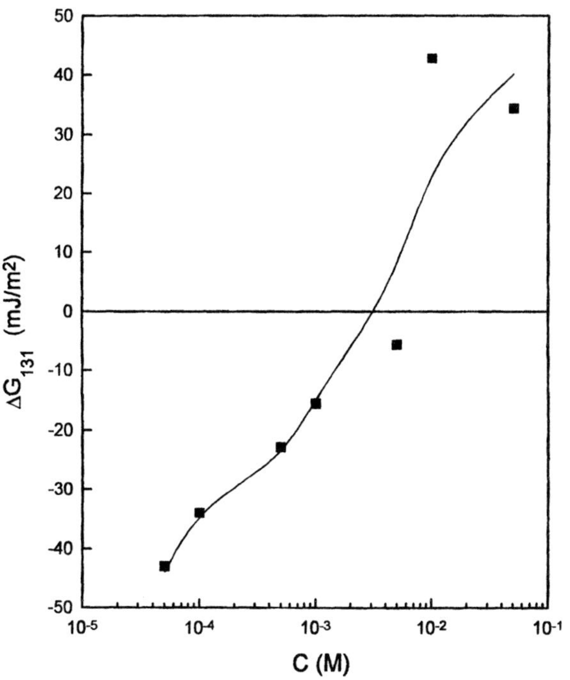

Figure 8 Free energy of interaction between particles of cholesterol immersed in solutions of sodium cholate of different concentrations. (From data in Ref. 47.)

above 10^{-3} M. These results should also have their consequence on the free energy of interaction between the particles of cholesterol with the adsorbed bile salt. That is shown in Fig. 8, where the significant fact is that the suspension would be unstable if [NaC] $< 5 \times 10^{-3}$ M and stable at higher concentrations. It must also be noted that this concentration is of the order of the CMC of the bile salt, 7.2×10^{-3} M (47).

Modification of the Lewis acid-base character of a surface can also be achieved by the presence of inorganic ions in the solution. In this case, however, a net correlation between the (more representative) value of γ_S^- and the entropy of hydration of the ions, ΔS_h, has been found (48) . In Fig. 9 a linear correlation is observed between those quantities for a series of cations and anions, with the only exception of K^+ and OH^- which show a special behavior. Although ΔS_h deals with ions in solution and γ_S^- describes the surface behavior, the two quantities should be related by the mechanisms of adsorption of the hydrated ions on

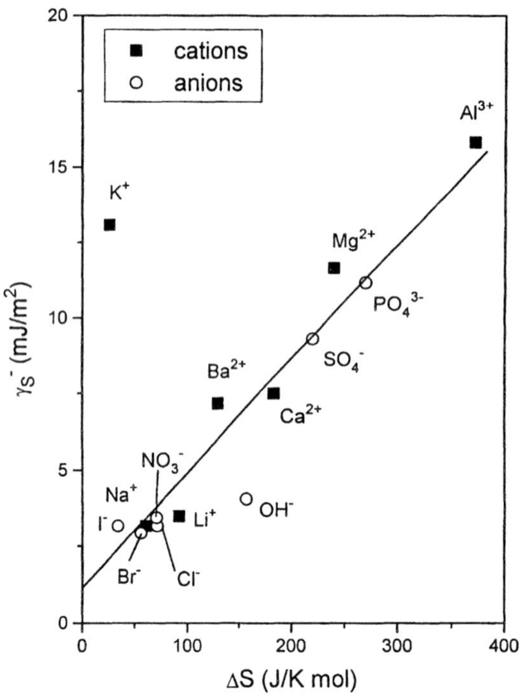

Figure 9 Relationship between the electron donor component, γ_S^- (mJ/m^2), of cholesterol surface with preadsorbed ions and the entropy of hydration of the ions, ΔS_h (J/K mol). (Taken from Ref. 48, with permission.)

the surface layer. The electron donor (also the electron acceptor) component of surface free energy originates mainly from the water molecules hydrating the adsorbed ions. Thus, ions being capable of producing a significant ordering of water molecules around them would also confer a high electron donicity after their adsorption on the surface. It is evident from the above results that adding ions in solution is another way to control the surface energetics of a solid in suspension and hence the global properties of the suspension.

V. ELECTRICAL PROPERTIES: THE ζ POTENTIAL

A. Origin of the Electric Charge at Solid–Liquid Interfaces

It is current experience that a solid material immersed in a liquid medium (water or aqueous solutions in most practical cases) shows some special properties that can be explained by the existence of a surface charge. The origin of such charge is strongly dependent on the nature of both the solid and the liquid phases. Briefly, the most frequent mechanisms are (49–51):

1. *Differential ion solubility*, as in the case of the well-studied system of dispersed AgI particles (50).
2. *Direct ionization of surface groups*. This is a common mechanism in ionizable polymeric surfaces that contain such groups in their molecules as carboxylic or sulfonic acids, sulfuric acid esters, or amino and quaternary ammonium groups.
3. *Isomorphous ion substitution*. It is a common mechanism in many minerals, like clays, oxides, etc., in which the structural ions are substituted by ions of valency different from the corresponding lattice ions.
4. *Specific ion adsorption*. In pharmaceutical applications, this is perhaps one of the most important mechanisms used in practice for imparting surface charge to the solid particles in a given formulation. In particular, the adsorption of surfactants is of great importance and will be treated below in more detail.
5. *Anisotropic crystals*. Some important materials have crystal structures that when cleaved can result in the production of both positively and negatively charged surfaces (49). This is the case of the aluminosilicate layers in clays.

B. The Electrical Double Layer at Equilibrium: ζ Potential

The electrical neutrality of a disperse system means that the surface charge density, σ_0, generated by any of the above mechanisms must be compensated by an opposite charge that will be distributed in the solution sorrounding the solid particle. The surface charge, together with its countercharge, constitute the *electrical double layer* (edl hereafter). A brief description of the edl will be given with the aim of clarifying the physical basis of the diverse techniques used in pharmaceutical research for investigating the electrical properties of disperse systems and, in particular, of understanding the meaning of the *electrokinetic* or ζ *potential*, because of its important role in characterizing double layers in practice.

1. The Gouy-Chapman Model of the Diffuse e.d.l.

Let us first describe the simple model of a homogeneously charged plain surface immersed in an aqueous liquid. The Gouy-Chapman theory (1915) describes the equilibrium distribution of charge and potential in the near-vicinity of the surface (the edl). It assumes that ions in solution have no volume (point charges) and move in liquid, considered as a continuous dielectric medium of constant permittivity, under the action of electric and thermal diffusive forces. In this situation, an asymmetrical distribution of ions originates a *diffuse* layer around the surface (see Fig. 10a). According to the Gouy-Chapman model for a plain surface (52,53) [other geometries can be found in more specialized books (8,18,51)], the electric potential along the edl diminishes with distance, from its value at the surface, ψ_0, to the value $\psi(x)$, at the distance x from the surface, where the volume charge density due to the ions in the solution is $\rho(x)$. The main equations governing the distribution of potential and ionic density in the edl are:

1) Poisson equation:

$$\frac{d^2\psi(x)}{dx^2} = -\frac{\rho(x)}{\epsilon\epsilon_0} \qquad [20]$$

giving the relationship between ψ and ρ as a function of the position (flat symmetry), with ϵ being the relative dielectric permitivity and ϵ_0 the permittivity of a vacuum.

2) Boltzmann equation:

$$n_i = n_i^0 \exp(-z_i e\psi/kT) \qquad i = 1, \ldots, N \qquad [21]$$

where n_i is the number of ions of type i per unit volume, n_i^0 is the number of ions of type i per unit volume in the bulk solution far from the surface, z_i is the valence of the ion i, e is the electron charge, k the Boltzmann constant, and T the absolute temperature.

3) The volume charge density in the neighborhood of the surface, given by:

$$\rho = \sum_i n_i z_i e \qquad [22]$$

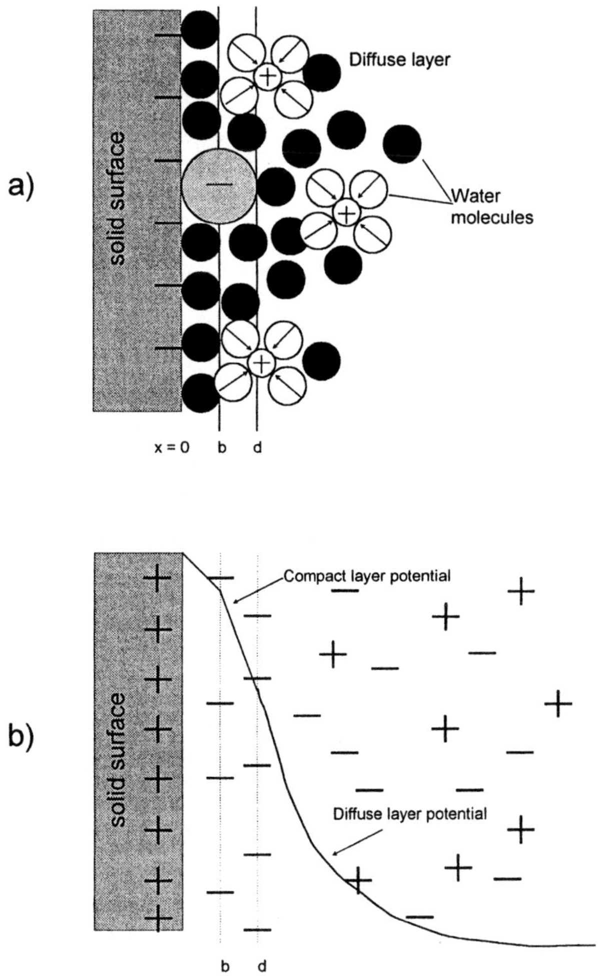

Figure 10 (a) The double-layer structure at the solid–solution interface according to the Gouy-Chapman-Stern-Grahame model. (b) The electrostatic potential in the compact and diffuse parts of the double layer. (Taken from Ref. 6, with permission.)

Combining Eqs. [20]–[22], the Poisson-Boltzmann equation is obtained:

$$\frac{d^2\psi(x)}{dx^2} = -\frac{1}{\epsilon\epsilon_0}\sum_i n_i^0 z_i e \exp(-z_i e\psi/kT) \qquad [23]$$

This equation has a simple solution when the electrical contribution is small compared to the thermal one: $(/z_i e\psi/ < kT)$—the Debye-Hückel approximation. In practice, when water is used as a liquid medium, at room temperature, this is valid for potentials below ~25 mV. This approximation allows to obtain the simple relationship between the electric potential and position (51) (Fig. 10b):

$$\psi = \psi_0\, e^{-\kappa x} \qquad [24]$$

with κ given by:

$$\kappa = \left(\frac{e^2 \sum n_i^0 z_i^2}{\epsilon\epsilon_0 kT}\right)^{1/2} \qquad [25]$$

This quantity is called the Debye-Hückel parameter and κ^{-1} is often referred to as the thickness of the double layer. At 25°C in water the value of κ is given by: $\kappa = 3.288\sqrt{I}$ (nm^{-1}), where I is the molar ionic strength of the solution: $I = 1/2 \sum (c_i z_i^2)$. As the ionic concentration increases, the thickness of the double layer decreases, a process referred to as *compression* of the double layer. Typical values of $1/\kappa$ range from 3.04 nm at $I = 10^{-2}$ M to 96.17 nm at $I = 10^{-5}$ M, in water at 25°C. The exact solution of Eq. [23] for symmetrical electrolytes, $z_+ = -z_- = z$, gives (51):

$$\tanh(z\psi^*/4) = \tanh(z\psi_0^*/4)\exp(-\kappa x) \qquad [26]$$

where $\psi^* = e\psi/kT$ is the reduced potential (dimensionless). For low potentials (lower than ≈25 mV), $\tanh p \approx p$, and Eq. [26] reduces to Eq. [24].

The surface charge density, σ_0, can be obtained from the integration of Eq. [20] and the condition of electroneutrality:

$$\sigma_0 = -\epsilon\epsilon_0\left(\frac{d\psi}{dx}\right)_{x=0} \qquad [27]$$

For symmetrical electrolytes this equation gives:

$$\sigma_0 = \frac{4n^0 ze}{\kappa}\sinh(z\psi_0^*/2) \qquad [28]$$

where n^0, the number of ions per unit volume in the bulk solution, is $n^0 = n_+^0 = n_-^0$. For very small potentials (Debye-Hückel approximation) a similar analysis leads to (51):

$$\sigma_0 = \epsilon\kappa\psi_0 \qquad [29]$$

and the integral capacity per unit area of the double layer, K, is given by:

$$K = \frac{\sigma_0}{\psi_0} = \epsilon\kappa \qquad [30]$$

Under these conditions, therefore, the system behaves like a parallel plate capacitor with a separation $1/\kappa$ between the plates, so the interpretation of κ^{-1} as the double-layer thickness is most obvious.

2. Modifications to the Gouy-Chapman Model: The Inner Stern Layer

The main assumptions of the previous model, i.e., that the electrolyte ions can be regarded as point charges and that the solvent can be treated as a structureless dielectric of constant permittivity, were demonstrated to be quite unsatisfactory. Simple calculations showed that even at modest potentials and electrolyte concentrations those hypotheses failed. On the one hand, the finite size of the ions, whether hydrated or not, determines their maximum concentration at the surface as well as their distance of closest approach to it. On the other hand, the high electric fields near the surface ($\sim 10^6 - 10^8$ V/m) must produce some ordering of solvent dipoles, especially for water (6). All of these circumstances were assumed by Stern (54) and later by Grahame (52), and their model (also named Gouy-Chapman-Stern-Grahame model) is the most commonly used to describe the electrical properties of colloidal suspensions.There are presently more sophisticated corrections to that model (8,55), but they are not of interest for the purposes of this work. According to the Stern-Grahame model, the electrical double layer should be divided into two main regions (see Fig. 10): an inner one (compact, or Stern layer), from $x = 0$ to $x = d$, and an external, diffuse layer, for $x > d$. The inner layer is again subdivided into two zones, limited by the inner Helmholtz plane (IHP) ($x = b$) and the outer Helmholtz plane (OHP) ($x = d$). Strongly adsorbed ions that have lost their hydration layer are located in the IHP. The charge density and potential at this plane are σ_i and ψ_i, respectively. The location of the outer Helmholtz plane is determined by the diameter of the hydrated adsorbed ions, d, the charge density and potential at this plane being σ_d and ψ_d, respectively. It is also assumed in this model that the dielectric permittivities in the two inner zones are different, ϵ_1 and ϵ_2, respectively, and both lower than the permittivity of the bulk solution, ϵ. One can visualize the structure of the edl, according to this model, as if it were composed of two parallel plane capacitors (for flat-plate geometry) with integral capacitances K_i and K_d for inner and outer Helmholtz layers, respectively, given by:

$$K_i = \frac{\epsilon_1}{b}, \; K_d = \frac{\epsilon_2}{d} \qquad [31]$$

Now we can write the potential drops across these capacitors:

$$\psi_0 - \psi_i = \sigma_0(b/\epsilon_1) \quad [32]$$

and

$$\psi_i - \psi_d = \sigma_d(d - b)/\epsilon_2 \quad [33]$$

Finally, the condition of global electroneutrality requires that

$$\sigma_0 + \sigma_i + \sigma_d = 0 \quad [34]$$

If ϵ_1 and ϵ_2 are assumed to be constant, Eqs. [32] and [33] predict a linear potential drop within each part of the inner double layer, as shown in Fig. (10b). The dependence of the electric potential with distance, in the region from $x = d$ to the bulk solution, i.e., in the diffuse part of the double layer, will be exponential if the surface potential is moderate, as predicted by Eq. [24]. In the case of higher surface potentials, the dependence is that shown in Eq. [26], substituting ψ_0 by ψ_d at $x = d$. Similarly, the surface charge density, σ_d, is related to ψ_d by an equation of the type [28].

3. The Electrokinetic or ζ Potential

The full description of the electrical double layer using the set of equations [24]–[34], requires the knowledge of six physical quantities: the charge densities at the three singular planes (σ_0, σ_i, and σ_d) and the corresponding values of the potential in those planes, ψ_0, ψ_i, and ψ_d. However, among these quantities, only σ_0 is experimentally accessible through, for example, potentiometric titration (18). Thus, indirect methods must be used for obtaining information on the other quantities. Concerning the values of the parameters in the inner layer, σ_i and ψ_i, information from adsorption experiments is needed. It must be noted that the origin of the interaction between the surface and the ions specifically adsorbed in the IHP is of nonelectrostatic nature. Nevertheless, for the purpose of calculation of the energy of interaction between particles in suspension, or between particles and additives in the suspension, it is the external part of the double layer that is important. Thus, the quantitative knowledge of σ_d and ψ_d is of practical interest. This is only possible if ψ_d is assumed to be identical to the electric potential responsible for the existence of the so-called electrokinetic phenomena, associated with the motion of both phases (solid and liquid) relative to each other. The potential at the slipping or shear plane, placed at the boundary between the mobile and inmobile phases, is known as electrokinetic or ζ potential. The most familiar electrokinetic phenomenon is electrophoresis, in which dispersed particles migrate under the action of an external electric field. Other phenomena currently used in the analysis of edl's include streaming potential (a potential difference is generated when the liquid phase moves relative to the solid by application of a pressure gradient), electroosmosis (a flow of the liquid is provoked by application of an external electric field), or sedimentation potential (a potential difference is generated when the solid phase moves relative to liquid by the action of gravity). Figure 11 shows schematically the fundamentals of the main electrokinetic techniques. In the case of electrophoresis, the quantity of interest is the electrophoretic mobility, μ_e, or particle velocity per unit applied field. From μ_e, the electrokinetic potential, ζ, can be estimated by using a suitable theoretical model of the phenomenon (see Sec. VI), but the main question about the relationship between ζ and the other potentials at the edl must still be considered. There are many interesting discussions in the bibliography about this matter (8,51). Shortly, the interpretation of all electrokinetic phenomena assumes that in the tangential liquid–solid displacement, a thin layer of unknown thickness (of the order of 10^{-8} cm) has zero fluidity, whereas beyond this stagnant layer the fluidity has its bulk value. The plane of the transition between zero and bulk fluidity is called the *slipping plane*, and ζ is identified as the potential at that plane. With all these simplifications, following Lyklema (50), the problem of identifying ζ can be rephrased as, where is the slipping plane located? At the present day experience, and according to this author, the answer is fairly simple and straight-

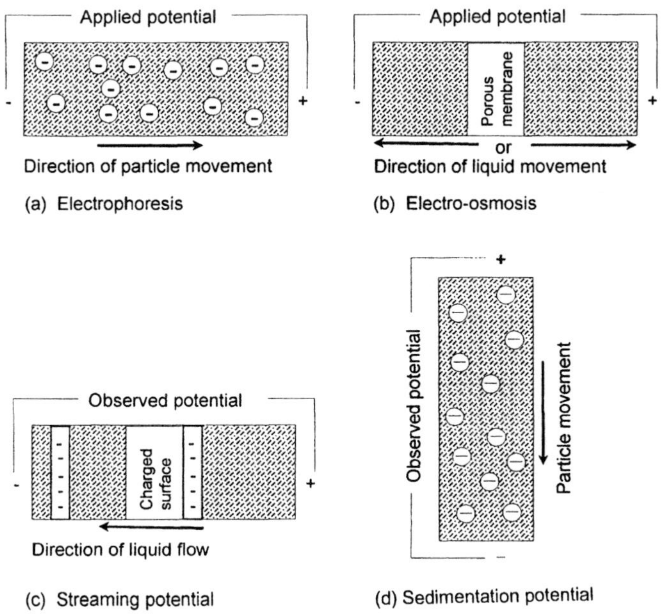

Figure 11 Schematic representation of the four main electrokinetic phenomena. (Taken from Ref. 49, with permission.)

forward: for smooth double layers in the absence of organic impurities, specially oligomeric or polymeric, the best choice is to identify the slipping plane with the OHP or, in other words, to identify ζ with ψ_d (50).

VI. EXPERIMENTAL DETERMINATION OF ζ POTENTIAL: SOME EXAMPLES OF PHARMACEUTICAL INTEREST

As mentioned in the previous section, electrophoresis is the most popular electrokinetic technique, and there are present in the market sophisticated instruments that take advantage of the latest technologies, such as laser light scattering. The procedure consists of analyzing the light scattered by the dispersed particles while they are moving by the action of an external applied electric field. The suspension is confined within a closed cylindrical capillary with platinum electrodes sealed at its ends (see Ref. 6). The complication arises that the electric field induces not only the movement of the particles but also of the liquid medium by electroosmosis at the capillary walls. There is, however, a distance from the wall of the cell where the flow of liquid in each direction cancels with the opposing flow (the cell is closed). This is the point at which one must measure the velocity of the particles because at that level (called the stationary level) they are moving at their true electrophoretic velocity with respect to the electrodes (or the laboratory reference frame).This level is usually found automatically in modern instruments.

It may happen that the nature of the suspensions is such that the light scattering method is not applicable. In such a case, a new method has been developed by the Australian group of Professors Hunter and O'Brien [Matec AcoustoSizer (6)], based on the emission of a sound wave generated by a short pulse of high frequency (in the MHz range) and its propagation across the suspension toward the detecting transducers. From the adequate analysis of the signals, the "dynamic mobility," and from it the ζ potential, of the particles can be obtained by using adequately modified forms of Smoluchowski's equation (see below; Eq. [35]) (6).

In order to obtain the ζ potential from electrophoretic mobility measurements, some theoretical model is needed.The simplest one was proposed by Smoluchowski:

$$\mu_e = \frac{\epsilon\epsilon_0}{\eta}\zeta \qquad [35]$$

where η is the viscosity of the liquid medium. This relationship is valid for rigid and nonconducting particles of arbitrary shape; it assumes that the electrical conductivity of liquid is the same within the edl and in the bulk, i.e., there is no "surface conductance." Also, the viscosity and dielectric permittivity is the same in all points of the liquid; finally, most important is that it is valid only if the double layer around the particle is very thin compared with its radius of curvature, a, at any point of the surface (i.e., $\kappa a \gg 1$). At the other extreme, when the double layer is very thick compared with the particle radius (i.e., for very small particles in dilute salt solution, where $\kappa a \ll 1$) the formula proposed by Hückel [in the 1920s (56)] is applicable:

$$\mu_e = \frac{2\epsilon\epsilon_0}{3\eta}\zeta \qquad [36]$$

For intermediate values of κa, the problem becomes more complicated. Henry (57) proposed the following solution:

$$\mu_e = \left(\frac{2\epsilon\epsilon_0}{3\eta}\zeta\right) f(\kappa a) \qquad [37]$$

where $f(\kappa a)$ is a function that varies smoothly from 1 to 1.5 as κa varies from 0 to ∞. Equation [37], however, gives satisfactory results only if ζ is low (less than about 25 mV). For higher values of ζ, the complexity increases because the equations involved do not have analytical solutions. Thus, numerical procedures are needed, and a very general one was proposed by O'Brien and White (58). The description of their procedure goes beyond the scope of this work. However, their results are worth to be shortly presented and are given in Fig. 12 (58) for suspensions of different κa values. In the figure, the dimensionless mobility $U(=3\eta e\mu_e/2\epsilon\epsilon_0 kT)$ is plotted as a function of the reduced ζ potential ($\zeta = e\zeta/kT$).

The estimation of ζ allows a follow-up of the electrical properties of the particles in a suspension as the conditions of the electrolytic medium (pH, valencies, concentrations, and mobilities of ions in solution) are modified. Also, the additives used in pharmaceutical suspensions, as thickeners or preservatives may affect the ζ potential of the particles and from such effects the action of these additives on the dispersed particles can be inferred. Let us show some results on the ζ potential of nitrofurantoin, a pharmaceutical drug already considered in Sec. IV of this work. Figure 13 shows the variation of ζ (obtained by both Smoluchowski's equation and the method by O'Brien and White) with the pH of the medium (water, without the addition of any electrolyte) (59). Prior to considering other aspects of this figure, it is worthwhile to note that the trends of variation of ζ are fairly similar no matter which model is assumed. Generally, $|\zeta|$ is always higher if estimated from the O'Brien and White method than if the Smoluchowski formula is used, as expected from the fact that the former includes the so-called electrophoretic retardation and relaxation effects (51). Only when the ionic strength of the solution is high (extreme values of pH) do both calculations of

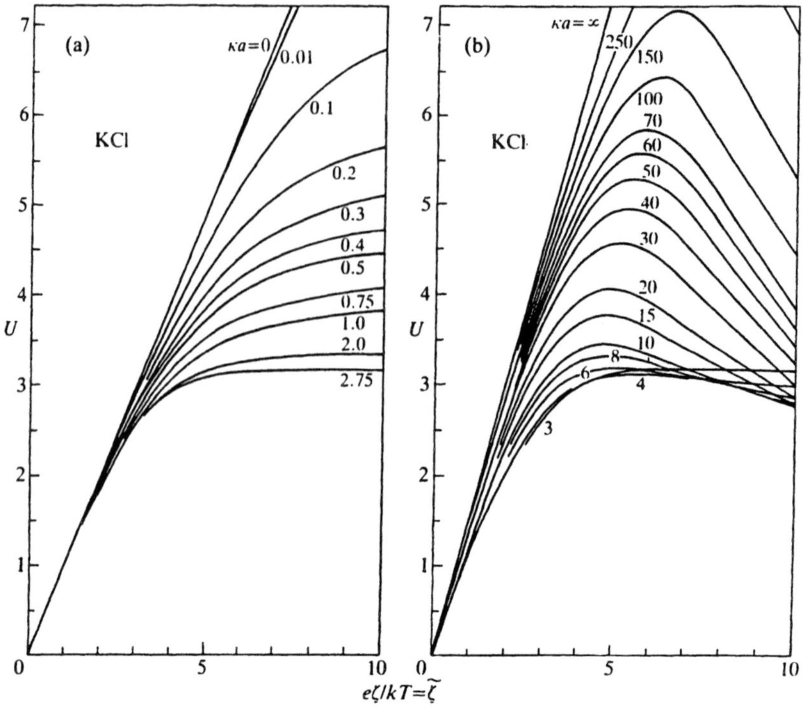

Figure 12 The dimensionless electrophoretic mobility, U, vs. dimensionless ζ potential for different values of κa. The electrolyte in solution is assumed to be KCl. (From Ref. 58, with permission.)

ζ become equal. A simple *charge generation mechanism* for nitrofurantoin can be proposed that is compatible with the data in Fig. 13. The net negative surface charge could be the consequence of the loss of the (acid in character) hydrogen atoms from the imide group of the nitrofurantoin molecule (60). Increasing pH of the medium would displace the dissociation equilibrium in the direction of increasing the release of H^+ and hence the negative charge of the particle, as observed.

The effect of the addition of inorganic electrolytes (NaCl, $CaCl_2$, $AlCl_3$) to the dispersion on the ζ potential of nitrofurantoin is shown in Figs. 14 and 15 (59). Concerning the effect of NaCl and $CaCl_2$ (Fig. 14), no important differences are observed between Smoluchowski and O'Brien and White methods of calculation of ζ. It can be observed that $|\zeta|$ decreases when the concentration of $CaCl_2$ is increased in the system, as would be expected due to the phenomenon of

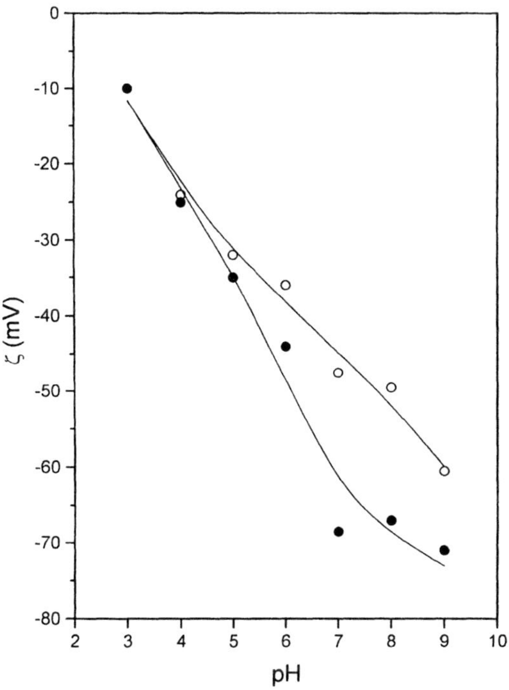

Figure 13 The ζ potential of nitrofurantoin as a function of pH. Open symbols: Smoluchowski's equation; full symbols: O'Brien and White method. (After Ref. 59, with permission.)

double-layer compression. However, $|\zeta|$ presents a maximum when the concentration of NaCl is $\sim 10^{-4}$ M. Hence, at low concentrations, $|\zeta|$ is lower for a given concentration of NaCl than for the same concentration of $CaCl_2$. This result can be considered unusual and could be attributed to the effect of a specific adsorption of Cl^- ions and the more effective screening of the surface charge by the Ca^{2+} counterions, which is more noticeable in the low concentration range (59). The effect is less apparent the lower the pH (lower concentration of OH^- competing with the Cl^- ions for the adsorption sites). On the other hand, the effect of higher acidity is a reduction of the absolute value of ζ potential, as corresponds to the neutralization of the negative charge of nitrofurantoin by the increasing amount of H+ ions. The important effect of trivalent Al^{3+} cations on the ζ potential of nitrofurantoin is clearly observed in Fig. 15, where ζ is plotted as a function of $AlCl_3$ concentration for three representative pH values. The fact that Al^{3+} is capable of changing to positive the sign of the net surface charge of

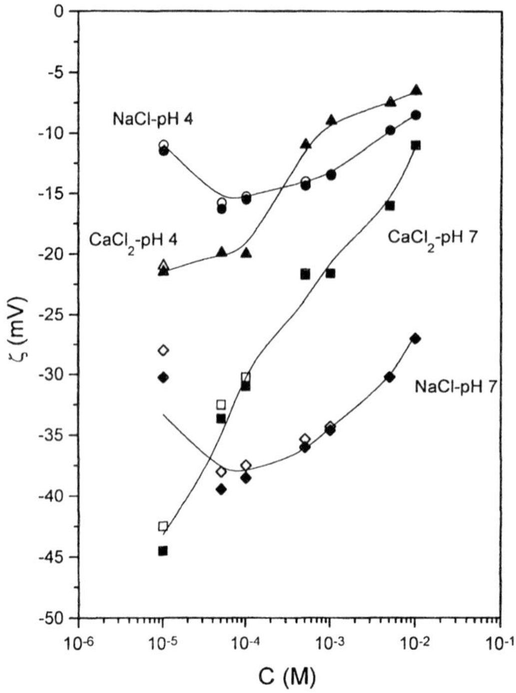

Figure 14 The ζ potential of nitrofurantoin as a function of NaCl and $CaCl_2$ concentrations at pH 4 and pH 7. Open symbols: Smoluchowski's equation; full symbols: O'Brien and White method (after Ref. 59, with permission.)

nitrofurantoin surface and even increasing it is indicative of the strong interaction of both the cation and its hydrolysis products with the drug.

Other common components in pharmaceutical suspensions are *thickeners*, substances that not only determine the bulk viscosity of the suspension but also enhance its physical stability by influencing the electrical properties of the double layer of the particles. The effect of a polymeric thickener (Carbopol 934) on the ζ potential of nitrofurantoin is shown in Fig. 16 (59). It is noticeable that the important increase of surface charge of nitrofurantoin particles after the adsorption of the polymer, in comparison with the values of ζ in the absence of it (see Figs. 14 and 15). Note also that the values of ζ predicted by the O'Brien-White model with NaCl as electrolyte are considerably higher than those predicted by Smoluchowski's formula; the high surface charge attained by nitrofurantoin in the presence of Carbopol at pH 7 would give rise to important relaxation and retardation effects (not included in Smoluchowski's model) and hence to larger

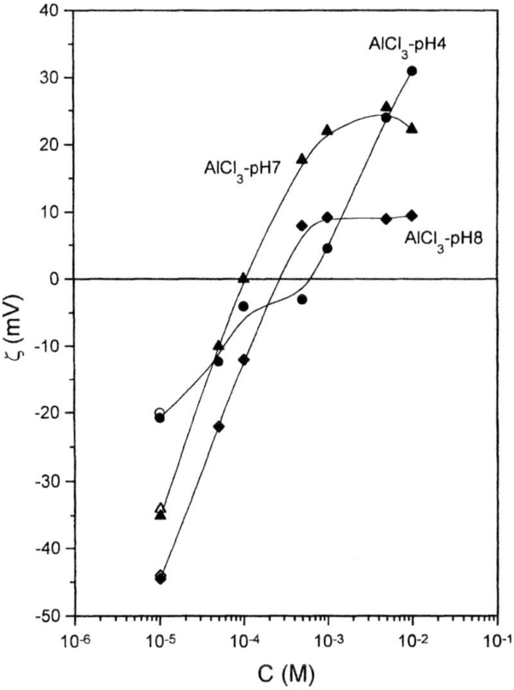

Figure 15 The ζ potential of nitrofurantoin as a function of $AlCl_3$ concentration at pH 4, 7, and 8. Open symbols: Smoluchowski's equation; full symbols: O'Brien and White method. (from Ref. 59, with permission.)

values of the magnitude of ζ. The origin of the high negative values of ζ potential of nitrofurantoin after the adsorption of Carbopol molecules is the existence of negative charge distributed along the molecular structure of the polymer, even at pH 7. It is the addition of a base to a Carbopol suspension to increase the pH that ionizes the -COOH groups of the chain, thus resulting in an increase of the negative charge in the polymeric chain. In turn, this is the responsible mechanism of the unwinding of polymer chains and hence the thickening action of this compound. Besides this effect, the increase in negative charge due to adsorption of the polymer also contributes to stabilization of the suspension by increasing electrostatic repulsion between the particles.

Like thickeners, other additives used in pharmaceutical suspensions can affect the ζ potential of the drug. Preservative substances are often used to prevent or at least minimize possible sources of deterioration in pharmaceutical preparations. Preservatives include antimicrobial agents to eliminate or inhibit the growth

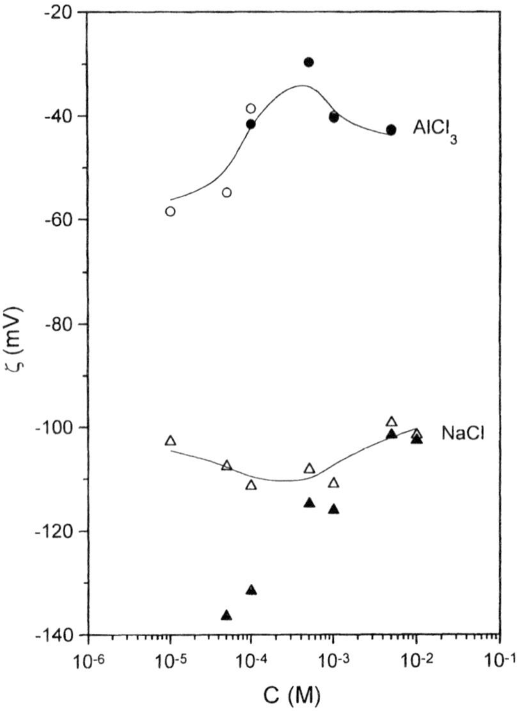

Figure 16 The ζ potential of nitrofurantoin with 0.1% Carbopol 934 at pH 7 as a function of NaCl and $AlCl_3$ concentration. (From Ref. 59, with permission.)

of microorganisms that may be present in the ingredients, or that may be added during manufacture or (more commonly) during use. Another significant source of alteration is oxidation of one or more of the ingredients upon exposure to atmospheric oxygen, or through "autooxidation," i.e., chain reactions initiated by ultraviolet radiation in the presence of traces of oxygen. These cases may require the addition of small amounts of antioxidants or photoprotectors, which in turn may have effects on the drug–aqueous solution interface and thus on the stability of the suspension.As an example, the usefulness of a group of pharmaceutical preservatives in aqueous nitrofurantoin suspensions was analyzed (61) and found that antipirin (ANT), benzoic acid (BA), and sodium metabisulfite (NaMB) were among the best choices as photoprotector, antibacterial, and antioxidant agents, respectively. The electrokinetic studies carried out with these additives on suspensions of nitrofurantoin showed (62) (see also Chapter 5, Sec. II, this book) that ANT had little effect on the surface charge of nitrofurantoin,

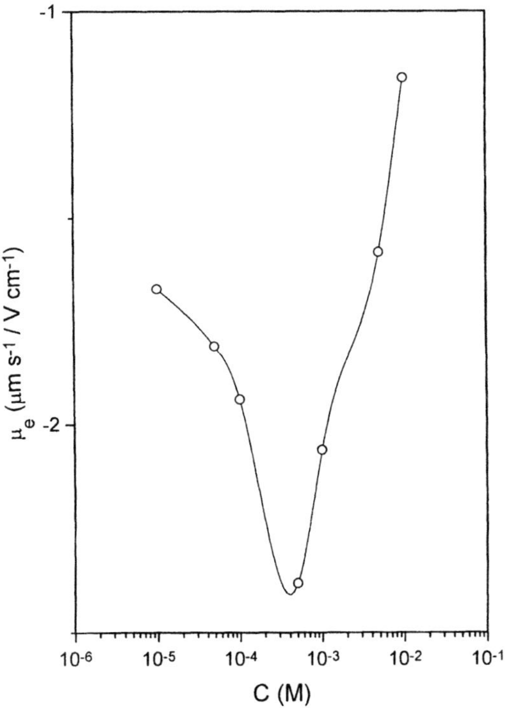

Figure 17 Electrophoretic mobility of nitrofurantoin as a function of preservative (sodium metabisulfite) concentration. (Taken from Ref. 62, with permission.)

although it may contribute to destabilization of the suspensions by lowering the electrophoretic mobility of the particles. However, BA, and to a greater extent NaMB [Fig. 17 (62)], may have considerable effects on the surface charge of the drug in aqueous suspension, mainly at intermediate concentrations where the negative electrophoretic mobility is largest.

VII. ADSORPTION OF SURFACTANTS AND POLYMERS

It has been emphasized throughout this work that one of the main tasks when dealing with the formulation of pharmaceutical dispersions (either drug suspensions or in cosmetic applications) is avoiding irreversible aggregation of suspended particles that cannot be redispersed after sedimentation. This effect, called *caking*, results in the formation of a hard, compact mass at the bottom of the container. There are several procedures to produce physically stable suspensions. Of course, a colloidally stable suspension would be the best option, as in this case the particles remain suspended in the medium by simple Brownian mechanism. This can be helped by adding dispersing agents to slow down particle aggregation. Perhaps the most widely used procedure to prevent caking is the controlled flocculation of suspended particles: a flocculated suspension contains a network of loosely aggregated particles held together by van der Waals forces or interparticle bridges. Sedimentation results in a large-volume, porous sediment that is easily redispersed by shaking to yield a homogeneous suspension. The achievement of this stability control normally requires addition of polyelectrolytes, surfactants, polymers, or other additives to the suspension. These substances can adsorb on the dispersed particles, conferring to them the specific properties required for a better stability of the suspension.

From a theoretical point of view, *adsorption at interfaces*, either solid–liquid (S/L), liquid–vapor (L/V), or liquid–liquid (L/L), is one of the topics of highest interest in the physical chemistry of colloidal systems due to the increasing number of important technological areas in which it finds application. We will only be here concerned with solid–liquid interfaces and, in particular, with the two main ways of favoring stabilization of pharmaceutical suspensions, which are the adsorption of surfactants and of polymers as additives in suspension. Let us first consider the fundamentals of these processes and later some results will be discussed.

A. Adsorption of Surfactants

Although there are numerous drug molecules with surfactant-like properties (63), relatively few surfactant molecules exhibit pharmacological effects that have been exploited for therapeutic purposes. Some ordinary characteristics of surfactants (detergency, emulsifying, solubilizing, etc.) could produce occasional nondesired effects if administered together with the drug, due to its ability to solubilize and remove lipids from biomembranes and, in general, react with proteins of all types (64). However, surfactants play an important role in pharmaceutical formulation. Thus, they are used in pharmaceutical technology as wetting adjuvants. Both ionic and nonionic surfactants are currently used with the aim of improving the wettability of essentially hydrophobic powders in aqueous suspensions. Other therapeutic and cosmetic applications (including their use as germicides, antiseptics, hair conditioners, etc.) are described for a detailed list of surfactants in, for example, Ref. 64.

The adsorption of molecules at an S/L interface creates a region of transition, of the order of molecular dimensions, in which the composition changes from the bulk solid phase to that of the bulk liquid. From a thermodynamic point of view, the adsorbed molecules modify the surface tension (surface free energy in the case of solids) of the surface as a result of its accumulation on the substrate. The ''excess'' of concentration of the adsorbed molecules on the solid substrate (see Sec.II), also named ''adsorption density,'' Γ_1, (strictly surface excess concentration, in mol/m^2; see Ref. 20) is related to the change of surface tension γ by means of the Gibbs adsorption isotherm:

$$\Gamma_1 = -\frac{a}{RT}\frac{d\gamma}{da} \qquad [38]$$

where a is the activity of solute in the solution and R is the universal gas constant. The spontaneous adsorption implies that $d\gamma/da$ is negative, and the sign in Eq. [38] means that in such a case, $\Gamma_1 > 0$, i.e., the concentration of solute in the interface is higher than in the bulk solution. In practice, Eq. [38] allows the determination of the density of adsorption, Γ_1, at L/V and L/L interfaces by the direct measurement of surface (interfacial) tension as a function of the concentration of electrolyte. In the case of S/L interfaces, the usual measurement is the density of adsorption and thus Gibbs' equation gives information on the changes of surface free energy resulting from the adsorption process. It should be noted here that not all of the molecules that are adsorbed at the interfaces and hence reduce their surface free energy can be considered surfactants. There are some well-defined characteristics of the molecules belonging to a substance that can be considered as surfactant. According to most authors, a ''functional surfactant'' is any substance of fairly high molecular weight that possesses a polar region, generally called ''head,'' and a nonpolar region, called ''tail,'' in the same molecule, which is hence of amphiphilic nature. Other common characteristics of surfactants are (64) as follows: (a) their solubility in at least one phase of the system; (b) the fact that they form oriented monolayers at the interfaces; (c) the fact that they exhibit interface equilibrium concentration higher than those in the bulk solution and form micelles at specific concentration; and (d) the fact that they exhibit one or more of the following characteristics: detergency, foaming, wetting, emulsifying, solubilizing, and dispersing.

It is very important to understand the nature of the interactions between surfaces and adsorbed surfactants in order to optimize their effects in the pharmaceutical and biomedical applications.The *mechanisms* of adsorption of surfactants at the S/L interface are diverse and can be one (or more) of the following (17): (a) ion exchange, i.e., substitution of previously adsorbed ions on the solid by surfactant ions of identical charge; (b) ion pairing, i.e., adsorption of surfactant ions on the surface sites of opposed charge not occupied by counterions; (c) acid–base interactions, mainly hydrogen bonds; (d) adsorption by polarization of π electrons, i.e., if the adsorbate has aromatic rings in its molecule and the adsorbant possess strongly positive sites; (e) adsorption by dispersion forces occurs via London-van der Waals dispersion forces between adsorbent and adsorbate molecules; (f) hydrophobic bonding, i.e., the molecules are adsorbed onto or adjacent to other molecules already adsorbed on the solid adsorbent. Fig. 18 (17) shows schematically the diverse mechanisms described above.

The quantitative analysis of the adsorption of a surfactant onto a solid surface is made by means of *adsorption isotherms*, which relate the amount of surfactant adsorbed per unit area of solid with the solution surfactant concentration required to produce a given surface coverage or degree of adsorption. It is also assumed that temperature and pressure are held constant. Of course, the adsorp-

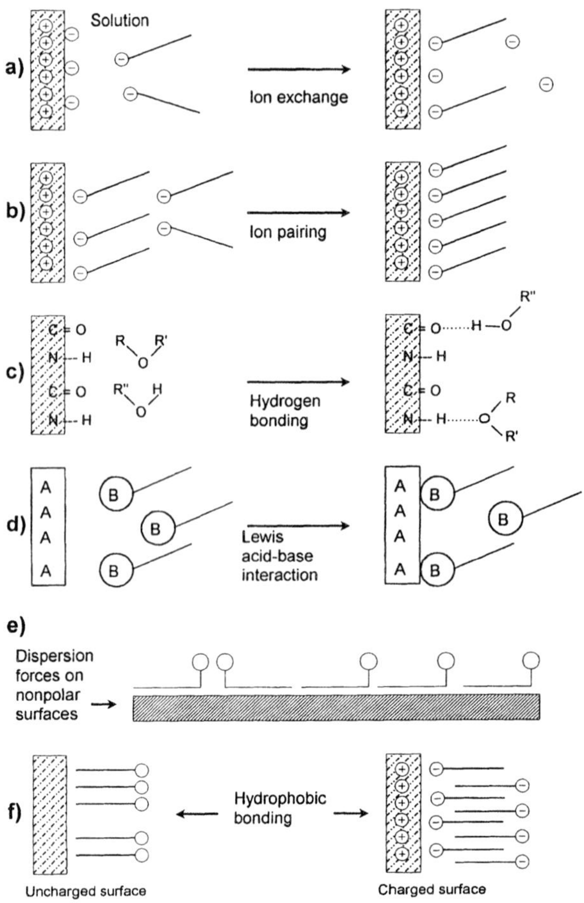

Figure 18 Schematic representation of the different mechanisms of adsorption of surfactants on solid surfaces. (Taken from Ref. 17, with permission.)

tion isotherm is obtained experimentally in each particular case and provides information on the specific mechanisms involved, which in turn are dependent on the nature of both the solid substrate and the molecules adsorbed. Thus, it is possible to get information on the orientation of the adsorbed molecules relative to the surface and solution, or on the surfactant concentration at which surface saturation occurs. A type of adsorption isotherm commonly observed in adsorption from solution of surfactants is the Langmuir-type isotherm, expressed by (65):

$$C_l^s = \frac{C_m^s C_l}{C_l + a} \qquad [39]$$

where C_l^s is the surface concentration (in mol/m^2), C_l the molar concentration of surfactant at adsorption equilibrium in the liquid phase, C_m^s the surface concentration of the surfactant (mol/m^2) at monolayer adsorption, and a is a constant (mol/L). This isotherm assumes that (17,66): (a) the adsorbent is homogeneous, (b) both solute and solvent have equal molar surface areas, (c) both surface and bulk phases exhibit ideal behavior (e.g., no solute–solute or solute–solvent interactions in either phase), and (d) the adsorption film is monomolecular. Many surfactant solutions show Langmuir-type behavior, even when these restrictions are not met, because of the mutual compensation of the several factors that affect the shape of the Langmuir isotherm. These factors are diverse and include (67) micellization of the surfactant, sign of the surface potential as compared to that of the adsorbed ion, and, most important, the heterogeneity of the solid adsorbent: the summation of adsorption onto sites of varying energy may yield an isotherm resembling a BET (multilayer isotherm) or a Freundlich isotherm ($n_l^s = kC_l^{1/n}$, where k and n are constant, with n generally greater than unity and n_l^s the number of moles of adsorbate per unit mass of adsorbent at adsorption equilibrium). Another current factor that determines the shape of the isotherm (S shape is usual) is the existence of lateral interactions between the chains of the adsorbed surfactant. It should be noted here that adsorption isotherms are currently classified according to the above characteristics and the catalog of their denominations is wide. The reader is referred to specialized texbooks (see Refs. 68–70).

1. Adsorption and Modification of Electric and Energetic Surface Properties

The ability of surface-active agents to adsorb on solid-liquid interfaces and produce a modification of the interfacial properties depends on the chemical nature of the three mutually interacting components of the system: the substrate, solid particles; the adsorbate, surfactant molecules; and the liquid medium, usually an aqueous solution. If the nature of the solid substrate is nonpolar, then the adsorption takes place by dispersion force interactions. The orientation of the adsorbed molecules is such that the hydrophobic groups of the chain become associated with the surface while the hydrophilic head orients toward the aqueous phase. The mechanism of this type of adsorption implies that in the first steps the hydrophobic chains will be lying on the surface much like trains or L's (see Fig. 18e). Further adsorption forces the molecules to a progressively perpendicular orientation to the surface such that when saturation is reached an approximated close-packed assembly will result (Fig. 18f). Saturation takes place at a concentration near the CMC of the surfactant. The changes of conformation of the adsorbed molecules, from L-shaped to a more perpendicular arrangement, may be followed up in practice by the appearance of some inflexion point in the isotherm. The adsorption process is usually limited to the formation of a single monolayer, given that the orientation of the hydrophilic groups toward the aqueous phase avoids the formation of a second adsorbed layer.

Adsorption may also take place on *polar, uncharged surfaces*, including natural and most synthetic polymeric materials (polyesters, polyamides, polyacrylates). This is a case of higher technological importance than the previous one. The molecular mechanism is also more complex and the orientation of the adsorbed molecules on the surface is the result of the balance of several forces: dispersion forces, dipolar interactions, and hydrogen bonding and other acid–base interactions. If dispersion forces predominate, then the adsorption mechanism is similar to that of nonpolar surfaces. If, on the contrary, polar interactions are the most intense, adsorption may take place in a reverse mode, i.e., with the hydrophilic head oriented to the surface and the hydrophobic chain toward the aqueous phase. In aqueous systems, the final orientation of the molecules is also dependent on the relative intensity of the solvent–adsorbent and solvent–adsorbate interactions.

The third type of adsorbent surfaces (*surfaces having discrete electrical charges*) is the most complex for several reasons. First, adsorption may take place through any of the mechanisms already cited. Second, adsorption involving electrostatic interactions is significantly more sensitive to external conditions such as pH, presence of polyvalent cations or of non-surface-active cosolutes. Materials with charged surfaces include a wide variety ranging from inorganic compounds of technological importance (oxides like silica, alumina, titania, zinc oxide, clay minerals, etc.), or latex polymers containing ionic comonomers, to many natural surfaces such as proteins, cellulose, etc. The predominant mechanisms of adsorption on those surfaces also range from ion exchange through ion bonding to dispersion or hydrophobic interactions. As a result, adsorption isotherms may be much more complex than those for the simpler systems. Further information on this topic may be found in Refs. 49 and 71.

Most substances usually employed in pharmaceutical formulations belong either to the group "polar, uncharged surfaces" or to the group of substances that ionize after immersion in the liquid medium, i.e., surfaces with discrete electrical charge. Frequently, the pharmaceutical component is designed to modify the surface properties of a given biological interface. This is the case of the system cholesterol–bile salt solution, already mentioned in this chapter (Sec. IV). In this system, the solid cholesterol surface plays the role of adsorbant substrate of "electrically uncharged nature." The important dipole moment of cholesterol molecules (6.6×10^{-30} C · m in dioxane at 25°C; see ref. 72) must play a fundamental role in the formation and structuring of the cholesterol–aqueous solution interface. After the adsorption of bile salts, cholesterol surface becomes modified as to its electrokinetic and surface free energy properties (see Sec. IV). The mechanism of adsorption of the bile salts is via dipole–dipole interactions and it is the combined effect of adsorbed water dipoles and ions coming from solution that gives rise to an electrical double layer, responsible for the electrical properties of the cholesterol–solution interface (73). In Fig. 19 is shown the effect of the adsorption of sodium cholate on the ζ potential of cholesterol at different pH values of the suspension. The figure makes clear the significant role played by H^+ and OH^- ions, which can be considered as potential determining ions for cholesterol

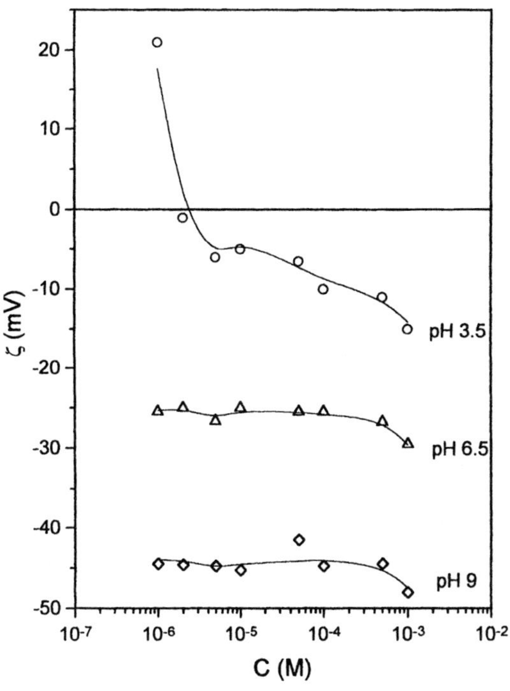

Figure 19 The ζ potential of cholesterol at 10^{-2} M NaCl aqueous solutions as a function of sodium cholate concentration at different pH values. (From Ref. 73, with permission.)

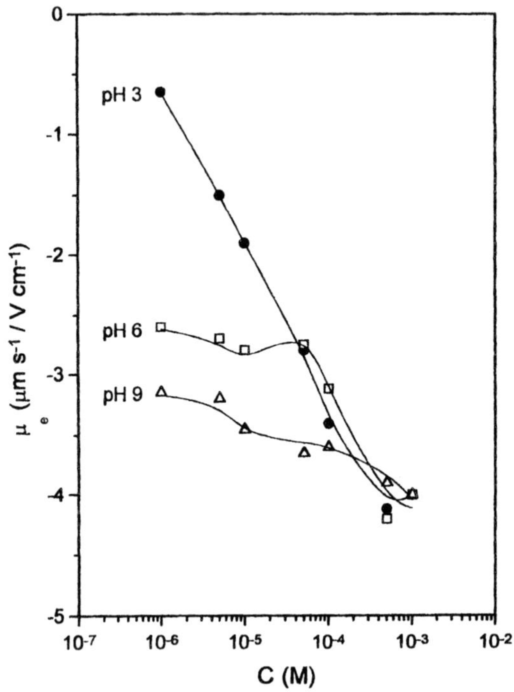

Figure 20 Electrophoretic mobility of a latex (Aquacoat) as a function of sodium dodecyl sulfate concentration at different pH values. (Taken from Ref. 77, with permission.)

surface. The modification produced by the adsorbed sodium cholate molecules is less evident, although a slowly increasing (in absolute value) trend of ζ with cholate concentration is observed. A competition between the adsorbed water and bile acid molecules [with dipole moments larger than that of water (74)] would have the net effect of increasing the negative electrokinetic charge associated with the presence of ion-determining ions around the resulting structure. The effect is more noticeable at low pH (Fig. 19) where, according to Fini et al. (75), at least 97% of the bile salt is in free acid form. Under such conditions the ionic (or free charge) contribution to the potential is negligible and the main effect in the modification of the ζ potential must be attributed to the presence of dipolar molecules of bile acid (73).

A more complex substrate is Aquacoat, a pharmaceutically used polymer obtained by etherification of natural cellulose in alkaline medium with ethyl chloride and further treatment with 1.3% sodium dodecyl sulfate (SDS) and 2.5% cetyl alcohol to enhance the colloidal stability of the resulting latex (76). The

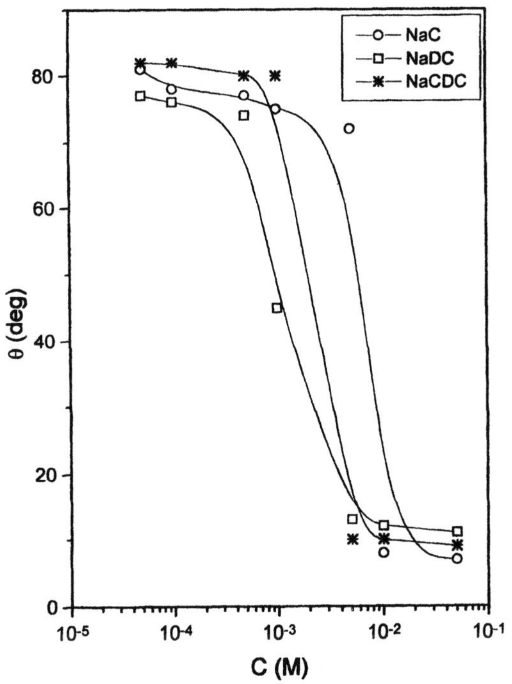

Figure 21 Contact angles, θ (degrees), of the bile salts (sodium cholate, NaC; sodium deoxicholate, NaDC, and sodium chenodeoxicholate, NaCDC) on the cholesterol surface precontacted with the solution, as a function of bile salt concentration. (After Ref. 47, with permission.)

surface of this polymer thus contains a mixture of strong and weak acid groups that are mainly responsible for the charge generation at the interface. The effect of adsorption of the surfactant SDS, on this polymer was studied. As shown in Fig. 20 (77), the electrophoretic mobility of Aquacoat is strongly affected by SDS concentration. Noticeable is the important role of H^+ and OH^- ions in the determination of its (negative) electrokinetic mobility. As observed, whatever the initial negative charge on the particles, the increase in SDS concentration (up to $\approx 5 \times 10^{-4}$ M) gives rise to a higher negative mobility of the polymer particles. At a concentration of surfactant [SDS] $\approx 10^{-3}$ M, a surface saturation is reached and the growing trend of μ_e with [SDS] seems to level off.

It has already been mentioned that, besides electrostatic effects, the interaction between a solid substrate and a surfactant has important consequences on the *wettability* of the surface. In particular, in pharmaceutical formulation it is common to use wetting agents that modify the surface free energy of the surface until it is made wettable by the liquid medium. Although this effect has been treated in Sec. IV, here it will be reconsidered from the more usual perspective of the contact angle, θ, between a liquid solution and a surface, after modification of the latter by adsorption of the wetting agent. Changes in θ can be understood from both Gibbs's and Young's equations (Eqs. [13] and [38]). The latter can be rewritten as $\cos \theta = (\gamma_S - \gamma_{SL})/\gamma_L$, where one can see that θ will decrease if either γ_{SL} or γ_L, or both, is reduced and γ_S remains essentially unchanged. The process of adsorption of bile salts on cholesterol provides a good example of how the hydrophobic cholesterol surface becomes hydrophilic after being wetted by these biological surfactants. This is shown in Fig. 21 (47), where the contact angle of drops of some bile salt solutions on cholesterol are plotted as a function of their concentration. The wetting effect is noticeable only after some critical concentration is reached, where a sharp decrease of θ occurs, changing from $\approx 80°$ to $\approx 10°$. Higher concentrations of bile salt do not modify substantially the wettability of the surface.

B. Adsorption of Polymers

Both synthetic and natural polymers can behave as complex surfactants, playing an important role in the mechanisms of stabilization or flocculation of suspended particles. Synthetic polymers are designed to fulfill some specified function. Thus, homopolymers (built from a single monomer) or random copolymers (obtained from different monomers randomly arranged within the polymer chains) are used as flocculant agents, while block or graft copolymers (with two types of polymeric chains, one insoluble in the dispersion medium, attached to the solid surface, and another soluble in the dispersion medium) are currently used as stabilizing agents (78,79). The study of the adsorption of polymers at solid–liquid interfaces forms a part of the general field of polymer chemistry. For the purpose of discussion of the effects of some currently used polymers in the formulation of pharmaceutical suspensions, only some specific points, dealing with adsorption of polymers on solids, will be presented here. First, the polymers should be *soluble*. This depends on the relative contribution of the interaction energies segment–segment of the polymer, ϵ_{22}, segment–solvent, ϵ_{12}, and solvent–solvent, ϵ_{11}. The difference between a good or a poor solvent may be quantified by means of the parameter χ, related to the enthalpy of polymer-solvent mixing, defined by Flory as (80,81):

$$\chi = \frac{[\epsilon_{12} - \frac{1}{2}(\epsilon_{11} + \epsilon_{22})]z}{kT} \qquad [40]$$

where z is the coordination number. It is accepted (18,80) that in a "good" solvent the polymer chain will expand to increase the polymer–solvent contact, whereas in a "poor" solvent the polymer coil will contract, reducing the number of polymer-solvent links. Of course, the extent of these effects will be also determined by the entropy of mixing, which depends on the polymer molecular weight *M* and concentration *c*. Another way to face this question is by considering the osmotic pressure, π, of the solvent, given, in a semidilute polymer solution by (18,80):

$$\frac{\pi}{c} = \frac{RT}{M}\left[1 + k\left(\frac{1}{2} - \chi\right)c\right] \quad [41]$$

where k is a constant. When $\chi = 1/2$ the polymer solution behaves ideally. For $\chi > 1/2$, at fixed c, π decreases and the polymer coil collapses (poor solvent), but for $\chi < 1/2$, π increases and the coil swells (good solvent). Thus, polymer chains can show either positive or negative deviations from ideal behavior. For dilute polymer solutions, there is a temperature at which the osmotic pressure behaves as in an ideal solution. This is the "θ temperature" (then, according to Eq. [41], $\chi = 1/2$). As Flory (81) pointed out, under "θ solvency" conditions, polymer molecules of, say, a million molecular weight behave as if they were ideally small molecules. Another important condition for an efficient adsorption of polymers on solid surfaces is avoiding "depletion" effects, i.e., those phenomena that tend to exclude the polymer segments from the interfacial region. The extent of these phenomena (adsorption vs. depletion) depends on the net enthalpy of adsorption parameter, χ_s, resulting from the balance of the surface–polymer and surface–solvent energies of interaction, schematically shown in Fig. 22. χ_s is defined as (80,82):

$$\chi_s = \frac{[\epsilon_{2s} - \epsilon_{1s} + ½(\epsilon_{11} - \epsilon_{22})]}{kT} \quad [42]$$

where ϵ_{1s}, ϵ_{2s} represent, respectively, the solvent-solid, and segment-solid interaction energies [see Fig. 22 (80)]. If $\chi_s > \chi_{sc}$ (critical enthalpy for adsorption) then adsorption will occur, otherwise polymer segments will be depleted at the interface with respect to the bulk solution.The problem arises in predicting the χ values from ϵ_{ij}, which warrants further research. Polydispersity of polymeric solutions is another point to be taken into account when dealing with their adsorption onto surfaces. In practice, the distribution of molecular weights makes the polymeric solution to be treated as a multicomponent system. Recent works have stressed the use of at least relatively narrow molecular weight fractions (20). Kinetic effects in polymer adsorption are also worthy of attention in practical applications. Due to the large number of possible configurations at the solid–solution interface, the adsorption equilibrium can be exceedingly slow to attain; adsorption that appears to have leveled off after an hour or two may actually continue for days or months. In practice, the above cited distribution of polymer

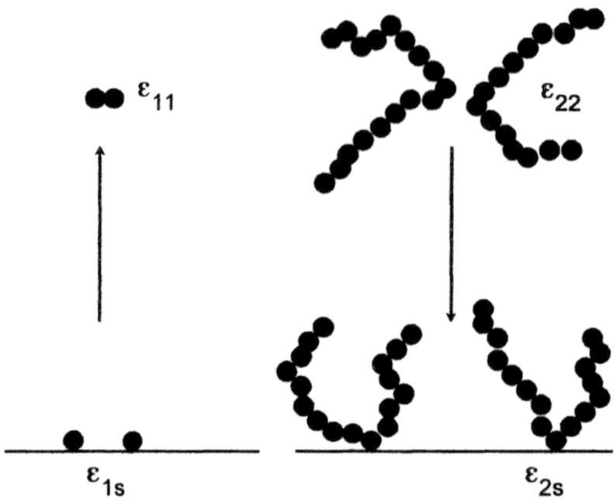

Figure 22 Schematic representation of polymer–polymer and polymer–substrate interactions. χ_S is the net energy difference between the process in the right-hand diagram and that in the left-hand one. (From Ref. 80, with permission.)

weights usually plays a fundamental role in this effect (83). Thermodynamically, the adsorption of the largest polymers is favored because of the higher number of surface contacts per molecule. The smaller polymers, on the contrary, have a higher diffusion coefficient and they will be the first to reach the surface. Thus, in the initial state, the shorter polymer molecules are preferentially concentrated at the surface while at equilibrium it is the larger polymers that reside at the surface. Furthermore, larger polymers can replace the shorter ones and this process is slow because it requires desorption of the shorter chains. Both the transport to the surface and surface rearrangements of polymeric chains justify the long time needed to reach the adsorption equilibrium.

The direct basic experiments to obtain information on the final arrangement of polymer on the substrate are the adsorption isotherms. A number of detailed approaches have been made to describe the adsorption and conformation of polymers at the interfaces, as a function of the several parameters involved in these processes, i.e., concentration of polymer in solution, molecular weight, polymer chain structure, polymer–solvent interactions, temperature, etc. There are treatments of different degrees of approximation, including those based on a lattice model (84–87), the so-called random walk approaches (88,89) and others of a mechanostatistical nature (90–92). In view of the sophistication of the various treatments, it is almost embarrassing, as mentioned by Adamson (20, p. 432), that most polymer adsorption data fit the simple Langmuir equation, within experimental error.

VIII. STABILITY OF SUSPENSIONS

A. Interactions Between Particles in Suspension: Classical and Extended DLVO Approaches

Once the particles have overcome the energetic barriers and attained the physical state of a dispersed system, the next question concerns the stability of such a system. Different kinds of forces can be considered between the particles, and between the particles and the solvent molecules, either of attractive or repulsive nature. Attractive interactions will favor the coagulation in the suspension, while repulsive ones will create energetic barriers between the particles, hindering their mutual approximation and thus coagulation. This means that one could control the stability of a suspension by means of the adequate knowledge of the nature and mechanisms of the interactions taking part between the different components in the suspension, as well as their dependence on such parameters as concentration, temperature, pH of the solution, etc.

The interactions between identical solid particles in a liquid medium can be one or more of the following: (a) attractive, Lifshitz-van der Waals (LW), (b) repulsive, electrostatic interactions (EL) between the electric double layers surrounding the particles in suspension, (c) solvation, or structural forces, coming from the more or less structured layer of solvent molecules around the solid particles. Interactions of this type can also be of either attractive (hydrophobic) or repulsive (hydration pressure, usually between hydrophilic particles) nature. Hydrophilic repulsion is usually of very short range and hydrophobic attraction can be of even higher range than the van der Waals one.

The theory that first accounted for the effect of van der Waals and electrostatic interactions on the stability of particles was proposed independently by Derjaguin and Landau and then by Verwey and Overbeek in the 1940s (93,94). The theory is now classical and is known as DLVO theory in their honor. During the last two decades, a number of experiments and new theoretical approaches (13,95,99,100) addressed the completion of the DLVO theory including the third type of interaction (c above), although their origin was not always easy to understand, as it quantitatively accounted for their effects. These approaches can be classified under the common name of extended DLVO theories.

From the surface thermodynamics point of view, the treatment proposed by van Oss et al. (11,13) gave the possibility of a quantitative estimation of solvation forces, either of hydrophobic or hydrophilic nature (see Sec. II). In this approach, the free energy of interaction (ΔG) between the particles consists of three components: electrostatic, Lifshitz-van der Waals, and acid-base:

$$\Delta G = \Delta G^{\mathrm{EL}} + \Delta G^{\mathrm{LW}} + \Delta G^{\mathrm{AB}} \qquad [43]$$

In addition to kinetic contributions, it is the dependence of each term of Eq. [43] on the distance H between the particle surfaces that determines the stability of the systems.

The electrostatic term was well established by the classical DLVO theory. Assuming constant, moderate surface potential, a reasonable expression for spherical particles is (18):

$$\Delta G^{\mathrm{EL}} = 2\pi\epsilon\epsilon_0 a\psi_0^2 \ln[1 + \exp(-\kappa H)] \qquad [46]$$

where a is the radius of the spherical particles, κ the reciprocal of the Debye length, ϵ the relative dielectric constant of the medium, ϵ_0 the permittivity of vacuum, and ψ_0 the surface electric potential, which can be assumed to be equal to the electrokinetic, or ζ, potential.

The Lifshitz-van der Waals attractive energy of interaction between the same spheres can be computed from (96):

$$\Delta G^{\mathrm{LW}} = -\frac{A}{6}\left[\frac{2a^2}{H(4a + H)} + \frac{2a^2}{(2a + H)^2} + \ln\frac{H(4a + H)}{(2a + H)^2}\right] \qquad [45]$$

where A is the Hamaker constant, dependent on the characteristics of both the particles and the liquid medium. Typical values of A range between 10^{-21} and 10^{-20} J for a large number of colloidal systems. It has been demonstrated, even experimentally (97,98), that there must be some relationship between the value of A for a given dispersed system and the LW contribution to the total interfacial energy of the particles. In fact, it was shown (99) that

$$A = 24\pi H_0^2\gamma_{12}^{\mathrm{LW}} \qquad [46]$$

where H_0 is the equilibrium separation distance between the interfaces. According to van Oss et al. (11), the most reasonable value estimated for H_0 is 1.58 ± 0.08 Å, as deduced from the application of Eq. [46] to a large number of systems. $\gamma_{12}^{\mathrm{LW}}$ is the LW contribution to the interfacial tension of the solid material (1)/solution (2) system, and it can be written in terms of the LW contributions to the surface free energy of the individual materials, both experimentally accessible (see Eq. [7]), as

$$\gamma_{12}^{\mathrm{LW}} = (\sqrt{\gamma_1^{\mathrm{LW}}} - \sqrt{\gamma_2^{\mathrm{LW}}})^2 \qquad [47]$$

The dependence of the structural forces of interaction on distance between the particles is made by most authors using the empirical exponential function (11,95,99,100):

$$F^{\mathrm{S}}(H) = F_0^{\mathrm{S}} \exp(-H/\lambda) \qquad [48]$$

where F_0^{S} is related to the wettability of the surface, positive in the case of repulsion (hydrophilic surfaces) and negative in the case of attraction (hydrophobic surfaces). The parameter λ is the correlation length of the water molecules. If the latter were not hydrogen-bonded to each other, the theoretical value for λ would be close to 2 Å, the order of magnitude of the radius of gyration of single water molecules (101). But it is admitted that about 10% of water molecules are engaged in hydrogen bonds, so that a reasonable estimate for the mean value of λ in the case of water appears to be close to 1 nm for hydrophilic surfaces and $\approx$0.6 nm for the hydrophobic case in ionic solutions (13,99). Nevertheless, the value of λ for water is not free of relative uncertainty, so values of λ between 0.6 and 1.0 nm have been reported for hydrophilic repulsion, and 1–2 nm (even higher), for hydrophobic attraction (95, p. 277–283).

According to van Oss et al.'s surface thermodynamic approach, these non-electrostatic forces must be related to the acid–base component of the interfacial tension between materials 1 and 2. The expression for the AB term in Eq. [43] is finally (between two spheres of radius a) (13, p. 80):

$$\Delta G^{\mathrm{AB}} = -2\gamma_{12}^{\mathrm{AB}} \pi a \lambda \exp\left(\frac{H_0 - H}{\lambda}\right) \qquad [49]$$

where $\gamma_{12}^{\mathrm{AB}}$ can be computed from Eq. [7]. The factor $\Delta G^{\mathrm{AB}}(H_0) \equiv -2\gamma_{12}^{\mathrm{AB}}$ measures the strength of the acid–base interaction between particles of material *1* immersed in the material 2, at the distance of equilibrium H_0. The higher its absolute value the stronger the repulsion [for $\Delta G^{\mathrm{AB}}(H_0) > 0$] or attraction [for $\Delta G^{\mathrm{AB}}(H_0) < 0$] between the particles.

B. Energy-Balance Diagrams

From Eqs. [43]–[49] and the values of the quantities relative to the particular conditions of a dispersed system, such as average particle size, ζ potential, ionic strength of the solution, and surface free energy components of dispersed solid and liquid, it is possible to build the diagrams showing the total free energy of interaction between the particles in a suspension as a function of the distance, H, between them.These diagrams allow the thermodynamic prediction of the stability of the suspension. A typical diagram including the electrostatic and van der Waals interactions (DLVO approach) between particles is shown in Fig. 23. Curves labeled a, b, and c correspond to increasing concentrations of electrolyte. It can be observed that the strength of the electrostatic interaction diminishes with increasing electrolyte concentration as a consequence of the compression of the electric double layer (see Eq. [44]): Curve a shows the existence of an energy barrier avoiding the coagulation of the particles; hence the suspension would be stable. On the contrary, in curve c there is a potential well, as the attractive van der Waals forces are more intense than those of repulsion between electrical double layers. This enhances coagulation and the suspension becomes

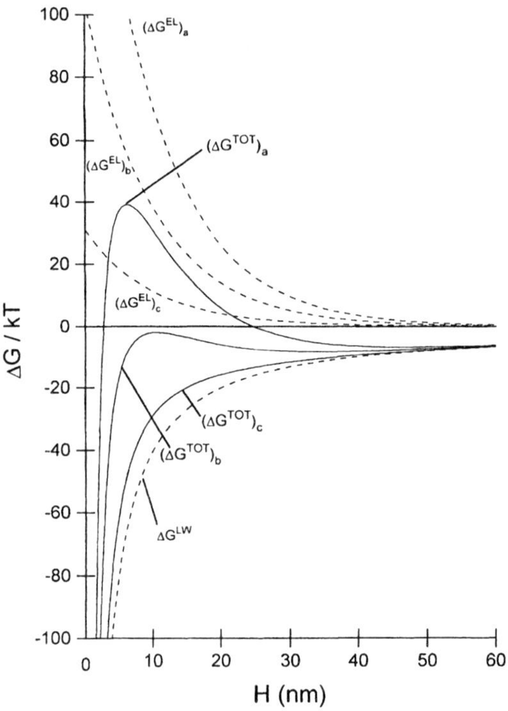

Figure 23 Total free energy of interaction (classical DLVO), ΔG (in kT units), for (a) a stable, (b) a slightly unstable, and (c) an unstable sol. Case (b) corresponds to the critical coagulation concentration.

unstable. In curve b the concentration of electrolyte is such that the total energy of interaction between particles is zero or negative for any distance of separation between them. This corresponds to the so-called critical coagulation concentration (CCC), above which stability of the suspension disappears (see below).

If non-DLVO contributions are included, the energy diagrams can show significant changes and predictions can be modified significantly. The shape of the curves will be as shown in Fig. 24. When hydrophilic acid–base repulsion exists (Fig. 24a), the extended theory predicts stability of the suspension, even at high electrolyte concentrations, contrary to classical DLVO calculations (see Fig. 23). The existence of hydrophobic acid–base attraction (Fig. 24b) between the particles manifests in instability of the suspension even for low electrolyte concentrations. The predictions of the extended DLVO approach have been verified in recent works on suspensions of hydrophilic particles of clays, like hectorite (99), or inorganic particles of zinc sulfide (100), and also in suspensions of hydrophobic particles of the ethylcellulose latex, Aquacoat (102). Experimental results obtained from direct measurements with the surface force apparatus have also confirmed the important effect of the "solvation" forces in the interactions between surfaces of either hydrophilic or hydrophobic nature on the stability of such suspensions (95, p. 277–286).

C. Kinetic Aspects of Suspension Coagulation: Stability Ratio

The velocity at which a suspension coagulates is also of interest in the formulation of a pharmaceutical suspension. It can be characterized by means of the stability ratio, W, a measure of the effectiveness of the potential barrier in preventing the particles from coagulating, defined by:

$$W = \text{(number of collisions between particles)}/\text{(number of collisions resulting in coagulation)}$$

The velocity of coagulation in the absence of a potential barrier (fast coagulation rate, R_f) is only limited by the velocity of Brownian diffusion of the particles. If the particles in suspension are under the effect of an energetic barrier, then the coagulation rate, R_s, would be slower. A simple theoretical expression for W is given by (18,103):

$$W = \frac{Rf}{R_s} = 2\int_2^\infty \exp(\Delta G/kT)\frac{ds}{s^2} \qquad [50]$$

where $s = (H + 2a)/a$ and ΔG is the total (either classical or extended) potential energy of interaction between spherical particles. Equation [50] assumes that coagulation is considered in the very early stages of the process, i.e., before formation of doublets, triplets, etc.

There are several experimental methods for the determination of W as a function of electrolyte concentration (103,104). One of the most frequently used procedures is based on the measurement of the time variation of the turbidity of the suspension, starting from the moment of addition of electrolyte to the suspension. According to Jayasuriya et al. (105), W can be determined by means of the following equation:

$$W = \frac{(d\tau/dt)_f}{(d\tau/dt)_s} \qquad [51]$$

where τ is the turbidity of the suspension, and the subscripts f and s stand for "fast" and "slow" coagulation respectively. In Eq. [51] the slopes of the turbidity vs. time curves must be computed for $t \to 0$, i.e., in the initial seconds of the coagulation process. A plot of log W as a function of electrolyte concentration

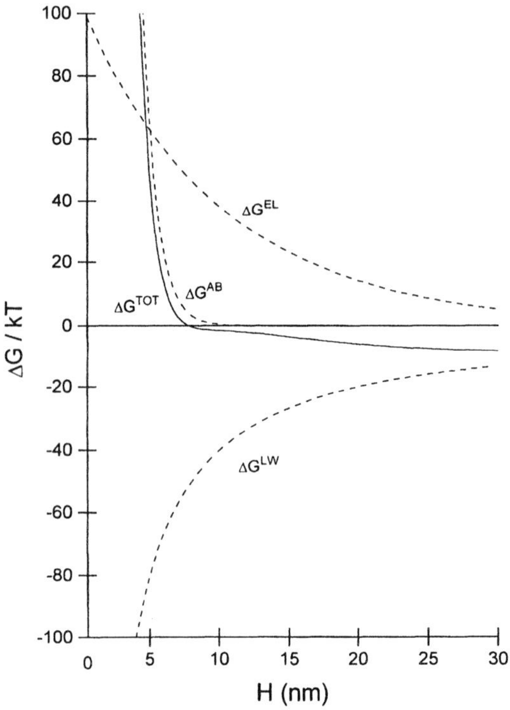

(a)

Figure 24 Total free energy of interaction between solid colloidal particles inmersed in solution, obtained as a sum of three contributions: electrostatic (EL), Lifshitz-van der Waals (LW), and acid–base (AB), following the extended DLVO model. (a) Spherical hydrophilic particles of radius 200 nm in 10^{-3} M solution of indifferent 1:1 electrolyte and neutral pH; ζ potential 22 mV; Hamaker constant $A = 10^{-19}$ J; and $\Delta G(H_0) = 55.4$ mJ/m^2. (b) Identical hydrophobic particles but in this case $\Delta G(H_0) =$ -30 mJ/m^2.

in the suspension clearly shows two regions (see below, Fig. 25): (a) for concentrations lower than the CCC, there exists a linear dependence of negative slope; that is the region where the suspension is stable ($W \gg 1$ for very stable suspensions; $W > 1$ for suspensions of low stability). (b) For electrolyte concentrations higher than or equal to the CCC, W is constant, equal to unity. The point of intersection between the two lines in the plots log W vs. concentration allows the determination of the experimental CCC; this value, in turn, can be compared with the predicted value obtained from Eq. [50], using the theoretical curves of interaction energy vs. distance between the particles, similar to those of Figs. 23 and 24.

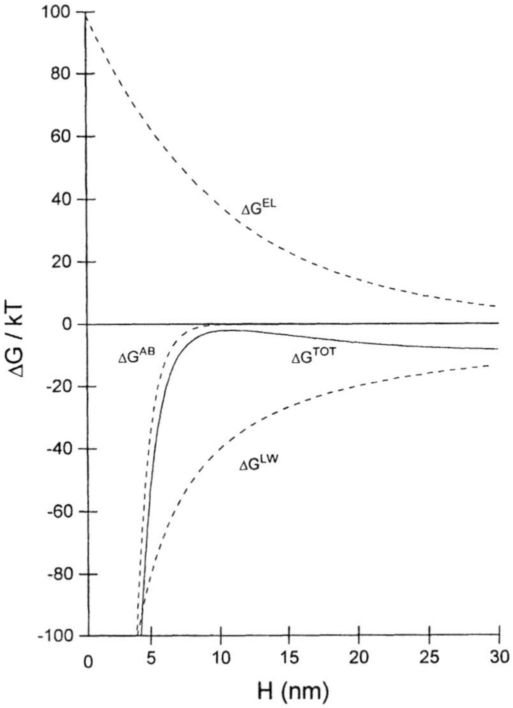

(b)

Another parameter of current use in pharmaceutical practice for the characterization of the stability of suspensions is the flocculation ratio, or the extent of flocculation, β, defined as (106):

$$\beta = \frac{F}{F_\infty} \quad [52]$$

F is the flocculation coefficient, calculated from $F = V/V_0$, where V is the sediment volume and V_0 is the total initial volume of the suspension. F_∞ is some reference, frequently taken as the value of F for a nonflocculated or stable suspension. According to its definition, the greater the value of β the more unstable is the suspension.

D. Some Results of Pharmaceutical Interest

In order to illustrate the above comments on the stability of suspensions and show the predictive power of the DLVO and extended-DLVO models, let us show

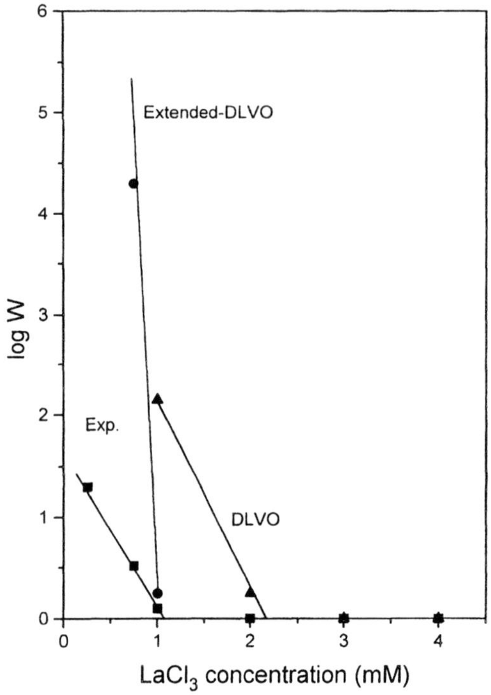

Figure 25 Experimental and theoretical (DLVO and extended DLVO approaches) stability ratios of Aquacoat suspensions as a function of $LaCl_3$ concentration at pH 6. (After Ref. 102, with permission).

some experimental results obtained with pharmaceutical suspensions. Stability properties of Aquacoat were studied in a wide range of ionic compositions of the dispersion medium, taking into account the interfacial properties of the particles and the different interactions existing between them (102). Figure 25 shows the variation of experimental stability ratio, W, and its theoretical values using both the DLVO and ext-DLVO approaches, with $LaCl_3$ concentration as the coagulating electrolyte. The effect of increasing electrolyte concentration is to decrease the stability of the latex dispersion until a given concentration, the CCC, is reached. Higher concentrations do not essentially affect the stability of the system; any potential barrier between particles is completely absent for concentrations greater than CCC. In order to obtain the theoretical values of W, using Eq. [50], the total interaction energy must be computed. For this purpose, the two approaches have been used and the results are shown in Fig. 26. The shape of the curves in this figure is well known, and only a comment on the AB contri-

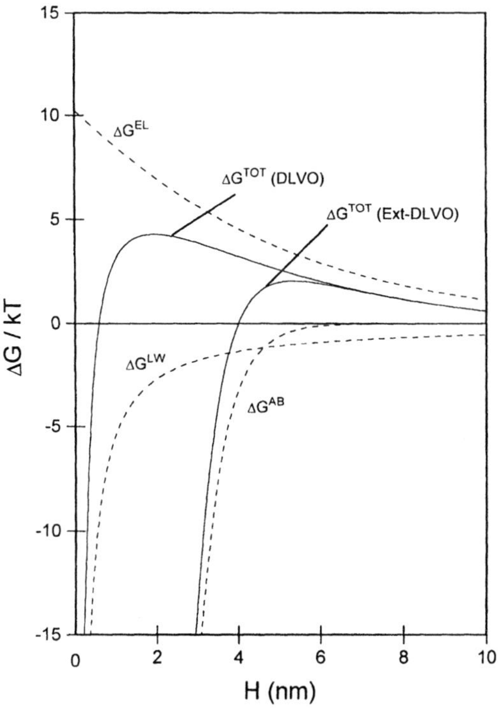

Figure 26 Free energy of interaction–distance plots for Aquacoat particles in 10^{-3} M $LaCl_3$ and pH 6. Equations [46]–[51] were used to compute ΔG^{LW}, ΔG^{EL}, and ΔG^{AB}. Parameters: $A = 0.156 \times 10^{-20}$ J, $\gamma_S^+ = 0.1$ mJ/m^2, $\gamma_S^- = 12.4$ mJ/m^2; $\gamma_S^{LW} = 31.2$ mJ/m^2, $a = 170$ nm; $\zeta = -9$ mV. (After Ref. 102, with permission.)

bution is about its order of magnitude: as observed, it is a rather short-ranged attractive interaction that can be much larger, in absolute value, than the LW components. Its presence significantly alters the shape of the ΔG-H classical (DLVO) curves, as also found by other authors (13,99,100) working with entirely different systems. Concerning the CCC values deduced from Fig. 25, an excellent agreement is found between the extended model and experimental data, whereas the stability of the suspension is clearly overestimated by the classical version of the theory. It should be noted, however, that except in the vicinity of the CCC, the results shown in Fig. 25 corresponding to the extended model overevaluate the stability of the suspension, due to the constant value of γ_s^- used for calculation of ΔG^{AB}, whatever the electrolyte concentration. The agreement obtained between W_{EXP} and $W_{ext\text{-}DLVO}$ is not so close in the case of mono- or divalent cations (102) although, in general, a better qualitative explanation of the stability of sus-

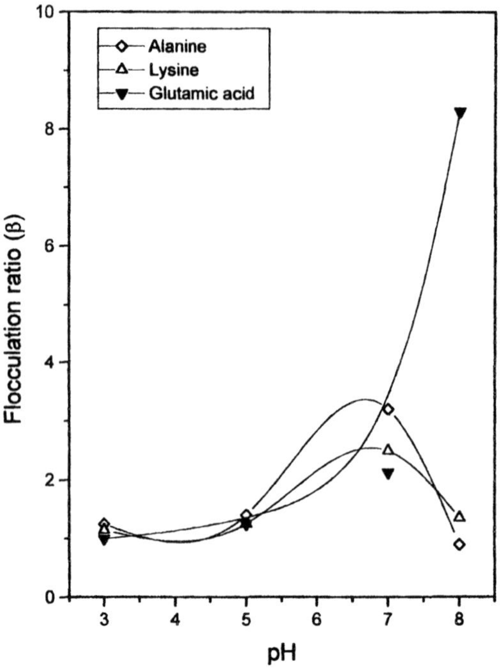

Figure 27 Flocculation ratio of nitrofurantoin suspensions as a function of pH in the presence of 10^{-2} M of amino acid (alanine, lysine, and glutamic acid) solutions. (Taken from Ref. 107, with permission.)

pensions is found if the classical DLVO theory is extended to include the effect of acid–base interactions between the particles. Quantitative improvements should be obtained if possible effects of the composition of the ionic medium on the strength of the acid–base interactions are considered.

Sometimes the stability conditions of the pharmaceutical suspension are discussed in terms of the extent of flocculation (β). Let us consider nitrofurantoin suspensions again. The effect of some amino acids that are potentially useful in the control of the stability of these suspensions has been studied in a series of works. In Fig. 27 (107), the dependence of β on pH for the three amino acids, alanine, lysine, and glutamic acid at constant (10^{-2} M) concentration is shown. In acidic conditions, β reaches relatively small values, and a similar effect is observed at alkaline pH. However, at pH 7 the β values were higher, which suggests that under these conditions the particles are aggregated forming flocculi that are easy to redisperse (108). The suspensions are more unstable when the pH is close to neutrality. The effect is especially remarkable when glutamic acid at pH 8 is used. At the concentration studied, β reaches values up to eightfold

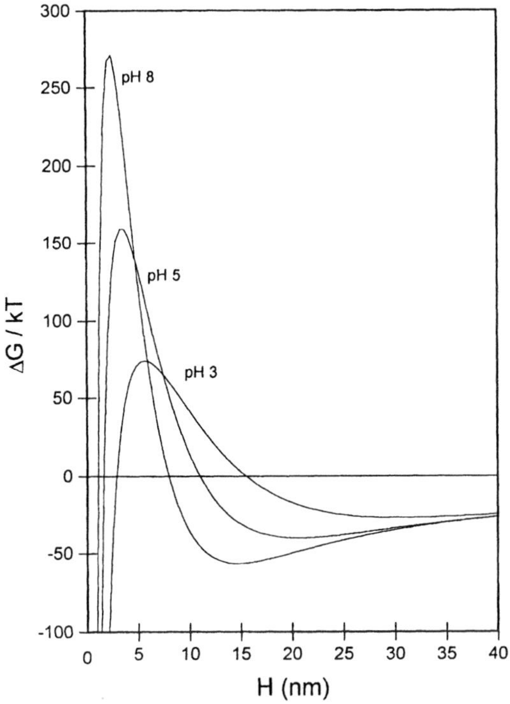

Figure 28 Free energy of interaction plotted as a function of distance for different pH values. Amino acid: glutamic acid. (Taken from Ref. 107, with permission.)

those found for other amino acids at any pH value. The increasing tendency to unstabilize nitrofurantoin suspensions with glutamic acid as the pH rises is qualitatively viewed in Fig. 28 (108), where total free energy of interaction is represented as a function of interparticle distance, H, for different pH values. The presence of a secondary minimum (Fig. 28: H = 15 nm; $\Delta G = -60$ kT), at pH 8 corresponds to the large sedimentation volumes (high β) obtained at this pH. At pH 3 and 5 the system should be more stable because of the less pronounced secondary minimum and thus low energy of attraction between particles. The results of Fig. 27 confirm this prediction.

IX. SUMMARY

The formulation of a suspension implies the use of a number of components (the active principle, plus wetting agents, dispersants, thickeners, flavoring additives, preservatives, etc.) which together determine the global properties of the suspension. Wetting, stability, and rheological properties of the suspension are the macroscopic result of the interfacial properties of the dispersed individual particles interacting with the liquid dispersion medium. Throughout this chapter, the main properties of the solid–liquid interfaces have been described, placing emphasis on the thermodynamic and electrical quantities governing the overall macroscopic behavior of a suspension. The methods for the experimental determination of quantities such as the solid surface free energy or the ζ potential of dispersed particles have been discussed and their modification by the presence of usual components of pharmaceutical suspensions studied. Hydrophobicity and hydrophilicity in pharmaceutical suspensions has been quantitatively estimated in terms of the thermodynamic free energy of interaction between particles dispersed in a liquid. The stability properties of the suspensions considered in this work have been discussed in the context of the classical Derjaguin-Landau-Verwey-Overbeek (DLVO) model, as well as of the extended model, introducing other solvation interactions not considered in the earlier approach. The application of all concepts introduced in this chapter has been made to a drug of pharmacological interest, nitrofurantoin, and to a biological substance of great importance, cholesterol, dispersed in aqueous solutions of different bile salts.

ACKNOWLEDGEMENTS

The authors thank the individuals who helped with the preparation of the manuscript: Drs. F. J. Arroyo and A. Ontiveros. Also, financial support given by the Ministerio de Educación y Ciencia (DGICYT) under Project no. MAT98-0940 is acknowledged.

REFERENCES

1. SR Byrn. Solid State Chemistry of Drugs. New York: Academic Press, 1982.
2. HR Brittain. Overview of physical characterization methodology. In: HR Brittain, ed. Physical Characterization of Pharmaceutical Solids. New York: Marcel Dekker, 1995, p. 1.
3. RA Nash. Pharmaceutical suspensions. In: HA Lieberman, MM Rieger, GS Banker, eds. Pharmaceutical Dosage Forms: Disperse Systems, Vol. 1. New York: Marcel Dekker, 1988, p. 151.
4. CM Ofner III, RL Schnaare, JB Schwartz. Oral aqueous suspensions.In: HA Lieberman, MM Rieger, GS Banker, eds. Pharmaceutical Dosage Forms: Disperse Systems, Vol. 2. New York: Marcel Dekker, 1989, p. 231.
5. RD Schoenwald, DR Flanagan. Bioavailability of disperse dosage forms. In: HA

Lieberman, MM Rieger, GS Banker, eds. Pharmaceutical Dosage Forms: Disperse Systems, Vol. 2. New York: Marcel Dekker, 1989, p.115.
6. RJ Hunter. Introduction to Modern Colloid Science. Oxford: Oxford University Press, 1993.
7. MJ Falkiewicz. In: HA Lieberman, MM Rieger, GS Banker, eds. Pharmaceutical Dosage Forms: Disperse Systems, Vol. 1. New York: Marcel Dekker, 1988, p. 13.
8. J Lyklema. Fundamentals of Interface and Colloid Science, Vol. 1. London: Academic Press, 1991.
9. SW Ip, JM Toguri. J Mater Sci 29: 688, 1994.
10. RJ Good. In: KL Mittal, ed. Contact Angle, Wettability and Adhesion. Utrecht: VSP, 1993, p. 3.
11. CJ Van Oss, MK Chaudhury, RJ Good. Chem Rev 88:927, 1988.
12. CJ Van Oss. Colloids Surfaces. 78:1, 1993.
13. CJ Van Oss. Interfacial Forces in Aqueous Media. New York: Marcel Dekker, 1994.
14. FM Fowkes. Ind Eng Chem 56: 40, 1964.
15. FM Fowkes, FL Riddle, WC Pastore, AL Weber. Colloids Surfaces 43: 367, 1990.
16. RJ Good, MK Chaudhury, CJ Van Oss. In: LH Lee, ed. Fundamentals of Adhesion. New York: Plenum Press, 1991, p.153.
17. MJ Rosen. Surfactants and Interfacial Phenomena, 2nd Ed. New York: John Wiley and Sons, 1989, Chapters 2 and 6.
18. R Hunter. Foundations of Colloid Science, Vol. 1. Oxford: Clarendon Press,1987.
19. J Lyklema. Fundamentals of Interface and Colloid Science, Vol. 2. London: Academic Press, 1995.
20. AW Adamson. Physical Chemistry of Surfaces, 5th Ed. New York: John Wiley and Sons, 1990.
21. RJ Good. In: RJ Good, RR Stromberg, eds. Surface and Colloid Science, Vol. 11. New York: Plenum Press, 1979, Chapter 1, p. 1.
22. W Duncan-Hewitt, R Nisman. In: KL Mittal, ed. Contact angle, Wettability and Adhesion. Utrecht: VSP, 1993, p.791.
23. JK Spelt, E Moy, DY Kwok, AW Newmann. In: AW Neumann, JK Spelt, eds. Applied Surface Thermodynamics. New York: Marcel Dekker, 1996, Chapter 6.
24. E Chibowski, F González-Caballero. J Adhesion Sci Technol 7: 1195, 1993.
25. E Chibowski, L Hołysz, A Zdziennicka, F González-Caballero. In: AK Chattopadhyay, KL Mittal, eds. Surfactants in Solution. New York: Marcel Dekker, 1996, p.31.
26. SN Omenyi, RS Snyder, CJ Van Oss, OR Absolom, AW Neumann. J Colloid Interf Sci 8: 402, 1981.
27. RF Giese, PM Costanzo, CJ Van Oss. Phys Chem Minerals. 17: 611, 1991.
28. CJ Van Oss, RF Giese, Z Li, K Murphy, J Norris, MK Chaudhury, RJ Good. In: KL Mittal, ed. Contact Angle, Wettability and Adhesion. Utrecht: VSP, 1993, p. 269.
29. E Chibowski, L Hołysz. Langmuir 8: 710, 1992.
30. L Holysz, E Chibowski. Langmuir 8: 303, 1992.
31. E Chibowski. In: KL Mittal, ed. Contact Angle, Wettability and Adhesion. Utrecht: VSP, 1993, p. 641.
32. E.Chibowski, F.González-Caballero. Langmuir 9:330, 1993.

33. TD Blake, JM Haynes. J Colloid Interf Sci 30: 421, 1969.
34. RJ Good. J Colloid Interf Sci 42: 473, 1973.
35. RJ Good, NY Lin. J Colloid Interf Sci 54: 52, 1976.
36. SN Harder, DA Suck, JA Wood. J Pharm Sci 59: 1787, 1970.
37. CF Lerk, AJM Schoonen, JT Fell. J Pharm Sci 65: 843, 1976.
38. JT Fell, E Efentakis. Int J Pharm 4: 153, 1979.
39. L Erhardt, L Pratt, E Schindler. Pharm Ind 35: 719, 1973.
40. M de Lourdes Costa, A Bazkin. J Pharm Pharmacol 37: 455, 1985.
41. E Chibowski, M Bolivar, F González-Caballero. J Colloid Interf Sci 154: 400, 1992.
42. E Chibowski, ML Kerkeb, F González-Caballero. J Colloid Interf Sci 155: 444, 1993.
43. A Arancibia, J López. Ciencia Ind Farm 10: 10, 1978.
44. Y Kumano, S Nakamura, S Tahara, S Ohte. J Soc Cosmet Chem 28: 285, 1977.
45. E Chibowski, M Bolívar, AV Delgado, F González-Caballero. Colloid Polym Sci 271: 688, 1993.
46. V Gallardo, AV Delgado, A Parera, J Salcedo. J Pharm Pharmacol 42: 225, 1990.
47. B Janczuk, ML Kerkeb, F González-Caballero, E Chibowski. Langmuir 10: 1012, 1994.
48. E Chibowski, ML Kerkeb, F.González-Caballero. Langmuir 9: 2491, 1993.
49. D Myers. Surfaces, Interfaces and Colloids. New York: VCH, 1991, p. 70.
50. J Lyklema. In: Th F Tadros, ed. Solid/Liquid dispersions. London: Academic Press, 1987, p. 63.
51. RJ Hunter. Zeta Potential in Colloid Science. London: Academic Press, 1981, Chapter 2.
52. DC Grahame. Chem Rev. 41: 441, 1947.
53. J Th Overbeek. In: H R Kruyt, ed. Colloid Science, Vol. 1. Amsterdam: Elsevier, 1952.
54. O Stern. Z Electrochem 30: 508, 1924.
55. WB Russel, DA Saville, WR Schowalter. Colloidal Dispersions. Cambridge: Cambridge Univ. Press, 1989, Chapter 4.
56. E Hückel. Physik Z 25: 204, 1924.
57. DC Henry. Proc Roy Soc (London) Ser A 133: 106, 1931.
58. RW O'Brien, LR White. J Chem Soc Faraday Trans 2 74: 1607, 1978.
59. AV Delgado, V Gallardo, A Parera, F González-Caballero. J Pharm Sci 79: 82, 1990.
60. DE Cadwallader. J Am Pharm Assoc N S 15: 413, 1975.
61. MA Ruiz-Martínez, V Gallardo, A Parera, AV Delgado. Ars Pharmaceutica 28: 315, 1987.
62. V Gallardo, J Salcedo, A Parera, AV Delgado. Int J Pharmaceutics 71: 223, 1991.
63. D Attwood, AT Florence. Surfactant Systems. London: Chapman and Hall, 1983, p. 108.
64. MM Rieger. In: HA Lieberman, MM Rieger, GS Banker, eds. Pharmaceutical Dosage Forms: Disperse Systems, Vol. 1. New York: Marcel Dekker, 1988, p. 285.
65. I Langmuir. J Am Chem Soc. 40:1361, 1918.
66. JJ Betts, BA Pethica. Trans Faraday Soc 56:1515, 1960.
67. JA Kitchener. J Photogr Sci 13:152, 1965.

68. W von Rybinski, MJ Schwuger. In: MJ Schick, ed. Nonionic Surfactants. Physical Chemistry. New York: Marcel Dekker, 1987, p. 45.
69. GH Giles, D Smith, A Huitson. J Colloid Interf Sci 47:755, 1974.
70. BT Ingram, RH Ottewill. In: DN Rubingh, PM Holland, eds. Cationic Surfactants. Physical Chemistry. New York: Marcel Dekker, 1991, p. 87.
71. R Aveyard. In: ThM Tadros, ed. Solid/Liquid Dispersions. London: Academic Press, 1987, p. 111.
72. WD Kumler. J Am Chem Soc 67: 1901, 1945.
73. J Salcedo, A Delgado, F González-Caballero. J Colloid Interf Sci 138: 10, 1990.
74. WD Kumler, IF Halverstadt. J Am Chem Soc 64:1941, 1942.
75. A Fini, A Roda, R Fugazza, B Grigolo. J Solution Chem 8:595, 1985.
76. ME Labib, AA Robertson. J Colloid Interf Sci 77:151, 1980.
77. V Gallardo, J Salcedo, P Vera, AV Delgado. Colloid Polym Sci 271:967, 1993.
78. B Vincent. In: THF Tadros, ed. Solid/Liquid Dispersions. London: Academic Press, 1987, p. 149.
79. J Vanderhoff, MS El-Aasser. In: HA Lieberman, MM Rieger, GS Banker, eds. Pharmaceutical Dosage Forms: Disperse Systems, Vol. 1. New York: Marcel Dekker, 1988, p. 93.
80. T Cosgrove. In: THF Tadros, ed. Solid/Liquid Dispersions. London: Academic Press, 1987, p. 131.
81. PJ Flory. Principles of Polymer Chemistry. New York: Ithaca Press, 1953.
82. A Silverberg. J Chem Phys 48: 2835, 1968.
83. DF Evans, H Wennerström. The Colloidal Domain: Where Physics, Chemistry, Biology, and Technology Meet. New York: VCH, 1994.
84. A Silverberg. Faraday Disc Chem Soc 59:203, 1975.
85. CAJ Hoeve. J Polym Sci Symp 61:389, 1977.
86. RJ Roe. J Chem Phys 60:4192, 1974.
87. JMHM Scheutjens, GJ Fleer. J Phys Chem 84:178, 1980.
88. R Simha, HL Frisch, FR Eirich. J Phys Chem 57:584, 1953.
89. E Di Marcio, R Rubin. Am Chem Soc Polym Prepr 11:1239, 1970.
90. SG Ash, D Everett, GH Findenegg. Trans Faraday Soc 66:708, 1970.
91. PG de Gennes. Compt Rend 291 Ser B:21, 1980.
92. HL Frisch. J Phys Chem 59:633, 1955.
93. BV Derjaguin, L Landau. Acta Physicochim 14:633, 1941.
94. EJW Verwey, JThG Overbeek. Theory of Stability of Lyophobic Colloids. Amsterdam: Elsevier, 1948.
95. J Israelachvili. Intermolecular and Surface Forces, 2nd Ed. London: Academic Press, 1992, Chapter 13.
96. J Gregory. J. Colloid Interf Sci. 83: 138, 1981.
97. VS Yushchenko, VW Yaminsky, ED Shuchkin. J Colloid Interf Sci 96: 307, 1983.
98. HK Christenson, PM Claesson, J Berg, PC Herder. J Phys Chem 93: 1472, 1989.
99. CJ van Oss, RF Giese, PM Constanzo. Clays Clay Min 38: 151, 1990.
100. JDG Durán, MC Guindo, AV Delgado, F González-Caballero. Langmuir 11: 3648, 1995.
101. DYC Chang, DJ Mitchell, BW Ninham, BA Pailthorpe. In: F Franks, ed. Water, Vol. 6. New York: Plenum Press, 1979 p. 239.

102. P Vera, V Gallardo, J Salcedo, AV Delgado. J Colloid Interf Sci 177:553, 1996.
103. B Bijsterbosch. In: ThF Tadros, ed. Solid/Liquid Dispersions. London: Academic Press, 1987, p. 91.
104. JS Laskowski, R J Pugh. In: JS Laskowski, J Ralston, eds. Colloid Chemistry in Mineral Processing. Amsterdam: Elsevier, 1992, Chapter 4, p. 115.
105. RM Jayasuriya, MS El-Aasser, JW Vanderhoff. J Polym Sci, Polym Chem Ed. 23: 2819, 1985.
106. L Dintenfass. Kolloid Z 163: 48, 1959.
107. V Gallardo, MA Ruiz, P Vera, F González-Caballero, AV Delgado. J Pharm Pharmacol 48:450, 1996.
108. V Gallardo, M Bolivar, J Salcedo, AV Delgado. Prog Colloid Polym Sci 84: 447, 1991.

5

Emulsions as Drug Delivery Systems

Katrin Buszello* and Bernd W. Müller
Christian Albrecht University, Kiel, Germany

* *Current affiliation*: Boehringer Ingelheim GmbH, Ingelheim, Germany.

I. INTRODUCTION

Colloidal particles, such as microparticles, nanospheres, emulsion particles, liposomes, and mixed micelles, have been investigated as potential carrier systems for the delivery or the targeting of drugs to specific sites in the body in recent years. Among them emulsion formulations have gained particular interest as a carrier of lipophilic drugs due to their biocompatibility and satisfactory long-term stability and the fact that they can easily be manufactured on an industrial scale using proven technology (1).

Emulsions can be defined as a mixture of at least two inmiscible liquid or liquid crystal phases with the interface being stabilized by an adsorbed monomolecular or micellar layer of surface active agents as shown in Fig. 1. Depending on the nature of the dispersed and the continuous phase, different types of emulsions can be distinguished: oil-in-water (O/W) or water-in-oil (W/O) emulsions. If the dispersed phase represents an emulsion itself the systems formed are called multiple emulsions.

Emulsions intended for total parenteral nutrition (TPN) are formulated using vegetable or neutral oils as dispersed phase and phospholipids as emulsifier with the objective that they are not recognized as foreign by the body. They can be considered as being artificial chylomicrons and enter the fat metabolism pathway through the adsorption of apolipoproteins and the subsequent action of lipoprotein lipase.

Extemporaneous emulsions were prepared by adding a concentrated drug solution to a commercially available parenteral emulsion. Examples are shown in Table 1.

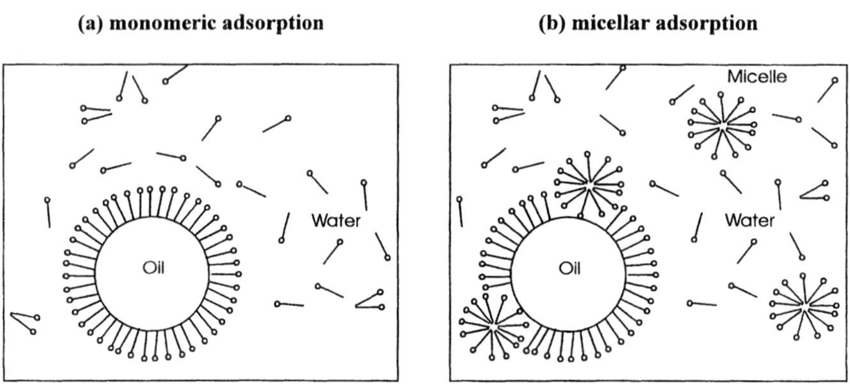

Figure 1 Schematic cross-sectional view of surfactant adsorption at oil/water interface (2). (a) Oil particle stabilized by condensed monolayer of surfactant (symbolized by circle for polar or ionic head and straight tail for hydrocarbon chain). (b) Oil particle stabilized partly by adsorbed single surfactant molecules and partly by micellar aggregates.

The limits of conventional drug delivery systems are evident. Following the administration of a drug, the final amount of drug reaching its target site may only be a small fraction of the administered dose. Accumulation at nontarget sites often leads to undesired side effects. The approach to apply the drug directly to its intended site of action is limited when the target tissue is inaccessible or widespread. Selective drug targeting would not only lead to an improved drug effectiveness and a reduction in adverse reactions but also offer the possibility of applying highly potent drugs.

If lipid-soluble drugs are to be administered intravenously they must be dissolved in an aqueous medium. The approach to use the water-soluble salt of those compounds sometimes gives unstable solutions or solutions at an unphysiological pH. The possibility of using biocompatible organic solvents or other principles of solubilization is limited (4).

Parenteral O/W emulsion formulations have also been adapted to the delivery of hydrophilic or poorly lipophilic drugs by synthesizing lipophilic prodrugs or drug derivatives that can be located in the oil phase of the emulsion, as for instance:

Dexamethasone palmitate (5)
Isocarbacyclin methyl ester (6)
Palmitoyl rhizoxin (7)

The use of parenteral emulsions may be apparently useful in overcoming solubility and/or stability problems of selected drug substances for which intravenous formulations are desired and difficult to develop (8). Interest in this concept has increased significantly with the greater understanding of the processes involved in drug delivery both at the cellular and subcellular level.

Table 1 Marketed Extemporaneous Emulsion Formulations Based on Intralipid® Formulations[a]

Emulsion preparation	Drug	Company	Indication
Diprivan	Propofol	Zeneca Pharmaceuticals, U.K.	General anaesthesia
Limethason	Dexamethasone palmitate (prodrug)	Green Cross, Japan	Chronic rheumatoid arthritis
Lipo-NSAID	Flurbiprofen axetil	Kaken Pharmaceutical Co., Japan	Postoperative and cancer pain
Vitalipid	Vitamins A, D_2, E, K_1	Kabi-Pharmacia, Sweden	Parenteral nutrition

[a] Composition of Intralipid® (Kabi-Vitrum, Sweden) (3): soybean oil, 10 or 20%; egg lecithin, 1.2%; glycerol, 2.5%.

II. BIODISTRIBUTION OF DRUG CARRIERS

General considerations for a good drug carrier are that it should be biocompatible, biodegradable, of fine and uniform particle size, have good stability, be suitable for targeting, and be pharmaceutically acceptable. The choice of a particular emulsion system to be used for drug delivery is generally dependent on the route of administration, the drug characteristics, and the effect required.

A certain specificity in biodistribution may be achieved passively by control of the physicochemical properties of the injected carrier system, such as particle size and dose, together with surface charge and surface characteristics. An influence of emulsion properties as well as the particle structure on the elimination rates from plasma and drug distribution in animals has been described after intravenous administration of emulsions containing coenzyme Q10 as shown in Fig. 2 (9). The possible influence of particle structures on the pharmacokinetics of incorporated drugs and the bioacceptibility of the carrier can be deduced.

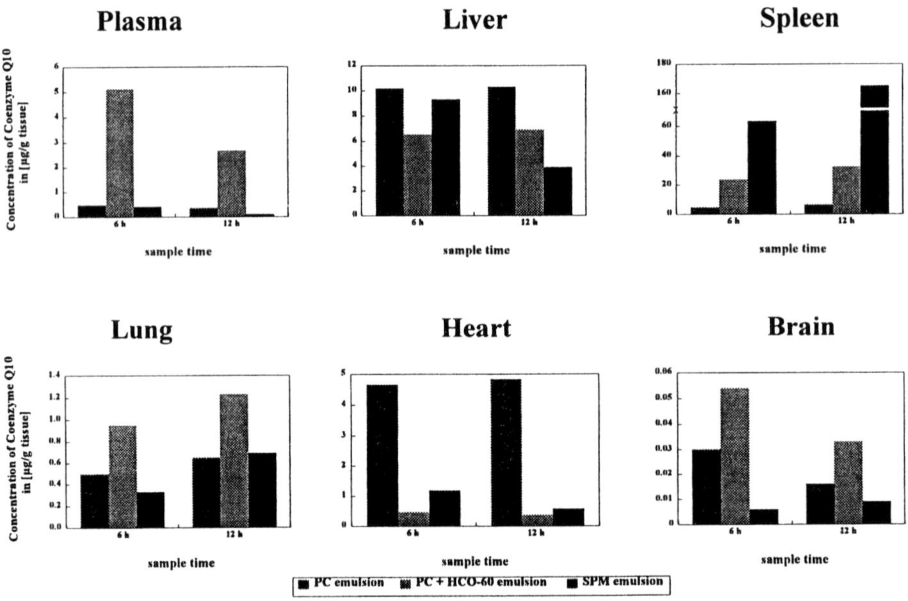

Figure 2 Coenzyme Q10 (CoQ10) concentrations (CoQ10 µg/g tissue) in various rat tissues at 6 and 12 h following the intravenous administration of 0.6 mg/kg of [^{14}C] CoQ10 in emulsion prepared with either PC (phosphatidylcholine), PC + HCO-60 (hydrogenated castor oil), or SPM (sphingomyeline). Values at 6 h represent the mean of three to six rats. Values at 12 h represent the mean of two rats; deviation from the mean was less than 20% in all cases.

The principal strategy for active targeting involves preparation of emulsion particles bearing ligands that will recognize determinants on the surface of the target cells. Iwamoto et al. investigated the influence of coating of the oil droplets in O/W emulsions with cell-specific polysaccharide derivatives on the target ability of those formulations. The saccharide moieties of glycolipids and glycoproteins on the cell surface are considered to play an important role in various intercellular recognition processes (10).

III. INFLUENCE OF PARTICLE SIZE AND PARTICLE DOSE

The biodistribution of colloidal systems can be related to various physiological processes as a function of particle size. Particles with a mean diameter larger than approximately 7 μm become entrapped in the capillary network of the lungs by mechanical filtration (11). However, smaller particles can also be deposited in the lung due to aggregation, positive surface change, or hydrophilic surface characteristics. Generally, particles smaller than 7 μm are retained by the phagocytotic mononuclear cells of the reticuloendothelial system (RES) in liver, spleen, and bone marrow (11). The smaller the particles the more likely they are to accumulate in the bone marrow. Uptake of small colloidal drug carriers in the liver can be exploited to improve the treatment of parasitic, fungal, and bacterial diseases such as, for instance, leishmaniasis, which mainly involves the macrophages of the RES (12), or to substitute specific liver enzymes.

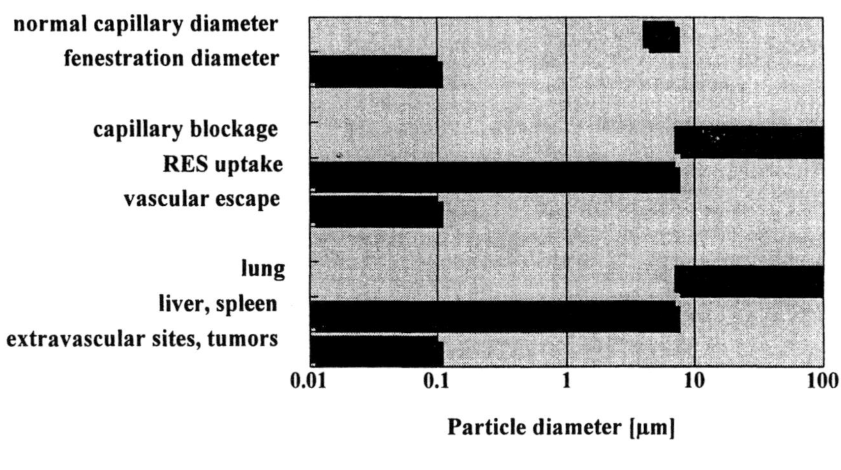

Figure 3 Relationship between particle size, physiological processes, and biodistribution of intravenously injected particles.

A difficulty with emulsion particles as drug carriers is to reach any site of action in the extravascular space, especially through intact endothelium. Particles below 100 nm in diameter are capable of escaping from the vascular system via fenestrated capillaries and discontinuous or sinusoidal capillaries because 100 nm is believed to be the cutoff value for the largest pores in the capillary endothelium. However, since the proportion of capillaries that can be passed by emulsion droplets is very low, the possibility of targeting extravascular sites of action is severely limited. However, tumors or inflamed tissues possessing a defective microvasculature are possible targets in this way. A schematic illustration is given in Fig. 3.

In order to reach organs or tissue sites other than liver, spleen, and bone marrow, it will be necessary to change the pattern of distribution of carrier systems in the organism. Kurihara et al. (7) demonstrated that by properly selecting

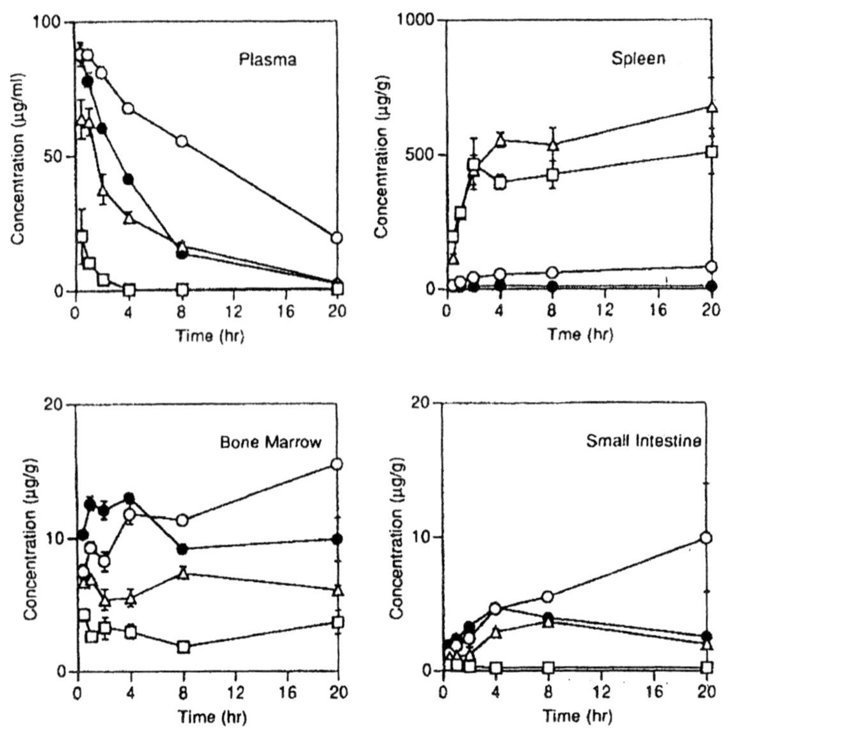

Figure 4 Tissue distributions of RS-1541 into the spleen, bone marrow, and the small intestine after i.v. injection of colloidal solution (●), 104 nm emulsion (○), 232 nm emulsion (△), an 550 nm emulsion (□) in rats at a dose of 4 mg/kg (7). Each value represents the mean and S.E. of three rats.

the particle size, lipid emulsions can control the behavior of a drug in the body, as shown in Fig. 4. Small-size emulsions (100–110 nm) showed the highest plasma concentrations of rhizoxin, though they were unable to suppress distributions of the drug in peripheral tissues. Emulsions larger than 200 nm in size, on the contrary, effectively inhibited the drug from entering the bone marrow, small intestine, and other non-RES organs, where many cytotoxic compounds showed undesired toxicities (7).

Initial clearance rate by the RES can be affected by the presence of large numbers of particles occupying available RES receptors or exhausting opsonizing factors (13). Overloading the RES by single large doses or repeated administration may lead to subsequent remain of injections in the circulation and could be used to affect distribution patterns.

Davis et al. confirmed that the infusion of fat emulsions into the rabbit can cause slight temporary impairment of the RES as determined by subsequent administration of a radiolabeled colloid probe. An infused emulsion can cause RES blockade by one or both of two mechanisms. Immediately after infusion some of the particles may be recognized as foreign and are then cleared by the RES (largely the Kupffer cells of the liver). In addition, or alternatively, the Kupffer cells of the liver may become overloaded by emulsion remnants that will result from normal metabolism of the emulsion by tissue lipases (14).

IV. SURFACE CHARACTERISTICS

When administered into the bloodstream, colloidal particles normally become opsonized with various blood components. Through the nature and the amount of this opsonization process such systems will be recognized as foreign very rapidly and will be removed by the various elements of the RES.

Alteration of tissue distribution has been made successful by modifying the surface structure of the emulsifier shell of the oil droplets. Small differences in the composition of the oil–water interface which affect emulsion particle hydrophobicity/hydrophilicity can alter RES uptake significantly. Hydrophobic particles are taken up by macrophages without the necessity of opsonization. Coating of the surface of the oil droplets with hydrophilic polymers reduces macrophage uptake significantly, as illustrated in Fig. 5 (15,16), not only due to an alteration of emulsion particle hydrophobicity but also by affecting droplet–macrophage adhesion processes. This is due to a steric stabilization effect together with a decrease of the uptake of opsonizing blood components onto the surface of emulsion droplets (17). Colloidal particles presenting hydrophilic surfaces with a low contact angle will be almost ignored by phagocytotic cells (15).

Surface coatings with hydrophilic polymers reduce macrophage uptake significantly. Studies have demonstrated that incorporation of polyethylene glycol

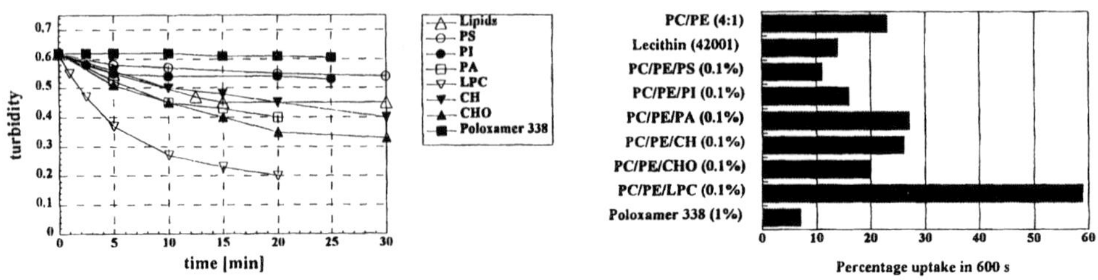

Figure 5 The effect of the nature of the emulsifier on the phagocytosis of fat emulsions by mouse peritoneal macrophages. PC (phosphatidylcholine), PE (phosphatidylethanolamine), PS (phosphatidylserine), PI (phosphatidylinositol), PA (phosphatidic acid), CH (cholesterol), CHO (cholesterol oleate), LPC (lysophosphatidylcholine).

(PEG)–modified phospholipids provides a hydrophilic shell that surrounds the lipophilic core and creates a steric barrier. PEG derivatives were shown to prevent serum protein binding, to reduce phagocytotic cell–mediated recognition, and to result in increased circulation lifetimes (18). Polyglycerolated phospholipids are said to inhere a similar potential of reducing opsonization processes while showing reduced toxicity (19).

PEG-PE (20)

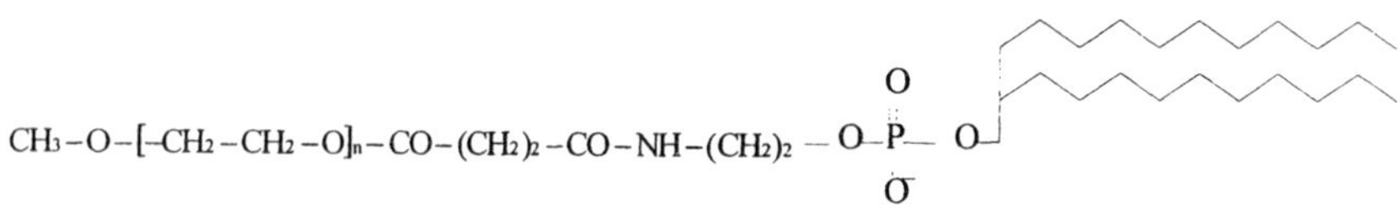

Poly(glycerol)phosphatidylglycerol (20)

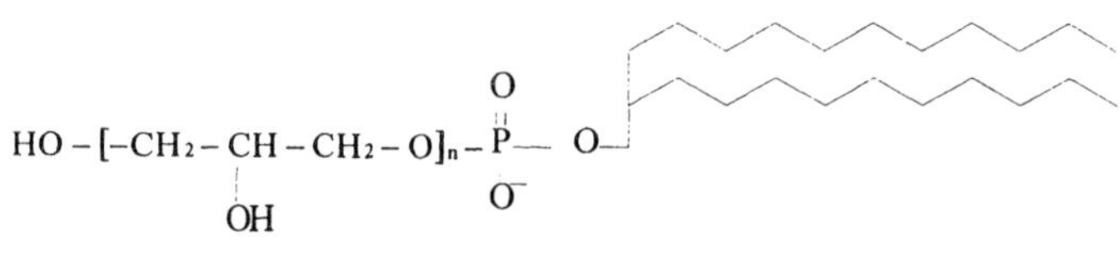

Liver and spleen uptake was almost totally eliminated if a nonionic coemulsifier, the block copolymer poloxamine (P 908), was employed as determined by gamma scintigraphy (17,21). This corresponded well with the blood clearance data that demonstrated a delayed clearance from the circulation as shown in Fig. 6. Moreover, nonionic block copolymers can effectively enhance antibody response to a variety of viral, parasitic, or bacterial antigens (22).

Stossel et al. investigated how the surface of a particle influences its accept-

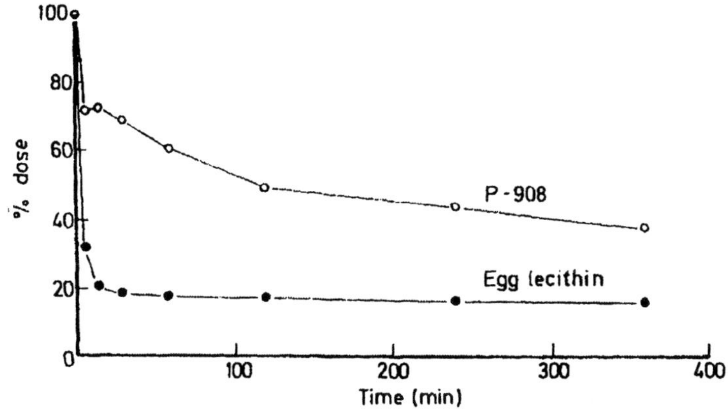

Figure 6 The effect of the nature of the emulsifying agent on the blood clearance of fat emulsions administered by i.v. injection to rabbit (n = 3) (21).

ability as a substrate for phagocytosis. It was shown that methyl groups as outermost substituents decrease droplet–macrophage adhesiveness. Emulsions prepared with Pluronics that have methyl side groups were not engulfed by the mononuclear phagocytotic cells (23).

Iwamoto et al. suggested that the tissue distribution of an O/W emulsion can be controlled by coating their surfaces with a cell-specific cholesterol-bearing polysaccharide, such as pullulan, amylopectin, and mannan (10,24).

Cholesteryl-Pullulan (CHP-55-2.1) (25)

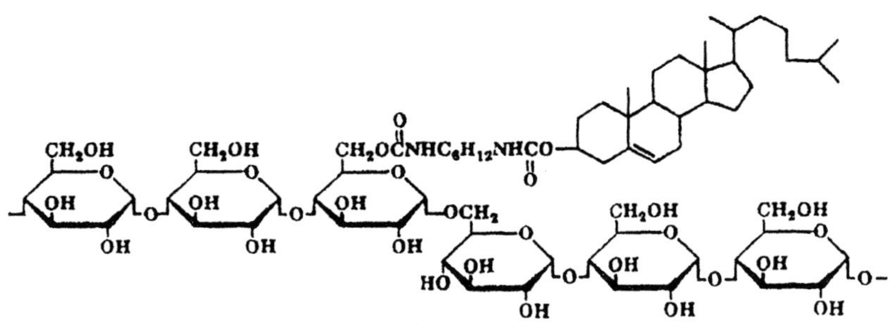

Reports that the coemulsification or incubation of fat emulsions with gelatin prior to intravenous injection enhanced its rate of clearance from the blood by fixed RES cells of the liver, lung, spleen, and bone marrow as illustrated in Fig. 7 (21,26), have led to the development of an artificial fat emulsion for RES func-

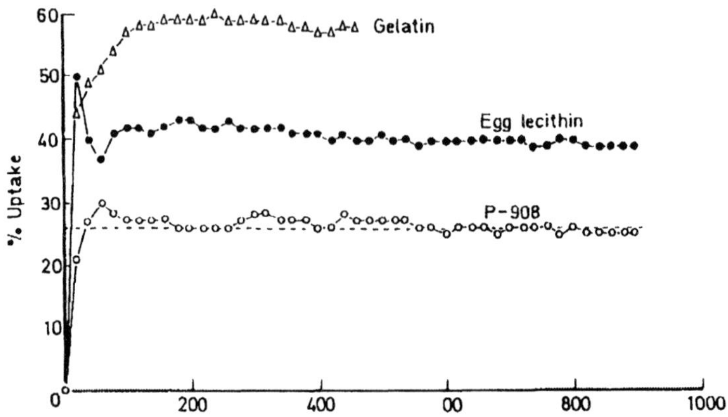

Figure 7 The effect of the nature of the emulsifying agent and added gelatin on the liver (and spleen) levels of fat emulsions administered by i.v. injection to rabbit (n = 3). The dotted line represents the liver blood-pool (21).

tionality tests. The mechanisms employed are interactions with circulating plasma opsonin proteins, thereby exploiting a receptor-mediated process involving fibronectin (17).

Takino et al. developed emulsion carrier systems for lipophilic drugs with the potential for prolonged circulation in the blood or hepatic targeting. A coating with sphingomyelin on the surface of the oil droplets resulted in avoidance of the RES (1). Takino et al. demonstrated further that [^{14}C]cholesteryl oleate administered with oil droplets of a conventional O/W emulsion formulation was rapidly taken up from the circulation by the RES cells, while those containing sphingomyelin survived in the circulation for a considerably longer period (27).

V. EFFECT OF SURFACE CHARGE

Particle surface charge has marketed effects on the clearance and deposition of colloids. Clearly, the connection between phagocytosis, RES uptake, and surface charge is far from simple and concomitant changes in other surface properties may override any effects produced by variations in surface charge. So the difference in organ distribution cannot be attributed to surface charge alone. Other factors need to be taken into account.

Typically, emulsion formulations reported in the literature are negatively charged. They are based on lecithin combined with nonionic or anionic emulsifiers. The approach to design an emulsion vehicle bearing a positive instead of a negative or a neutral charge is based on the results published in the liposome research literature (28). The positive charge of the colloidal carriers is conferred on them by the incorporation of a cationic lipid, mostly stearylamine (29). There have been extensive studies using other cationic compounds to charge-reverse the particle surface such as 1,2-dimyristyloxypropyl-3-dimethylhydroxyethyl bromide (DMRIE) (30), 2,3-dioleyloxypropyl-1-trimethylammonium bromide (DOTMA, Lipofectin) (30,31), or dimethylaminoethane-carbamoyl cholesterol (DC-cholesterol) (32). However, due to toxicity considerations they have not yet been established in colloidal formulations for parenteral purposes except for gene delivery.

DC-Cholesterol

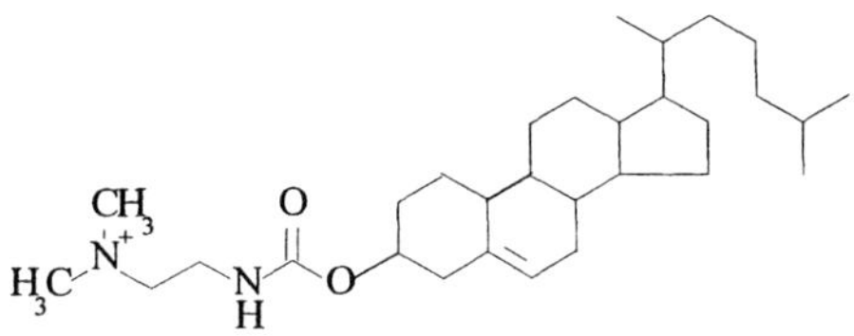

DOTMA (Lipofectin)

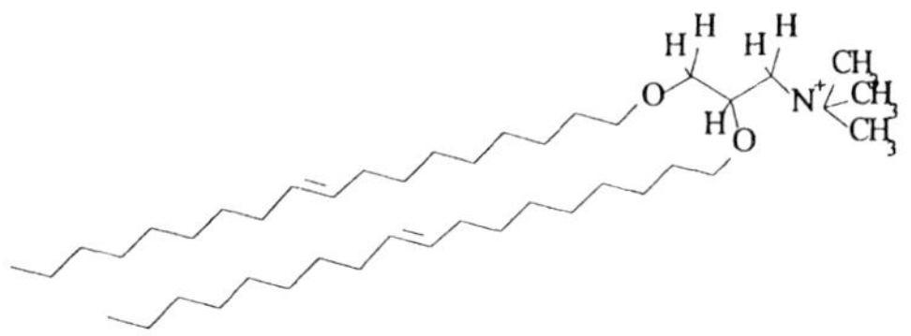

Surface charge affects the phagocytosis of emulsion particles by leukocytes and has a subsequent impact on the type of opsonin. It has already been shown that emulsions with neutral surface charges are taken up more slowly by macrophages than those bearing charged surfaces (15,23). This is in accordance with the results of Stossel et al. who investigated the effects of several drugs and hormones on the initial rate of phagocytosis in dependence on the surface characteristics of the emulsion droplets. It appeared that a particle with a strong net charge, either positive or negative, is ingested more readily then an uncharged or weakly charged particle (23).

Negatively charged emulsions, as compared with neutral or positively charged ones, showed a faster fate of clearance and higher liver and spleen uptake, while positively charged colloids showed an initial accumulation in the lungs and subsequent relocation to liver and spleen. Davis et al. indicated that surface charge might be an important factor and that a proper choice of the phosphatide emulsifier combination could be used to minimize the uptake of fat emulsions by the RES. But they did not find any clear correlation between ζ potential and in vitro macrophage phagocytosis for soybean oil emulsion systems prepared with a range of phosphatide emulsifiers (15).

When an emulsion formulation is administered into the bloodstream it comes into contact with positively charged electrolytes (e.g., sodium or calcium ions) that might cause a reduction in surface charge and can even display a charge reversal (14). While there is evidence that such systems behave satisfactorily, it can be suggested that charged emulsions could behave differently when introduced into the bloodstream with respect to the uptake of plasma blood components and opsonic factors and the possible sequestration by cells of the RES (14) as illustrated in Fig. 8.

It is believed that positively charged emulsions may alter the pharmacokinetic profile of the incorporated selected drugs resulting in enhanced localization of higher drug contents in targeted organs. This hypothesis has already been proposed by other authors who investigated charge-reversed emulsions (14,34).

Since the corneal surface is negatively charged it can be assumed that positively charged emulsions will be attracted by the negative charge of the cornea, resulting in a more prolonged residence time and leading to drug flux enhancement. Such a hypothesis is supported by data reported by Singh and Mezei for liposomal and emulsion preparations (35,36).

The major obstacle to parenteral application of positively charged emulsion formulations are toxicity concerns. No clear conclusion could be drawn from the literature on the toxic effects of formulations containing stearylamine. Most authors refer to the publication of Adams et al. (37). Adams et al. have shown the toxicity of stearylamine-loaded liposomes in mice after intracerebral injection. However, Klang et al. assessed the potential induced toxicity of a positively charged emulsion using a combination of emulsifiers consisting of phospholipids, Poloxamer-188, and stearylamine, which was proved suitable for parenteral use and for ocular application. The studies indicated that hourly administration of a positively charged emulsion vehicle containing stearylamine was well tolerated without any evidence of any toxic or inflammatory response to the ocular surface

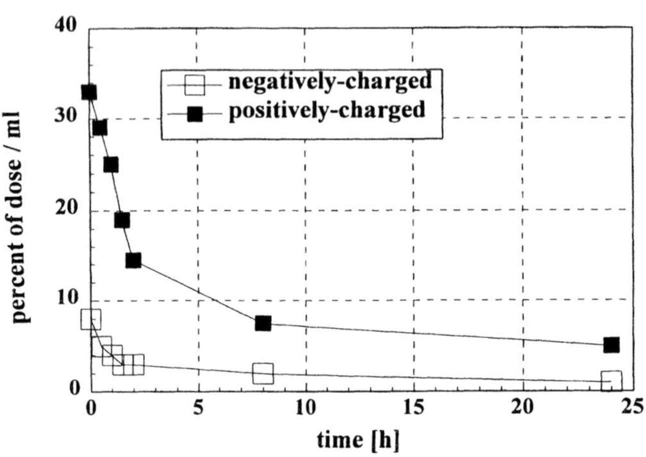

Figure 8 Effect of submicron emulsion surface charge on plasma concentration of radiolabelled cholesteryl oleate following i.v. administration (33).

during the 5 days of the study. No marked acute toxicity effect, no hepatotoxicity, no nephrotoxicity, were observed when the positively charged emulsion was injected intravenously in mice. This is probably due to the fact that the doses administered were well below the ones used in animal studies with liposomal formulations. Furthermore, positively charged emulsion vehicles did not penetrate through the blood–brain barrier and therefore did not reach the brain, which is considered to be the organ where stearylamine side effects were produced (34).

Davis et al. assessed the effect of charge reversal on the uptake of fat emulsions by the RES by in vivo and in vitro methods. No differences could be seen between emulsions that are infused as such or those whose charges have been reversed by added calcium (14).

VI. STABILITY CONCERNS

Emulsion droplets are normally stabilized by enhancing the mechanical strength of the interfacial film formed around the oil droplets (38), by steric stabilization effects, and/or by the presence of charged surfactants which create an electrostatic barrier. The stabilizing factor of the latter is the electrostatic repulsion of similarly charged droplets. The emulsion stability can be considerably improved with the use of mixed emulsifying agents (39).

The addition of a drug to the oil phase often negatively influences the emulsion stability (40). The destabilizing effects associated with the incorporation of drugs into fat emulsions limit the therapeutic applications of these interesting dosage forms (41,42). It was demonstrated that the electrical charge caused by the phospholipids which stabilized the emulsions is relatively low and depends on the exact phospholipid mixture composition used (43). Attempts to enrich the phospholipid mixture with negatively charged phosphatides to provide stabilizing electrical charges were disappointingly unsatisfactory (44,45).

Coating of emulsion particles with cholesterol-bearing polysaccharides, such as pullulan, amylopectin, and mannan, led to a reduction of the divalent metal ion–induced aggregation of the emulsion droplets. The cholesteryl group behaves as a hydrophobic anchor in the surface lipid layer resulting in a lower fluidity of the surface of the emulsion droplets (24).

The number of antiflocculants suitable to stabilize emulsions is limited and the effect is less distinct compared with polymeric particles. Selection of an antiflocculant requires consideration of net charge increase, effects of pH shift, and toxicological acceptance (38).

An attempt was made to elucidate the role of oleic acid in the stabilizing process of a diazepam submicron emulsion using a new approach of emulsion interface characterization. The study provided evidence that the close-packed film resulting from molecular interactions of the various film-forming components at the oil–water interface of the emulsion efficiently prevented coalescence of adjacent droplets upon collisions. This was due to a combined stabilizing mechanism involving steric stabilization accompanied by an electrostatic repulsive surface charge together with improved mechanical properties of the film (44). These results were confirmed by studies from Yamaguchi (45) and Washington (46) who suggested that oleic acid prevents flocculation and coalescence.

VII. STRUCTURE OF EMULSION CARRIERS

The influence of particle structures and constituents on the pharmacokinetics of incorporated drugs and the bioacceptance of the carrier is obvious (47) (Table 2).

A wide variety of surface-active agents can be used to prepare emulsion formulations. This variety is tremendously narrowed in the case of parenteral formulations. One of the major obstacles is the hemolytic effect of the majority of emulsifiers, which excludes their use for intravenous application (49) as seen in Fig. 9. The emulsion droplet surface can be charged negatively or positively by the incorporation of charged emulsifiers or enriched with reactive groups to which ligands can be covalently linked.

Jeppsson and Rössner studied whether different emulsifying agents influence the clearance of the lipid particles from the bloodstream. A difference in the soybean oil content had only a minor influence on the fractional elimination rate. The addition of phosphatides promoted the elimination of the fat particles.

Table 2 Factors That May Affect the Tolerability of Drug Emulsions (48)

Variable	Potential effect	Comment	Ref.
Lysophospholipids	Hemolysis	Only at very high nontherapeutic doses (25–30%)	49
Additional synthetic phospholipids	Unexpected toxicity	May vary depending on phospholipid	50
Degree of saturation of phospholipid/fat	Increase in total and LDL-cholesterol effects on membrane proteins	May occur on continuous infusion	51, 52
Content of MCT	Increase in total and LDL-cholesterol CNS effects	May occur on continuous infusion and/or at high doses	52, 53
Cationic phospholipids	Increased immunogenicity of transported drug	Most likely with liposomal forms	28, 34

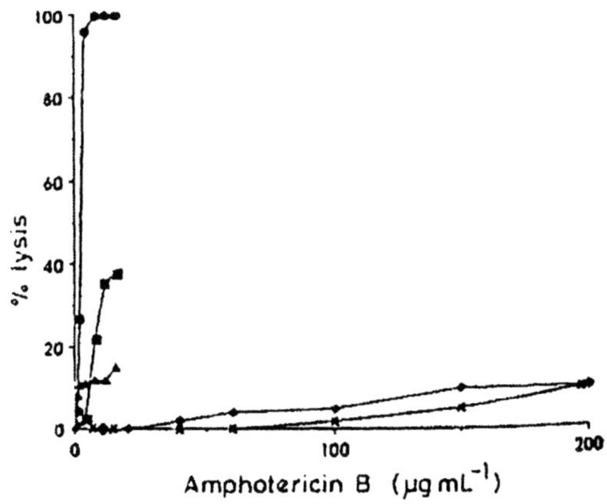

Figure 9 Erythrocyte lysis caused by emulsion formulations. All data normalized to 100% lysis. Key: × control, ◆ amphotericin B emulsion/lecithin, ● Fugizone, ▲ amphotericin B emulsion/Pluronic L92, ■ amphotericin B emulsion/Pluronic F127 (105).

Addition of co–block polymers such as Pluronics considerably impaired the removal of the fat particles (40).

The results of Sasagawa indicated that hydrogenated castor oil (HCO-60) emulsions, compared with conventional lecithin-stabilized emulsions, are more stable to lipoprotein lipase and show low uptake by RES organs, long circulations in the plasma, and high distribution in tumors. Thus, these sterically stabilized emulsions could show potential as effective carriers for highly lipophilic antitumor agents to enhance the drug delivery in tumors (54). This was confirmed by Sakaeda et al. who found that the rate of selective delivery of Sudan II to liver, lungs, and spleen could be suppressed by using HCO-60 (55). Lin et al. confirmed that HCO-60 was a good emulsifier for the preparation of lipiodolized emulsions with better stability and prolonged and selective delivery properties. Furthermore, an O/W emulsion containing HCO-60 was shown to be superior in the selective distribution of adriamycin-HCl to the liver and in decreasing concentration in heart and kidney (45).

Certain L-isomeric phospholipids with their gel transition temperature near body temperature can produce neurotoxic side effects as reported in the early literature on lysophospholipids, which suggests that the choice of lipids for the formulation of emulsions intended for drug delivery should be carefully considered (50).

The behavior of an emulsion is largely dependent on the lipid that constitutes the oil phase. If an emulsion is composed of unsaturated fatty acids, an absorption-promoting action may be obtainable (51). Polyunsaturated fatty acids have been found to have strong cytotoxicity against several types of cancer cells (25). Fatty acids themselves can influence the immune system. Both ω-3 and ω-6 fatty acids have been reported to possess immunosuppressive properties (48).

Ljungberg demonstrated for cholesteryl oleate and prednimustine that the emulsification process is dependent to a large extent on the physicochemical properties of the oil phase. These compounds gave coarse and unstable emulsions, but solubilization in castor oil resulted in proper emulsions (58).

Klang et al. showed that the selection of an appropriate composed oil phase allowed a tremendous increase in drug load of emulsion formulations due to an optimization of drug solubility within the inner oil phase (28) as shown in Fig. 10.

Deckelbaum et al. explored how enzyme affinities and enzyme activities (lipoprotein lipase, hepatic lipase) regulate hydrolysis of phospholipid-stabilized emulsions of medium-chain (MCT) vs. long-chain triacylglycerols (LCT). It was shown that MCT emulsions are more readily hydrolyzed by lipoprotein and hepatic lipases than LCT emulsions because of greater MCT solubility and mobility at the oil–water interface. The major factors involved are the affinity of the lipase for the interface and the accessibility of individual substrate molecules to the lipase (53). In mixtures of LCT and MCT emulsions, a higher affinity for the LCT-containing particles results in partitioning of the lipase away from the MCT emulsion with consequently diminished MCT hydrolysis.

The use of saturated MCT was the most effective way to increase blood concentration of Sudan II, resulting in higher distribution to liver, lungs, spleen, and brain (52).

Moriwaki et al. reported that the oral adsorption of vitamin A was enhanced

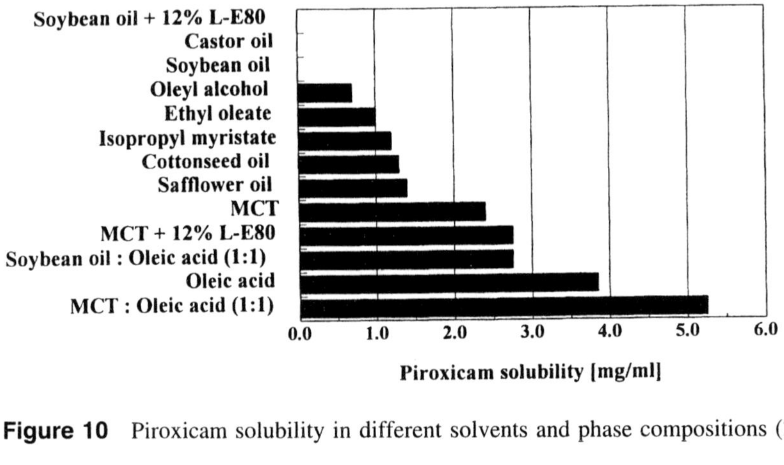

Figure 10 Piroxicam solubility in different solvents and phase compositions (28).

in milk fat globule membrane (MFGM) emulsions when butter oil was used as the oil phase but not when triolein was used instead. This suggests the involvement of oil in the enhancement mechanism (59).

The pharmacokinetics of incorporated drugs can be influenced by the partition coefficient. Sakaeda suggested that an increase in partition coefficient (experimentally more than 108) would be required for targeted drug delivery (55).

The composition of emulsion formulations plays an important role concerning intercellular cell recognition processes. Emulsion vehicles coated with cholesterol-bearing polysaccharide are very promising carriers. The saccharide moieties of glycolipids and glycoproteins on the cell surface are considered to play an important role in various intercellular recognition processes (10). Cell recognizability is also being improved by incorporation of apoproteins or galactoproteins in the lipid particles to enhance their specificity (60).

VIII. IN VIVO BEHAVIOR OF EMULSION SYSTEMS

Administration of emulsions has been carried out by every conceivable enteral and parenteral route. The oral route is problematic since many emulsion formulations are rapidly destabilized in the gut following interactions with bile salts.

The most important route for a wide range of therapeutic applications, the intravenous injection of emulsions, is normally followed by their interaction with plasma proteins. These are the so-called opsonins, which by adsorbing onto the surface of oil droplets mediate their endocytosis by macrophages of the RES, and the high-density lipoproteins (HDLs), which remove phospholipid molecules from the interface leading to droplet disintegration and release of drug. It was observed that an increase of the packing density of the interface led to drastic reduction of phospholipid removal by HDLs. However, the effect of lipid composition on vesicle half-life is not as substantial since vesicle size overrides interface fluidity in determining uptake by the RES. Moreover, the transfer of apolipoproteins is reported to provide a mechanism of identifying and marking plasma lipoproteins and lipoprotein-like colloidal systems for systematic metabolism by enzymes and tissues (15,61).

Prolonged residence of emulsions in the circulation is required when these are designed to reach non-RES tissues in the vascular system, extravascular sites of action, or to act as circulating drug reservoirs. Recent work has shown that a hydrophilic droplet surface achieved by the use of co–block polymers, PEG derivatives, and other hydrophilic molecules substantially prolong the half-life of emulsion droplets (11).

Regardless of the residence time of emulsions within the vascular system, much of an injected dose is taken up via endocytosis by RES cells to end up in the lysosomal apparatus.

When given by other parenteral routes, e.g., intraperitoneally, subcutaneously, or intramuscularly, the majority of emulsion droplets enter the lymphatic system and eventually the blood circulation where particles behave as if given intravenously. Liver, spleen, and bone marrow uptake is significantly lower. Indeed, relative to their size, lymph nodes take up a much greater (over 100-fold) proportion than any other RES tissue. Lymphatic transport was predominantly associated with chylomicron-based transport (62).

Efforts to ascertain whether emulsions given orally facilitate absorption of drugs in the gastrointestinal tract have given mixed results. Although agents such as insulin or vitamins administered via emulsions do reach the blood circulation, their absorption is minimal and unpredictable. It is nevertheless apparent that emulsions of a lipid composition that renders them resistant to attack by bile salts or phospholipases survive long enough to facilitate absorption of some of their contents, probably through the lymphatics.

Engel et al. have attempted to enclose insulin in a W/O/W emulsion to improve intestinal absorption (63). These data are in agreement with those presented by Matsuzawa et al. who stated that W/O/W multiple emulsions stabilized by gelatin can improve ileal and colonic absorption of insulin (64). Recent studies have shown that the oral or intestinal absorption of vitamin A, vitamin D_3, and insulin was improved by administering them as milk fat globule membrane (MFGM) emulsions compared with the administration of simple solution or emulsions using synthetic emulsifiers (65).

Decker et al. defined two fundamental properties dominating the delivery of drugs from emulsions: first, the concentration of the compound in the lipid phase of the emulsion is directly proportional to the concentration of the compound in the cell at equilibrium, and second, the rate of transfer is directly proportional to the concentration of particles in contact with cells. The transfer is consistent with direct partitioning from the lipid phase of the emulsion to cells and occurs by the direct collision of emulsion particles with cells (66).

IX. CLINICAL APPLICATIONS OF EMULSION SYSTEMS

An obvious major consideration in the use of emulsions in drug delivery is that a given emulsion-drug formulation should exhibit clear benefits over and above those seen with the conventional formulation of the drug. Encouraging results with emulsion-drug carriers in the treatment or prevention of a wide spectrum of diseases in experimental animals and in humans indicate that emulsion-based products for clinical and veterinary application may be forthcoming. These could include anticancer and antimicrobial therapy, vaccines and diagnostic imaging, artificial blood substitution, and treatment of ophthalmic diseases (Table 3).

Emulsions are well accepted as delivery systems for either hydrophilic or

Table 3 Advantages of Emulsion Formulations in Comparison with Conventional Formulations

Drug	Advantages	Ref.
Amphotericin B in Intralipid 20%	Fewer side effects, incl. renal impairment or chills equivalent anticandiasis activity	67, 68, 69
Cyclosporine	Reduced nephrotoxicity	57, 70
Diazepam (Diazemuls)	Reduced thrombophlebitis improved LD_{50} (mice)	4, 71, 72
Propofol in Intralipid 10% (Diprivan)	Reduced local pain on IV injection	73, 74

lipophilic drugs. A nonexhaustive survey of drugs in emulsion formulations for various routes of administration is given in Table 4.

The pharmacological activity of barbiturates was shown to be prolonged by incorporating these drugs into a soybean oil emulsion administered intravenously (4,40). Ljungberg and Jeppsson compared pentobarbital and thiopental emulsions to their corresponding sodium salts dissolved in water with respect to their effects in mice. Both emulsion preparations gave a prolonged duration of sleep as opposed to the aqueous solutions. No side effects were registered (4).

Mizushima et al. indicated that corticosteroids incorporated in lipid emulsions are taken up by the RES and some inflammatory cells much more than are free corticosteroids; thus, they have a stronger antiinflammatory effect (5) (Fig. 11).

Grolier et al. developed stable and sterile lipid carriers that could entrap large quantities of carotenoids and direct them to cultured rat hepatocytes (60).

Prostaglandin E_1 incorporated in emulsions is more stable than free PG E_1 against enzymatic inactivation in the body and causes less irritation (86). A PG E_1 emulsion is expected to be highly efficient for a drug delivery in clinical treatment since even by diluting the lipid vehicle with excess of transfusion fluid of different pH a significant amount of PG E_1 was retained in the lipid particles (91).

Kamperman and Sallis investigated the feasibility and benefits of entrapping phosphocitrate into liposomes and multiple emulsions. Due to the limited phosphocitrate payload in liposomes, attention was focused on the possibility of using multiple emulsions. A considerable increase in circulation half-life and a favorable anticalcifying response could be achieved with the comparatively low-dose emulsion preparation over the free drug (92).

A new approach to resolve the side effects associated with conventional formulations of diazepam consisted of dissolving the drug in the oil phase of an

Table 4 List of Drugs Under Investigation for Incorporation in Emulsion Formulations

Drug	Route of administration	Indication	Ref.
Amphotericin B	IV	Systemic antifungal agent, use in neutropenic patients	69, 75
Barbiturates	IV	Sedation, anaesthesia	4, 40
Cyclosporine	IV	Immunosuppressant	57, 70
Deoxycholate–amphotericin B	IV	AIDS-associates cryptococcal meningitis	76
Diazepam	Topical	Sedation	77
Epirubicin	Transfemoral cannulation, intrahepatic arterial infusion	Chemotherapy, hepatocellular carcinoma	78
Etomidate	IV	General anaesthesia	79
Isocarbacycline methyl ester (PG-1_2)	IV	Thrombotic disorders	6
Miconazole	IV and topical	Antifungal	80
NSAID	Topical	Analgesia, antiinflammatory	81, 82
Palmitoylrhizoxin	IV	Cytotoxic	54
Perfluorooctyl bromide: oxygent	IV	Blood substitutes	83
Perfluorooctyl bromide: imagent	IV SC	Contrast agent	84, 85
PG-E_1	IV	Antiplatelet and vasodilatory actions	86
Physostigmine	Oral	Anticholinesterase, Alzheimer's disease	87
Pilocarpine	Ocular	Glaucoma treatment	88
Pregnanolone	IV	General anaesthesia	89
Tetrahydrocannabinol	Ocular	Glaucoma treatment	35
Vitamin E	Topical	Antioxidant	90

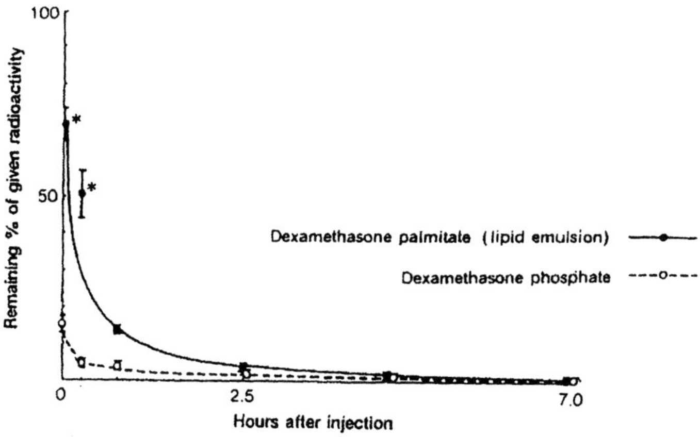

Figure 11 Accumulation in the spleen of [3H] dexamethasone palmitate given as a lipid emulsion and [3H] dexamethasone palmitate as an aqueous solution in rats (5).

O/W emulsion (42,44). Von Dardel et al. found that by administration of diazepam in an emulsion formulation, a highly significant reduction in the incidence of local vascular side effects can be achieved with no alteration in the therapeutic effect in comparison with the conventional preparative form (93). Ljungberg and Jeppsson (4) have shown in pharmacological studies that the LD_{50} of diazepam is three times larger in the emulsion form than in the aqueous form, but there is no difference in the therapeutic effect. This caused a threefold increase in the therapeutic index in the case of the new lipid emulsion form.

The efficacy of diazepam applied topically in emulsion creams depends greatly on the oil droplet size and to a lesser degree on the formulation composition and the oil type. Emulsions as vehicles for transdermal delivery of diazepam generate significant systemic activity of the drug as compared with conventional ointments. The efficacy of this formulation was shown to be as effective as in the range of parenteral delivery against convulsive effects induced by pentamethylenetetrazole in mice (77).

Other authors have used submicron emulsion delivery systems to improve the biological availability of various active ingredients following oral administration. However, problems associated with the oral administration, such as the low pH and high ionic strength of stomach environment, and their negative effect on emulsion stability limited the therapeutic application of this interesting dosage form (88).

Recent studies have shown that the oral or intestinal absorption of vitamin A, vitamin D_3, and insulin was improved by administering them as MFGM emulsions compared with the administration of simple solution or emulsions using synthetic emulsifiers (65). The MFGM, which consists mainly of lipids (55%) and proteins (44%) (95), is known to stabilize milk as an emulsion by enclosing the fat droplets in milk. MFGM could be a good alternative to a synthetic emulsifier for oral use considering its natural origin.

Yuasa et al. found that the oral absorption of α-linolenic acid tended to be increased when it was administered as MFGM emulsion compared with Tween-80 emulsions. In the intestinal loop the study failed to demonstrate an obvious enhancement of α-linolenic acid absorption by the use of MFGM emulsion compared with Tween-80 emulsion. However, the oral absorption process is limited by the emptying of the stomach and, hence, any change in the intestinal absorption process would only modestly affect its oral absorption (96). However, the present study strongly suggests that a decrease in emulsion droplet size could lead to a decrease in the intestinal absorption of amphiphatic compounds (97). Sato et al. have shown that MFGM can be used as an intestinal absorption enhancer of cyclosporines. After intraduodenal dosing in rats the blood and lymphatic fluid concentrations of a cyclosporine derivative increased significantly (98).

The use of emulsion for oral application is limited since other attractive alternatives such as the self-emulsifying oil delivery systems (SEDDs) are avail-

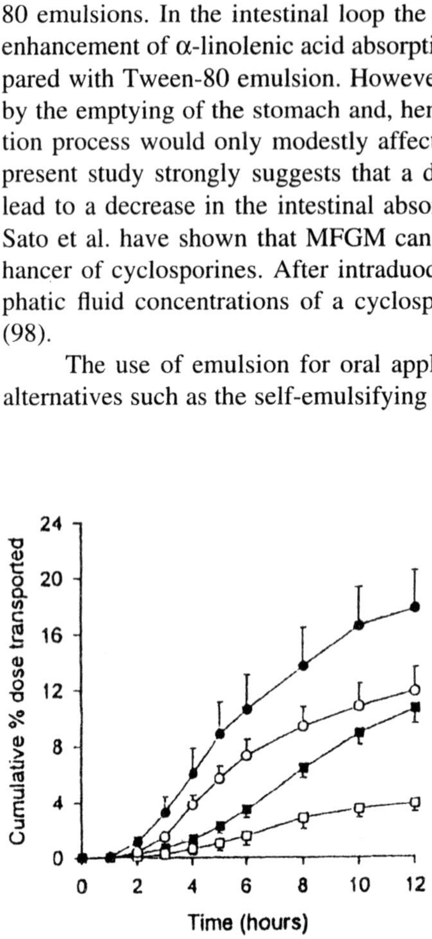

Figure 12 Cumulative percent dose of HF (Halofantrine) collected in intestinal lymph as a function of time and lipid vehicle in the triple-cannulated anaesthetized rat model. The administered dose was 2 mg of HF base and the lipid vehicles were a micellar lipid formulation (●), an emulsified formulation (○), and a lipid solution formulation that resulted in two data sets: lipid solution, group A (■) and lipid solution, group B (□) (62).

Table 5 Marketed Emulsion Preparations Based on Innovative Emulsion Formulations

Emulsion preparation	Drug	Company	Indication
Diazemuls	Diazepam	Kabi-Pharmacia (Sweden)	Status epilepticus, sedation, convulsion, tetanus, delirium tremens
Liple	Alprostadil ($PG\text{-}E_1$)	Green Cross (Japan)	Peripheral vascular disorders and maintenance of patent ductus arteriosus
Fluosol-DA	Perfluorodecalin Perfluorotripropylamine	Green Cross and Alpha Therapeutics (Japan)	Artificial blood substitute

able. They are isotropic mixtures of oil and surfactants, and can improve oral availability of drugs poorly soluble in water. SEDDs are physically stable solutions and easy to manufacture. Taken in gelatin capsules these formulations will disperse in the gastrointestinal tract to form a fine emulsion, upon dilution with gastrointestinal fluids.

Halofantrine, a highly lipophilic antimalarial agent, was formulated in lipid vehicles that were either a lipid solution, an emulsion, or a micellar system. Lymphatic transport was a major contributor to oral bioavailability. The rank order effect of the vehicles for the promotion of lymphatic transport was micelles > emulsion > lipid solution (62) (Fig. 12). These overall results underline the promising properties of emulsion drug carriers as therapeutic delivery systems for a variety of drugs (Table 5).

X. CANCER CHEMOTHERAPY

Modern chemotherapy focuses on the specific delivery of anticancer drugs to the desired target with a minimum of systemic side effects (56).

There is evidence that drug concentrations administered intravenously via emulsion particles in certain tumors are higher than in neighboring normal tissues. This might be due to a higher endocytotic activity of some tumor cells combined

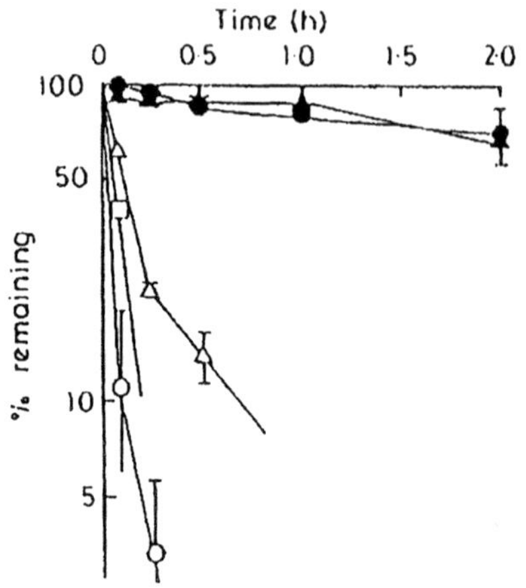

Figure 13 Disappearance of Mitomycin C (MMC) or nonyloxycarbonyl MMC from the rat thigh muscle after intramuscular injection in liposomes and O/W emulsion. □, MMC in saline; ○, MMC in liposomes; △, MMC in O/W emulsion; ●, nonyloxycarbonyl MMC in liposomes; ▲, nonyloxycarbonyl MMC in O/W emulsion (100).

with augmented local permeability of capillaries allowing the passage of small oil droplets. Eskey et al. provided an accurate model of mass transport within the tissues for evaluation of rates of tumor blood flow with the help of small, rapidly diffusing tracers (93).

Small droplet size of emulsion particles has a favorable influence on the efficacy of an encapsulated cytostatic drug. Kurihara et al. demonstrated that small-size emulsions (100–110 nm) showed the highest plasma concentrations of rhixozin, though they were unable to suppress distributions of the drug in peripheral tissues (7).

Many attempts have been made to deliver cytotoxic drugs to the tumor site by means of drug delivery systems that either utilize physical devices or chemically transform the drug molecules to prodrugs. Sasaki et al. [100] designed a promising delivery system for a prodrug of the anticancer agent mitomycin C by combining physical and chemical approaches. They suggested that the combination of lipid carrier devices with lipophilic prodrugs may be a useful adjunct to cancer chemotherapy, as illustrated in Fig. 13.

Wheeler et al. anticipated that the therapeutic activity of taxol, a cell cycle–specific anticancer drug, will be improved significantly when administered in an appropriately designed drug carrier system (18).

Experiments with cytostatic drugs entrapped in emulsions have clearly shown reduced toxicity compared with controls receiving the free drug. Fukui

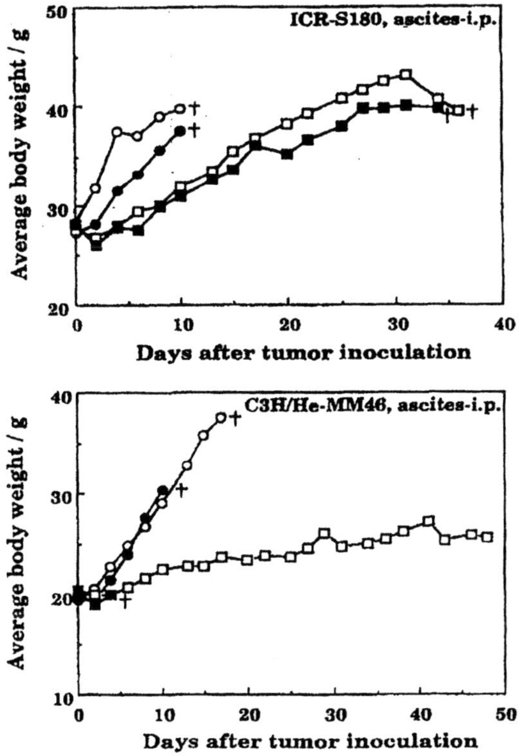

Figure 14 Effect of intraperitoneally administered CHP/ALA-emulsion on the change in body weight of ICR mice after S180 inoculation (top) and that of C3H/He mice after MM46 inoculation (bottom). The CHP/ALA-emulsion and conventional CHP self-aggregate were intraperitoneally injected for 5 consecutive days starting 1 day after tumor inoculation. Dose: ○, control; ●, 5.0 mg CHP per mouse; □, 2.5 mg of CHP plus 5.0 mg of ALA per mouse; ■, 5.0 mg of CHP plus 10.0 mg of ALA per mouse; n = 5 (25).

et al. investigated the in vivo anticancer effect of an O/W emulsion of linolenic acid (LA) and α-linolenic acid (ALA) stabilized with cholesterol-bearing pullulan (CHP) on the proliferation of human colon cancer cells and mouse fibroblasts in vitro. It was found that LA promoted the growth of these cancer cells, whereas ALA showed a strong growth inhibition (25). Fukui et al. showed further that the cytotoxic effect of the CHP-ALA-trioctanoylglyceride emulsion against human colon cancer was much higher than that of free ALA (101). Intraperitoneal injection of the CHP/ALA emulsion effectively prolonged the survival time of mammary tumor cell–bearing mice (25) (Fig. 14).

Proprietary emulsion-drug formulations that are stable and provide a high drug-to-lipid mass ratio have been facilitated. Tibell et al. compared five different formulations of cyclosporine using emulsions or liposomes as carriers, to that of the commercially available intravenous formulation Sandimmun in the heterotopic heart transplant model in rats. The new formulations tested did not reduce the immunosuppressive effect of cyclosporine but reduced the acute renal side effects following intravenous administration of Sandimmun (57).

Valinomycin, an antitumor agent, was formulated in emulsion form utilizing the commercially available Intralipid 10%. Animal testing of this system showed that the emulsion formulation produced similarly shaped dose–response curves to that of an aqueous suspension, but the emulsion formulation required a 30-fold lower dose than the suspension to produce similar effects (102).

XI. ANTIMICROBIAL THERAPY

The approach to use emulsion-drug carriers against microbial diseases is superior to free antimicrobial agents both in terms of distribution to the relevant intracellular sites and in treating disseminated disease states effectively. As already discussed, emulsion droplets are able to localize in liver and spleen, where many pathogenic microorganisms reside. For instance, current treatment of immunocompromised patients (e.g., those with AIDS) is ineffective probably because the microorganisms reside intracellularly, mostly in monocytes. Entrapment of antibiotics into emulsions offers a means to introduce drugs.

Another group of microbial diseases, severe fungal infections, has benefited substantially from the use of emulsions as a drug delivery system. Amphotericin B is the current drug of choice for the treatment of deep-seated mycoses, such as those occurring in transplant or AIDS patients (103). Nevertheless, its significant side effects to human cells such as central nervous system damage and nephrotoxicity has limited its potential as a therapeutic agent (104). Forster et al. showed that the emulsion formulation was less toxic to human erythrocytes than solubilized amphotericin B (105), as shown in Fig. 15.

Extemporaneous emulsion-amphotericin preparations were also capable of reducing drug toxicity in clinical studies. Joly et al. showed that an Intralipid carrier reduced the infusion-related toxicity of amphotericin B deoxycholate without altering its antifungal efficacy but did not offer substantial benefit against renal toxicity (106).

A large number of clinical trials have now confirmed the effectiveness of emulsion-amphotericin formulations as a safe alternative to the conventional formulations of amphotericin B in a majority of patients with invasive or superficial fungal infections. Levy et al. showed that, irrespective of the formulation used, the incorporation of amphotericin B in an emulsion formulation extended the

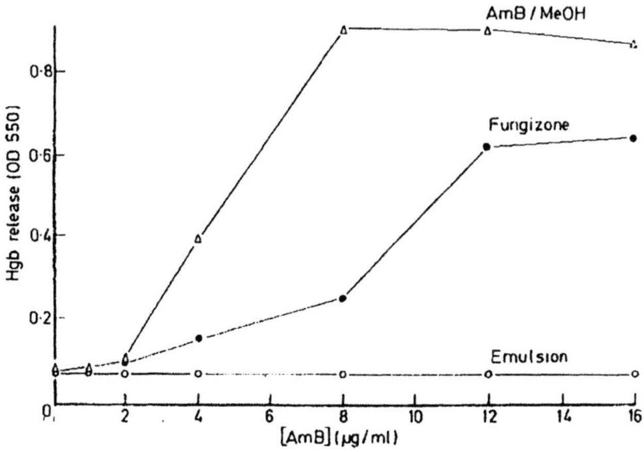

Figure 15 The lysis of human erythrocytes by Amphotericin B formulations, the effect of Amphotericin B concentration. Hemoglobin release determined at a wavelength of 550 nm. Legend: (△) Methanolic solution, (●) Fungizone, and (○) Emulsion (21).

survival time of mice infected with *Candida albicans* when compared to Fungizone, the marketed corresponding product (107).

XII. IMMUNOLOGICAL ADJUVANT

There is a demand for novel immunological adjuvants that are both acceptable and effective in potencing hormonal and cell-mediated immunity.

Emulsions are one of the few major nontoxic candidate adjuvants for human use currently under investigation. There is evidence that emulsion adjuvanticity might be due to its lipoid nature. Fatty acids can themselves influence the immune system. For instance, both ω-3 and ω-6 fatty acids have been reported to possess immunosuppressive properties (57).

Immunotoxins composed of ricin A chain coupled to a cell-specific monoclonal antibody have continued to show promise as a new class of antitumor agents. Griffin and Raso investigated the use of the carboxylic ionophor monensin formulated in an O/W emulsion for the in vitro and in vivo potentiation of immunotoxins. Monensin in this pharmacologically available form significantly improved the in vivo efficacy of both anti–human transferrin receptor immunotoxin and anticarcinoembryonic antigen immunotoxin when used as regional therapy.

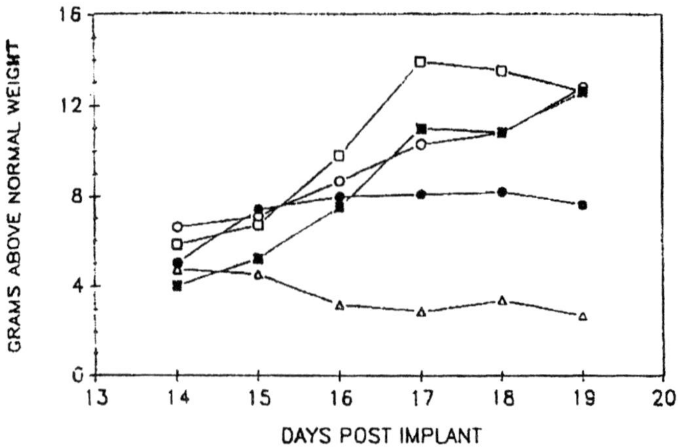

Figure 16 Mean weights of tumor-bearing mice treated with immunotoxin. Animals (n = 4/group) were weighted daily, and data were plotted as mean weight gain over time in days. SEM ≤ ± 7% for all data points. □, vehicle control; ■, control (anti-CEA) immunotoxin and monensin in emulsion; ○, anti-TfR immunotoxin; ●, anti-TfR immunotoxin and monensin in buffer; △, anti-TfR immunotoxin and monensin in emulsion (108).

Monensin emulsion greatly increased the tumoricidal action of immunotoxins coadministered in vivo (108) (Fig. 16).

Yoshioka et al. formulated a nonionic surfactant vesicle-in-water-in-oil (V/W/O) system for potential use in vaccine delivery. Initial studies of the system in vivo, as an adjuvant for tetanus toxoid, using cottonseed oil as the external oil phase, showed enhanced immunological activity over the free antigen or vesicles (109).

XIII. DIAGNOSTIC IMAGING

Emulsions containing contrast agents have been used in diagnostic imaging, including computed tomography, magnetic resonance imaging, and radionuclide imaging.

The use of neutron-activatable nuclides for scintigraphic imaging has immensely expanded the possibilities of radiolabeling (110,111). Low amounts of stable isotopes, mostly found in the lanthanide series, can be incorporated during normal manufacturing processes. For monitoring the in vivo behavior of parenteral O/W emulsions in a noninvasive manner, radiolabeling by neutron activation appeared to be quite promising.

Fatty acid salts of samarium could be incorporated as neutron-activatable excipients in emulsion formulations (111) (Fig. 17).

At the tissue level, images of the RES can be obtained very easily with radioactive tracer–containing colloids and emulsions by following their passive biodistribution pattern. Selective tissue imaging is likely to benefit from the use of small particles that can be designed to have long half-lives in the circulation. Thus it was possible to formulate emulsions that distribute radioactive agents in non-RES tissues, e.g., myocardial infarcts, lymph nodes, and tumors. Bone marrow imaging with emulsions in patients with malignant lymphoma has now been used to measure bone marrow suppression during chemotherapy and radiotherapy.

Bakan et al. developed a chylomicron remnant–like emulsion to serve as a hepatocyte-selective delivery system. The particle size profile (less than 200 nm) and high degree of incorporation of iodized triglycerides into the emulsion suggest that the formulation holds substantial promise as a hepatocyte-selective imaging agent for computed tomography of the liver (112). Incorporation of cholesterol into the monolayer was of critical importance in generating a stable emulsion near the targeted size (113).

To optimize the use of iodized oil for diagnostic computed tomography and for interstitial radiation therapy, De Baere et al. quantified the distribution of iodized oil after injection of different formulations of iodized oil into the hepatic artery. The ratio of tumor to nontumorous liver uptake of iodized oil was significantly higher with large droplet O/W emulsion than with any other preparation (114).

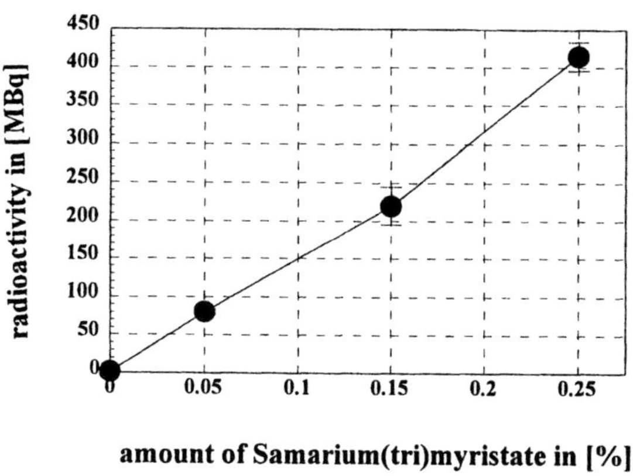

Figure 17 Influence of the amount of Samarium fatty acid on the radioactivity of neutron activable parenteral O/W emulsions (111).

XIV. PERFLUOROCHEMICAL EMULSIONS

Perfluorochemical (PFC) emulsions are attracting increasing interest for biomedical use as, for example, vehicles for respiratory gas transport, contrast agents for diagnostic tissue imaging, or drug delivery systems (115,116) (Fig. 18). Riess developed stable, ready-to-use, concentrated, injectable, fluorocarbon-based emulsions that are highly efficient in tissue oxygenation (117). Fujita et al. evaluated fluorocarbon emulsion as a candidate for artificial blood. They stated that fluorocarbon emulsions should consist mainly of particles smaller than 100 nm as a candidate for red blood cell substitute (118).

Fluosol-DA (Green Cross and Alpha Therapeutics, Japan), which consists of perfluorodecalin and perfluorotripropylamine, has already been marketed as an artificial blood substitute worldwide.

Composition of Fluosol DA (119)

Oil phase	
Perfluorodecalin	14.0
Perfluorotripropylamine	6.0
Emulsifier	
Pluronic F-68	2.72
Egg lecithin	0.4
Potassium oleate	0.03

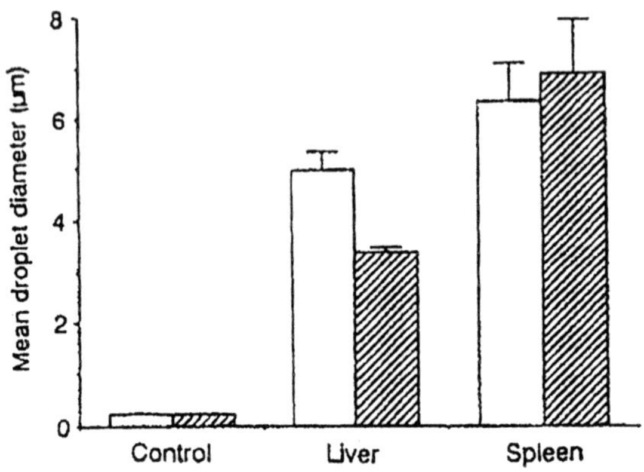

Figure 18 Mean diameter of PFC droplets extracted from rat liver or spleen at 72 h after injection of low dose Fluosol (group VI □) or 30% (w/v) FDC emulsion (group III, ■). Control mean values are droplet diameters of emulsions exposed to the extraction procedure (115).

Kryoprotector
Glycerol 0.8
Ionic balance, pH, and osmotic control
$NaCl$, KCl, $MgCl_2$, $CaCl_2$, $NaHCO_3$, dextrose

However, since this formulation has to be kept frozen and has to be mixed from three separate preparations after thawing prior to use, a new generation of fluorocarbon-based injectable carriers has been developed. Oxygent and Imagent (Alliance Pharmaceutical Corp.) are currently under clinical testing. Those formulations have been proved to be stable for at least 12 months at room temperature, to be autoclavable, and to be ready for administration as it is (119).

Zarif et al. investigated the so-called dowel molecule $C_6F_{13}CH{=}CHC_{10}H_{21}$ ($F_6H_{10}E$) as stabilizer in injectable fluorocarbon emulsions intended to serve as oxygen carriers, contrast agents, or drug delivery systems. It was proven to be well tolerated and was excreted reasonably fast without metabolism (120).

XV. EMULSIONS FOR OCULAR DELIVERY OF DRUGS

Drugs are normally applied to the eye as solutions (eye drops) and more rarely as suspensions or as ointments. However, the loss of drug either by tear flow or by drainage into the nasolacrimal duct as well as the impairment of vision by viscous ophthalmic ointments are major drawbacks for optimum delivery. As a submicron emulsion is biologically compatible, stable, and sterilizable, it appeared to be of interest as an ocular vehicle. Emulsion formulations have some advantages in providing a reservoir for slow release of drug in a nonviscous medium.

Naveh et al. showed that pilocarpine incorporated into a submicron emulsion might serve as a long-acting form of pilocarpine requiring just a single daily application. A single-dose application of pilocarpine emulsion induced an unexpectedly prolonged progressive decrease in intraocular pressure in rats that was maintained during a 29-h period whereas the pressure decreasing effect of pilocarpine aqueous eye drops persisted during the initial 5 h post instillation (88).

Muchtar et al. (35) have demonstrated that submicron emulsions can also be used as an ocular delivery system for the lipophilic antiglaucoma drug, Δ^8-tetrahydrocannabinol. They showed that the emulsion was able to achieve a long-lasting antidepressant effect on the intraocular pressure of rabbits following a single instillation.

The residence time of the formulation depends both on the particle size and on the nature of the surface charge. A positively charged O/W emulsion containing the antifungal agent miconazole was developed for ophthalmic use

(121). Positively charged emulsions were shown to have an affinity for corneal surfaces, binding for sufficient periods to optimize the release of the encapsulated drug. Wehrle et al. (121) confirmed the importance of positive charge on corneal retention of emulsions presumably as a result of association with the polyanionic corneal and conjunctival mucoglycoproteins.

XVI. MULTIPLE EMULSIONS

Multiple emulsions are complex systems wherein droplets of the dispersed phase contain additional but smaller droplets, identical to or different from the continuous phase. Even so there are many interesting fields of potential research and applications, such as vaccine formulations, enzyme immobilization, and drug overdose treatment (122). The disadvantages of current multiple emulsions are obvious. Internal droplet growth and release as well as degradation to heterogeneous O/W or W/O formulations results in shear sensitivity and bad long-term multiplicities. Besides, those systems are problematic to produce at industrial scale.

However, there is particular interest in the development of slow-release drug delivery systems (123,124) seen in Fig. 19 or in selectively transferring drugs into the lymphatic system using W/O/W emulsions. W/O/W emulsions

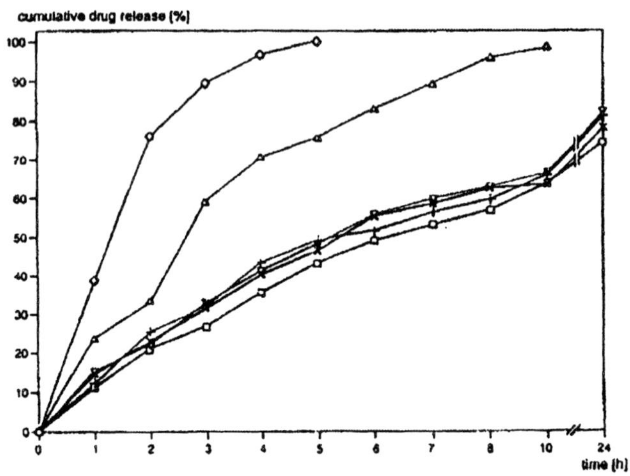

Figure 19 Drug release studies from various Rifampicin formulations W/O/W multiple emulsions uncoated (×), coated (□), uncoated formulation with serum (▽), coated formulation with serum (+), plain drug (◇), drug with albumin (△) (125).

enter the lymphatic system easily and reach regional lymph nodes selectively (56).

Multiple emulsions have already been used orally (98). Fusion has been shown to take place in a dopamine W/O/W emulsion in an acid environment (99). However, they can be stabilized by gelatinization of the internal aqueous phase.

Engel et al. have attempted to enclose insulin in a W/O/W emulsion to improve intestinal absorption (63). W/O/W multiple emulsions stabilized by gelatin can improve ileal and colonic absorption of insulin (64).

Oba et al. used a W/O/W multiple emulsion composed of oleic acid as a carrier of carboxyfluorescein via the enteral route as a model for future drug transport. The W/O/W emulsion was considered to be superior to micelles because it maintained a higher blood level of carboxyfluorescein over long periods and transferred it to the lymphatic system (51).

A W/O/W multiple emulsion containing rifampicin as encapsulant was coated with an *o*-palmitoyl derivative of mannan to assess its toxicity in vitro and its distribution behavior in vivo. A significant enhancement in lung uptake and a decreased internalization by spleen was noticed. The multiple emulsion was found to be nontoxic (125).

XVII. CONCLUSIONS

Research into the use of emulsions in drug delivery has led to remarkable improvements in the design of formulations for specialized tasks, in emulsion long-term stability, and scaled-up production.

Because of the structural versatility of the dispersed phase in terms of size, composition, surface charge, and ability to incorporate almost any drug regardless of solubility, or to carry on their surface cell-specific ligands, emulsion formulations can potentially be tailored to ensure optimum clinical use. There is evidence that emulsion-entrapped drugs exhibit superior pharmacological properties to those observed with conventional formulations.

Remarkable advances have been made in understanding and controlling emulsion behavior in vivo. This includes controlled retention of entrapped drugs in the presence of biological fluids, controlled particle residence in the blood circulation or other compartments in the body, and enhanced uptake of oil droplets by target cells. This has facilitated the application of a wide range of emulsion-entrapped drugs in the treatment and prevention of diseases in experimental animals and clinically, particularly in areas such as cancer chemotherapy, antimicrobial therapy, vaccines, diagnostic imaging, and the treatment of ophthalmic disorders.

In addition, it seems reasonable to use O/W emulsions as vehicles in the screening and evaluation of experimental drug substances that have poor water solubility and/or that in the early stages of evaluation/development do not merit extensive, costly, and time-consuming formulation studies (126).

Several emulsion preparations have already been licensed and a number of others are likely to follow soon. The future of emulsions as drug delivery systems appears to be secure.

REFERENCES

1. Takino, T., Konishi K, Takakura, Y., and Hashida, M., Biol. Pharm. Bull., 17(1), 121 (1994).
2. Rowe, E.L., J. Pharm. Sci., 54, 260 (1965).
3. Hansrani, P.K., Davis, S.S., and Groves, M.J., J. Parent. Sci. Technol., 37, 145 (1983).
4. Ljungberg, S. and Jeppsson, R., Acta Pharm. Suecica, 7, 435 (1970).
5. Mizushima, Y., Hamano, T., and Yokoyama, K., Ann. Rheum. Diseases, 41, 263 (1982).
6. Hoshi, K., and Mizushima, Y., Prostaglandins, 40, 155 (1990).
7. Kurihara, A., Shibayama, Y., Mizota, A., Yasuno, A., Ikeda, M., and Hisaoka, M., Biol. Pharm. Bull., 19(2), 252 (1996).
8. Silvestri, S., Wu, L.L., and Bowser, B., J. Pharm. Sci., 81, 413 (1992).
9. Kimura, A., Yamaguchi, H., Watanabe, K., Hayashi, M., and Awazu, S., J. Pharm. Pharmacol., 38, 659 (1986).
10. Iwamoto, K., Kato, T., Kawahara, M., Koyama, N., Watanabe, S., Miyake, Y., and Sunamoto, J., J. Pharm. Sci., 80, 219 (1991).
11. Illum, L., Davis, S.S., Müller, R.H., Mak, E., and West, P., Life Sci., 40, 367 (1987).
12. Alving, C.R., Pharmacol. Ther., 22, 407 (1983).
13. Saba, T.M., Arch. Intern. Med. 126, 1031 (1970).
14. Davis, S.S., Illum, L., Washington, C., and Harper, G., Int. J. Pharm., 82, 99 (1992).
15. Davis, S.S. and Hansrani, P., Int. J. Pharm., 23, 69 (1985).
16. Takada, M., Yururika, T., Katayama, K., Iwamoto, K., and Sunamoto, J., Biochim. Biophys. Acta, 802, 237 (1984).
17. Illum, L., West, P., Washington, C., and Davis, S.S., Int. J. Pharm., 54, 41 (1989).
18. Wheeler, J.J., Wong, K.F., Ansell, S.M., Masin, D., and Bally, M.B., J. Pharm. Sci., 83, 1558 (1994).
19. Eibl, H.J., personal communication.
20. Torchilin, V.P., and Trubetskoy, V.S., Adv. Drug Del. Rev., 16, 141 (1995).
21. Davis, S.S., Washington, C., and West, P., Ann. N.Y. Acad. Sci., 507, 75 (1987).
22. Brey, R.N., Pharm. Biotechnol., 6, 297 (1995).
23. Stossel, T.P., Mason, R.J., Hartwig, J., and Vaughan, M., J. Clin. Invest., 51, 615 (1972).
24. Carlsson, A., Sato, T., and Sunamoto, J., Bull. Chem. Soc. Jpn., 62, 791 (1989).
25. Fukui, H., Akiyoshi, K., Sato, T., and Sunamoto, J., J. Bioactive Comp. Polym., 8, 305 (1993).

26. Tonaki, H., Saba, T.M., Mayron, L.W., and Kaplan, E., Exp. Mol. Pathol., 25, 189 (1976).
27. Takino, T., Nakajima, C., Takakura, Y., Sezaki, H., and Hashida, M., J. Drug Targeting, 1, 117 (1993).
28. Klang, S.H., Baszkin, A., and Benita, S., Int. J. Pharm., 132, 33 (1996).
29. Elbaz, E., Zeevi, A., Klang, S., and Benita, S., Int. J. Pharm., 96, R1 (1993).
30. Oku, N., Tokudome, Y., Namba, Y., Saito, N., Endo, M., Hasegawa, Y., Kawai, M., Tsukada, H., and Okada, S., Biochim. Biophys. Acta, 1280, 149 (1996).
31. Felgner, P.L., Gadek, T.R., Holm, M., Roman, R., Chan, H.W., Wenz, M., Northrop, J.P., Ringold, G.M., and Danielsen, M., Proc. Natl. Acad. Sci. USA, 84, 7413 (1987).
32. Gao, X. and Huang, L., Biochem. Biophys. Res. Commun., 179, 280 (1991).
33. Klang, S.H., Parnas, M., and Benita, S., Proceedings of the CDC Conference, 3rd Expert Meeting, Berlin, 1997.
34. Klang, S.H., Frucht-Pery, J., Hoffman, A.S., and Benita, S., J. Pharm. Pharmacol., 46, 986 (1994).
35. Muchtar, S., Almog, S., Torracca, M.T., Saettone, M.F., and Benita, S., Ophth. Res., 24, 142 (1992).
36. Singh, K. and Mezei, M., Int. J. Pharm., 19, 263 (1984).
37. Adams, D.A., Richardson, G.J., Ryman, B.E., and Wisniewski, H.M., J. Neurol. Sci., 31, 173 (1977).
38. Lucks, J.S., Müller, B.W., and Müller, R.H., Int. J. Pharm., 63, 183 (1990).
39. Buszello, K. and Müller, B.W., Third European Congress of Pharmaceutical Sciences, Edinburgh (1996).
40. Jeppsson, R., and Rössner, S., Acta Pharmacol. Toxicol., 37, 134 (1975).
41. Collins-Gold, L.C., Lyons, R.T., and Bartholow, L.C., Adv. Drug Del. Res., 5, 189 (1990).
42. Levy, M.Y. and Benita, S., Int. J. Pharm., 54, 103 (1989).
43. Washington, C., Chawla, A., Christy, N., and Davis, S.S., Int. J. Pharm., 54, 191 (1989).
44. Levy, M.J., Schutze, W., Fuhrer, C., and Benita, S., J. Microencapsulation, 11(1), 79 (1994).
45. Yamaguchi, T., Nishizaki, K., Itai, S., Hayashi, H., and Ohshima, H., Pharm. Res., 12(9), 1273 (1995).
46. Washington, C., and Davis, S.S., Int. J. Pharm., 39, 33 (1987).
47. Westesen, K., and Wehler, T., J. Pharm. Sci., 81(8), 777 (1992).
48. Parnham, M.J., Proceedings of the CDC Conference, 3rd Expert Meeting, Berlin, 1997.
49. Bock, T.K., and Müller, B.W., Pharm. Res., 11, 589 (1994).
50. Yanez, A.M., Wallace, M., Ho, R., Shen, D., and Yaksh, T.L., Anaesthesiology, 82(5), 1189 (1995).
51. Oba, N., Sugimura, H., Umehara, Y., Yoshida, M., Kimura, T., and Yamaguchi, T., Lipids, 27(9), 701 (1992).
52. Sakaeda, T. and Hirano, K., J. Drug Target., 3(3), 221 (1995).
53. Deckelbaum, R.J., Hamilton, J.A., Moser, A., Bengtsson-Olivecrona, G., Butbul, E., Carpentier, Y.A., Gutman, A., and Olivecrona, T., Biochemistry, 29, 1136 (1990).

54. Kurihara, A., Shibayama, Y., Mizota, A., Yasuna, A., Ikeda, M., Sasagawa, K., Kobayashi, T., and Hisaoka, M., Biopharm. Drug Dispos., 17(4), 331 (1996).
55. Sakaeda, T., Takahashi, K., Nishihara, Y., and Hirano, K., Biol. Pharm. Bull., 17(11), 1490 (1994).
56. Lin, S.-Y., Wu, W.-H., and Lui, W.-Y., Pharmazie, 47, 439 (1992).
57. Tibell, A. and Norrlind, B., Transpl. Int., 7(6), 438 (1994).
58. Lundberg, B., J. Pharm. Sci., 83, 72 (1994).
59. Moriwaki, T., Yoshikawa, H., Takada, K., and Muranishi, S., Yukuzaigaku, 50, 263 (1990).
60. Grolier, P., Azais-Braesco, V., Zelmire, L., and Fessi, H., Biochim. Biophys. Acta, 1111, 135 (1992).
61. Connelly, P.W., and Kuksis, A., Biochim. Biophys. Acta, 666, 80 (1981).
62. Porter, C.J., Charman, S.A., and Charman, W.N., J. Pharm. Sci., 85(4), 351 (1996).
63. Engel, R.H., Riggi, S.J., and Fahrenbach, M.J., Nature, 219, 856 (1968).
64. Matsuzawa, A., Morishita, M., Takayama, K., and Nagai, T., Biol. Pharm. Bull., 18(12), 1718 (1995).
65. Honda, T., Okamoto, M., Maeda, M., Kawagoshi, T., Nihonmatsu, I., Yamashita, I., Yoshida, M., Tazawa, K., and Fujimaki, M., Drug Del. Systems, 7, 43 (1992).
66. Decker, D.E., Vroegop, S.M., Goodman, T.G., Peterson, T., and Buxser, S.E., Chem. Phys. Lipids, 76(1), 7 (1995).
67. M.J., Polacheck, I., Barenholz, Y., and Benita, S., J. Med. Vet. Mycol., 31, 207 (1993).
68. Ryan, K.A., J. Intraven. Nurs., 19, 123 (1996).
69. Sorkine, P., Nagar, H., Weinbroum, A., Setton, A., Israitel, E., Scarlatt, A., Silbiger, A., Rudick, V., Kluger, Y., and Halpern, P., Crit. Care Med., 24, 1311 (1996).
70. Tibell, A., Lindholm, A., Sawe, J., Chen, G., and Norrlind, Pharmacol. Toxicol., 76, 115 (1996).
71. Sarubin, J., Fortschr. Med., 100, 2135 (1982).
72. Levy, M.J., Langerman, L., and Gottschalk-Sabag, S., Pharm. Res., 6, 510 (1989).
73. Klement, W., and Arndt, J.O., Br. J. Anaesth., 67, 281 (1991).
74. Doenicke, A.W., Roizen, M.F., Rau, J., Kellermann, W., and Babl, J., Anaesth. Analg., 82, 472 (1996).
75. Ayestaran, A., Lopez, R.M., Montoro, J.B., Estibalez, A., Pou, L., Julia, A., Lopez, A., and Pascal, B., Antimicrob. Agents Chemother., 40(3), 609 (1996).
76. Joly, V., Geoffray, C., Reynes, J., Goujard, C., Mechali, D., Maslo, C., Raffi, F., and Yeni, P., J. Antimicrob. Chemother., 38(1), 117 (1996).
77. Schwarz, J.S., Weisspapir, M.R., and Friedman, D.I., Pharm. Res., 12, 687 (1995).
78. Yoshikawa, M., Saisho, H., Ebara, M., Iijima, T., Iwama, S., Endo, F., Kimura, M., Shimamura, Y., Suzuki, Y., and Nakano, T., Cancer Chemother. Pharmacol., 33 Suppl., S149 (1994).
79. Altmayer, P., Grundmann, U., Ziehmer, M., and Larsen, R., Anaesthesiol. Intensivmed. Notfall,ed. Schmerzther., 28, 415 (1993).
80. Levy, M.J., Polacheck, I., Barenholz, Y., and Benita, S., Pharm. Res., 12, 223 (1995).
81. Friedman, D.I., Schwarz, J.S., and Weisspapir, M., J. Pharm. Sci., 84, 324 (1995).

82. Treffel, P., Gabard, B., Skin Pharmacol., 6, 268 (1993).
83. Keipert, P.E., Artif. Cells Blood Subst. Immobil. Biotechnol., 23, 381 (1995).
84. Flaim, S.F., Hazard, D.R., Hogan, J., and Peters, R.M., Invest. Radiol., 26, S122 (1991).
85. Steinbach, G.C., Baker, K.G., Lim, G.L., Mattrey, R.F., and Arellano, R.S., Artif. Cells Blood Substit. Immobil. Biotechnol., 22, 1501 (1994).
86. Mizushima, Y., Yanagawa, A., and Hoshi, K., J. Pharm. Pharmacol., 35, 667 (1983).
87. Rubinstein, A., Pathak, Y.V., Kleinstern, J., Reches, A., and Benita, S., J. Pharm. Sci., 80, 643 (1991).
88. Naveh, N., Muchtar, S., and Benita, S., J. Ocul. Pharmacol., 10, 509 (1994).
89. Gray, H.S., Holt, B.L., Whitaker, D.K., and Eadsforth, P., Br. J. Anaesth., 68, 272 (1992).
90. Bekyarova, G., Yankova, T., and Galunska, B., Artif. Cells Blood Substit. Immobil. Biotechnol., 24, 629 (1996).
91. Yamaguchi, T., Fukushima, Y., Itai, S., and Hayashi, H., Biochim. Biophys. Acta, 1256, 381 (1995).
92. Kamperman, H., and Sallis, J.D., J. Pharm. Pharmacol., 47, 802 (1995).
93. von Dardel, O., Mebius, C., and Mossberg, T., Acta Anaesth. Scand., 20, 221 (1976).
94. Benita, S., Friedman, D., and Weinstock, M., Int. J. Pharm., 30, 1986, 47–55.
95. Kanno, C., Jap. J. Zootech. Sci., 51, 74 (1980).
96. Yuasa, H., Sekiya, M., Ozeki, S., and Watanabe, J., Biol. Pharm. Bull., 17, 1262 (1994).
97. Yuasa, H., Sekiya, M., Ozeki, S., and Watanabe, J., Biol. Pharm. Bull., 17, 756 (1994).
98. Sato, H., Liu, H.X., Adachi, I., Ueno, M., Lemaire, M., and Horikoshi, I., Biol. Pharm. Bull., 17, 1526 (1994).
99. Eskey, C.J., Wolmark, N., McDowell, C.L., Domach, M.M., and Jain, R.K., J. Natl. Cancer Inst., 86, 293 (1994).
100. Sasaki, H., Kakutani, T., Hashida, M., and Sezaki, H., J. Pharm. Pharmacol., 37, 461 (1985).
101. Fukui, H., Akiyoshi, K., and Sunamoto, J., J. Biomater. Sci. Polym. Ed., 7, 829 (1996).
102. Repta, A.J. in Breimer, D.D. and Speiser, P. (Eds.), Topics in Pharmaceutical Sciences, Elsevier/North-Holland, Amsterdam, 1981, p. 131 (1981).
103. Medoff, G., Brajtburg, J., Kobayashi, G.S., and Bolard, J., Annu. Rev. Toxicol., 23, 303 (1983).
104. Washington, C., Taylor, S.J., and Davis, S.S., Int. J. Pharm, 46, 25 (1988).
105. Forster, D., Washington, C., and Davis, S.S., J. Pharm. Pharmacol., 40, 325 (1988).
106. Joly, V., Aubry, P., Ndayiragide, A., Carriere, I., Kawa, E., Mlika-Cabanne, N., Aboulker, J.P., Coulard, J.P., Larouze, B., and Yeni, P., Clin. Infect. Dis., 23, 556 (1996).
107. Levy, M.J., Polacheck, I., Barenholz, Y., and Benita, S., J. Med. Vet. Mycol., 31, 207 (1993).
108. Griffin, T., and Raso, V., Cancer Res., 51, 4316 (1991).

109. Yoshioka, T., Skalko, N., Gursel, M., Gregoriadis, G., and Florence, A.T., J. Drug Target., 2(6), 533 (1995).
110. Digenis, G.A., Parr, A.F. and Jay, M., in Davis, S.S. and Wilson, C.G. (Eds.), Drug Delivery to the Gastrointestinal Tract. Hardy, J.,G., Ellis Horwood Limiteds, Chichester, 111 (1989).
111. Buszello, K., and Müller, B.W., Proceedings of the 24th International Symposium on Controlled Release of Bioactive Materials, Stockholm, 1997, p. 811.
112. Bakan, D.A., Longino, M.A., Weichert, J.P., and Counsell, R.E., J. Pharm. Sci., 85(9), 908 (1996).
113. Longino, M.A., Bakan, D.A., Weichert, J.P., and Counsell, R.E., Pharm. Res. 13(6), 875 (1996).
114. De Baere, T., Zhang, X., Aubert, B., Harry, G., Lagrange, C., Ropers, J., Dufaux, J., Lumbroso, J., Rougier, P., Ducreux, M., and Roche, A., Radiology, 201(3), 731 (1996).
115. Bentley, P.K., Johnson, O.L., Washington, C., and Lowe, K.C., J. Pharm. Pharmacol., 45, 182 (1993).
116. Magdassi, S., and Siman-Tov, A., Int. J. Pharm., 59, 69 (1990).
117. Riess, J.G., Artif. Cells Blood Substit. Immobil. Biotechnol., 22(2), 215 (1994).
118. Fujita, T., Sumaya, T., and Yokayama, K., Eur. Surg. Res., 3, 436 (1971).
119. Riess, J.G., Vox Sang, 61, 225 (1995).
120. Zarif, L., Postel, M., Trevino, L., Riess, J.G., Valla, A., and Follana, R., Artif. Cells Blood Substit. Immobil. Biotechnol., 22(4), 1193 (1994).
121. Wehrle, P., Korner, D., and Benita, S., Pharm. Dev. Technol., 1(1), 97 (1996).
122. Chiang, C., Fuller, G.C., Frankefeld, J.W., and Rhodes, C.T., J. Pharm. Sci., 67, 63 (1978).
123. Benoy, C.H., Elson, L.A., and Schneider, R., Br. J. Pharmacol., 45, 135 (1972).
124. Brodin, A.F., and Frank, S.G., Acta Pharm. Suec., 15, 111 (1978).
125. Khopade, A.J., Mahadik, K.R., and Jain, N.K., Pharmazie, 51(8), 558 (1996).
126. El-Sayed, A.A., and Repta, A.J., Int. J. Pharm., 13, 303 (1983).

6

Parenteral Fat Emulsions: Structure, Stability, and Applications

Jörn-Stefan Lucks and Bernd W. Müller
Christian Albrecht University, Kiel, Germany

Karsten Klütsch
B. Braun Melsungen AG, Melsungen, Germany

I. INTRODUCTION

Parenteral fat emulsions can be defined as oil-in-water emulsions with mean droplet diameters ranging from 200 to 500 nm. For intravenous applications it is necessary that there be no droplets larger than the diameter of the capillaries (about 5 μm) to avoid blockage. However, the discussion about the maximum particle size is at present in a state of controversy. More recent limits for oil droplets in fat emulsions and total parenteral nutrition (TPN) regimes permit an amount of 0.4% of the droplets to be larger than 5 μm. Recent investigations demonstrated the coexistence of lecithin-coated oil droplets and liposomal structures in the aqueous phase. Their contribution to emulsion stability is yet unknown. Due to the large interface parenteral fat emulsions are thermodynamically instable. Several theories are used to explain the mechanisms of stabilization. The interaction of all forces is very difficult to describe. A complex theory covering all aspects is still lacking. Despite this theoretical shortcoming parenteral fat emulsions are widely employed for (total) parenteral nutrition and as drug carriers. In addition to numerous experimental formulations (e.g., for cytostatics) there are several commercially available emulsions containing hypnotics such as diazepam, etomidate, or propofol. In most cases, however, the incorporation of drugs drastically reduces the physical stability of the emulsion. Therefore, only a few products are on the market despite considerable endeavors. The application of parenteral fat emulsions as carriers for nasal and ocular delivery of drugs is almost neglected but might be promising. A moderate increase in droplet size could be tolerated much better compared to the intravenous route. Parenteral fat emulsions are eliminated from the bloodstream similarly to chylomicrons. Their artificial character may lead to an undesirable interaction with the immune system and the reticuloendothelial system (RES). Careful selection of the emulsion system (e.g., emulsifiers) can circumvent overloading the RES.

II. DEFINITIONS

In the literature on parenteral fat emulsions there are many different terms that are used synonymously and sound alike but that describe different physical systems. The International Union of Pure and Applied Chemistry (IUPAC) generally defines emulsions as *dispersion of liquids or liquid crystals in another liquid* (41,71). In general, depending on the outer phase it is possible to distinguish between water-in-oil and oil-in-water emulsions. Parenteral fat emulsions (PFEs) belong to the oil-in-water emulsions and have a mean droplet size of 200–500 nm. Parenteral fat emulsions are milky in appearance. When diluted with water they change by degrees to a clear solution with a blue shimmer caused by diffraction of light on the oil droplets. Due to their particle size parenteral fat emulsions are suitable for *intravenous application*. Therefore they are sometimes referred to as intravenous emulsions (3,12,27,59,192,194). Mizushima's research group (Kawasaki, Japan) uses the term *lipid microspheres* in his publications (116–118,124). However, in this case these are also parenteral fat emulsions. Under certain conditions (high pressure, low amount of oil, high amount of lecithin) it is possible to produce emulsions with a droplet size under 100 nm (132,191). These are called *nanoemulsions*, but differ from parenteral fat emulsions only in their size. Due to the large interface—and the high interface energy that goes with this—parenteral fat emulsions and nanoemulsions are *thermodynamically unstable systems*.

Between these systems and the so-called *microemulsions* a clear differentiation must be made. The microemulsions are often wrongly described as parenteral fat emulsions (142) and represent—from the physical point of view—a completely different system. Microemulsions are thermodynamically stable, single phase, transparent or opalescent, liquid, and in unsheared state are optically isotropic systems consisting of several liquids that cannot be mixed together and are produced with the aid of surfactants or surfactant mixtures (88). The physical structure of these systems is the object of many investigations (51,56,64, 84,103,152), in which they are described either as swollen micelles or as critical solutions (47).

III. PHYSICAL STRUCTURE OF PARENTERAL FAT EMULSIONS

Emulsions are defined as dispersions of liquids and/or liquid crystals in other liquids (41,71). It is noticeable that an *excess of phospholipid* is used to stabilize the emulsions. Theoretical calculations (55) show that the amount to achieve a monomolecular film covering the oil–water interface is only 50% of the phospholipid that is really needed for the stabilization. In fact, nearly all commercial emulsions are stabilized with 1.2% lecithin irrespective of the amount of oil used. However, in more recent preparations lesser amounts (0.6–0.8%) are used. Friberg (13,48) in the 1970s and Rydhag (153,154) in the early 1980s proved in their studies that for stabilization of emulsions with lecithins as emulsifiers the formation of liquid crystalline structures on the interface is absolutely necessary. The soya lecithin used can form lamellar *liquid crystalline structures*, especially in the presence of further negatively charged phospholipids which results in an upstake of water with the uptake of water. According to this, these structures, which can be clearly distinguished from the oil and water phase, so that it can be said that they have their own emulsifier phase, consisted of several phospholipid layers. This structure concept also found its way into the textbooks (104). Shortly after this electron microscopic liposomal structures could be detected in commercial fat emulsions. Groves et al. (55) discovered that the oil droplets were surrounded by a single phospholipid layer and showed free multilamellar structures, corresponding to *multilamellar liposomes*. With the existence of these phospholipid reservoirs the distinct good heat stability (e.g., when autoclaving) of the commercial fat emulsions could be explained without any other mechanism being suggested.

However, Westesen's recent investigations (193,194) give a different view. Her electron microscopic studies show that at suitable freezing speed soya oil too forms crystals and lamellar structures that cannot be distinguished from lamellar phospholipid layers, or that these are possibly simulated. She found that most of the particles above 100 nm were filled with oil and only a few multilamellar vesicles were present. Very occasionally W/O/W structures could be observed. After freezing of the samples in melting propane multilayered structures on the interface could not be detected. Most of the oil droplets were surrounded by a single layer.

In addition, she found many small particles under 60 nm whose structure could not be explained by electron microscopy. With the aid of nuclear magnetic resonance (NMR) measurements and x-ray small-angle investigations bilayer structures could be detected. Thirty percent of all particles are (supposedly) small unilamellar vesicles (SUVs); altogether the number of particles under 100 nm is distinctly greater than the number of particles above 100 nm. However, the usual method for characterization of fat emulsions, photon correlation spectroscopy (PCS), is not able to trace these small vesicles beside the larger particles that are also present. If the larger particles are centrifuged the PCS measurement of the subordinate liquid shows a mean PCS diameter of 38 nm.

The consequences arising from these new observations are considerable. Due to the free phospholipid in fat emulsions used for parenteral nutrition there is an increase in serum cholesterin level (5) and abnormal lipoproteins are formed (61,115). Both of these phenomena are more pronounced with 10% emulsions that have a higher amount of free phospholipid. As a result of these findings in the meantime emulsions are being produced that are stabilized with *less lecithin* (0.6–0.8%) (19).

IV. PHYSICAL STABILITY

Emulsions are thermodynamically instable systems. The work ΔW which is necessary for dispersion is determined by the *interface tension* γ and the interface magnification of the disperse phase ΔA:

$$\Delta W = \Delta A \times \gamma.$$

Fjellestad (44) discovered that for lecithin-stabilized soja oil–water–systems

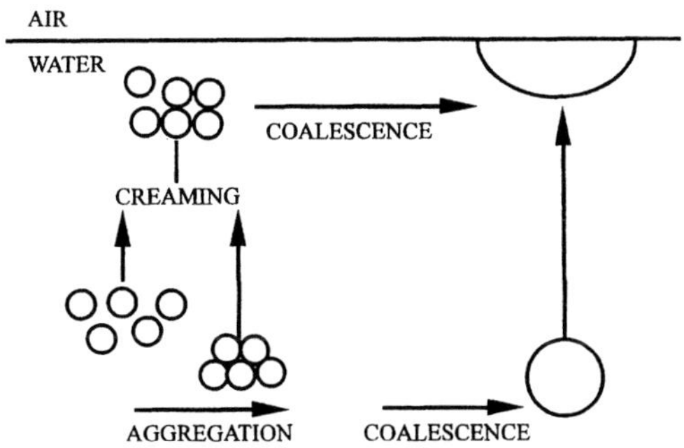

Figure 1 Diagram of creaming and coalescence of an O/W emulsion (38).

there were interface tensions of 0.3 and 1.5 mN/m. If 1 mL soja oil is distributed in fine droplets with a radius of 250 nm, then the interface increases, assuming a spherical form from 4.86 cm^2 to 2.4 $\times$ 10^5 cm^2, which corresponds to 24 m^2. Consequently, the interface energy is between 0.0072 and 0.036 J. Due to this *high interface energy* the system tries to reach a thermodynamically more favorable level by reducing the interface. A possible entropy increase of the oil is overcompensated by the high-order state of the water near the oil droplets (e.g., hydrate covering) and the entropy loss of the water molecules that is connected with this.

Macroscopically the phase separation can be described as follows (92): As a result of the difference in density the oil droplets float on the surface. Reversible formation of agglomerates (grape-like flakes) occurs and simultaneously or following this coalescence of drops (Fig. 1) (38) takes place. Whereas the formation of a creamy layer (due to floating agglomerates) does not represent a decrease in quality, as it can be eliminated by simple shaking, the coalescence of oil droplets is irreversible and represents a decrease in quality.

The stability of colloidal emulsions—also due to the unexplained structure—is very difficult to foresee, as particularly in mixtures for total parenteral nutrition (TPN) there are many interactions. In a series of articles Washington attempted to put the prediction of stability of emulsions in TPN regimes on a rational basis (183–189). To explain the stability of emulsions the following effects and theories are referred to:

DLVO theory
Mechanical stability of the emulsifier film
Hydration forces
Steric (= entropic or osmotic) stabilization
Marangoni effect (with surfactants that are soluble in water).

To determine the charge, the *electrophoretic mobility*, i.e., the mobility of the charged particles in the electric field, is determined. From this the so-called ζ potential can be calculated, which is a parameter for the stabilization by means of electrostatic forces. The *mechanical stability* of the emulsifying film is also of considerable importance. On the one hand, it must be sufficiently *rigid* to stabilize the interface, but on the other hand must also be sufficiently *flexible* that collision of emulsion droplets does not lead to rupture of the emulsifying film and to resulting coalescence.

The macroscopically perceivable mechanical properties (such as resistance to shear forces) of a gel system are described by the concept of *viscosity*. In analogy to this the concept *microviscosity* describes the viscosity of narrowly limited areas such as the emulsifying layer at the interface. By using fluorescence markers whose properties are dependent on the immediate surroundings of the molecules, an attempt is made to quantify these effects. First attempts to transfer this technology, which up to now was only used to investigate biological cells and liposomal structures, especially to parenteral fat emulsions, were promising (37). A correlation between reduced microviscosity and low stability of fat emulsions containing MCT as compared to fat emulsions with soja oil was found. However, there is also a series of investigations that show that MCT/LCT emulsions provide better stability in TPN preparations than LCT emulsions (112).

The term *steric stabilization* describes the stabilization of colloid systems by molecules with large hydrophilic areas (62,138) such as ethoxylated surfactants (e.g., poloxamers). The hydrophilic areas consist of ethylene oxide chains which have free mobility and which extend far into the aqueous dispersion medium. If particles come nearer together, then direct contact is not possible due to steric reasons. When particles come nearer to each other, the ethylene oxide chains penetrate each other and thus decrease their mobility, which results in a loss of entropy. However, the water that is released achieves an increase in entropy. Due to the osmotic drop between free water phase and the space that is filled with the ethylene oxide chains the water tries to dilute these latter chains. A driving force for this repulsion is thus the *osmotic pressure*. In production of emulsions with surfactants which are soluble in water the *Marangoni effect* plays a role (85,150). Before reaching adsorption equilibrium, surfactant molecules diffuse continuously from the water phase to the interface. Between two droplets that are directly next to each other a reduction of surfactant molecules in the water phase between them takes place. This means that the interface tension decreases more slowly than in the rest of the droplet surface. Due to this gradient in the interface tension a movement of the adsorption layer occurs along the interface into this intermediate space. A thin layer of water is also transported with this. This means that the thinning out of the aqueous intermediate layer is prevented and coalescence of the oil droplets does not occur (Fig. 2) (92). However, the two forces of *hydration* (the force that is necessary to get rid of the

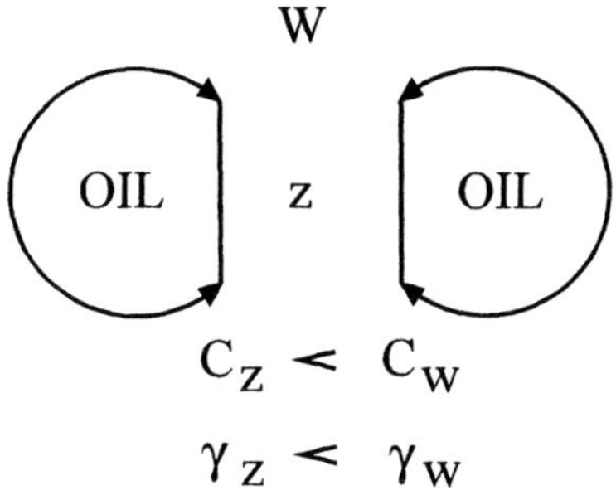

Figure 2 Stabilization by means of the Marangoni effect. c, Emulsifier concentration; γ, interface tension; z, intermediate space.

water layer surrounding an oil droplet) and the Marangoni effect are difficult to detect in experiments but should not be ignored.

V. PARENTERAL FAT EMULSIONS

Emulsions for oral or topical application are among some of the oldest formulations of drugs. As a result of their physical instability, especially due to interaction with drugs and the resulting toxicity, it is only in the middle of our century that it has become possible to produce emulsions that can also be applied parenterally (28,30). These systems are subject to strict criteria regarding the *maximum particle size*. Regulations of the latest editions of USP XXIII and BP 88 do not limit the particle size explicitly, but the diameter of the smallest blood vessel (5 μm) is still the limit (16,30). However, some authors (199) want the maximum value to be 1 μm. But it should also be taken into consideration that the oil droplets are *biologically degradable* and thus a small number of larger particles would seem tolerable. So in 1997 it was suggested that the larger particle size fraction be limited to 0,4% above 5 μm. Connections can be established between the tolerability of the emulsion and the maximum droplet size (50). Parenteral fat emulsions are used today for parenteral nutrition and as drug carrier systems. Similar systems are employed as blood substitutes (e.g., perfluorocarbon emulsions) and as contrast medium (iodized fat phases).

VI. PARENTERAL NUTRITION

The idea of giving patients who are in a critical nutrition state (patients in intensive care or in convalescence) nutrition, especially calories, by parenteral route, has been in existence for a long time. Parenteral nutrition is indicated when adequate nutrition is not possible orally or extraorally. This is particularly the case with (52):

Traumas of the gastrointestinal tract
Diseases of the gastrointestinal tract (mainly cancer, infections)
Burns
Radiation expositions
Psychic disorders (e.g., bulimia)

The earliest recorded experiments were carried out in England by Courten in 1679 (24). He injected a dog intravenously half an ounce of olive oil that had not been emulsified. The result was that the dog died of a fat embolism. At the end of the nineteenth century experiments were carried out giving campher oil subcutaneously to patients with spinal tuberculosis. At this time attempts were once again made to administer fat intravenously (63,157,162). However, these preparations could not be used clinically.

The first fat emulsions that could be applied clinically were developed at the beginning of the 1920s. Kleinberger and Pamperl (87) divided the development of parenteral fat emulsions into three generations. The first emulsions were developed in the 1920s in Japan. Under the commercial name of Yanol a 3% castoreum emulsion which was stabilized by egg lecithin was put on the market (139,200, 201). These emulsions were submitted to intensive testing in the 1930s in the USA, but had to be withdrawn due to side effects such as shaking fits, fever, and shock reactions (177). The use of *purified phosphatides* at the end of the 1940s led to the second generation of fat emulsions (113). The best known product was on the market under the name *Lipomul* (in Europe: *Infonutrol*). Cottonseed oil was used as the fat component, and soja phosphatides and poloxamer-188 were the emulsifiers (see Table 1). Isotonicity was produced by glucose (87).

Table 1 Composition of Cottonseed Oil Emulsions of the Second Generation (mOsm)

	Lipomul	Lipiphysan	Lipofundin		Lipophysan
Cottonseed oil	15.0	15.0	10.0	20.0	15.0
Soja lecithin	1.2	2.0	—	—	2.0
Hydrated soja phosphatide	—	—	0.75	1.5	—
Poloxamer 188	0.3	—	—	—	—
Glucose	4.0	—	—	—	—
Sorbitol	—	5.0	5.0	5.0	—
Glycerol	—	—	—	—	2.5
Water for injection purposes	ad 100.0	100.0	100.0	100.0	100

The more widespread clinical use of these emulsions led to the observation of many *side effects* (161). As an early reaction (usually at the beginning of the infusion), reddening of the skin, feelings of heat, sickness, vomiting, and headaches were observed. In severe cases there was also shaking fits, tachycardia, drop in blood pressure, feelings of anxiety with difficulties in breathing, and shock-like conditions. For these symptoms fat embolism was not held to be responsible, but incorrectly the colloid character of the emulsions, as similar observations for dextran solutions had been made (203).

An *endogenic pyrogen release* was thought to be due to *impurities* in the emulsifier (especially in the case of poloxamer-188) (72). This result was confirmed in a more recent investigation (9). After a longer period of administration sometimes the so-called overloading syndrome occurred, which was characterized by hyperlipidemia, fever, anorexia, pains in the upper stomach, hemolysis, and anemia, sometimes with lethal end (hepatosplenomegaly and insufficiency of the liver) (54,114,178). In addition the cottonseed oil which was used contained the steroid *gossypol*, which is also toxic (95). For these reasons the Food and Drug Administration decreed a general prohibition on fat emulsions for parenteral use in 1964.

The name Intralipid stands for emulsions of the third generation. These emulsions were introduced in Sweden at the beginning of the 1960s. *Fractionated soja bean oil* was used as fat phase and *lecithin* as emulsifier. From investigations of Wretland's research group (163) in the 1950s egg lecithin is supposed to have *better tolerability* than soja lecithin. Cats were given soja bean oil emulsions which had different lecithins as emulsifiers. Systems stabilized with egg phosphatide were tolerated without any clinical abnormalities, whereas with the formulations containing soja phosphatide severe falls in blood pressure and apnea were observed. Other authors (202) think that these effects are dependent on the degree of purity of the lecithins that were used. However, more recent studies (57) have shown that emulsions stabilized with soja lecithin can also be tolerated without clinical complications.

Since the beginning of the 1980s emulsions have been developed with safflower oil (dyer's safflower oil) (11,15,22,171,198), as this has more linoleic acids and less saturated fatty acids as compared to soja oil. However, the formulations of the fourth generation replace parts of the soja oil with middle chain triglycerides (MCTs) (57a,83,86). The explanation for this is provided by the *different* metabolism of middle chain triglycerides as compared to the long chain derivatives (4,39). Middle chain triglycerides are oxidized more quickly and more completely, prevent lipolysis, and thus lower the plasma level of free fatty acids and cholesterin synthesis (197). For this reason in the last years the amount of soja bean oil (and thus the amount of unsaturated fatty acids) was reduced, as an unfavorable influencing of the liver and lung function and immune reactions were observed due to the high linolenic acid content (35,43,148,159,182). This also occurred in lipofundin MCT/LCT as the first fat emulsion parts of the soja oil were replaced by middle chain triglycerides (from cocoa oil), thus improving the tolerability. New developments aim to further optimize the composition of these fat emulsions not only by lowering the LCT amount but also by optimizing the proportion of ω-3 to ω-6 fatty acids by adding fish oil. Fatty acids are the preliminary stages for certain lipid mediators (prostaglandin, thromboxane, leukotrienes) that control the course of infectious reactions in the body and have an influence on the tonus of the blood vessels. The ω-6 fatty acids are the preliminary stage of Eicosanoides which have a proinflammatory and prothrombotic effect. These effects are considerably weaker with the derivatives of the ω-3 fatty acids. A balanced supply of ω-3 and ω-6 fatty acids is therefore particularly important in cases of sepsis, SIRS, and chronic cases of infection in order to prevent the spreading of an inflammatory reaction.

The use of structured lipids in fat emulsions is also possible. In contrast to the physical mixtures of MCT and LCT these lipids contain a mixture of middle chain and long chain fatty acids within a tryglyceride. Distinctions are made between chemically defined and randomized structured lipids.

The chemically defined structured lipids are produced by a complicated enzymatic reesterification from various oils, e.g., MCT oil and soja bean oil. The enzymatic reesterification permits exactly defined triglycerides with a special placement of the fatty acids on the different positions of the glycerol framework to be produced. In animal experiments a favorable influencing of the fat clarification, the protein synthesis, and the bacteria clarification by the liver could be observed after the use of these emulsions (145,197).

When structured lipids of the randomized type (produced by heat reesterification) were used the different fatty acids were randomly distributed on the positions of the triglyceride framework. In clinical examinations with structured lipids of this type similar advantages over the pure LCT emulsions could be shown, as could also be observed with the physical mixtures with the reduced LCT amount (68,109,140,155). The oil-emulsifier composition of selected parenteral fat emulsions is given in Table 2.

VII. TOTAL PARENTERAL NUTRITION

Patients who can only be fed parenterally over a longer period of time need not only calories but also amino acids, electrolytes, trace elements, and vitamins. However, if separate infusions of the different solutions are applied very large volumes (up to 5 L) must be given. For this reason so-called all-in-one solutions are aimed at. A popular standard formulation (≈2500 kcal, 2600 mL) consists of 1000-mL amino acid solution (isoleucine, leucine, lysine, methionine, phenylalanine, threonine, tryptophane, valine, alanine, arginine, histidine, proline, tyro-

Table 2 Selected Parenteral Fat Emulsions Currently on the Market

Product	Producer	Oil component	Concentration	Emulsifier
Abbolipid/Liposyn	Abbott	Soja oil/safflower oil	10%/20%	Egg-PL
Intralipid	Pharmacia	Soja oil	10%/20%/30%	Egg-PL
LipofundinN/Endolipide	B. Braun	Soja oil	10%/20%	Egg-PL
LipofundinMCT/LCT/ Medialipide/Vasolipid	B. Braun	Soja oil/MCT	10%/20%	Egg PL
Lipovenos	Fresenius	Soja oil	10%/20%	Egg-PL
Ivelip/Salvilipid	Clintec/Baxter	Soja oil	10%/20%	Egg PL
Clinoleic	Clintec/Baxter	Soja oil/olive oil	20%	Egg-PL
Intralipos	Green Cross	Soja oil	10%/20%	Egg-PL

sine, glycine), 500 mL glucose 50% plus 500 mL glucose 5% and 500 mL of a fat emulsion of 20%. In this the electrolytes (sodium acetate, sodium chloride, potassium acetate, potassium chloride, potassium phosphate, calcium gluconate, and magnesium sulfate), the trace elements (iron, molybdenum, chrome, copper, manganese, selenium, zinc, fluoride, and iodide), and the water- and oil-soluble vitamins (folic acid, vitamin B complex) are dissolved.

The *physical stability* of such TPN regimes is the object of many investigations (6,10,16,29,60,69,94,143,158,166,172,186,195). Due to the numerous components there are often unexpected interactions. Thus there is sometimes precipitation of calcium phosphate in the mixture (89) if by inattention calcium chloride and sodium or potassium phosphate are to be dissolved in the system. Special vitamins, e.g., ascorbic acid, are sensitive to different application conditions (temperature, light, oxygen, etc.). For this reason in practice it is recommended that the vitamins only be added to the all-in-one mixture immediately before the infusion is given. Most of the publications deal with the physical stability of the used fat emulsions, whose shelf life can be reduced by additives from years to days. New developments in industry show standardized TPN preparations that contain carbohydrates, amino acids, fat, and electrolytes in a multicompartment bag. Due to the separate storage in different compartments these preparations have a stability of 2 years and can be stored at room temperature. Before the application the contents of the separate compartments are mixed and the TPN preparation can then be given extra vitamins and trace elements. Examples for TPN regimes are given in Table 3.

VIII. DRUG CARRIER SYSTEMS

There are several advantages that favor the use of parenteral fat emulsions as *carriers for drug applications* (147):

- Poor solubility in water and good solubility in oil
- Stabilization of drugs that are sensitive to hydrolysis
- Avoidance of sorption of drugs on the plastic material of the infusion set
- Reduction of side effects and toxicity of intravenously applied drugs
- Possible retardation of release
- Concentration in special compartments (or organs) and drug release at that site (drug targeting)

There are generally two types of drug emulsions, those in which the drug is already *dissolved in the oil phase* before production (this is described as de novo production) and those in which the drug is dissolved in a commercial fat emulsion for parenteral nutrition, usually with the aid of organic solvents such as ethanol or dimethyl acetamide *shortly before application*. However, with the

Table 3 Example for a TPN Regime

Moderately increased requirements: 16 g nitrogen, 2550 kcal (2150 nonprotein kcal)
Osmolarity: 1195 mOsm/L (without additions)
Shelf life: 90 days

Preparation	Volume (ml)	Amino acids (g)	Nitrogen (g)	Glucose g	Lipids (g)	Energy (kcal)	Electrolytes (mmol) Na^+	K^+	Mg^{2+}	Ca^{2+}	Phosphate	Chloride	Acetate
Aminoplasmal™ −10% E	1000	100	16			400	43	25	2.6		9	57	59
Glucose 10%	500			50		200							
Glucose 50%	500			250		1000							
Lipofundin MCT/LCT 20%	500				100	955					7		
Sodium chloride 5.85%	77						77					77	
Potassium chloride 15%	15							30				30	
Magnesium sulfate 50%	2								4.0				
Calcium gluconate 20%	15									6.8			
Potassium phosphates	20							20			12		
Total	2629	100	16	300	100	2555	120	75	6.6	6.8	28	164	59
Maximum further electrolytes							130	75	6.4	6.2	15		
Maximum total electrolytes							250	150	13	13	43		
Further additions	10 ml Tracutil™ (9 trace elements). Compatible vitamin preparations.												

second method, the in situ production, there is the danger of *drug precipitation* in the aqueous phase or the possibility of *breaking* of the carrier emulsion. This way is often possible for clinical research but not for commercial products that are to gain registration.

The first successful experiments with emulsions as parenteral, in particular intravenous drug carriers were published in the early 1970s by Jeppsson's research group (University of Uppsala and Vitrum AB, Sweden) (72–80). Very soon the potential of this drug formulation was realized (27,28) and numerous examples were published in the literature. In the meantime several articles giving a summary of this subject were published (21,147,170).

Diazepam is used in anaesthesia intravenously and in the treatment of epileptic convulsions. Due to its low solubility in water the drug had to be dissolved in a propylene glycol–water mixture. When this solution is diluted with blood there is a precipitation of diazepam in the vein and pain as well as a *vein infection* is often observed. In addition to this, the drug is often susceptible to *adsorption* on the infusion equipment and on plastic syringes (196). Due to the importance of this drug much research was carried out on the development of an emulsion formulation that contained diazepam (25,76,78,97–99,196). The frequency of the above mentioned side effect could be drastically reduced by the use of an emulsion formulation (26, 90,156). Further hypnotic drugs on which work was carried out were barbiturate acids (73,80,106), propofol (169), and etomidate (174).

For therapy of the following severe illnesses emulsion formulations were developed: *chronic arthritis* [corticoids (119,120,123) and, nonsteroidal antirheumatic drugs (122,125,164,167)], *Alzheimer's disease* [inhibition of central cholinesterase by physostigmine (7,8,49,144,151), *vascular diseases* [e.g., therapy of thrombosis with prostaglandine E_1 (65,121,126,168,176)], *cancer* [praclitaxel (175), perilla keton (141), rhizoxine (173), penclomedine (134,135,146), mitomycin C (136), bleomycin (137), CCNU (46,105), hexamethylmelamine (2), NSC No. 278214 (40)], and *systemic micoses* (e.g., by immune suppression after organ transplantations or HIV infections, amphotericin B (45,70,91,190), imidazole derivatives (127). Recently, parenteral fat emulsions have also been used as adjuvants for the application of vaccinations (101,102).

In spite of considerable research activities only the parenteral fat emulsions with drugs that are listed in Table 4 are commercially available. However, as many drugs are soluble in neither water nor oil, attempts have recently been made to use emulsion suspension systems (160). The drug that is suspended in the oil phase must be suitably finely milled before emulsifying. This should be possible with a Dyno Mill (Bachofen, Basel, Switzerland) if the production parameters are appropriately selected (100). However, it remains open as to whether partial dissolution of the drug occurs due to increases in temperature (e.g., autoclaving) and that whether recrystallization takes place when the temperature cools down to room temperature.

Table 4 Registered Emulsions Containing Drugs

Brand name	Producer	Drug	Oil phase	Applications
Diazepam-Lipuro	Braun Melsungen	Diazepam	Soja oil + MCT	IV
Disoprivan	Zeneca	Propofol	Soja oil	IV
Etomidat-Lipuro	Braun Melsungen	Etomidate	Soja oil + MCT	IV
Lipotalon	Merckle	Dexanethaso-nepalmitate	Soja oil + MCT	i. arthr.
Stesolid	Dumex	Diazepam	Soja oil, acet. Monoglycerides	IV

IX. FURTHER USES

A. Contrast Media

Emulsions for parenteral application can also be used for contrast radiographs of organs of the reticuloendothelial system (RES). Here lipophilic phases that are marked with elements with a high periodic number (brome, iodine) are used. These make it possible to show liver and spleen in computer tomography and in ultrasonic examinations. Iodized oils (93,180) are used, along with *bromized perfluorocarbons* (107,111).

B. Perfluorocarbon Emulsions

Perfluorized hydrocarbons are able to dissolve oxygen and carbon dioxide to a considerable extent. For this reason they have been developed in emulsion form as *blood substitute* (50,149). As in the last decades increasingly infectious diseases (especially HI viruses) have been transmitted by blood conserves, perfluorocarbon emulsions are now the object of intensive research (53,81,108). For the disperse phase perfluorotributylamine, perfluorobutyltetrahydofuran, perfluorodecaline, perfluoromethyldecaline, and perfluoromethylcyclohexane are used. The three last-named substances seem to be eliminated via the skin and lungs (30). Very often poloxamer-188 (Pluronic F68) is used. In Japan there is now on the market a product with the name Fluosol DA (Green Cross, Osaka).

X. IN VIVO BEHAVIOR OF INTRAVENOUS FAT EMULSIONS

Soja emulsions are eliminated from the bloodstream with a half-life period of 7–10 min (1). For this two mechanisms are held responsible: the metabolic elimination analogous to the chylomicrons which are produced during feeding, and the elimination according to the colloidal foreign body character. Chylomicrons consist of 86–95% fat, 2–7% phospholipids, 3–7% cholesterin, and 1–2% proteins, the so-called apolipoproteins (85).

The *apolipoproteins*, along with the lipids which are bonded over hydrophilic interactions, form the so-called lipoproteins. Chylomicrons are responsible for the uptake in the blood and transport of the fat contained in food. In the blood apolipoprotein CII is taken up from high-density lipoproteins (HDLs) which are activated by the lipoprotein lipase. With their help the fats are hydrolyzed in the chylomicrons. The free fatty acids are now available for the body cells or bind with the albumins of the blood. The remaining particles are completely degraded in the liver.

The VLDLs and LDLs that are produced in the liver are responsible for the transport of fats (and cholesterin). HDLs contain less fat and cholesterin as compared with chylomicrons, VDLDs and LDLs, but they contain considerably more proteins (40–55%), especially apolipoprotein C II, which on one hand is at the disposal of the chylomicrons and LDL for the stimulation of lipoproteinlipase, but on the other hand can be absorbed again by the chylomicrons which are largely degraded.

After injection of a fat emulsion an associate formation with apolipoproteins (18) could be detected, which makes the analogy to chylomicrons (96, 110,128,129) seem logical. On the other hand, the foreign body character of the fat emulsions must be taken into consideration.

According to Juliano (82), several mechanisms can be held responsible for the particle clearance:

- Particle aggregation
- Binding of particles to the cells which are circulating in the blood
- Deposition at the endothelial cells
- Leaving the blood vessels
- Uptake by the mononuclear phagocyting system (MPS)

The formation of larger oil droplets by coalescence of the fat emulsion due to a possibly reduced physical stability which is caused by interaction with the blood components is comparable to an aggregation. Particles larger than 5–15 μm are mechanically *filtered* out of the blood through the nearest *capillary bed* (in the case of intravenous application, the lung). The binding to blood cells or endothelial cells is usually not specific and in the case of fat emulsions has not yet been published. Parenteral fat emulsions with a droplet size of approximately 200–500 nm are too big to pass through the capillary endothelium. They must leave the bloodstream via the liver and spleen, as these possess fenestrated endothelials. Capillaries in *infected tissue* or *tumors* have higher permeability so that it seems possible that in these cases penetration of the droplets in the surrounding tissue is possible.

The interaction between the parenteral fat emulsions and the MPS, to which the RES belongs, is of crucial importance (150). It consists of circulating *blood monocytes* and *macrophages* which are in the tissues (e.g., Kupffer cells of the liver) or in the endothelial of the blood capillaries. Their job is to remove substances from the body that have been recognized as foreign to it. By sorption of opsonines and proteins of the complement system the fat droplets are marked as foreign and so they are phagocytized by the cells of the RES.

The extent to which phagocytosis occurs is markedly dependent on the physicochemical properties of the fat emulsion droplets such as size, charge, and surface hydrophobicity (20). The absorption of fat emulsions is therefore also of importance because by a blockade of the RES (whether this occurs due to saturation of the phagocyting cells or by exhaustion of the opsonin supply in the blood) the immune system is weakened.

The literature on the interaction of fat emulsions with the RES is contradictory. In previous years fat emulsions were used as test systems for the function of the RES (36,179,181). Waddel (181) reported that the fat emulsions that he used were mainly absorbed by the liver and spleen and were similar in effectivity to the colloidal carbon particles. Later a more differentiated picture emerged. The extent of absorption in the RES was now correlated with the *type of emulsifier* used (31, 79). The higher the charge, the greater was the absorption that occurred. For a series of polyoxamer emulsifiers the reverse relationship to the molecular weight resulted. By using polyoxamine-908 it is even possible to almost completely avoid an RES uptake (70). The research group of Illum and Davis recently published the results of three gamma camera studies (32–34), according to which the RES is only slightly influenced by the fat emulsion. Here the products in question are the commercial products Intralipid and Lipofundin as well as experimental fat emulsions. However, with higher normal clinic doses an influence of the fat components on the RES function could be observed. Of patients who were on parenteral nutrition and who were given either the LCT or the the MCT/LCT emulsions, who took the LCT group showed a reduced clearance rate of RES dependent on the infusion speed. In contrast, the administration of the same doses of MCT/LCT did not lead to any disturbance of the RES function.

REFERENCES

1. D. H. Adamkin, K. N. Gelke, and B. F. Andrews. Fat emulsions and hypertriglyceridemia. J. Parent. Ent. Nutr., 8; 563–567, 1984.
2. M. M. Ames and J. S. Kovach. Parenteral formulation of hexamethylmelanine potentially suitable for use in man. Cancer Treat. Rep., 66; 1579–1581, 1982.

3. D. W. Andersen, L. J. Filer, and L. D. Stegink. Intravenous lipid emulsions in the treatment of essential fatty acids deficiency: Studies in young pigs. Pediatric Res., 18; 1350–1355, 1984.
4. A. C. Bach, A. Frey, and O. Lutz. Clinical and experimental effects of medium chain triglyceride based fat emulsions–a review. Clin. Nutr., 8; 223–235, 1989.
5. A. C. Bach, J. Ferezou, and A. Frey. Phospholipid-rich particles in commercial parenteral fat emulsions. An overview. Prog. Lipid Res., 35; 133–153, 1996
6. A. C. Barat, K. Harrie, M. Jacob, T. G. Diamantidis, and N. L. McIntosh. Effect of amino-acid solutions on total nutrient admixture stability. J. Parent. Ent. Nutr., 11; 384–388, 1987
7. S. Benita, D. Friedman and M. Weinstock. Pharmacological evaluation of an injectable prolonged release emulsion of physostigmine in rabbits. J. Pharm. Pharmacol., 38; 653–658, 1986.
8. S. Benita, D. Friedman, and M. Weinstock. Physostigmine emulsion: a new injectable controlled release delivery system. Int. J. Pharm., 30; 47–55, 1986.
9. P. K. Bentley, S. S. Davis, O. L. Johnson, K. C. Lowe, and C. Washington. Purification of Pluronic F 68 for perfluorochemical emulsification. J. Pharm. Pharmacol., 41; 661–663, 1989.
10. F. S. Bettner and D. J. Stennett. Effects of pH, temperature, concentration, and time on particle counts in lipid-containing total parenteral nutrition admixture. J. Parent. Ent. Nutr., 10, 375–380, 1986.
11. B. A. Bivins, R. P. Rapp, K. Record, H. C. Meng, and W. O. Griffen. Parenteral safflower oil emulsions. Ann. Surg., 191, 307–315, 1980.
12. C. D. Black and N. D. Popovich. A study of i. v. emulsion compatibility: effects of dextrose, amino acids, and selected electrolytes. Drug Intell. Clin. Pharm., 15, 184–193, 1981.
13. D. Blohm, C. Bollschweiler, and H. Hillen. Pharmaproteine. Angew. Chem., 100, 213–231, 1988.
14. H. Bolderman, M. Wicklmayr, K. Rett, P. Banholzer, G. Dietze, and H. Mehnert. Changes in hepatic morphology during parenteral nutrition with lipid emulsions containing LCT or MCT/LCT quantified by ultrasound. J. Parent. Ent. Nutr., 15, 601–603, 1991
15. G. J. Burckart, P. F. Whitington, D. K. Halbrehder, and R. A. Helms. Triglyceride and fatty acid clearance in neonates following safflower oil emulsion infusion. J. Parent. Ent. Nutr., 7, 251–253, 1983.
16. W. R. Burnham, P. Hansrani, C. E. Knott, J. A. Cook, and S. S. Davis. Stability of a fat emulsion based i.v. feeding mixture. Int. J. Pharm., 13, 9–22, 1983.
17. P. A. Cairns, D. C. Wilson, J. Jenkins, D. McMaster, and B. G. McClure. Tolerance of mixed lipid emulsion in neonates: effect of concentration. Arch. Dis. Child., 75; F113–F116, 1996
18. L. A. Carlson. Studies on the fat emulsion Intralipid. I. Association of serum proteins to Intralipid triglyceride particles. Scand. J. Clin Lab. Invest., 40, 139–144, 1980.
19. H. Carstensen, B. W. Müller, and R. H. Müller. Characterization of the properties of parenteral fat emulsions related to their uptake by the RES. Proc. 18th Int. Symp. Contr. Rel. Bioact. Mat., Amsterdam, 1991, p. 473–474.
20. H. Carstensen, R. H. Müller, and B. W. Müller. Particle size, surface hydropho-

bicity and interaction with serum of parenteral fat emulsions and model drug carriers as parameters related to RES uptake. Clin. Nutr., 11; 289–297, 1992.

21. L. C. Collins-Gold, R. T. Lyons, and L. C. Bartholow. Parenteral emulsions for drug delivery. Adv. Drug Deliv. Rev., 5; 189–208, 1990.
22. A. G. Coran, R. Drongowski, T. M. Sarahan, and J. R. Wesley. Comparison of a new 10% and 20% safflower oil fat emulsion in pediatric parenteral nutrition. J. Parent. Ent. Nutr., 5, 236–239, 1981.
23. R. Cotter, L. Martis, F. Cosmas, C. Taylor, S. Young, B. Rowe, and R. Johnson. Comparison of the elimination of 10 and 20% Travamulsion lipid emulsion from the blood of beagle dogs. Am. J. Clin. Nutr., 41; 994–1001, 1985.
24. W. Courten. Experiments and observations of the effect of several sorts of poison upon animals. Experiments made at Montpellier in the years 1678–1679. Phil. Trans. Reg. Soc. (Lond.), 27; 488–499, 1710.
25. O. v. Dardel, C. Mebius, and T. Mossberg. Diazepam in emulsion form for i.v. usage. *Acta Anaesth. Scand.,* 20; 221–224, 1976.
26. O. v. Dardel, C. Mebius, T. Mossberg, and B. Svensson. Fat emulsion as a vehicle for diazepam. A study of 9492 patients. Br. J. Anaesth., 55; 41–47, 1983.
27. S. S. Davis. Pharmaceutical aspects of i.v. fat emulsions. J. Hosp. Pharm., 32; 149–170, 1974.
28. S. S. Davis. The emulsion—obsolete dosage form or novel drug delivery system and therapeutic agent ? J. Clin. Pharm., 1; 11–27, 1976.
29. S. S. Davis and M. Galloway. Studies on fat emulsions in combined nutrition solutions. J. Clin. Hosp. Pharm., 11; 33–45, 1986.
30. S. S. Davis, J. Hadgraft, and K. J. Palin. Medical and Pharmaceutical Applications of Emulsions, Vol. 2. New York, Marcel Dekker, 1983.
31. S. S. Davis and P. Hansrani. The influence of emulsifying agents on the phagocytosis of lipid emulsions by macrophages. Int. J. Pharm., 23; 69–77, 1985.
32. S. S. Davis, L. Illum, R. H. Müller, F. Landry, J. Wright, and G. Harper. The effect of infused fat emulsions on reticuloendothelial function in the rabbit. Clin. Nutr., 9; 260–265, 1990.
33. S. S. Davis, L. Illum, C. Washington, and G. Harper. Studies on the interaction of charge-reversed emulsions with the reticuloendothelial system. Int. J. Pharm., 82; 99–105, 1992.
34. S. S. Davis, L. Illum, P. West and M. Galloway. Studies on the fate of fat emulsions following i.v. administration to rabbits and the effect of added electrolytes. Clin. Nutr., 6; 13–197, 1987.
35. A. R. Dennison, M. Ball, L. J. Hands, P. C. Crowe, R. M. Watkins, and M. Kettlewell. Total parenteral nutrition using conventional and medium chain triglycerides: effect on liver function tests, complement, and nitrogen balance. J. Parent. Ent. Nutr., 12; 15–19, 1987
36. N. R. Di Luzio and S. J. Riggi. The development of a lipid emulsion for the measurement of reticuloendothelial function. J. Reticuloendothelial Soc., 1; 136–149, 1964.
37. J. E. Diederichs, H. Weyhers, and R. H. Müller. Parenteral fat emulsions 1: Effect of emulsion composition on the microviscosity of the emulsifier film. Proc. 18th. Int. Symp. Control. Rel. Bioact. Mat., 1991, pp. 471–472.

38. B. Dobias. Emulsionen (Teil 1). Tenside Surf. Det., 15; 225–232, 1978.
39. J. Eckart, M. Adolph, U. Mühlen, and V. van der Naab. Fat emulsions containing medium chain triglycerides in parenteral nutrition of intensive care patients. *J. Parent. Ent. Nutr.*, 4; 360–366, 1980.
40. A. A. A. El-Sayed and A. J. Repata. Solubilization and stabilization of an investigational antineoplastic drug (NSC no. 278214) in an i.v. formulation using an emulsion vehicle. Int. J. Pharm., 13; 303–312, 1983.
41. D. H. Everett. Pure Appl. Chem., 31; 577–638, 1972.
42. G. Fischer, Total Parenteral Nutrition, Boston, Little, Brown, 1976.
43. E. Fiaccadori, G. Torotella, G. L. Gonzi, C. Beghi, D. Albertini, S. Picolini, L. Belli, and A. Avogadro. Hemodynamic and respiratory effects of medium-chain and long-chain triglyceride fat emulsions: a prospective, randomized study. Riv. Ital. Nutr. Parent. Ent., 15; 6–14, 1997.
44. A. G. Fjellestad. Untersuchungen der Grenzflächenspannung zur Optimierung von Emulsionen als intravenöse Arzneistoffträgersysteme. MS thesis, Universities of D-Kiel and N-Oslo.
45. D. Forster, C. Washington, and S. S. Davis. Toxicity of solubilized and colloidal amphotericin B formulations to human erythrocytes. J. Pharm. Pharmacol., 40; 325–328, 1988.
46. C. L. Fortner, W. R. Grove, D. Bowle, and M. D. Walker. Fat emulsion vehicle for i.v. administration of an aqueous insoluble drug. Am. J. Hosp. Pharm., 32; 582–584, 1975.
47. H. J. Franzky. Mikroemulsionen—Mizellen oder Lösungen? PhD thesis, Christian-Albrecht- University, D-Kiel, 1986.
48. S. Friberg and P. O. Jansson. Surfactant association structure and emulsion stability. *J. Colloid Interf. Sci.,* 55; 614–623, 1976.
49. D. Friedman, S. Benita, and M. Weinstock. Controlled release of injectable emulsion of physostigmine. Proc.12th Int. Symp. Contr. Rel. Bioact. Mat., 1985, pp. 330–331.
50. T. Fujita, T. Sumaya, and K. Yokoyama. Fluorocarbon emulsions as a candidate for artificial blood. Eur. Surg. Res., 3; 436–453, 1971.
51. J. Georges and J. W. Chen. Microemulsion studies: correlation between viscosity, electrical conductivity and electrochemical and fluorescent probe measurements. Colloid Polyrn. Sci., 264; 896–902, 1986.
52. R. P. Geyer. Parenteral nutrition. Bull. Parent. Drug Assoc., 21; 215–225, 1967.
53. U. Gross, H. Reichelt, and E. Döring. Stabile Perfluorcarbon-Emulsionen. Tenside Surf. Deter., 25; 284–292, 1988.
54. G. Grotte, S. Jacobsen, and A. Wretlind. Lipid emulsions and technique of peripheral administration in parenteral nutrition, pp. 335–362, in Ref. 42.
55. M. J. Groves, M. Wineberg, and A. P. R. Brain. The presence of liposomal material in phospholipid stabilized emulsions. J. Dispers. Sci. Technol., 6; 237–243, 1985.
56. D. Guest and D. Langevin. Light-scattering study of a multiphase microemulsion system. J. Colloid Interf. Sci., 112; 208–220, 1986.
57. S. Hailer and G. Wolfram. Influence of artificial fat emulsions on the composition of serum lipoproteins in humans. Am. J. Clin. Nutr., 43; 225–233, 1986.

57a. S. Hailer, G. Wolfram, and N. Zöllner. Changes in serum lipoproteins in humans following the infusion of a fat emulsion containirig medium- and long-chain triglycerides. *Eur. J. Clin. Invest.,* 17; 402–407, 1987.
58. T. Hajri, J. Ferezou, and C. Lutton. Effects of intravenous infusions of commercial fat emulsions (Intralipid 10 or 20%) on rat plasma lipoproteins: phospholipid in excess are the main precursors of lipoprotein-X-like particles. Biochim. Biophys. Acta, 1047; 121–130, 1990.
59. P. K. Hansrani, S. S. Davis, and M. J. Groves. The preparation and properties of sterile i.v. emulsions. J. Parent. Sci. Tech., 37; 145–150, 1983.
60. K. R. Harrie, M. Jacob, D. McCormick, J. S. Reid, and N. L. McIntosh. Comparison of total nutrient admixture stability using two intravenous fat emulsions, Soyacal and Intralipid 20%. J. Parent. Ent. Nutr., 10; 381–387, 1986.
61. D. Haumont, R. J. Deckelbaum, M. Richelle, W. Dahaln, E. Coussaert, B. E. Bihain, and Y. A. Carpentier. Plasma lipid and plasma lipoprotein concentrations in low birth weight infants given parenteral nurition with twenty or ten percent lipid emulsion. J. Pediatr. 115, 787–793, 1989.
62. W. Heller and T. L. Pugh. Steric stabilization of colloidal solutions by adsorption of flexible macromolecules. J. Polym. Sci., 47; 203–217, 1960.
63. E. Hodder. Practitioner, 10; 14, 1873,
64. H. Hoffmann, G. Platz, and W. Ulbricht. From micellar solutions to microemulsions—a kinetic study. Ber. Bunsenges. Phys. Chem., 90; 877–887, 1986.
65. K. Hoshi, Y. Mizushima, S. Kiyokawa, and A. Yanagawa. Prostaglandin E1 incorporated in lipid microspheres in the treatment of peripheral vascular disease and diabetic neuropathy. Drugs Exp. Clin. Res., 12; 681–685, 1986.
66. K. Huth, W. Schoenborn, and J. Börner. Zur Pathogenese der Unverträglichkeitserscheinungen bei parenteraler Fettzufuhr. Med. Ernähr., 8; 146–148, 1967.
67. K. Huth. Parenterale Ernährung und Sondenkost, in Lehrbuch der Ernährungstherapie, G. Thieme Verlag, Stuttgart, 1986.
68. A. Hyltander, R. Sandström, and K. Lundholm. Metabolic effects of structured triglycerides in humans. *NCP*, 10; 91–97, 1995.
69. L. Illano, M. Delanghe, H. van den Baviere, and A. Lauwers. Effects of electrolytes in the presence of some trace elements on the stability of all-in-one emulsion mixtures for total parenteral nutrition. J. Clin. Hosp. Pharm., 9; 87–93, 1984.
70. L. Illum, P. West, C. Washington, and S. S. Davis. The effect of stabilizing agents on the organ distribution of lipid emulsions. Int. J. Pharm., 54; 41–49, 1989.
71. International Union of Pure and Applied Chemistry. Manual on Colloid and Surface Science. Boston, Butterworth, 1972.
72. R. I. Jeppsson. Effects of barbituric acids using an emulsion form intraperitoneally and subcutaneously. Acta Pharm. Suecica, 9; 199–206, 1972.
73. R. I. Jeppsson. Effects of barbituric acids using an emulsion form i.v. Acta Pharm. Suecica. 9; 81–90, 1972.
74. R. I. Jeppsson. Comparison of pharmacological effects of some local anaesthetic agents when using water and lipid emulsion as injection vehicles. Acta Pharmacol. Toxicol., 36; 299–311, 1975.
75. R. I. Jeppsson. Parenteral administration of lipid-soluble drugs in oil emulsion vehicles. Acta Pharm. Suecica Suppl., 13; 43, 1976.

76. R. I. Jeppsson, M. J. Groves, and H. S. Yalabik. The particle size distribution of emulsions containing diazepam for i.v. use. J. Clin. Pharm., 1; 123 -127, 1976.
77. R. I. Jeppsson, and S. Ljungberg. Intraarterial administration of emulsion formulations containing cyclandelate and nitroglycerin. Acta Pharm. Suecica, 10; 129–140, 1973.
78. R. I. Jeppsson and S. Ljungberg. Anticonvulsant activity in mice of diazepam in an emulsion formulation for i.v. administration. Acta Pharmacol. Toxicol., 36; 312–320, 1975.
79. R. I. Jeppsson and S. Rössner. The influence of emulsifying agents and of lipid soluble drugs on the fractional removal rate of lipid emulsions from the bloodstream of the rabbit. Acta Pharmacol. Toxicol., 37; 134–144, 1975.
80. R. I. Jeppsson and G. L. Schoefl. Electron microscopic observations on cerebral and pulmonary blood vessels after the i.v. injection of artificial lipid emulsions containing barbituric acids. Aust. J. Exp. Biol. Med. Sci., 52; 703-709, 1974.
81. O. L. Johnson, C. Washington, and S. S. Davis. Thermal stability of fluorocarbon emulsions that transport oxygen. Int. J. Pharm., 59; 131–135, 1990.
82. R. L. Juliano. Factors affecting the clearance kinetics and tissue distribution of liposomes, microspheres, and emulsions. *Adv. Drug Deliv. Rev.,* 2; 31–54, 1988.
83. U. Julius and W. Leonhardt. Elimination and metabolism of a fat emulsion containing medium chain triglycerides (Lipofundin MCT 10%). J. Parent. Ent. Nutr., 12; 116–120, 1988.
84. M. Kahlweit, R. Strey, D. Haase, K. Kunieda, T. Schmeling, B. Faulhaber, M. Borkovec, H. F. Eicke, G. Busse, F. Eggers, T. Funck, H. Richmann, L. Magid, O. Söderman, P. Stilbs, J. Winkler, A. Dittrich, and W. Jahn. How to study microemulsions. J. Colloid Interf. Sci., 118; 436–453, 1987.
85. P. Karison. Kurzes Lehrbuch der Biochemie für Mediziner und Naturwissenschaftler, Vol. 12. G. Thieme Verlag, Stuttgart, 1984.
86. D. P. Katz and J. L. Knittle. The effect of medium and long chain triglyceride on human adipose tissue metabolism. Lipids, 22; 435–441, 1987.
87. G. Kleinberger and H. Pamperl. Allgemeine Charakteristika und Fragen zur Galenik von Fettemulsionen. Infusionstherapie, 10; 108–117, 1983.
88. P. K. Kleinebudde. Untersuchungen an pharmazeutisch verwendbaren, flüssigen Mehrkomponentensystemen mit nicht mischbaren Flüssigkeiten. PhD thesis, Christian-Albrecht-University, D-Kiel, 1987.
89. M. J. C. Koorenhof and J. G. Timmer. Stability of total parenteral nutrition supplied as all-in-one for children with chemotherapy-linked hyperhydration. Pharm. Weekblad Sci. Ed., 14; 50–54, 1992.
90. T. Kronevi and S. Ljungberg. Sequale following intra-arterially injected diazepam formulations. Acta Pharm. Suecica, 20; 389–396, 1983.
91. K. A. Lamb, C. Washington, S. S. Davis, and S. Denyer. Toxicity of amphotericin B emulsion to cultured canine kidney cell monolayers. J. Pharm. Pharmacol., 43; 522–524, 1991.
92. H. Lange and C. P. Kurzendörfer. Zum Mechanismus der Stabilisierung von Emulsionen. Fette Seifen Anstrichmittel, 76; 120–126, 1974.
93. A. M. Laval-Jeantet, M. Laval-Jeantet, and C. Bergot. Effect of particle size on

the tissue distribution of iodized emulsified fat following intravenous administration. Invest. Radiol. 17; 617–620, 1982.

94. R. I. Lawrence, W. K. Flukes, V. J. Rust, and P. A. Bralthwate. Total parenteral nutrition using a combined nutrient solution. Aust. J. Hosp. Pharm., 11; S40–S42, 1981.
95. H. H. Le Veen, P. Glordano, and J. Spletzer. The mechanism of removal of i.v. injected fat. Arch. Surg., 831; 169–179, 1961.
96. N. P. Lenzo, L. J. Martins, B. C. Mortimer, and G. Redgrave. Effects of phospholipid composition on the metabolism of triacylglycerol, cholesteryl esters and phosphatidylcholine from lipid emulsion injected intravenously in rats. Biochim. Biophys. Acta, 960; 111–118, 1988.
97. M. Y. Levy and S. Benita. Design and characterization of a submicronized o/w emulsion of diazepam for parenteral use. Int. J. Pharm., 54; 103–112, 1989.
98. M. Y. Levy and S. Benita. Drug release from submicronized o/w emulsions: a new in vitro kinetic evaluation model. Int. J. Pharm., 66; 29–371, 1990.
99. M. Y. Levy, L. Langerman, S. Gottschalk-Sabag, and S. Benita. Side-effect evaluation of a new diazepam formulation: venous sequela reduction following i.v. injection of a diazepam emulsion in rabbits. Pharm. Res., 6; 510–516, 1989.
100. A. Lich. Intravenös applizierfähige O/W-Emulsionen als Vehikel für schlechtlösliche mikronisierte Wirksubstanzen. PhD thesis, Philipps University, D-Marburg/Lahn, 1991.
101. D. M. Lidgate, R. C. Fu, N. E. Byars, L. C. Foster, and J. S. Fleitman. Formulation of vaccine adjuvant muramyldipeptides. Processing optimization, characterization, and bioactivity of an emulsion vehicle. Pharm. Res., 6; 748–752, 1989.
102. D. M. Lidgate, T. Trattner, R. M. Shultz, and R. Maskiewicz. Sterile filtration of a parenteral emulsion. Pharm. Res., 9; 860–863, 1992.
103. B. Lindman, P. Stilbs, and M. E. Moseley. Fourier transform NMR self-diffusion and microemulsion structure. J. Colloid Interf. Sci., 83; 569–582, 1981.
104. P. H. List. Arzneiformenlehre, 4th Ed. Stuttgart, Wiss. Verlagsges, 1986.
105. C. Litterst, E. G. Mimnaugh, A. C. Cowles, T. Gram, and A.M. Guarino. Distribution of C14 lomustine (14C-CCNU) derived radioactivity following intravenous administration of three potential clinical formulations to rabbits. J. Pharm. Sci., 63; 1718–1721, 1974.
106. S. Ljungberg and R. I. Jeppsson. Intravenous injection of lipid soluble drugs. Acta Pharm. Suecica, 7; 435–440, 1970.
107. D. Long, A. Burgan, R. Long, R. Mattrey, J. Riess, and R. Follana. Preparation and evaluation of F-octylbromide emulsions for contrast agents and synthetic oxygen carrier. J. Fluor. Chem., 35; 235 ff, 1987.
108. S. Magdassi and A. Siman-Tov. Formulation and stabilization of perfluorocarbon emulsions. Int. J. Pharm., 59; 69–72, 1990.
109. I. K. Magnusson Borg, L. G. Sandberg, A. K. Wennberg, L. Lindmark, and E. G. Ekman. Effects of a fat emulsion containing medium chain fatty acids and long chain fatty acids on protein and energy metabolism in partially hepatectomized rats. Clin. Nutr., 14; 23–28, 1995.
110. L. J. Martins, N. P. Lenzo, and T. G. Redgrave. Phosphatidylcholine metabolism after transfer from lipid emulsions injected intravenously in rats. Implications for

high-density lipoprotein metabolism. Biochim. Biophys. Acta, 1005; 217–224, 1989.

111. R. Mattrey, P. Hilpert, C. Long, D. Long, R. Mitten, and T. Peterson. Hemodynamic effects of intravenous lecithin-based perfluorocarbon emulsions in dogs. Crit. Care Med., 17; 652–656, 1989.
112. W. Messerschmidt. Fettemulsionen in TPN Regimen—Untersuchung zur Stabilität. Krankenhauspharmazie, 12; 331-336, 1991.
113. J. M. McKibbin, R. M. Ferry, and F. J. Stare. The utilization of emulsified fat given intravenously. J. Clin. Invest., 25; 679–686, 1946.
114. H. C. Meng. Fat emulsions in parenteral nutrition, Seiten 305–334, in Ref. 42.
115. B. Messing, J. Peynet, J. Poupon, A. Pfeiffer, F. Thuillier, O. Chazouilleres, and A. Legrand. Effect of fat-emulsion phospholipids on serum lipoprotein profile during 1 mo of cyclic total parenteral nutrition. Am. J. Clin. Nutr., 52; 1094–1100, 1990.
116. Y. Mizushima. Lipid microspheres as novel drug carriers. Drugs Exp. Clin. Res., 11; 595–600, 1985.
117. Y. Mizushima. Application of lipid microspheres for targeting therapy. J. Bioact. Compat. Polym., 3; 148–156, 1988.
118. Y. Mizushima. Lipid microspheres for targeting therapy. Proc. 17th Int. Symp. Contr. Rel. Bioact. Mat. 1990, pp. 8–9.
119. Y. Mizushima, T. Hamano, and K. Yokoyama. Tissue distribution and anti-Inflammatory activity of corticosteroids incorporated in lipid emulsion. Ann. Rheum. Dis., 41; 263–267, 1982.
120. Y. Mizushima, T. Hamano, and K. Yokoyama. Use of a lipid emulsion as a novel carrier for corticosteroids. J. Pharm. Pharmacol., 34; 49–50, 1982.
121. Y. Mizushima, K. Hoshi, H. Aihara, and M. Kurachi. Inhibition of bronchoconstriction by aerosol of a lipid emulsion containing prostaglandin E1. J. Pharm. Pharmacol., 35; 397, 1983.
122. Y. Mizushima, R. Igarashi, K. Hoshi, A. K. Sim, M. E. Cleland, H. Hayashi, and J. Goto. Marked enhancement in antithrombotic activity of isocarbacyclin following its incorporation into lipid microspheres. Prostaglandins, 33; 161–168, 1987.
123. Y. Mizushima, K. Kaneko, and K. Hoshi. Targeting steroid therapy in rheumatoid arthritis. Ann. Rheum. Dis., 42; 479–480, 1983.
124. Y. Mizushima, Y. Shoji, M. Kato, T. Fukushima, and S. Kurozumi. Use of lipid microspheres as a drug carrier for antitumor agents. J. Pharm. Pharmacol., 38; 132–134, 1986.
125. Y. Mizushima, Y. Wada, Y. Etoh, and K. Watanabe. Antiinflammatory effects of indomethacin ester incorporated in lipid microspheres. J. Pharm. Pharmacol., 35; 398–399, 1983.
126. Y. Mizushima, A. Yanagawa, and K. Hoshi. Prostaglandin E1 is more effective, when incorporated in lipid microspheres, for treatment of peripheral vascular diseases in man. J. Pharm. Pharmacol., 35; 666–667, 1983.
127. T. Morita, Y. Ishizuka, T. Arioka, Y. Mito, H. Yaginuma, and Y. Nozawa. Anticandidal activity of GBR 14206, a new imidazole derivative intravenous administration in lipid emulsion form. Jpn. J. Med. Mycol., 30; 143–148, 1989.
128. B. C. Mortimer, W. J. Simmonds, S. J. Cockman, R. V. Stick, and T. G. Redgrave. The effect of monostearoylglycerol on the metabolism of chylomicron-like lipid

emulsions injected intravenously in rats. Biochim. Biophys. Acta, 1046; 46–56, 1990.
129. B. C. Mortimer, W. J. Simmonds, C. A. Joll, R. V. Stick, and T. G. Redgrave. The effect of added monoacylglycerols on the removal from plasm of chylomicron-like emulsions injected intravenously in rats. Biochim. Biophys. Acta, 1002; 359–364, 1989.
130. R. H. Müller and S. Heinemann. Stability of TPN regimens based on Lipofundin/ Intralipid containing a high electrolyte load. Arch. Pharm. 324, 694, 1991.
131. R. H. Müller, S. Heinemann, and J. E. Diederichs. Parenteral fat emulsions II: charge and rigidity of emulsifier film related to the long-term stabillty. Proc. 18th Int. Symp. Control. Rel. Bioact. Mat., 1991, pp. 469–470.
132. M. Mütsch. Chylomicron-Remnants als mögliche Träger von Leberlherapeutika– Evaluierung von weiteren Nanoemulsionen mit vielseitigeren Anwendungsmöglichkeiten. PhD thesis No. 9097, Eidgenössische Techniche Hochschule Zürich, 1990.
133. R. A. Myers and V. J. Stella. The effects of various oil in water emulsions on the svstemic bioavailability and lymphatic transport of a lipophilic compound penclomedine. Pharm. Res, Suppl., 6; S92, 1989.
134. R. A. Myers and V. J. Stella. Factors affecting the lymphatic transport of peaclomedine (NSC 338720), a lipophilic cytotoxic drug: comparison to DDT and hexachlorobenzene. Int. J. Pharm., 80; 51–62, 1992.
135. R. A. Myers and V. J. Stella. Systemic bioavailability of penclomedine (NSC 338720) from oil in water emulsions administered intraduodenally to rats. Int. J. Pharm. 78; 217–226, 1992.
136. Y. Nakamoto, M. Fujiwara, T. Noguchi, T. Kimura, S. Muranishi, and H. Sezaki. Studies on pharmaceutical modification of anticancer agents. 1. Enhancement of lymphatic transport of mitomycin C by parenteral emulsions. Chem. Pharm. Bull., 23; 2232–2238, 1975.
137. Y. Nakamoto, M. Hashida, S. Muranishi, and H. Sezaki. Studies on pharmaceutical modification of anticancer agents. II. Enhanced delivery of bleomycin into lymph by emulsions and drying emulsions. Chem. Pharm. Bull., 23; 3125–3131, 1975.
138. D. H. Napper. Polymer Stabilization of Colloidal Dispersions. London, Academic Press, 1983.
139. T. Nomura. J. Exp. Med., 13; 51 ff, 1929.
140. J. Nordenström, A. Thörne, and T. Olivecrona. Metabolic effects of infusion of a structured triglyceride emulsion in healthy subjects.
141. M. Paborji, C. M. Riley, and V. J. Stella. A novel use of Intralipid for the parenteral delivery of perilla ketone (NSC 348407), an investigational cytotoxic drug with a high affinity for plastic. Int. J. Pharm., 42; 243–249, 1988.
142. A. Panaggio, C. T. Rodes, and L. R. Worthen. The possible use of autoclaving microemulsions for sterilization. Drug Dev. Ind. Pharm., 5; 169–173, 1979.
143. V. A. Parry, K. R. Harrie, and N. L. McIntosh-Lowe. Effect of various nutrient ratios on the emulsion stability of total nutrient admixtures. Am. J. Hosp. Pharm., 43; 3017–3022, 1986.
144. Y. Pathak, A. Rubinstein, and S. Benita. Enhanced stability of physostigmine salicylate in submicron o/w emulsion. Int. J. Pharm., 65; 169–175, 1990.

145. E. Pescheidl, C. Winzer, S. Richter, M. Hedwig-Geising, and A. Rügheimer. Influence of structured lipids on reticuloendothelial-system function and liver morphology in an endotoxin rat model. Clin. Nutr. 12, P33, 1993.
146. R. J. Prankrerd, S. G. Frank, and V. J. Stella. Preliminary development and evaluation of a parenteral emulsion formulation of penclomedine: a novel, practically water insoluble cytotoxic agent. J. Parent. Sci. Technol., 42; 76–81, 1988.
147. R. J. Prankrerd and V. J. Stella. The use of oil in water emulsions as a vehicle for parenteral drug administration. J. Parent. Sci. Technol., 44; 139–149, 1990.
148. P. Radermacher, B. Santak, H. Strobach, K. Schrör, and J. Tarnow. Fat emulsions containing medium chain triglycerides in patients with sepsis syndrome: effects on pulmonary hemodynamics and gas exchange. Intensive Care Med. 18; 231–234, 1992.
149. J. G. Riess. Fluorocarbon based blood substitutes: what progress? Int. J. Artif. Organs, 14; 255–258, 1991.
150. I. M. Roitt, J. Brostoff, and D. K. Male. Kurzes Lehrbuch der Immunologie. Stuttgart, G. Thieme Verlag, 1987.
151. A. Rubinstein, Y. V. Pathak, J. Kleinstern, A. Reches, and S. Benita. In vitro release and intestinal absorption of physostigmine salicylate from submicron emulsions. J. Pharm. Sci., 80; 643–647, 1991.
152. E. Ruckenstein. The surface of tension, the natural radius, and the interfacial tension in the thermodynamics of microemulsions. J. Colloid Interface Sci., 114; 173–179, 1986.
153. L. Rydhag. The importance of the phase behaviour of phospholipids for emulsion stability. Fette Seifen Anstrichmittel, 81; 168–173, 1979.
154. L. Rydhag and I. Wilton. The function of phospholipids of soybean lecithin in emulsions. J. Am. Oil Chem. Soc., 58; 830–837, 1981.
155. R. Sandström, A. Hyltander, U. Körner, and K. Lundholm. Structures triglycerides were well tolerated and induced inreased whole body fat oxidation compared with long-chain triglycerides in postoperative patients. J. Parent. Ent. Nutr. 19; 381–386, 1995
156. J. Sarubin. Verminderte Thrombophlebitis-Häufigkeit nach i.v. Gabe von Diazepam in neuem Lösungsmittel (Sojabohnenöl-Emulsion). Fortschr. Med., 100; 2135–2136, 1982.
157. P. Saxl and F. Donath. Klinische, experimentelle und pharmakologische Studien über die Abfangfunktion des retikulo-endothelialen Systems. Wien. Arch. Int. Med., 13; 7 ff, 1926.
158. F. A. Sayeed, M. G. Tripp, K. B. Sukumaran, B. Mikrut, H. A. Stelmach, and J. A. Raihle. Stability of various total nutrient admixture formulations using Liposyn 11 and Aminosyn 11. Am. J. Hosp. Pharm., 44; 2280–2286, 1987.
159. P. C. Sedman, S. S. Somers, C. W. Ramsden, T. G. Brennan, and P. J. Guillou. Effects of different lipid emulsions on lymphocyte function during total parenteral nutrition. Br. J. Surg. 78; 1396–1399, 1991
160. P. C. Schmidt, H. Perschbacher, K. J. Steffens, and H. P. Kraemer. Development of a suspension-emulsion system for parenteral application in animals. Acta Pharm. Technol., 35; 34–37, 1989.
161. H. Schön, W. Zeller, and U. Krause. Intravenöse Ernährungstherapie mit Fettemulsionen. Pharmakotherapia, 1; 311–325, 1963.

162. E. Schott. Versuch einer vollständigen parenteralen Ernährung. Dt. Arch. klin. Med., 112; 403–441, 1913.
163. O. Schuberth and A. Wretlind. Intravenous infusion of fat emulsions, phosphatides and emulsifying agents. Acta Chir. Scand. Suppl., 278; 1–21, 1961.
164. J. F. Scieszka, T. J. Vidmar. L. C. Haynes, and M.J. Cho. Biochemical stability in serum of a lipid-soluble probe molecule entrapped in an o/w emulsion as a carrier for passive drug targeting. Int. J. Pharm., 45; 165–168, 1988.
165. L. E. Scriven and C. V. Sternling. The Marangoni effect. Nature, 187; 186–188, 1960.
166. J. C. Shelton, A. Smith, and I. R. Thrussell. Factors affecting the stability of nutritional emulsion mixtures. Clin. Nutr., 8; 167–171, 1989.
167. Y. Shoji, Y. Mizushima, A. Yanagawa, T. Shiba, M. Fujii, and M. Amino. Enhancement of antiinflammatory effects of biphenylyl acetic acid by its incorporation into lipid microspheres. J. Pharm. Pharmacol., 38; 118–121, 1986.
168. A. K. Sim, A. P. McCraw, M. E. Cleland, H. Aihara, and K. Hosoda. The effect of prostaglandin E1 incorporated in lipid microspheres on thrombus formation and disaggregation and its potential to target to vascular sites. Arzneim. Forsch. Drug Res., 36; 1206–1209, 1986.
169. P. J. Simons, L. D. Cockshott, E. J. Douglas, E. A. Gordon, K. Hopkins, and M. Rowland. Disposition in male volunteers of a subanaesthetic intravenous dose of an oil in water emulsion of ^{14}C Propofol. Xenobiolica, 18; 429–440, 1988.
170. M. Singh and L. J. Ravin. Parenteral emulsions as drug carrier systems. J. Parent. Sci. Tech. 40; 36–41, 1986.
171. P. Somani, W. D. Leathem, and A. L. Barlow. Safflower oil emulsion: single and multiple infusions with or without added heparin in normal human volunteers. J. Parent. Ent. Nutr., 4; 307–311, 1980.
172. K.J. Steffens, R. Schreiber, and B. Meibohm. TPN-Regime und parenterale Fettemulsionen-Untersuchungen zur physikalischen Stabilität. Krankenhauspharmazie, 13; 426–430, 1992
173. V. J. Stella and W. N. Waugh. Development of parenteral formulations of experimental cytotoxic agents: I. Rhizoxin (NSC 332598). Int. J. Pharm., 43; 191–199, 1988.
174. H. Suttmann, A. Doenicke, J. Kugler, and M. Laub. Eine neue Zubereitung von Etomidat in Lipidemulsion–Bioverfügbarkeit und Venenreizung. Anaesthesist, 38; 421–423, 1989.
175. B. D. Tarr, T. G. Sambandan, and S. H. Yalkowsky. A new parenteral emulsion for the administration of taxol. Pharm. Res., 4; 162–165, 1987.
176. D. L. Teagarden, B. D. Anderson, and W. J. Petre. Dehydration kinetics of prostaglandin E1 in a lipid emulsion. Pharm. Res., 6; 210–215, 1989.
177. S. W. Thompson. The Pathology of Parenteral Nutrition with Lipids. Springfield, IL: Charles C. Thomas, 1974.
178. S. W. Thompson, L. D. Jones, J. F. Ferrell, R. D. Hunt, H. T. Meng, H. Sasaki, F. Schaffner, and W. S. Singleton. Testing of fat emulsions for toxicity. III. Toxicity studies with new fat emulsions and emulsion components. Am. J. Clin. Nutr., 16; 43–61, 1965.
179. H. Tonaki, T. M. Saba, L. W. Mayron, and E. Kaplan. Phagocytosis of gelatinized r. e. test lipid emulsion by Kupffer cells: electron microscopic observations. Exp. Mol. Pathol., 25; 189–201, 1976.

180. M. Vermess, R. H. Adamson, J. L. Doppman, and M. Girton. Computed tomographic demonstration of hepatic tumor with the aid of i.v. iodinated fat emulsion. Radiology, 125: 711–715, 1977.
181. W. R. Waddel, R. P. Geyer, E. Clarke, and F. J. Stare. Function of the reticuloendothelial system in removal of emulsified fat from blood. Am. J. Physiol., 177; 90–94, 1954.
182. D. L. Waitzberg, R. Bellinati-Pires, M. M. Salgado, I. P. Hypolito, G. Colleto, O. Yagi, E. M. Yamamuro, J. Gama-Rodrigues, and H. W. Pinotti. Effect of total parenteral nutrition with different lipid emulsions on human monocyte and neurophil functions. Nutrition, 13; 128–132, 1997.
183. C. Washington. The electrokinetic properties of phospholipid-stabilized fat emulsions. II. Droplet mobility in mixed electrolytes. Int. J. Pharm., 58; 13–17, 1990.
184. C. Washington. The electrokinetic properties of phospholipid-stabilized fat emulsions III. Interdroplet potentials and stability ratios in monovalent electrolytes. Int. J. Pharm., 64; 67–73, 1990.
185. C. Washington. The stability of intravenous fat emulsions in total parenteral nutrition mixtures. Int. J. Pharm., 66; 1–21, 1990.
186. C. Washington. The electrokinetic properties of phospholipid stabilized fat emulsions. VI. Zeta potentials of Intralipid 20% in TPN mixtures. Int. J. Pharm. 87; 167–174, 1992.
187. C. Washington, A. Athersuch, and D. J. Kynoch. The electrokinetic properties of phospholipid stabilized fat emulsions. IV. The effect of glucose and of pH. Int. J. Pharm., 64; 217–222, 1990.
188. C. Washington, A. Chawla, N. Christy, and S. S. Davis. The electrokinetic properties of phospholipid-stabilized fat emulsions. Int. J. Pharm., 54; 191–197, 1989.
189. C. Washington, M. A. Connolly, R. Manning, and M. C. L. Skerratt. The electrokinetic properties of phospholipid stabilized fat emulsions. V. The effect of amino acids on emulsion stability. Int. J. Pharm., 77; 57–63, 1991.
190. C. Washington, O. Lutz, and S. S. Davis. Stability of an amphotericin B emulsion. J. Pharm. Pharmacol. Suppl., 43; 93p, 1991.
191. H.-G. Weder, and M. Mütsch. Möglichkeiten der Herstellung von physikalisch-chemisch definierten Liposomen und Nanoemulsionen, in Liposomen und Niosomen in Pharniazie und Kosmetik. APV- Kurs 502, Nürnberg, 1990.
192. K. Westesen. Development and characterisation of novel intravenous emulsion systems. Arch. Pharm., 324; 639, 1991.
193. K. Westesen and T. Wehler. Investigations of the structure and particle size distribution of a model intravenous emulsion. 6th Int. Conf. on Pharm. Technol. (APGI), Paris, 1992, pp. 164–173.
194. K. Westesen and T. Wehler. Physicochemical characterization of a model intravenous oil-in-water emulsion. J. Pharm. Sci., 81; 777–786, 1992.
195. T. L. Whateley, G. Steele, J. Urwin, and G. A. Smail. Particle size stability of Intralipid and mixed total parenteral nutrition mixtures. J. Clin. Hosp. Pharm., 9; 113–126, 1984.
196. M. Winsnes, R. I. Jeppsson, and B. Sjöberg. Diazepam adsorption to infusion sets and plastic syringes. Acta Anaesth. Scand., 25, 93–96, 1981.

197. G. Wolfram. Der Einsatz von Fettemulsionen in der postoperativen Ernährung. Chir. Gastroent. 10; 173–176, 1994.
198. K. H. Wong and M. Deitel. Studies with a safflower oil emulsion in total parenteral nutrition. Can. Med. Assoc. J., 125; 1328–1334, 1981.
199. H. S. Yalabik-Kas, S. Eryilmaz, and A. A. Hincal. Formulation, stability and toxicitv studies of intravenous fat emulsions. S.T.P. Pharma, 1; 12 ff, 1985.
200. S. Yamakawa. J. Jpn. Soc. Internal Med., 17; 1–22, 1920.
201. S. Yamakawa and T. Nomura. Jikken Iho, 14; 523 ff, 1928.
202. D. A. Yeadon, L. A. Goldblatt, and A. M. Altschul. Lecithin in oil/water emulsions. J. Am. Oil Chem. Soc., 35; 435–438, 1958.
203. N. Zöllner and H. Schumacher. Klinische Beobachtungen bei der parenteralen Zufuhr von Fettemulsionen. Med. Welt., 13; 2669–2680, 1962.

7

Use of Emulsions as Topical Drug Delivery Systems

Eric W. Smith
Ohio Northern University, Ada, Ohio

Howard I. Maibach
University of California Medical School, San Francisco, California

Christian Surber
University Hospital, Basel, Switzerland

I. INTRODUCTION

Drug administration methods for the treatment of skin diseases remained virtually unchanged for decades. From a patient viewpoint, topical drug delivery is a popular method of dosing as it is a noninvasive and painless way of administering active drugs and has the added component of being a very manual and tactile experience. Early practitioners apportioned as much importance to the nature of the compounded topical mixture as to any drugs or extracts that were included in the product. Together with the psychological aspects of the healing process that would occur on inunction of the formulation, this holistic approach was believed to be essential for the clinical effect. Following the explosion in the knowledge of pharmacology and pharmaceutical principles that occurred midcentury, the drug entity was deemed to be the most important aspect of the clinical efficacy of any drug product. Thereafter nearly the total research effort was applied to the development and discovery of newer and more potent drug entities. These drugs were mixed into rather crude products with little emphasis placed on the efficacy of drug delivery from the formulation to the biosystem. Thereafter the concepts of pharmacodynamics, bioavailability, and pharmacokinetics were nurtured to the extent that these principles now rule the formulation development process. It was soon evident for topical formulations that incorporating the same concentration of a specific drug into different formulations often produced markedly different clinical results, and the problem of bioequivalence was born.

As we approach the end of the century we are just beginning to understand the intricate and complex array of events that take place when a topical formulation is rubbed onto the skin by a patient. We now know that the age-old tradition of dispersing drug crystals in a bland cream formulation seldom produces optimal results and in most cases is no more efficacious to the patient than the cream without the drug. Similarly, it is becoming clearer that conventional topical drug delivery formulations that have been on the market for many years rarely achieve optimal results since little attention was paid 30 years ago to the topical availability of the drug from the formulation. Since then, many investigators have been engaged in the research, development, and optimization of topical drug delivery systems. The main problem facing the formulator concerns the development of a topical delivery system that will demonstrate the characteristics of optimal drug delivery to treat the clinical condition, display negligible toxicity, retain drug and excipient stability, and comply with the aesthetic necessities of the formulation. New products that are emerging have been more thoroughly tested and have been designed specifically to deliver the drug in a known manner. Even so, many industries still approach this development process as a trial-and-error experience because of incomplete knowledge of the events that are thought to occur when a complex topical delivery matrix (such as an emulsion) is applied to the relatively hostile environment of the stratum corneum.

II. THE SKIN

The popularity of the skin as a drug delivery portal has been exemplified by the increasing number of topical products that are available on the market. Although formulations were initially intended for localized delivery of drugs to treat cutaneous pathologies, we now have a broad array of systemically delivered drugs that enter the body through the skin. The limitation in this regard is the exceptional barrier that the stratum corneum presents to exogenous chemicals in the course of the body's normal defense mechanism. This barrier has limited the scope of drugs that may be delivered to the systemic circulation, but newer delivery techniques such as iontophoresis, electroporation, liposomal and penetration enhancer formulation have vastly expanded the range of chemical structures that may now be delivered in clinically effective amounts.

The skin is a multilayered organ composed of several distinct layers: the epidermis (stratum corneum and the viable epidermis), the dermis, and the subcutaneous tissue. The stratum corneum functions as a protective physical and chemical barrier and is only slightly permeable to water. This tissue comprises 10–15 layers of flattened, keratinized cells which, together with the intercellular lipid matrices (mainly ceramides, cholesterol, and fatty acids), form the principle barrier to the ingress of exogenous substances (1). The epidermis is more aqueous in nature and cell replication takes place at the epidermal–dermal mitotic layer. The dermis is highly perfused by blood and lymph vessel networks that provide an efficient means of drug removal into the systemic pool. The hypodermis comprises adipose cells that serve principally as an energy source, a mechanical cushion for the outer skin layers, and an insulation layer in the temperature regulation function of the skin.

III. ABSORPTION THROUGH THE SKIN

Dermatological drug products have been applied topically for many years to achieve a localized pharmacological action in the skin tissues. The drug molecule is considered to diffuse from the applied vehicle to a target tissue in the proximity of drug application site in order to produce its therapeutic effect prior to systemic clearance (of parent drug or its cutaneous metabolites) and elimination. Antiinflammatory corticosteroids have been employed in this fashion for four decades. The transfer of drug across the stratum corneum in this manner is by passive diffusion and for most drugs this process occurs very slowly. This tissue consists of aggregates of closely packed cells and contains both lipid and aqueous regions. Lipid-soluble substances pass readily through the lipid components of the intercellular bilayers whereas water-soluble drugs are able to pass through the more hydrated environments between the bilayer polar headgroups.

Passage through the stratum corneum can be via an intercellular route along the lipid molecule bilayers or transcellular, successively crossing keratinocytes and bilayers in a transverse fashion. It is believed that the former is most favorable from an energetic viewpoint, even though the pathway is longer and more tortuous. Further drug diffusion through the skin may occur through the skin appendages. The continuous barrier generated by the stratum corneum is circumvented somewhat by the eccrine sweat glands and hair follicles, although the stratum corneum layers do penetrate some distance into each of these appendageal openings. The net result is a relatively higher rate of ingress for some types of molecules via these shunt pathways. However, the total surface area of the skin that presents as appendageal opening to the applied drug vehicle is negligibly small (2) when compared to the continuous surface area of stratum corneum cells in contact with the applied formulation. For most drugs, therefore, diffusion through the stratum corneum is the rate-limiting step to penetration, although shunt pathways may be important for the penetration of aqueous (charged) molecules.

In recent years, the skin has served as the portal for administration of systemically active drugs. In this case the agent applied topically has no target receptor in the dermal strata and should theoretically interact with these strata in as minimal a way as possible during its passage to the systemic circulation via the dermis. Substances that pass through the stratum corneum are generally able to diffuse freely through the lower epidermal strata and into the dermis, from which the compound has access to the systemic circulation by diffusion through the walls of the blood or lymphatic vessels.

The biochemistry of the stratum corneum intercellular matrix has been the focus of much research since this is the locus of the rate-limiting factor in the diffusion of topically applied drugs. Any abnormal alteration of the lipid bilayer composition or conformation is usually accompanied by an increase in the permeability of the pathway, allowing drug molecules to pass through the barrier at a greater rate. Penetration enhancer chemicals (3) that are formulated into topical formulations are believed to disrupt the normal alignment of lipid molecules in the bilayers and thereby facilitate the concomitant diffusion of coapplied drugs. Similarly, many physical methods of penetration enhancement (4) (e.g., electroporation) aim to reversibly alter the homeostasis of the lipid bilayers in order to decrease the barrier potential generated. Penetration enhancer development and use has become a science in its own right. A factor that is often overlooked is that the topical formulation constituents, however ubiquitous, may also contribute in a reversible manner to the disruption of the lipid intercellular bilayers and thereby may also serve to enhance the permeation of formulated drugs. Even water or glycols that are often included in topical formulations as drug solubilizers will penetrate the stratum corneum and interact with the biochemicals in this tissue. Formulators often inadvertently (or purposefully) include ''solubilizers'' in topical formulations which will have marked effects on the stratum corneum once applied to the skin.

IV. USE OF EMULSION FORMULATIONS IN DERMATOLOGY

Emulsions are liquid or semisolid mixtures of lipophilic and hydrophilic constituents that have been stabilized into a homeogeneous dispersion by the inclusion of an emulsifying agent (surface-active agent). The purpose of the emulgent molecules is to align at the interface between the oil and water phases and to lower the surface interfacial energy between these two hostile regions. Thereby the macrodispersion is stabilized and coalescence of the phases (cracking of the emulsion) is prevented. Other classes of emulgent increase the viscosity of the continuous phase of the emulsion and thereby prevent coalescence by physical means, but these agents are not used as widely as the interfacial stabilizers. Topical creams are viscous semisolids and are usually oil-in-water (O/W) emulsions (aqueous creams) or water-in-oil (W/O) emulsions (oily creams). Lotions are generally less viscous dispersions that have a greater water fraction.

From a pharmacological viewpoint, emulsions are used to apply solutions or suspensions of drugs to the skin for therapeutic purposes where a highly occlusive effect is not necessary. Drug manufacturers will usually formulate a specific drug (especially corticosteroids) into several types of formulations to meet different patient or clinician needs. Aqueous creams and lotions tend to be more patient-acceptable since they are easily applied and spread on the skin, have a colling effect, and are not obvious in use. Oil emulsions, on the other hand, tend to have a greasy feel on application, are somewhat occlusive, (tending to create a warming sensation at the application site), and often leave a detectable sheen on the skin. Clinicians may prescribe different formulation types depending on the dermatological condition to be treated. Weeping lesions are usually treated with O/W emulsions because these bases tend to spread easily on moist tissue, whereas oily creams are often prescribed for dry, proliferative conditions because the occlusive properties of W/O bases tend to rehydrate the skin and increase its emolliency. Additionally, oily creams usually generate a marked increase in drug permeation through the stratum corneum because of the hydration of the superficial layers. Therefore, the clinical response (and potential toxicity) of any topically applied drug is expected to be greater when occlusive, W/O formulations are used.

The objective when selecting a suitable base for delivery of a therapeutic drug is usually to optimize drug delivery through the skin. In reality, optimum delivery is rarely achieved as many formulations currently on the market were developed long before the current understanding of drug delivery and thermodynamic leaving potential was refined. Newer formulations have tended to follow the general development scheme of products already on the market in terms of balancing the therapeutic potential of the formulation with its aesthetic appeal, its conditions and convenience of use, the stability and safety of the product. There has been little real innovation in terms of designing an emulsion delivery vehicle that can optimally and totally deliver the drug it contains in a predefined manner. Perhaps the lack of innovation in this field has been fostered by the current requirements for the registration of bioequivalent formulations, which require that the innovator formulation be a pharmaceutical equivalent of the reference formulation already on the market. This requires that the innovator product must contain the *same concentration* of the same drug as the reference product. This requirement is still in place even though it is now possible to achieve higher percutaneous delivery, and therefore, better clinical efficacy, with vehicles that have been optimized in terms of delivery potential and can therefore contain much lower total drug concentrations. In essence, the formulator of today is forced to adopt inferior delivery principles that were commonplace two decades ago, simply to comply with the requirements for registration of the product. The cost saving of this practice that is lost to the manufacturer and to the patient is obvious.

V. DRUG DIFFUSION ACROSS THE STRATUM CORNEUM

When compounding a formulation for transdermal drug delivery, the physical and chemical characteristics of each excipient must be determined with respect to the active ingredient, other components in the formulation, and the skin. Since emulsions are complex matrices composed of several chemical components, the potential for interaction and instability is marked. This poses an increased burden on the formulator since each chemical in the topical delivery vehicle has the ability to profoundly alter (or negate) the rate at which the incorporated drug is delivered to the stratum corneum.

A. Diffusant Solubility

The solubility of the drug in the applied emulsion vehicle is of primary importance in determining the rate at which the diffusant traverses the rate-limiting membrane. Diffusion is a passive kinetic process that takes place along a concentration gradient from a region of high diffusant concentration to one of lower concentration. Therefore, the higher the concentration of drug *in solution* in the applied emulsion, the greater will be the diffusion gradient (since theoretically the stratum corneum contains no diffusant molecules prior to administration of the dosage form). It is emphasized that the drug must be in solution to influence the concentration gradient; drug that is contained in the delivery vehicle in a suspension state (in equilibrium with its saturated solution) does not contribute to the concentration gradient. However, drug crystals in suspension may act as the reservoir of drug, which maintains saturation for an extended time and thereby maintains the zero-order delivery kinetics. The solubility of a drug can also be influenced by the presence of a cosolvent in the formulation. Cosolvents that increase the solubility of the active drug produce greater concentration gradients across the vehicle–skin interface. However, as the solubility of the drug in the solvent increases, the partitioning of the drug between the membrane and the vehicle may decrease and, hence, the net effect of the cosolvent on drug delivery may be minimal or may be retardive.

Similarly, one has to consider the biphasic nature of the emulsion matrix: lipophilic drugs would preferentially dissolve in the lipid phase of the emulsion. If one is formulating an O/W emulsion, then drug dissolution will be maximal in the inner, dispersed phase and minimal in the outer, hydrophilic, continuous phase. The latter is in contact with the stratum corneum and the dispersed phase is essentially suspended away from the skin by the continuous hydrophilic phase. The majority of drug partitioning across the vehicle–stratum corneum interface is therefore taking place from the aqueous environment of the vehicle where the drug is at lowest concentration. The effect of dispersed phase globule size in this regard has already been addressed (5).

B. Partition Coefficient

The different solubilities of the diffusant molecule in the delivery vehicle and in the stratum corneum biosystem will establish a partitioning potential across the formulation–skin interface. The degree of partitioning of the drug into the stratum corneum is dependent on its relative affinity for the vehicle and for the intercellular environment. The higher the partition coefficient of the drug for the skin, the greater the concentration of drug established in the proximal layer of corneocytes. Vehicle to stratum corneum partitioning often constitutes the rate-limiting step in transdermal drug delivery. Theoretically, the drug should have a high solubility in (affinity for) the stratum corneum and a low solubility in the delivery vehicle. This will facilitate partitioning across the interface into the skin (maximizing thermodynamic leaving potential). Drug–vehicle interaction plays a significant role in determining the rate of drug molecule diffusion within the base and the partition coefficient of a given drug between the vehicle and skin. A high affinity of the drug for the base is not desirable as drugs that bond with components of the vehicle are released into the skin very slowly. Hence, the release of drug is favored by formulation (dissolution) in a hostile solvent environment from a potential energy viewpoint.

Determinations of drug solubility in the vehicle can assist in estimating the extent of drug affinity for the formulation. This is often a complex task where emulsions are concerned because of the distinct environments of lipophilicity and hydrophilicity, especially when one considers multiple emulsion where there is more than one dispersed region in the formulation of a specific type, or microemulsions (6) where discrete physical influences of neighboring molecules may affect drug solubility (7). While the overall determination of the solubility of a drug in an emulsion may indicate the total fraction of drug present in solution, it is only the drug dissolved in the external continuous phase, which is in contact with the skin, that will provide the partitioning potential.

C. Diffusion Coefficient

Drug diffusion phenomena are important in the delivery vehicle and in the skin environments. In the formulation, drug in solution must diffuse from loci of high concentration to replenish molecules lost at the vehicle–skin interface by partitioning, thereby reattaining the equilibrium in the formulation. Since drug is continually lost at the interface on application of the vehicle to the skin, this replenishment diffusion is a continuous process, often involving partitioning of drug molecules across the continuous–dispersed phase interface and dissolution of drug from the reservoir system of drug in suspension. The diffusion coefficient of a drug, either in the topical vehicle or in the skin, depends on the physicochemical properties of (and the extent of interaction between) the drug and the diffusion medium. Diffusion decreases as the viscosity of the diffusion medium increases and as the potential for transient bonding interactions between the diffusant and formulation molecules increases. Theoretically the drug and vehicle molecules should be as dissimilar as possible to minimize the possibility for bonding between the two molecules during the diffusion process. Again, the dissimilar environments of the emulsion formulation matrices make the estimation of diffusion interactions a difficult process.

D. pH Variation

In general, only un-ionized molecules pass across lipid membranes because unionized drugs are more lipid soluble than their ionized (hydrophilic) forms. Most drugs are either weak acids or weak bases whose solubility in the aqueous phase of a given vehicle is determined by the state of ionization (determined by the pH) in that medium. Alteration of pH is only of concern in the aqueous phase of an emulsion formulation. Altering the pH of the hydrophilic phase may markedly increase the total concentration of drug in solution but may not supply drug molecules in the optimal un-ionized state for partitioning and dissolution in the stratum corneum. In this case, changing the pH to improve dissolution will not enhance the overall process of transdermal delivery. Furthermore, when changing the pH of a formulation the chemical stability of the drug and of the overall preparation must be considered. The formulator must be aware that the pH of optimum drug solubility is not always the pH of maximum stability of the formulation and a balance must often be achieved between the two.

E. Thermodynamic Activity

The thermodynamic activity of a drug in a particular vehicle is the potential for the active substance to be released from the delivery system to the skin with which it is in contact. The foregoing discussion suggests that maximum release of a drug would occur when it exists in the vehicle in maximal (un-ionized) concentration, in a dissimilar (energetically hostile) chemical environment. A saturated solution is therefore preferable as a topical drug delivery system as the maximum concentration gradient and maximum thermodynamic activity (leaving potential) is achieved with such a system. Saturated solutions, by implication, have exhausted the attractive bonding interactions between the drug and vehicle molecular groups. Drug molecules at saturation solubility are in a state of dynamic equilibrium, continually precipitating out of solution and reentering the solution state. The molecules in this state undergo the least degree of vehicle bonding interactions and are most likely to escape from the formulation by partitioning across the vehicle–skin interface. Drug release from subsaturated solutions is retarded with respect to saturated systems.

With respect to thermodynamic activity, the solubility of a drug can also be influenced by the presence of cosolvent chemicals in the formulation. Cosolvents that increase the solubility of the active drug produce greater concentration gradients across the vehicle–skin interface. However, as the solsubility of the drug in the solvent increases, because the number of attractive bonding interactions between the drug and cosolvent molecules increases, the partitioning of the drug between the membrane and the vehicle may decrease for the same reason. Hence, the net effect of the cosolvent on the rate and extent of drug delivery may be minimal or may be retardive. Since emulsions are complex, multicomponent matrices, there is a good potential that several of the formulation constituents may act as cosolvents for the drug molecules in this way. Therefore, adding propylene glycol, for example, to an emulsion formulation to increase the solubility of the incorporated drug may inadvertently improve the dissolution characteristics of the environment to such an extent that the higher proportion of drug molecules entering into solution at equilibrium have a lesser tendency to leave the vehicle and partition into the stratum corneum.

Therefore, from a diffusion viewpoint, the solubility of the drug in the vehicle should be maintained as near to saturation as possible. However, it is undesirable to have a base in which the drug is highly soluble as the release of the permeant will be retarded from this favorable environment. Preformulation studies on the solubility of the drug in all the components used in the preparation of the emulsion are essential for ensuring optimal drug release.

VI. EFFECT OF THE EMULSION ON TRANSDERMAL DELIVERY

It is a relatively complex matter to attempt to predict the dynamic changes that will take place in the delivery vehicle with each change in chemical composition or concentration outlined above. However, the problem is not solved when a stable emulsion has been generated that will (theoretically) allow maximal delivery of the incorporated drug to the stratum corneum. The factor that many formulators do not consider is that profound changes take place in the delivery vehicle and in the skin on application of the vehicle; the term *metamorphosis of the vehicle* has recently been introduced (8) to describe this phenomenon. Volatile components of the emulsion evaporate quickly on spreading (and rubbing) the formulation on the skin. Water is the obvious volatile constituent that will be lost, but it should also be remembered that even ''cosolvents'' such as propylene glycol are volatile and are lost to the atmosphere. This loss by evaporation drastically changes the dissolution and partitioning environment that has been created by the formulator and may result in both rapid precipitation of drug molecules from solution and altered partitioning potential between the applied film of emulsion and the stratum corneum.

Similarly, emulsion vehicle components may enter the stratum corneum layers, since all chemicals are free to diffuse from the applied film into the skin. The entry of these formulation constituents into the skin may markedly affect the dissolution and diffusion environment of this tissue. Water entering the skin from the formulation will hydrate and swell the stratum corneum layers and allow partitioning drug molecules to diffuse more easily through these layers. Occlusive vehicles such as W/O emulsions may have a similar effect. Chemicals that penetrate the skin, such as propylene glycol, glycerol, or fatty acids, may disrupt the ordered bilayer lipid structure of the intercellular spaces and allow facilitated drug diffusion. Penetration enhancers such as azone (9) are incorporated into topical formulations for exactly this purpose; what is often not understood is that *all* chemical constituents in the emulsion have the potential to interact with and alter the diffusive transfer that takes place across the barrier layer. Similarly, chemicals entering the skin may influence the development and maintenance of a drug reservoir in the strata. Prolonged binding of the drug to the molecules in the stratum corneum reduces the absorption of the drug into the viable epidermis and dermis, but forms a depot for the long-term delivery of the partitioned drug to these structures that may last for several days or weeks. Vehicles specifically formulated to maximize depot formation in the skin may therefore allow less frequent application regimens to be implemented.

One should also take into account that disease, age of the skin, and site of topical drug application usually violate the constraints of the simple diffusion theory and may require modification of the vehicle formulation. Intact, healthy skin presents a barrier to exogenous absorption that can be reduced considerably or nullified when the skin is damaged or is in a diseased state. To produce a clinically superior dermatological preparation the formulator must consider how the site and condition of the skin will affect drug absorption through the stratum corneum. In addition, the skin is now understood to be a metabolically active tissue system that has the ability to metabolize several classes of drugs that are diffusing through it—dermal first-pass metabolism. The possible influence of the emulsion constituents on this system should also be considered.

VII. CONCLUSION

In the formulation of an emulsion vehicle for topical drug delivery, many factors must be considered. Drug stability, specific product use, site of application, and product type must be combined in a multiphase dosage form that will readily allow drug release when the formulation is applied to the skin. The potential of the vehicle to release the drug is dependent on the physicochemical properties of the specific diffusant and the matrix of chemicals that have been combined together to form the emulsion. Safety, stability, and effective preservation must be combined with optimum drug delivery in the total formulation, which must retain aesthetic acceptability to the patient for compliance considerations. Although consumers may apply less stringent acceptability criteria to dermatologicals, there is still general preference for preparations that are easy to apply to the skin and easy to remove by washing. Emulsion formulations fulfill this role perfectly, together with generating a cooling sensation on application and providing some degree of skin rehydration and emolliency. With the recent advances in the quality and scope of chemical constituents that are available to the formulator to use in the development of topical emulsions, and with the expanding knowledge of the microstructure effects of dispersed systems, this appears to be an exciting pharmaceutical field for the future.

REFERENCES

1. P.M. Elias and G.K. Menon. Structural and lipid biochemical correlates of the epidermal permeability barrier. Adv. Lipid Res., 24, 1991, 1–26.
2. R.F. Bronaugh, R.F. Stewart, and E.R. Congdon. Methods for in vitro percutaneous

absorption studies II. Animal models for human skin. Toxicol. Appl. Pharmacol., 62, 1982, 481–488.
3. R.B. Walker and E.W. Smith. The Role of Percutaneous Penetration Enhancers. Adv. Drug Deliv. Rev., 18, 1996, 295–301.
4. M.R. Prausnitz. Reversible permeabilization for transdermal delivery of macromolecules. Crit. Rev. Ther. Drug Carr. Sys., 14, 1997, 455–483.
5. J.M. Haigh, E.W. Smith, E. Meyer, and R. Fassihi. Influence of the oil phase dispersion in a cream base on the in vivo release of betamethasone 17-valerate. S.T.P. Pharm. Sci., 2, 1992, 259–264.
6. C. Laugel, A. Baillet, M.P. Youenang Piemi, J.P. Marty, and D. Ferrier. Oil-water-oil multiple emulsions for prolonged delivery of hydrocortisone after topical application: comparison with simple emulsions. Int. J. Pharm., 160, 1998, 109–117.
7. C. Malcolmson, C. Satra, S. Kantaria, A. Sidhu, and M.J. Lawrence. Effect of oil on the level of solubilization of testoterone propionate into nonionic oil-in-water microemulsions. J. Pharm. Sci., 87, 1998, 109–122.
8. E.W. Smith, C. Surber, and H.I. Maibach. Topical dermatological vehicles. A holistic approach. In: Percutaneous Absorption. Drugs—Cosmetics—Mechanisms—Methodology, 3rd Ed., R.L. Bronaugh, and H.I. Maibach (eds.), Marcel Dekker, New York (1999), 779–787.
9. B.B. Michniak, M.R. Player, J.M. Chapman, and J.W. Sowell. In vitro evaluation of a series of azone analogs as dermal penetration enhancers. 1. Int. J. Pharm., 91, 1993, 85–93.

8
Vesicles as a Tool for Dermal and Transdermal Drug Delivery

J. A. Bouwstra
Leiden/Amsterdam Center for Drug Research, Einsteinweg, and Leiden University, Leiden, The Netherlands

I. INTRODUCTION

The natural function of the skin is the protection of the body against the loss of endogenous substances such as water and undesired influences from the environment caused by exogenous substances. This implies that the skin acts as a barrier for diffusion of substances through the underlying tissue. Overcoming this natural barrier is the main problem in dermal/transdermal administration of drugs. Methods to decrease the diffusional barrier include the use of penetration enhancers, sonophoresis, iontophoresis, and incorporation of drugs in vesicles that can serve as special drug carrier systems.

The diffusional barrier for most substances is located in the upper layer of the skin, the stratum corneum (SC). The SC consists of unnucleated cells, the corneocytes, entirely surrounded by lamellar lipid regions (Fig. 1). The lipid regions are required for a competent skin barrier and are the only continuous structure in the SC.

In the basal layer of the epidermis there is a continuous renewal of cells. These cells are subsequently transported to the upper layers in the epidermis. The composition of lipids changes markedly during apical migration through successive epidermal layers. In the basal and spinous layers the plasma membranes contain a large fraction of phospholipids. However, in the SC when the differentiation process is accomplished, the lipid composition changes markedly. The main constituents of the SC lipids are cholesterol, free fatty acids, and ceramides. SC ceramides represent a unique, heterogeneous group of at least six ceramides that differ from each other by the headgroup architecture and by the mean fatty acid chain length. The fatty acid esterfied to the amide of the sphingosine headgroup can be either α-hydroxy or nonhydroxy fatty acids. In pig SC the fatty acid chain length varies between 16 (ceramide 5) and 33 C atoms (1) (ceramide 1).

In recent years the lipid organization of human, pig, and mouse SC was studied using small and wide-angle x-ray diffraction (SAXD and WAXD, respectively) and electron microscopy in combination with RuO_4 post fixation. X-ray diffraction studies revealed the presence of crystalline lipid lamellar phases

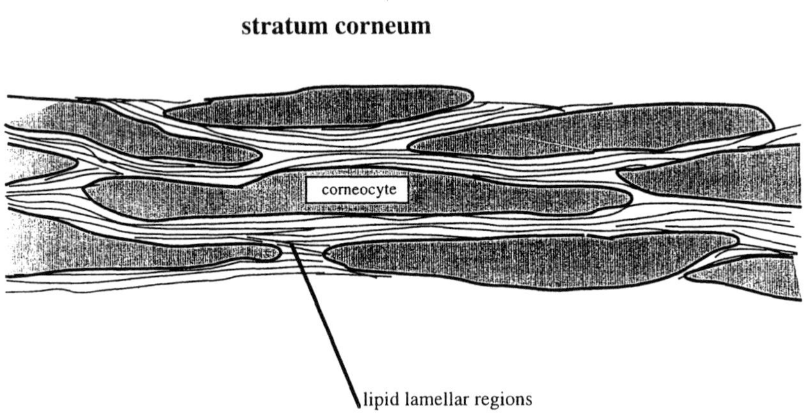

Figure 1 Stratum corneum structure. Corneocytes are embedded in lipid lamellar regions. The transport route is either transcellular or intercellular.

Figure 2 The lipid lamellae between the corneocytes with the characteristic broad–broad–narrow sequence, bar = 100 nm.

(2–7). Using RuO_4 post fixation an unusual lipid bilayer arrangement of a repeating pattern of two electron translucent broad bands and one electron translucent narrow band, which are not equal in width (8–12) was found (Fig. 2). In this chapter first a brief summary will be given of the lipid phase behavior found is SC of various species.

II. LIPID ORGANIZATION IN STRATUM CORNEUM

In Fig. 3 the small-angle x-ray diffraction curves of human SC measured at room temperature is drawn. This diffraction curve is characterized by a strong and a weak peak. Both peaks consist of a main position and a shoulder on the right-hand side. After recrystallization from 120°C the diffraction curves revealed a series of peaks that are located at the same interpeak distance, which is a characteristic diffraction pattern of a lamellar phase. From the positions of the various peaks it could be concluded that a lamellar phase with a periodicity of 13.4 nm was present. Comparing the peak positions in the diffraction curve after recrystallization with those present in the diffraction curve from untreated SC revealed

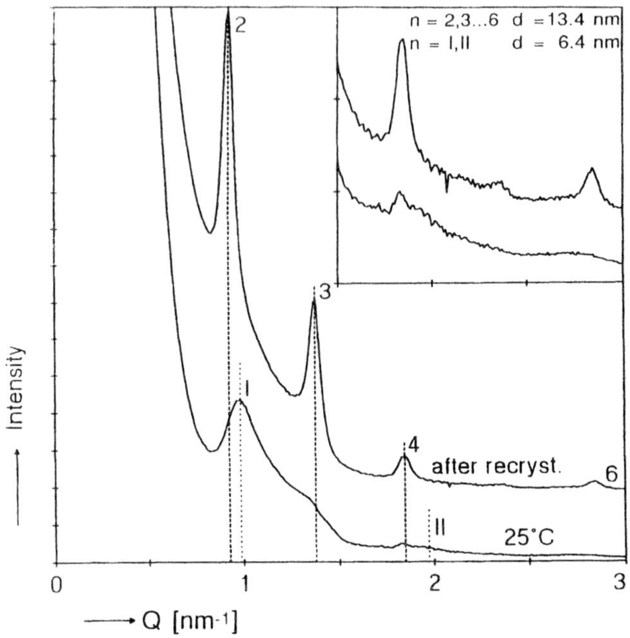

Figure 3 The diffraction curves of the human stratum corneum at a hydration level of approximately 20% w/w. The diffraction curve after recrystallization from 120°C revealed a series of sharp peaks based on a 13.4-nm lamellar phase. The diffraction pattern at room temperature indicates two lamellar phases with periodicities of 13.4 and 6.4 nm, respectively.

that the latter could be interpreted by the presence of two lamellar phases in untreated SC (3) with a periodicity of 13.4 and 6.4 nm, respectively.

The wide-angle x-ray diffraction pattern revealed two strong diffraction rings at approximately 0.412 and 0.373 nm, indicating that at least a crystalline packing (orthorhombic) of the lipids is present. Whether in addition to this crystalline packing a hexagonal or even a fluid phase is also present cannot be concluded from this pattern since the diffraction ring attributed to an hexagonal lateral packing is obscured by the strong diffraction peak at approximately 0.415 nm assigned to the orthorhombic phase. The presence of soft keratin in the corneocytes results in a two broad rings at a spacing of approximately 0.45 and 0.96 nm. From the diffraction patterns of intact SC we cannot determine whether a liquid lateral packing coexists in intact SC, since the diffraction ring that can be attributed to a liquid lateral packing is also present at a spacing of 0.45 nm. However, information on the SC lipid phase behavior can also be obtained with mixtures prepared from either synthetic or isolated ceramides mixed with fatty acids and cholesterol. Recently, lipid mixtures were examined that were prepared from isolated pig SC ceramides, free fatty acids, and cholesterol (13). These studies revealed that these mixtures exhibit a similar phase behavior as the lipids present in intact pig SC, namely, the presence of two lamellar phases with repeat distances of approximately 5.6 and 13.1 nm. In these mixtures no indications for the existence of a lateral liquid phase were found. However, when cholesterol sulfate was added to the mixtures, the solubilization of cholesterol in the lipid structure was increased and a liquid lateral phase (14) was formed. In earlier studies by Kitson et al. (15) it was observed that in mixtures of bovine brain ceramide 3, cholesterol, and palmitic acid at least 80% of the lipids are in a crystalline or gel state but that a small fraction of lipids was also in a liquid state. Therefore, it is likely that despite the presence of long chain fatty acids in the SC, a small fraction of lipids in SC forms a fluid phase.

From the above-mentioned studies we can conclude that the SC lipids form a very exceptional organization not observed in other biological membranes. Most probably this is caused by the presence of the long chain ceramides and the absence of phospholipids. Furthermore, the headgroups of the ceramides contain several hydroxyl groups that can form lateral hydrogen bonds with adjacent lipids. As a result of this, lipid bilayers are formed with a very tight packing. Furthermore, the lipids are very hydrophobic and the lamellae do not swell upon hydration. In this way the lipid lamellae that are surrounding the corneocytes form a hydrophobic mortar between the cells. The consequence of this is a very low natural hydration level in the SC, being only 10–20%.

In comparing the lipid organization in SC isolated from skin of pig, human, or mouse (see Table 1), some differences are observed. While the 13-nm lamellar phase is always present in the SC of the various species, the 6-nm phase is only occasionally present in mouse SC. Furthermore, the absence of an orthorhombic lateral packing in pig SC is also very striking. Of course, the lipid organization is not the only parameter that varies between the various species. In a study by Lavrijsen et al. (16), the Trans Epidermal Water Loss (TEWL) was measured as function of the number of strippings and a comparison was made between hairless mouse and human skin. Although in intact mouse and human skin the TEWL values were similar, the TEWL values as a function of the number of strippings was different. In mouse SC, the TEWL increased already after 1–2 strippings, while this occurs in human SC only after 10–12 strippings. Furthermore, the number of hair follicles per unit area in mouse, human, and pig skin differs tremendously between the various species and between different regions in one specie. This means that results obtained with animals can not be extrapolated to humans, although trends in interaction might be similar.

The presence of these crystalline lamellae between the corneocytes and the impermeable corneocyte envelope that forces drugs to diffuse through the

Table 1 A Comparison Between the Lipid Organization in Pig, Human, and Mouse Stratum Corneum

Skin specie	Lateral packing[a]	Lamellar phases	Crystalline cholesterol
Human	*Orthorhombic *Hexagonal? *Liquid?	Two lamellar phases with periodicities of 6.4 and 13.4 nm	Frequently present
Pig	*Hexagonal *Liquid	Two lamellar phases with periodicities of 6.0 and 13.2 nm	Always present
Mouse	*Orthorhombic *Hexagonal? *Liquid	One 13.4-nm lamellar phase, occasionally a 6-nm lamellar phase	Only occasionally present

[a] A question mark indicates that it is not known whether or not this phase is present at room temperature.

crystalline lamellae results in an excellent skin barrier. This means that the main problems for dermal/transdermal administration of drugs is overcoming this natural permeability barrier. In order to do so, there are several methods available. While chemical penetration enhancers or supersaturated of the drug concentrations in the formulation are both often used as cheap methods to increase the penetration rate of drugs, the problem is often the limited enhancement in skin penetration rate or the toxicity of the penetration enhancers itself. By using physical methods, such as iontophoresis, often much higher skin penetration enhancement can be achieved; however, since cost aspects play a negative role, this method will only be commercially attractive for those substances, such as peptides, that cannot be administered in another noninvasive way.

A third method to increase the skin transport is encapsulation of drugs in vesicles. Although several studies have been performed, the interactions between vesicles and skin is still largely unknown. In this chapter a review will be given on the state of the art in skin vesicle research.

III. LIPOSOMES AFFECT DRUG TRANSPORT THROUGH THE SKIN

A. Introduction

The first publications on interactions between liposomes and skin appeared in 1980 and 1982 (17,18). In these publications it was reported that liposomes applied in vivo to white rabbits skin favored the disposition of drugs in the epidermis and dermis, while the amount of drug found in the various organs was reduced. In these studies it was strongly suggested that vesicles penetrated the skin intact. This suggestion resulted in a considerable amount of scepticism in the scientific literature. That scepticism was even further amplified when for commercial reasons the cosmetic industry in their advertisement also claimed that vesicles could penetrate the skin as intact entities, with the aim of improving the skin appearance. These campaigns certainly did not harm this research field. This scepticism was demonstrated in a very amusing paper published by Gregoriadis (19). After the first papers by Mezei and Gulasekharam, a large number of studies were carried out, a review of which will be given below. This chapter will focus on (a) studies on in vitro skin penetration and drug disposition, (b) drug disposition when applied to skin in vivo, (c) studies designed to target drugs to the hair follicles, and, finally, (d) studies performed to obtain insight in the mechanisms involved in the mode of actions of vesicles.

B. Vesicles Affect Skin Penetration in Vitro

1. Vesicle Composition

After the introduction of liposomes as drug delivery systems for the transdermal route by Mezei and Guleksaharam (17,18), the first reports to appear on the same subject were those of Ganesan and Ho (20,21). In the studies reported in these papers, liposomes were prepared from dipalmitoyldiphosphatidylcholine (DPPC), which forms gel state bilayers. In their studies three (model) drugs were used, namely, glucose, progesterone, and hydrocortisone. These studies were performed to obtain detailed information on the mechanisms involved in the skin penetration rate when the drug was applied in vesicles. The authors proposed that the drug could be transported either (a) as free solute, (b) as free solute but also associated with the liposomes, or (c) by a direct transfer of the drug from liposomal bilayer to SC without partitioning into the water phase. They assumed that the vesicles neither absorbed intact nor fused with the SC surface. They performed permeation studies in vitro with mouse skin. Two observations let them conclude that liposomes did not penetrate as intact entities across the skin; (a) DPPC-radiolabeled liposomes did not result in measurable amounts of radioactivity in the receiver of the diffusion cell and (b) glucose entrapped in liposomes resulted in lower skin permeability then compared to glucose in normal saline. Furthermore, release studies on the more lipophilic drugs revealed almost no drug release from the liposomes to the saline solution. However, when applied on the skin, the drug penetrated across the skin, strongly suggesting that a direct transfer from vesicles to SC occurred. Very convincing the results of these studies did not confirm intact liposome penetration as suggestions by Mezei and Guluksaharam.

One of the first studies in which the vesicle composition varied systematically was carried out by Knepp et al. (22,23). In these studies the effect of liposome composition on skin penetration of progesterone was studied. In their first study (22) the release rate of progesterone from the delivery system consisting of liposomes in an agarose gel was examined. These studies revealed that the release from the agarose gel alone was very fast compared to the release from liposomes embedded in the agarose gel. Moreover, it was shown that the vesicles reduced the progesterone transport rate across hairless mouse skin. In their succeeding study (23), progesterone fluxes across mouse skin were studied in more detail over a period of 24 h. The drug was either applied in only the gel that served as control, or in egg-phosphatidylcholine, dioleoylphosphatidylcholine (DOPC) liposomes, DMPC dimyristoylphosphatidylcholine (DMPC) liposomes, or in dipalmitoylphosphatidylcholine (DPPC) liposomes. The liposomes were again applied in the agarose gel. In all cases the skin transport of progresterone from a liposome based formulation was lower than from the agarose gel. Furthermore, the DPPC vesicles forming gel state bilayers resulted in a slower skin penetration of progesterone compared to the egg-phosphatidylcholine liposomes forming liquid state bilayers. When *cis*-unsaturated fatty acid (oleic acid) was intercalated in vesicle bilayers prepared from saturated acyl chain phospholipids the drug transport rate across the skin was increased by an order of magnitude. From the results of this systematic study the authors concluded that a liposome-based reservoir system can modulate drug transport. The studies of Knepp et al. were performed with liposomes that were immobilized in an agarose gel. In two more recent studies (24,25) it was found that liposome incorporation in a gel can affect the interactions between the liposome and the skin. For example, due to the immobilization of the liposomes, a direct transfer of the drugs from liposomes to SC might be inhibited, while this direct transfer was considered to be an important mechanism in the papers by Ho and Ganesan (20,21). However, most likely the trend observed in the studies of Knepp et al. that incorporation of drugs in liquid state liposomes results in higher flux rates across the skin than the drug incorporation in gel state liposomes will also hold for gel-free formulations (see below).

Several other studies were carried out to evaluate the effect of liposome composition on skin penetration of drugs. In the early 1990s Dowton et al. (26) compared the effect of liposomal composition on the disposition of encapsulated cyclosporine A in mouse skin. They applied the vesicles nonocclusively in vitro. Two nonionic surfactant vesicles were compared with bovine brain ceramide liposomes and liposomes prepared from saturated and unsaturated phosphotidylcholine. The various liposomes were saturated with cyclosporine A. In this way an equal thermodynamic activity of the cyclosporine A was achieved in all formulations. The distribution of cyclosporine A after 24 h of application was determined and it was found that when the drug was applied in nonionic surfactant vesicles prepared from glyceryl dilaurate/cholesterol/polyoxyethylene-10 stearyl ether, the amount of cyclosporine A in deeper skin strata and receiver compartment was highest compared to the other formulations. In fact, these studies confirmed the results of the studies of Knepp et al. (22,23), although in the studies by Dowton et al. (26) the vesicles were applied nonocclusively without a gel. The findings of Knepp et al. (22,23) were also confirmed by studies of Hofland et al. (27). They found that when estradiol was incorporated in gel state nonionic surfactant vesicles prepared from long chain saturated surfactants and cholesterol, the drug transport across human SC was very low compared to estradiol skin transport when applied in liquid state vesicles (Fig. 4). These results were obtained with saturated estradiol formulations. However, in contrast to the studies of Knepp et al., the studies of Hofland et al. (27) revealed that a drug applied in liquid vesicles resulted in higher penetration rates then when applied in a phosphate-buffered saline solution. The differences found between these two studies might be based on the difference between the experimental setups such as the presence of agarose in the studies of Knepp et al. (22,23). However, another important difference in these two studies is an equal *thermodynamic activity* of estradiol in all formulation in the studies performed by Hofland et al., while in the progesterone diffusion studies, the progesterone *concentration* in the various formulations was kept equal, resulting in different driving forces from the formulations to the skin. In a subsequent study, van Hal et al. (28) found that a decrease in cholesterol content, which results in vesicles with more fluid and more flexible bilayers, resulted in an increase in estradiol transport across SC. Meuwissen et al. (29) compared the diffusion of the bilayer label fluorescein-dipalmitoylphosphatidylethanolamine (fluorescein-DPPE) in the skin when applied in gel state liposomes and in liquid state liposomes, and found using confocal laser scanning microscopy that fluorescein-DPPE applied in liquid state liposomes penetrated deeper in the skin than the label applied in gel state liposomes. Du Plessis (30) compared application of a lipophilic and hydrophilic drug after pretreatment with skin lipid liposomes and phospholipid liposomes and did not find large differences in drug permeation. However, this concerned pretreatment and not encapsulation of the drug in the vesicles. From all of the above-mentioned studies it seems very clear that the *thermodynamic state of the bilayers* of the vesicles in which the drugs are applied plays a crucial role in the effect of vesicles on drug transport rate across skin in vitro. Furthermore, it seems that at least in liquid state vesicles prepared from single-chain surfactants and cholesterol (28), an increased cholesterol/surfactant ratio inhibits the transport of lipophilic drugs across the skin.

2. Application Method

In a few studies occlusive application was compared to nonocclusive application and it was found that occlusive application of vesicles to the skin resulted in a

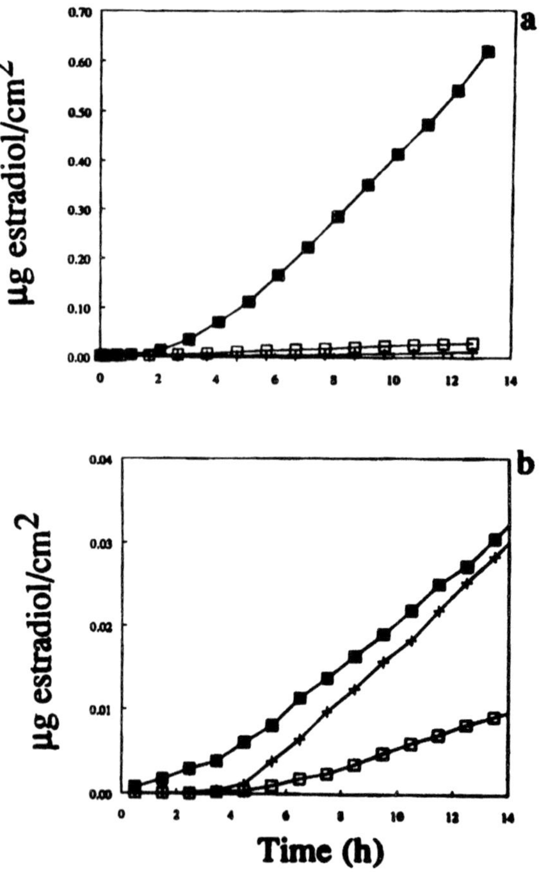

Figure 4 Estradiol flux through human stratum corneum when the drug is applied occlusively in liquid or gel state vesicles prepared from polyoxyethylene alkyl ethers and cholesterol. All formulations are saturated with respect to estradiol. (A) Plus estradiol applied in PBS; ■, estradiol applied in liquid state vesicles prepared from decaoxyethylene oleyl ether ($C_{9=9}EO_{10}$) and cholesterol; □, estradiol applied in phosphate buffered saline solution (PBS) after pretreatment of vesicles prepared from $C_{9=9}EO_{10}$ and cholesterol. (B) Plus estradiol applied in PBS; ■, estradiol applied in gel state vesicles prepared from trioxyethylene alkyl octadecyl ether ($C_{18}EO_3$) and cholesterol; □, estradiol applied in PBS after pretreatment of vesicles prepared from $C_{18}EO_3$ and cholesterol.

lower amount of drug in SC and deeper skin strata than when applied nonocclusively (31,32). Since water is an effective permeation enhancer, the results were somewhat unexpected. At least two reasons can be given for these findings, namely, (a) a more profound interaction between the liposomal constituents and the skin or (b) the absence of a hydration gradient in the skin, in case of occlusive application. According to Cevc et al. (31), the water gradient is an important driving force for diffusion and makes even intact vesicle penetration in the skin possible, when flexible vesicles, the so-called transfersomes, are used (see below). The effect of formulation volume on drug transport was also assessed. In a study of Niemic et al. (33), the effect of the formulation volume on skin disposition of cyclosporine A was examined. For these studies nonionic surfactant vesicles prepared from glyceryl dilaurate/cholesterol/polyoxyethylene-10 stearyl ether were selected and applied nonocclusively. It was found that an increase in the vesicle application volume resulted in a decrease in the amount of drug disposited in the lower strata of the skin. Possibly this is due to a longer dehydration period of the formulation when the application volume increased. It was also found that an increase in application time from 2 to 24 h resulted in an increase in the amount of drug in the skin. Du Plessis et al. (30) found that nonocclusive application of a drug encapsulation in vesicles was more effective than pretreatment of the skin with the empty liposomes, after which the buffer solution containing the drug was applied. The drugs used in these studies were hydrocortisone and inulin. In particular, the amount of drug in the deeper SC was increased when applied in liposomes. Hofland et al. (27) and van Hal et al. (28) studied the skin penetration of estradiol through human SC applied in vesicles occlusively and applied in a buffer solution after pretreatment using empty vesicles. The vesicles were prepared from polyoxyethylene alkyl ethers and cholesterol. Since they performed their studies with saturated estradiol formulations, the thermodynamic activity in all formulations was equal. The results of both studies confirmed the findings of du Plessis et al. (30), namely, that pretreatment was less effective than application of drugs in vesicles.

3. Physical Characteristics of Liposomes

The size and lamellarity of liposomes may affect the skin penetration rate or the disposition of the drug in the various skin layers. Some studies have been carried out to elucidate the effect of these parameters. In a study by du Plessis et al. (34), the effect of vesicle size on the disposition of the drug was evaluated using skin from various species. They stated that if intact penetration of liposomes would occur, the smaller vesicles would lead to a higher drug disposition in lower skin strata than when the drug was applied in larger vesicles. For this purpose vesicles were applied nonocclusively. The vesicles were prepared from phosphatidylcholine, cholesterol, and cholesterol sulfate. Two lipophilic drugs were chosen, namely, cyclosporine A and cholesterol sulfate. The liposome size varied between 60 and 600 nm. Pig skin, hamster skin, and mouse skin were used. After 24 h of application they found that the disposition of the drug in lower skin strata and the receiver compartment was not favored by the smaller vesicles, from which they concluded that intact penetration of vesicles would be very unlikely to occur. This conclusion was in agreement with that of Ho and Ganesan (20,21). Hofland et al. (27) also examined the effect of vesicle size on estradiol transport through human SC. They also found no increased estradiol transport when using smaller unilamellar vesicles and again it was concluded that vesicles most likely do not transport intact across the human SC. Finally, it was found that the effect of lamellarity on the disposition of cholesterol and cholesterol sulfate in the various skin strata had little effect (35). It seems that the physical parameters as vesicle size and lamellarity are less important than the application method and the thermodynamic state of the bilayers. These findings are in favor of the absence of intact vesicle penetration across the skin.

4. Various Skin Species

Du Plessis et al. (34,35) evaluated the effect of liposome size and lamellarity on the disposition of the drug into the skin. For this purpose vesicles were applied nonocclusively. As stated before, pig skin, hamster skin, and mouse skin were used. They observed more drug in the lower pig skin strata compared to the lower mouse skin strata and suggested that this might be due to a higher density of follicles in the pig skin. If this is the case, then for lipophilic substances the follicular route plays an important role. However, the amounts are given relative to the total amount applied, so that no information is available on the concentration in the various skin strata. As is generally known, pig skin is much thicker than mouse and hamster skin. Therefore, their conclusions are subject to some debate.

5. Liposomes Compared to Other Formulations

There are a number of studies in which liposomes were compared to lotion, creams, or penetration enhancers. Although information can be obtained about the relative effectiveness of liposomes compared to other formulations, no information can be obtained about mechanisms, since in these studies the thermodynamic activity of the drug in the various formulations might differ quite extensively. Furthermore, the different components in the liposomes and lotions or creams might have different interactions with the skin. Therefore, obtained differences in drug penetration rate cannot be correlated with the special shape of the liposomes. Nevertheless, these studies are of interest because they show whether or not drug application in liposomes will be in favor compared to other formulations that might already be on the market.

Lash and Wohlrab (36,37) encapsulated either cortisol or hydrocortisone in egg lecithin liposomes. The loaded liposomes were applied on human skin and the drug distribution was compared to application of an equal drug concentration in an ointment. In both studies it was found that the drug distribution after application in liposomes was a function of depth less steep than after application in an ointment, which might be important for a reduction of side effects. Michel et al. (38) performed a very extensive study and compared the effect of liposomes on the penetration of two lipophilic drugs in skin, DL-α-tocopherol nicotinate (TN) and the antiinflammatory substance L440, with application in conventional galenical formulations. The vesicle formulations with the drugs were tested on their stability, size, and appearance (freeze-fracture electron microscopy). TN was applied in various vesicle suspensions and its absorption was examined in human SC in vivo. These studies revealed that TN in multilamellar vesicles resulted in a higher uptake of TN in human SC than applied in small unilamellar vesicles, which is surprisingly in contrast to results of other in vitro studies described above. Furthermore, occlusive application was more effective than nonocclusive application, which was also in contrast to earlier studies. L440 was mainly used to compare liposomes with the galenic formulation, and it appeared that when equal doses were used the vesicles were more effective than the galenic formulations. Moreover, the liposomes were also more effective in the treatment of arachidonic acid–induced ear eudema, suggesting that the liposomes might be a promising drug delivery system for these types of drugs. Touitou et al. (39) compared penetration enhancers with liposomes. Interestingly, they found that liposomes prepared from phospholipon-90 and a small amount of cholesterol disposited caffeine mainly in the skin but did not increase the flux that much, while the penetration enhancers, being a combination of transcutol and oleic acid, increased the flux of caffeine through hairless mouse skin tremendously. In vivo the enhancers resulted also in significantly higher plasma levels of caffeine. From these results they concluded that liposomes mainly disposit the drug in the skin and therefore can act as an excellent reservoir, while the penetration enhancers increased the drug transport across the skin.

C. Vesicles Affect Skin Penetration in Vivo

In 1986, Komatsu et al. (40) applied butylparaben to the skin in a liposome formulation prepared from egg phosphatidylcholine, cholesterol, and diacetyl phosphate. They varied the amount of lipids as well as the amount of butylparaben in their formulations. The liposomes were applied occlusively to the dorsal skin of guinea pig in vivo. The fate of the liposomes was determined by autoradiography. While the radioactive DPPC was mainly found on the application site, even after 48 h, the applied ^{14}C-butylparaben radioactivity was found in small intestine, feces, gall bladder, and urinary bladder. In an additional in vitro study (41) the mechanisms involved in ^{14}C-butylparaben penetration was examined very systematically using a flow through diffusion cell. They found that (a) the percentage of ^{14}C-butylparaben transported across the skin decreased as function of increasing amount of lipids (Fig. 5) and (b) the transport rate of radioactive phospholipids (DPPC) across guinea pig skin was much lower than that of ^{14}C butylparaben, indicating that the lipids are mainly staying on the skin surface or present in the SC and that copenetration of butylparaben and phospholipids does not occur. This is in agreement with the results of the in vivo experiments (40). Again it was concluded that most likely the liposomes do not penetrate as intact entities across the skin. (c) Pretreatment with empty liposomes did not further increase the flux of ^{14}C-butylparaben. From the observation that ^{14}C-butylparaben flux decreased at increasing lipid content the authors concluded that ^{14}C-butylparaben solubilized in the water phase mainly contributed to the percutaneous penetration. However, another mechanism may also play a role. The thermodynamic activity of butylparaben in the bilayers most probably reduces when increasing the lipid content, which might lead to a decrease in the partitioning from the lipid phase to either the water phase or the SC. This might result in a reduced penetration through the skin.

In another excellent study (42), double labeling was carried out. ^{3}H-Phosphatidylcholine and ^{14}C-tretinoin were intercalated in soybean lecithin liposomes and applied on female hairless rat in vivo and on human skin in vitro. After 6 and 12 h of in vivo application the distribution of the labels was determined in SC, epidermis, and dermis. These studies revealed that the ratio of the labels in

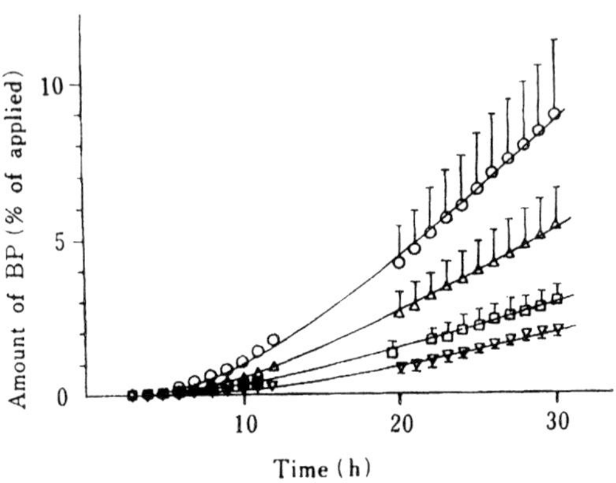

Figure 5 Percultaneous penetration of butylparaben from liposome suspensions in vitro (mean ± SD, $n = 4$). Liposomes were prepared from egg PC, cholesterol, and diacetyl phosphate in a ratio of 4:2:1. The amount of egg PC in the formulations in (∇) 32 mmol, (□) 16 mmol, (Δ) 8 mmol, and (○) 16/3 mmol.

SC was approximately constant as a function of the number of strippings but that the ratio was different from the ratio in the liposomes (Fig. 6). Furthermore, the $^3H/^{14}C$ ratio in epidermis was lower and decreased steeply until a skin depth of approximately 200 μm was reached. From these results the authors concluded that copenetration of a drug–liposome bilayer can be envisaged and that this is accompanied by migration of free drug. With liposomes epidermal and dermal concentrations are significantly greater than with a gel. Based on the reduced $^3H/^{14}C$ ratio in epidermis and dermis, the drug diffuses separately from liposomal bilayers in deeper skin strata.

Gesztes and Mezei (43) compared a tetracaine liposome formulation with commercial available pontocaine cream. After 1 and 2 h of application the anesthesia was tested by the pinprick method. It was found that the liposomal tetracaine produced a longer lasting anesthesia than the pontocaine cream. However, although these experiments are important for possible commercial use of liposomes, no mechanistic conclusions can be drawn from these studies. In a series studies carried out in the early 1990s (44), NAT-106 liposomes (see below), which induce large changes in the lipid organization in SC, were applied on pig skin occlusively. These liposomes were loaded with a 20,000 MW protein and the penetration behavior of the protein was followed by monoclonal antibody detection. Tissue samples were obtained only after 20 min of application. While application in a phosphate-buffered saline solution did not result in antibody activity, protein loaded in these liposomes resulted in antibody activity in all layers of the skin, clearly demonstrating that the transport of protein across the skin was facilitated by the liposomes. Although clearly demonstrated, these results are very surprisingly taking into account the size of the protein being 20,000 Da. In a second study (45), the effect of liposomes on percutaneous penetration was

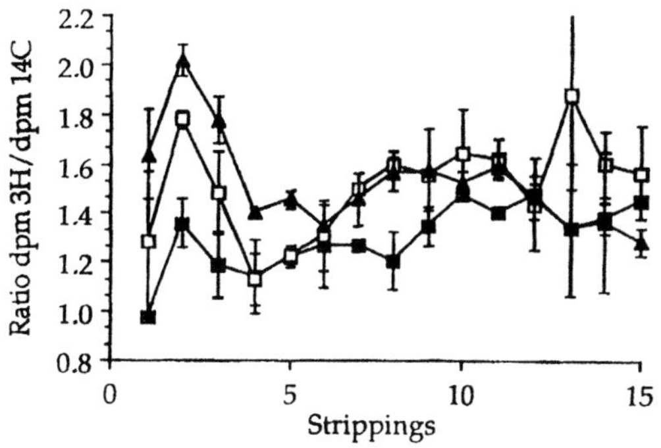

Figure 6 Evolution of the $^3H/^{14}C$ in the stratum corneum during in vivo distribution after 3 h (■), 6 h (□), and 12 h (▲), $n = 3$. In the applied liposomes the $^3H/^{14}C$ is 2.35.

investigated using three tracers; palmitoyl-^{14}C linonyl-phosphatidylcholine, sodium pertechnetate (^{99m}Tc), and heparin-^{35}S sulfonate. These labels were applied again in NAT-106 liposomes or a buffer solution (except for the radiolabeled phospholipid) to the skin. During 5 h of application blood samples were taken. The SC was stripped and samples were taken from the underlaying skin. The results revealed much higher concentrations of ^{35}S-heparin and ^{99m}Tc in blood and renal elimination, while a 3-fold increase in skin tissue was observed when applied in the NAT-106 liposomes again indicating that the liposomes increased the percutaneous penetration of hydrophilic drugs. In a related study it was shown that three liposome suspensions (NAT-106, NAT-50, and NAT-89, see below) also resulted in a decrease in the corneal blood supply as measured by laser Doppler flowmetry (46), indicating that *empty* liposomes induce changes in the lower regions of the skin.

Weiner et al. (47) used liposomally encapsulated interferon in a cutaneous herpes guinea pig model and compared the effect of preparation method and liposomal compositions on the therapeutic efficiency. In this study multilamellar vesicles, large unilamellar vesicles, and vesicles prepared by the dehydration–rehydration (DRV) method were compared. Only interferon encapsulated in vesicles prepared by the DRV method was effective, while no differences in therapeutic effect were observed with vesicles that differed in composition. Since in this study vesicles were prepared from egg lecithin (liquid state vesicles) as well as from ceramides (gel state vesicles), it was surprising that the composition of the vesicles had no effect because most studies, despite their being carried out in vitro, revealed that gel state vesicles are less efficient in drug transport enhancement than liquid state vesicles. The authors of this study suggested that the DRV method facilitates the interferon to be associated with the bilayers and that this was the reason for a better transfer of the drug into the lipophilic lipid regions of the SC, resulting in an improved therapeutic effect.

Skalko et al. (48) prepared multilamellar vesicles from soya lecithin or hostaphate. The size of the vesicles ranged between 0.77 and 8 μm, which is rather large. As a drug clindomycin hydrochloride was used. The concentration of the drug in all formulations was equal. They applied the various formulations to 30 patients all suffering from agnes vulgaris and observed that clinical treatment with a lotion containing liposomes during 2 and 4 weeks showed a much better efficacy than a nonliposomal form. Therefore, the results of these studies are very encouraging regarding the use of liposomes of this treatment.

The group of Cevc published a series of papers on the so-called transfersomes. Transfersomes are vesicles prepared from lipids and an edge activator which might be a single-chain lipid or surfactant, which makes the vesicles very deformable. In their first paper it is claimed that the presence of a water gradient exerts a force in the order of 10^{11} N per vesicle of a radius of approximately 60 nm. It is stated that as a result of this force deformable vesicles (which is a property of the transfersomes) can be squeezed through lipid bilayer regions (31). They compared occlusive with nonocclusive application of transfersomes and found nonocclusive application more effective than occlusive application (see above). In order to detect the mass flow across the skin and the disposition in the animal, they used tritiated DPPC as a radioactive compound. They reported that the transfersomes were much more effective than the "conventional" liposomes when applied nonocclusively (Fig. 7). After 8 h of transfersome application in mice significant amounts were found in blood (approximately 8% of the *applied* dose) and liver (20% of the *recovered* dose). They claimed that 50–90% of the dermally deposited lipids can be transported beyond the level of the SC. Although their transfersomes have certainly advantages with respect to increasing the transport of material across the skin, in these studies it was claimed that the deformable transfersomes penetrated intact into skin. This statement in particular brought renewed skepticism into this research field, since the statement was and still is unproven. The skepticism is also caused by the knowledge that such large associates are extremely difficult to transport across the skin as intact entities, especially since the SC consists of a very tight structure that is designed to act

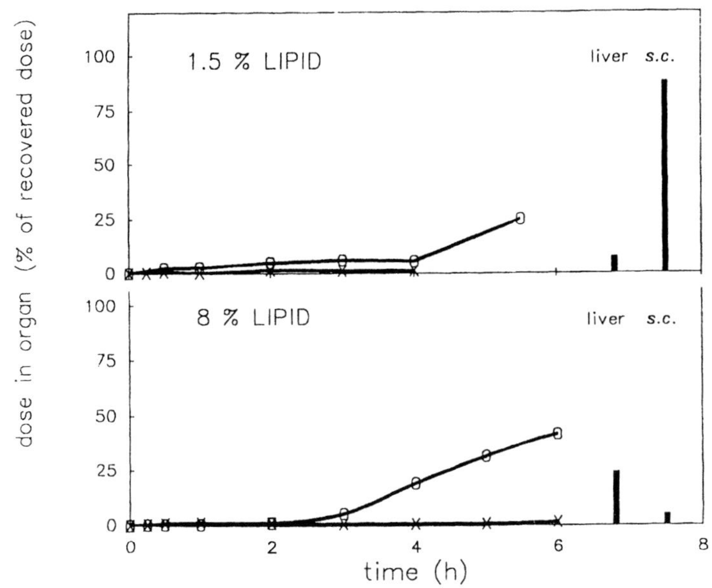

Figure 7 Systemic availability of vesicle-associated lipids as a function of time after dermal application of 8 and 1.5 mg/cm² in the form of transfersomes (O) or "standard" liposomes (×). The percentage of the recovered dose is given. Tritiated labeled phosphatidylcholine was used as marker.

as a barrier (see Sec. I). In more recent studies of this group several drugs were associated with the vesicles. In their first published study on drugs associated with transfersomes (49), phosphatidylcholine was combined with sodium cholate (as edge activator) and small amounts of ethanol, most probably required to lower the interfacial tension of the transfersomes, were used. A low interfacial tension makes these vesicles more deformable. The investigated transfersome suspension contained relatively high amounts of lidocaine or tetracaine (49). The transfersomes were tested on rats and on humans, and in both studies the transfersomes appeared to be more efficient than the conventional liposomes or solutions. However, on humans the transfersomes were applied occlusively resulting in an absence of the water gradient across the tissue. The water gradient was considered to be very important for the efficiency of the transfersomes. In another report, insulin was associated with transfersomes (50). Blood glucose levels could be lowered after about 3 h of application. A clear difference in glucose level in blood was observed between application of insulin associated with micelles, conventional liposomes, or transfersomes in the sense that transfersomes resulted in a significantly lower glucose level. This was observed using mice, but in more recent studies in humans also a decrease in glucose level was observed (51). Not only relatively large molecules as insulin are claimed to be transported across the skin associatied with the transfersomes, but even big molecules such as fluorescein isothiocyanate (FITC)-Bovine Serum Albumin (BSA) (52) or I (Iodide)-BSA can be transported across the skin when associated with transfersomes. The biodistribution of the derived radioactivity after 8 h application of mice skin of I-BSA associated with transfersomes is very similar to that of transfersome-associated I-BSA injected subcutaneously (22). These results are very encouraging and certainly show that transfersomes have advantages over vesicles prepared from only double-chain phospholipids and cholesterol. It would be of extreme interest to study the mechanisms involved in these types of vesicles in more detail.

In another study (53), sodium fluorescein was applied in liposomes prepared from soya lecithin as formulation, as pellet, as supernatant, and in a buffer solution. When applied on the skin for 30 min the skin was stripped and in each strip the amount of fluorescein was measured. Although the authors claimed a weak enhanced penetration of the label, the differences observed were very small. Furthermore, since only one time interval was chosen and not 100% of the label was recovered, some doubts arise as to whether the conclusion of the authors that vesicles enhance the penetration of the fluorescent label in the skin is justified.

Recently, a very extensive and well-designed study was described by Ogiso et al. (54). They compared gel systems in which *d*-limonene or laurocapram was present as penetration enhancer and studied the transport of β-histine with and without liposomes in these gels. The liposomes were prepared from either egg phosphatidylcholine or hydrogenated soybean phosphatidylcholine. In vivo studies were carried out with Wistar rat. Blood samples were taken after 10 h of application. The bioavailability of the drug was in the presence of the egg phosphatidylcholine liposomes higher than in the absence of the egg phosphatidylcholine liposomes. They also compared stripped and intact skin with and without liposomes in the formulation and found that in the presence of the liposomes the barrier of the SC was strongly reduced for the diffusion of β-histine compared to application in the same formulation in the absence of the liposomes. They also found that the increase in drug transport in the presence of gel state vesicles was smaller than in the presence of liquid state vesicles. Thus the fluidity of the membranes is an important factor for the penetration enhancement, which confirmed the results of earlier studies. In addition, they performed liquid analysis and found that the amount of ceramides in the SC decreased dramatically after application of a *d*-limonene-containing formulation, which led them to suggest that this penetration enhancer extracts lipids from the skin. Furthermore, the triglyceride content also decreased, suggesting that the follicular lipids were also extracted by this formulation. The lipid analysis also revealed that the phospholipid content increased after liposome treatment, suggesting that the liposomes replace the ceramides in the SC. However, cleaning surface might be a problem in determining the amount of lipids penetrated in the SC. Finally, from fluorescence microscopy no real evidence was found for the presence of intact vesicles in the SC and in the follicles, but they did not fully exclude the possibility of intact vesicle penetration. They assumed that in their formulations the replacement of ceramides by phospholipids play an important role in the mechanisms involved in the increased penetration of β-histine.

D. Follicular Drug Delivery

Only a few available papers focus on targeting to the hair follicles. It seems that the importance of this research area has always been underestimated. However, some pioneering studies have been undertaken to examine targeting of drugs to the follicles. Lieb et al. (55,56) used multilamellar vesicles prepared from phosphatidylcholine, cholesterol, and phosphatidylserine. As model drug, carboxyfluorescein was used, which was encapsulated in the vesicles but was also present in the external buffer phase. Vesicles were compared with an aqueous solution, ethanol, and propylene glycol. The formulations were applied to Hamster ear under nonoccluded conditions and the fluorescein was determined by quantitative fluorescence microscopy. Although in the pilosebaceous units more carboxyfluorescein was present than when using organic and surfactant solutions, the amount of label in the follicular unit was only increased from 0.1–0.4% for the organic and surfactant solutions to 1.2% (of applied dose) for the vesicles, which is a factor 3 increase. Although the authors concluded that liposomes targeted the fluorescein to the follicles, most of the label was still either in the donor phase of the in vitro diffusion setup or in the epidermis. Interestingly, when applied in vesicles the amount of fluorescein in epidermis increased from about 4% for the nonliposomal solutions to 18% for the liposome formulations, which is a factor 4.5 increase. In another study by the same group (57), the effect of liposomal composition on targeting to the various skin strata including to the pilosebaceous units was studied. Two drugs were used; α-interferon and cyclosporine. In case of the hydrophilic drug (α-interferon), the concentration in all of the formulations was kept equal, while in case of the lipophilic drug the formulations were saturated. Vesicles were prepared from three different surfactant compositions. For the hydrophilic drug the control was a buffer solution, while for the cyclosporine the control was a 40% ethanolic solution. The experiments were performed in vivo using male golden hamsters. The vesicles were applied nonocclusively. After 12 h the hamsters were sacrificed and drug distribution was determined by scraping. The radioactivity in various layers, such as the pilosebaceous units, dermis, cartilage, and dorsal, was determined. It was found that when the nonionic surfactant vesicles contained glycerol dilaurate, much higher values were found in the pilosebaceous units than when using the other vesicles. Although the authors claim that this vesicle suspension is more efficient, if the distribution is calculated as relative amounts in the various strata the aqueous solution, PC liposomes, and other nonionic vesicles are much more efficient. In case of the lipophilic cyclosporine, we observed the same phenomenon. In all skin strata and cartilage, much more label was disposited when applied in the vesicles based on glycerol dilaurate than when applied in the other vesicles. However, calculated as relative amounts the other formulation deposit relatively more in the hair follicles. Therefore, their conclusions are the subject of debate. Striking is the similarity in distribution of the hydrophilic and lipophilic drug applied in the liposomes, which was not expected, since it was often assumed that drug association to lipid bilayers would facilitate the transport of the drug in the tissue. In another study (58), cimetidine was delivered in formulations at pH 5.5, 8.3, and 7.4. At a pH value of 5.5 cimetidine is ionized, whereas it is neutral at pH 8.3. Surprisingly, when the drug was delivered in an aqueous solution at pH 8.3 more drug was driven into the pilosebaceous unit then when applied in liposomes or nonionic surfactant vesicles. At pH 8.3 the drug is very close to its saturation point in an aqueous solution, which was suggested by the authors to be the reason for the higher uptake in the pilosebaceous unit.

Yarosh et al. (59) used liposomes prepared from phosphatidylcholine, cholesterol hemisuccinate, and oleic acid to the skin with a DNA repair enzyme, T4 endonuclease V. Before and after exposure to UV-B light the enzyme was applied on the skin. After 6 h the animals were sacrificed and the DNA dimers counted. It appeared that post- and pretreatment with DNA repair–loaded liposomes resulted in a decrease in the number of dimers. Repeated application did not further enhance the repair. These results suggest that the DNA repair at least entered the cells. In a second study, Yarosh et al. (60) studied the fate of the di-I-labelled liposomes by bright-field fluorescence and by transmission electron microscopy. Liposomes containing a DNA repair enzyme were active in vivo. After isolation and sectioning of the mouse skin and staining with an antibody the enzyme could be detected in the epidermal cells and not in the dermis. Following Yarosh vesicles were present in the cytoplasmic membranes of epidermal cells and along the hair follicles. Furthermore, liposomes were found in cells of lymph nodes. Although intact penetration of liposomes across mouse SC seems to be very unrealistic, it seems that liposomal materials and their content are transported in sufficient amounts in the epidermis. These results are quite challenging, taking into account that an enzyme of approximately 4500 Da is involved. Interestingly, these vesicles contain an edge activator, i.e., oleic acid, which might increase the flexibility of the liposomes. By definition the composition of these liposomes approaches that of transfersomes (see below). Bernard et al (61) studied the targeting of an antiandrogen (RU 58841) to the hair follicles by liposomes and an ethanolic solution. They compared drug penetration through normal and scar hairless rat skin and found that penetration across normal skin was increased compared to that across scar skin, indicating that the follicles play a role in the transport of this drug through the skin. Furthermore, the concentration of the drug in normal skin was much higher than in scar skin. They also performed qualitative autoradiography and found localization of the drug in the sebaceous glands.

Finally, the group of Hoffman (62) made an extensive study of the fate of liposomes and targeting to the hair follicles. Liposomes were prepared from egg phosphatidylcholine. Melanin was encapsulated in the vesicles. Pieces of skin from the white-haired mouse were histocultured. The liposomes were incubated for 12 h. As control a melanin solution was taken. Histological studies and fluorescence microscopy revealed that the melanin was mainly present in the hair follicles when applied in liposomes. When applied as free solution almost no melanin was present in the hair follicles, indicating that by using liposomes tageting to hair follicles can be achieved. In two other related studies (63,64), similar results were obtained with entrapped calcein and a DNA-label fraction. They concluded that liposomes have an ability to target drugs to the follicles.

E. Mechanistic Studies

Several studies were performed to obtain a detailed insight into the interactions between vesicles and skin. Most studies focused on penetration characteristics of vesicles or vesicle constituents. In a study by Gabrijelcic et al. (65), a spin probe (ASL) was loaded in liposomes and the liposomes were applied to the skin. The translational mobility of the probe was detected. The mobility of the probe was measured with electron paramagnetic resonance imaging (EPRI). Gabrijelcic et al. compared encapsulated of the probe in liquid state and gel state vesicles and examined the effect of the preparation method. They observed that the label was only mobile when applied in liquid vesicles prepared by the reversed phase evaporation method. However, their experimental conditions were very unclear. Although the method seems to be very elegant, the consequence of their results is not clearly discussed. For example, it is not clear as to whether with this method (a) it is possible to distinguish between probe diffusion on the skin surface and probe diffusion in to the skin, and (b) it is possible to distinguish between diffusion in the viable epidermis/dermis and in the SC.

In 1993 the fate of liposomes after application to the skin was studied by $\gamma\gamma$ angular correlation spectroscopy with indium encapsulated in the vesicles. Liquid state vesicles were prepared from egg lecithin. The vesicles were applied to pig skin nonocclusively. From these measurements it was concluded that the vesicles disintegrate on the surface of the skin and that only a negligible amount of marker penetrated to the deeper layers of the SC. These results are in agreement with the results of studies carried out by Lasch et al. (66). They performed studies using fluorescence spectroscopy, in which the label was intercalated in the bilayers of the vesicles, and found that the label remained mainly on the surface of the SC, suggesting that no intact vesicle penetration occured. In a more recent study, vesicle–skin interactions were examined by confocal laser scanning microscopy (CLSM) (67). A large number of liposome compositions were examined. These studies revealed that the liposome constituents were only present in the outermost layers of the SC and that the constituents only acted as penetration enhancers.

In additional studies (68) the interactions between liposomes and skin were examined by freeze-fracture electron microscopy and Fourier transform infrared spectroscopy (FTIR). In the visualization study human SC was treated with liposomes prepared from either NAT-106, NAT-50, or NAT-89. These lipids consists of phosphatidylethanolamine (PE), phosphatidic acid (PA), and phosphatidylcholine including lysophosphatidylcholine (PC). The liposomes prepared from NAT-50 fused on the surface of the SC. No changes in the lipid organization were observed in deeper regions of the SC. Vesicles prepared from NAT-89 revealed intact vesicles between the upper SC cells. In addition, quite frequently rough structures were observed indicating a mixing between the SC lipids and the liposomes. Finally, application of vesicles prepared from NAT-106 revealed large changes in the lipid organization in the SC (Fig. 8). Almost no intact lipid bilayers were observed between the corneocytes; often flattened, isolated regions were visualized. This was also observed in deeper regions in the SC. These studies clearly show that liquid state liposomes prepared from different lipid mixtures result in different interactions with the SC. Interestingly, the liposomes prepared from the largest fraction PC, i.e., NAT-106, resulted in the strongest interactions with the SC. This can be explained by the presence of the single-chain lipid, lysophosphatidylcholine. Lysophosphatidylcholine can act as an edge activator

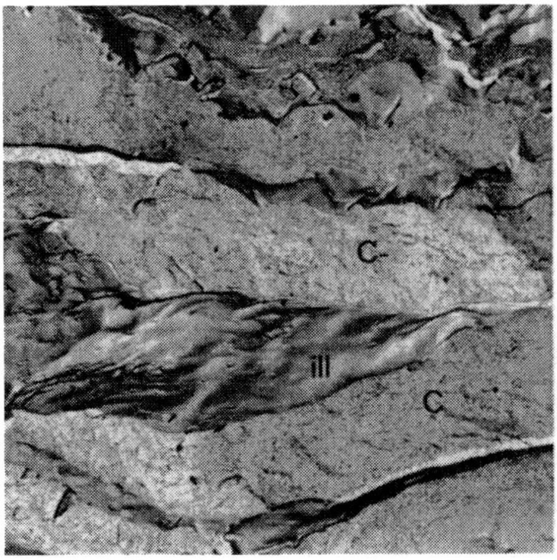

A

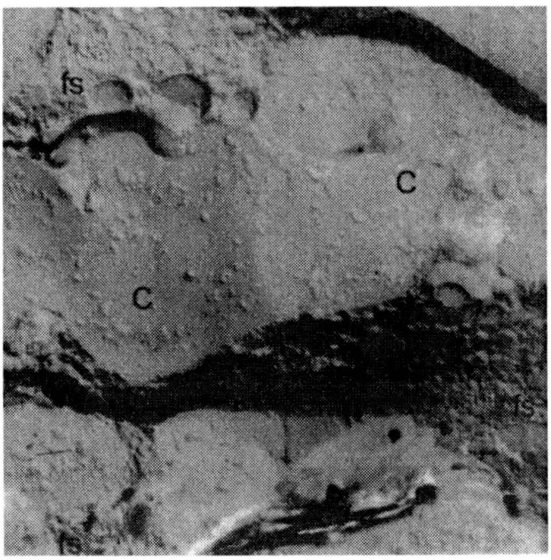

B

Figure 8 (A) PBS-treated human stratum corneum (SC). The micrograph depicts the corneocytes (C) between which the intercellular lipid lamellae (ILL) are located. The intercellular lipid lamellar regions are depicted as smooth surface planes. Sharp edges occur where the fracture plane crosses the lipid bilayers (scale bar = 0.1 mm). (B) NAT-106 liposomes had a very strong influence on the microstructure of the SC. The corneocytes (C) were considerably swollen and the smooth ultrastructure of the intercellular lipid bilayers showed flattened spherical structures (FS). The flattened structures were often circular (scale bar = 0.1 mm).

and increase the flexibility of the liposomal bilayers. In a related study (69), in vivo skin hydration was monitored either after occlusive application of NAT-106 liposomes prepared in D_2O or after application of pure D_2O. The hydration and the lipid penetration were monitored by FTIR. The liposomes proved to be superior compared to pure D_2O in driving D_2O into the skin. However, the amount of hydrogen oxide in the SC was lower compared to application of the pure D_2O. Finally, phospholipid components could be detected in deeper layers of the SC. Between the second and third strip the phospholipid content decreased dramatically; however, even after 16 strips phospholipids were still detectable (Fig. 9). Phospholipids were not observed in the control skin. The presence of phospholipids deep in the SC correlates with the results observed with freeze fracture, in which also in deeper layers of the SC interactions between phospholipid layers and SC were observed. The observed increase in D_2O content in SC and the decrease in accumulation of H_2O in the SC were suggested to be caused by an

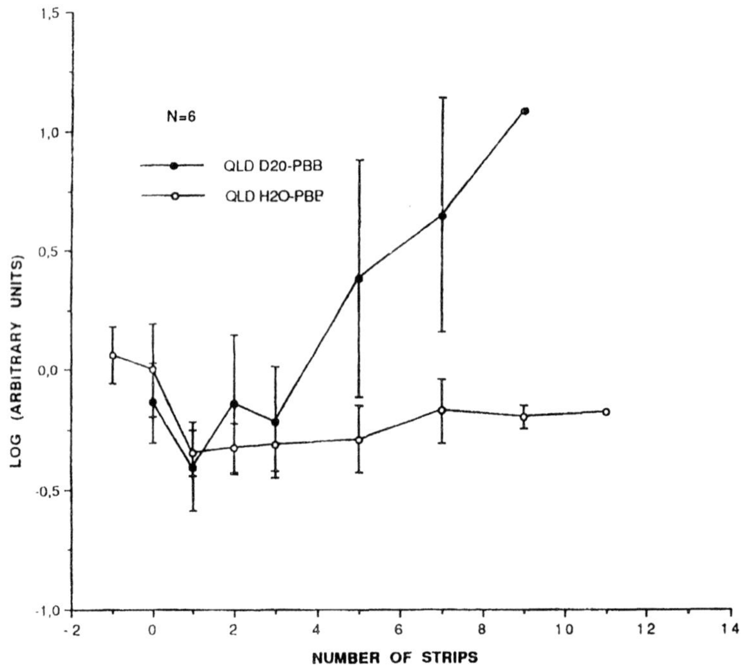

Figure 9 Closed circles: Quotient of peak-to-background ratios of OD-stretch peaks of liposome-treated skin divided by those of D_2O-treated skin. Open circles: Quotient of peak-to-background ratios of OH-stretch peaks of liposome-treated skin divided by those of D_2O-treated skin.

increase in the kinetics. In other words, the phospholipids may act as penetration enhancers for D_2O. In a study by Korting et al. (70), liposomes prepared from soya lecithin were prepared and applied on reconstructed epidermis. After application the skin was visualized on an ultrastructural level using OsO_4 staining and fixation. They observed no intact liposomes in the lower layers of the SC but rather deposition of lipids derived from liposomes between and within the corneocytes. These findings are in agreement with those found by Hofland et al. (68). Korting et al. found that only when liposomes were applied in high concentrations were they detected between the cells, and then only in the upper layer of the SC very close to the SC surface. These authors did not agree with the interpretations of Foldvari et al. (71), who claimed that intact liposomes penetrate in the skin.

In another study, vesicles prepared from cholesterol and surfactants were investigated (72). Vesicles were applied occlusively to human skin and the interactions studied with electron microscopy and CLSM. These studies revealed that vesicles applied occlusively to human skin are adsorbed on the SC surface. Furthermore, changes in the lipid organization in the superfacial layers of the SC were often observed. That vesicles can be adsorbed on corneocytes and corneocyte envelopes was demonstrated by Abraham and Downing (73). Hofland et al. (72) occasionally observed changes in the deeper layers of the SC, such as the presence of vesicular structures. The authors asserted that penetration of intact vesicles in the SC would be very unlikely to occur and explained the presence of vesicles by the penetration of vesicle constituents that might be capable of reforming vesicles in the SC. Furthermore, water pools not only in the corneocytes but also between the corneocytes, indicating that phase separation between SC lipids and water occurs. In a more recent study (74), it was found that the vesicles were only very occasionally present in the deeper layers of the skin and therefore are not expected to affect the diffusion characteristics of drugs across the skin. The studies of Hofland et al. (72) also revealed that a lipophilic label applied in gel state vesicles remains on the surface, while that applied in liquid state vesicles penetrates partly in the SC. Two recent studies appeared in which CLSM was used to study the penetration of fluorescent labels across the skin in more detail. Schätzlein and Cevc (75,76) intercalated rhodamine-DHPE (1,2-dihexadecanoyl-*sn*-glycero-3-phosphatidylethanolamine-*N*-lissamine rhodamine B sulfonyl, triethylamonium salt) in the bilayers of transfersomes prepared from soya phosphatidylcholine and sodium cholate. The suspensions were applied non-occlusively on mice skin and after 4–12 h the skin was examined ex vivo using CLSM. They claimed two vesicle transport routes in the SC, the so-called intercluster pathway between groups of cells, which should be a preferred route of transport between corneocytes that only partly overlap. However, the skin surface is not flat but contains many wrinkles. In recent experiments from our own group we also found ''clifts'' with a bright fluorescent appearance. However, in our studies these clifts may correspond to wrinkles in the skin (32). Intracluster pathways are also observed by Schätzlein and Cevc (75,76), since regions are observed with high fluorescence intensity across the lipid bilayer regions. The authors interpret these regions as being virtual pores between the corneocytes. In all interpretations, one should realize that CLSM is a technique that *cannot* be used to visualize the transport of vesicles as intact entities; it can only be used to visualize the transport of the label, which is not necessarily associated with the vesicle. Very recently, a paper appeared (71) in which liposomal transport was measured across human skin in vitro by time-resolved IR attenuated total reflection spectroscopy. In these studies, the transport of small unilamellar vesicles were compared with the transport of transfersomes, and it was found that transfersomes resulted in a faster penetration than liposomes. However, whether intact vesicles penetrate or liposomal fragments is not clear from these experiments either, since in this case only the lipids are detected and not the structure formed by the lipids.

IV. CONCLUSIONS

Considering all of the studies performed we can generally conclude that the drug transport can be adjusted on demand by association of the drug with vesicles. One of the central parameters is the thermodynamic state of the bilayers that affects quite dramatically drug permeation across the SC and in addition the interactions with the SC. It seems that the size and the multilamellarity of the vesicles only affects drug transport slightly. Furthermore, many of the studies revealed that it is very unlikely that intact vesicles transport across SC occurs. This is not only concluded from the results of visualization studies, but also from transport studies and studies using biophysical techniques. Concerning both, the gel and liquid state vesicles, the drug transport is often increased when single-chain surfactants are intercalated in the vesicles. These single-chain surfactants can act as penetration enhancers, especially in case of gel state vesicles. However, when incorporated in liquid state vesicles there might be an additional important effect, since incorporation of such a single-chain surfactant into the bilayers of liquid state vesicles results in a decrease in interfacial tension and thus in deformable vesicles (31,75,76). These deformable vesicles have been shown to increase the drug transport across the skin even further. However, questions arise about the mechanism involved in this increased drug transport. To examine the interaction of these deformable vesicles with the skin is a relatively new and interesting field. If it were possible that the lipid particles diffuse as intact entities across the skin, the question arises as to what occurs at the interface between SC–stratum granulosum. Do these particles remain the SC because of their amphiphilic nature or do these vesicles partition into the viable epidermis? And if these particles are taken up in the epidermis, is the drug then released and taken up by the small blood capillaries in the dermis, or are the vesicles, being too large to be taken up by the small blood vessels, taken up by the lymphatic nodes? Then the delivery system may reach the systemic blood circulation, similar to what occurs after subcutaneous injection. This latter possibility might be rather a utopie than something that occurs in reality, since many barriers have to taken by the vesicles before the blood circulation is reached.

Although many questions have already been answered concerning the mechanisms involved in the change in drug transport when associated with vesicles, some very fundamental questions concerning transport mechanisms involved in newly designed vesicles have to be answered. However, when using vesicles drug transport through the skin can be adjusted on demand by the physical characteristics of the vesicles, and even an inhibition of drug transport can be achieved, which might be of interest for the development of barrier creams.

REFERENCES

1. P. W. Wertz and D. T. Downing, Ceramides of pig stratum epidermis, structure determination. J. Lipid Res. 24 (1983) 753–758.
2. J. A. Bouwstra, G. S. Gooris, M. A. Salomons-de Vries, J. A. van der Spek, and W. Bras, Structure of human stratum corneum as function of temperature and hydration. A wide angle x-ray diffraction study. Int. J. Pharm. 84 (1992) 205–216.
3. S. H. White, D. Mirejovsky, and G. I. King, Structure of lamellar lipid domains and corneocytes envelopes of murine stratum corneum. An x-ray diffraction study. Biochemistry 27 (1988) 3725–3732.
4. J. A. Bouwstra, G. S. Gooris, J. A. van der Spek, and W. Bras, Structural investigations of human stratum corneum by small angle x-ray scattering. J. Invest. Derm. 97 (1991) 1004–1012.
5. J. A. Bouwstra, G. S. Gooris, W. Bras, and D. T. Downing, The lipid and keratin structure of pig stratum corneum. J. Lipid Res. 36 (1996) 685–695.
6. J. C. Carson, J. Doucet, J. L. Leveque, and G. Tsoucaris, Oriented structure in human stratum corneum revealed by x-ray diffraction. J. Invest. Derm. 96 (1991) 43–49.
7. J. A. Bouwstra, G. S. Gooris, J. A. van der Spek, S. Lavrijsen, and W. Bras, The lipid and protein structure of mouse stratum corneum: a wide and small angle diffraction study. Biochim. Biophys. Acta 1212 (1994) 183–192.
8. S. Y. Hou, A. K. Mitra, S. H. White, G. K. Menon, R. Ghadially, and P. Elias, Membrane structures in normal and essential fatty acid-deficient stratum corneum: characterization by ruthenium tetroxide staining and x-ray diffraction. J. Invest Dermatol 96 (1991) 215–223.
9. J. A. Bouwstra, O. Sibon, M. A. Salomons-de Vries, and F. Spies, Freeze substitution as tool for visualization of interactions between vesicles and skin. Proc. Controlled Rel. Soc. Meet., Orlando (1992), 410–419.

10. D. T. Downing, Lipid and protein structure in the permeability barrier of mammalian epidermis. J. Lipid Res. 33 (1992) 301–314.
11. M. Fartasch, I. D. Bassakas, and T. L. Diepgen, Disturbed extruding mechanism of lamellar bodies in dry non-eczematous skin of atopics. Br. J. Dermatol. 127 (1992) 221–227.
12. D. C. Swartzendruber, Studies of Epidermal Lipids Using Electron Microscopy. Semin. Dermatol. 11 (1992) 157–161.
13. J. A. Bouwstra, G. S. Gooris, K. Cheng, A. Weerheim, W. Bras, and M. Ponec, Phase behaviour of isolated skin lipids. J. Lipid Res., (1996) 999–1011.
14. J. A. Bouwstra, G. S. Gooris, F. E. R. Dubbelaar, A. M. Weerheim, and M. Ponec, pH and cholesterol sulfate affect stratum corneum lipid organization. J. Invest. Dermatol., 3 (1998) 69–74.
15. N. Kitson, J. Thewalt, M. Lafleur, and M. Bloom, A model membrane approach to the epidermal permeability barrier, Biochemistry, 33 (1994) 6707–6715.
16. A. P. M. Lavrijsen, Stratum corneum barrier in healthy and diseased skin, thesis, Leiden University, 1997, p 29.
17. M. Mezei and V. Gulasekharam, Liposomes-a selective drug delivery system for the topical route of administration. I. Lotion dosage forms. Life Sci. 26 (1980) 1473–1477.
18. M. Mezei and V. Gulasekharam, Liposomes—a selective drug delivery system for the topical route of administration. II. Gel dosage form. J. Pharm. Pharmacol. 34 (1980) 1473–1477.
19. G. Gregoriadis, Liposomes and anti-ageing creams: the facts beneath the face. The biochemist (1994).
20. N. F. H. Ho, M. G. Ganesan, and G. L. Flynn, Mechanism of topical delivery of liposomally entrapped drugs. J. Controlled Rel. 2 (1985) 61–65.
21. M. G. Ganesan, N. D. Weiner, G. L. Flynn, and N. F. H. Ho, Influence of liposomal drug entrapment on percutaneous absorption. Int. J. Pharm. 20 (1984) 139–154.
22. V. M. Knepp, R. S. Hinz, F. C. Szoka, and R. H. Guy, Controlled drug release from a novel liposomal delivery system, i. Investigations of transdermal potential. J. Controlled Rel. 5 (1988) 211–221.
23. V. M. Knepp, F. C. Szoka, and R. H. Guy, Controlled drug release from a novel liposome delivery system. II transdermal delivery characteristics. J. Controlled Rel. 12 (1990) 25–30.
24. J. Lasch and J. A. Bouwstra. Interactions of external lipids (lipid vesicles) with the skin. J. Liposome Res. 5 (1995) 543–569.
25. M. Foldvari, B. Jarvis, and C. J. N. Ogueijofor. Topical dosage form of liposomal tetraciane: effect of additives on the in vitro release and in vivo efficacy. J. Controlled Rel. 27 (1993) 193–205.
26. S. M. Dowton, Z. Hu, C. Ramachandran, D. F. H. Wallach, and N. Weiner, Influence of liposomal composition on topical delivery of encapsulated cyclosporin A, I. An in vitro study using hairless mouse skin. STP Pharma 3 (1993) 404–407.
27. H. E. J. Hofland, R. van der Geest, H. E. Boddé, H. E. Junginger, and J. A. Bouwstra. Estradiol permeation from nonionic surfactant vesicles through human stratum corneum in vitro. Pharm. Res. 11 (1994) 659–664.

28. D. van Hal, A. van Rensen, T. de Vringer, H. Junginger, and J. Bouwstra, Diffusion of estradiol from non-ionic surfactant vesicles through human stratum corneum in vitro. STP Pharma 6 (1996) 72–78.
29. M. M. Meuwissen, L. Mougin, H. E. Junginger, and J. A. Bouwstra, Transport of model drugs through skin in vitro and in vivo by means of vesicles. Proc. Int. Symp. Control. Rel. Bioact. Mater. 23 (1996) 303–304.
30. J. Du Plessis, N. Weiner, and D. G. Muller. The influence of in vivo treatment of skin with liposomes on the topical absorption of a hydrophilic and a lipophilic drug in vitro. Int. J. Pharm. 103 (1994) R1–R5.
31. G. Cevc and G. Blume, Lipid vesicles penetrate into intact skin owing to the transdermal osmotic gradients and hydration force. Biochim. Biophys. Acta 1104 (1992) 226–232.
32. M. E. M. J. van Kuyk-Meuwissen, H. E. Junginger, and J. A. Bouwstra, Interactions between liposomes and human skin in vitro, a confocal laser scanning microscopy study. Biochim. Biophys. Acta, 1371 (1998) 31–39.
33. S. M. Niemic, Z. Hu. C. Ramachandran, D. F. H. Wallach, and N. Weiner, The effect of dosing volume on the disposition of cyclosporin A in hairless mouse skin after topiucal application of a non-ionic liposomal formulation. An in vitro study. STP Pharma 4 (1994) 145–149.
34. J. Du Plessis, C. Ramanchandran, N. Weiner, and D. G. Muller, The influence of particle size of liposomes on the disposition of drugs into the skin. Int. J. Pharm. 103 (1994) 277–282.
35. J. du Plessis, K. Egbaria, and N. Weiner, Influence of formulation factors on the disposition of liposomal components into the different strata of the skin. J. Soc. Cosmet. Chem. 43 (1992) 93–100.
36. J. Lasch and W. Wohlrab, Liposome-bound cortisol: a new approach to cutaneous therapy. Biomed. Biochim. Acta 45 (1986) 1295–1299.
37. W. Wohlrab and J. Lasch, Penetration kinetics of liposomal hydrocortisone in human skin. Dermatologica 174 (1987) 18–22.
38. C. Michel, T. Purmann, E. Mentrup, E. Seiller, and J. Kreuter, Effect of liposomes on percutaneous penetration of lipophilic materials. Int. J. Pharm. 84 (1992), 93–105.
39. E. Touitou, F. Levi-Schaffer, N. Dayan, F. Alhaique, and F. Riccieri, Modulation of caffeine skin delivery by carrier design: liposomes versus permeation enhancers. Int. J. Pharm. (1994) 131–136.
40. H. Kumatsu, K. Higazi, H. Okamoto, K. Miyakwa, M. Hashida, and H. Sezaki, Preservative activity and in vivo percutaneous penetration of butylparaben entrapped in liposomes. Chem. Pharm. Bull. 34 (1986) 3415–3422.
41. H. Komatsu, H. Okamoto, K. Miyagawa, M. Hashida, and H. Sezaki, Percutaneous absorption of butylparaben from liposomes in vitro. Chem. Pharm. Bull. 34 (1986) 3423–3430.
42. V. Masini, F. Bonte, A. Meybeck, and J. Wepierre, Cutaneous bioavailability in hairless rats of tretinoin in liposomes or gel. J. Pharm. Sci. 82 (1993) 17–21.
43. A. Gesztes and M. Mezei, Topical anestesthesia of the skin by liposome encapsulated tetracaine. Anesth. Analg. 67 (1988) 1079–1081.

44. C. Artmann, J. Roding, M. Ghyczy, and H. G. Pratzel, Liposomes from soya phospholipids as percutaneous drug carriers. Artzneim.-Forsch/Drug Res. 40(II), 12 (1990), 1363–1365.
45. C. Artmann, J. Roding, M. Ghyczy, and H. G. Pratzel, Liposomes from soya phospholipids as percutaneous drug carriers. Artzneim.-Forsch/Drug Res. 40(II), 12 (1990), 1365–1368.
46. W. Gehring, M. Ghyzhy, M. Gloor, Ch. Heitzler, and J. Roding, Significance of empty liposomes alone and as drug carriers in dermatology, Artzneim.-Forsch/Drug Res. 40(II). 12 (1990) 1368–1371.
47. N. Weiner, N. Williams, G. Birch, C. Ramachandran, C. Shipman, Jr., and G. Flynn, Topical delivery of liposomally encapsulated interferon evaluated in a cutaneous herpes guinea pig model, Antimicrob Agents Chemother 33 (1989) 1217–1221.
48. N. Skalko, M. CajikovAc, and I. Jalsenjak, Liposomes with clindamycin hydrochloride in the therapy of acne vulgaris. Int. J. Pharm. 85 (1992) 97–101.
49. M. E. Plana, P. Gonzalez, L. Rodriquez, S. Sanchez, and G. Cevc, Noninvasive percutaneous Induction of Topical Analgesia by a New type of drug carrier, and prolongation of local pain insensitivity by anesthetic liposomes. Anesth. Analg. 75 (1992) 615–621.
50. G. Cevc, Dermal insulin. In: Frontiers in insulin pharmacology. M. Berger, A. Gries, eds. Georg Thieme, Stuttgart (1993) pp 61–74.
51. G. Cevc, Transfersomes, Liposomes and other lipid suspensions on the skin: permeation enhancement, vesicle penetration, and transdermal drug delivery. Crit Rev Ther Drug Carrier Syst. 13 (1996) 257–388.
52. A. Paul and G. Cevc, Noninvasive administration of protein antigens: transdermal immunization with the bovine serum albumin in transfersomes. Vaccine Res. 4 (1995) 145.
53. L. Coderch, M. Oliva, M. Pons, A. de la Maza, A. M. Manich, and J. L. Parra, Percutaneous penetration of liposomes using the tape stripping technique. Int. J. Pharm. 139 (1996) 197–203.
54. T. Ogiso, N. Niinaka, and M. Iwaki, Mechanism for enhancement effect of lipid disperse system on percutaneous absorption. J. Pharm. Sci., 85 (1996) 57–64.
55. L. Lieb, C. Ramanchandran, K. Egberia, and N. Weiner, Topical delivery enhancement with multilamellar liposomes into pilosebaceous units: I. In vitro evaluation using fluorescent techniques with the hamster ear model. J. Invest. Dermatol 99 (1992) 108–113.
56. L. Lieb, C. Ramanchandran, K. Egberia, and N. Weiner, Liposomes encapsulated active ingredients penetrate through the follicle. In: Liposome Dermatics, O. Braun Falco, H. C. Korting, eds. Springer-Verlag, Wlen, (1992) pp 200–205.
57. S. M. Miemic, C. Ramanchandran, and N. Weiner, Influence of non-ionic liposomal composition on topical delivery of peptide drugs into pilocebaceous units: an in vivo study using the hanster ear model. Pharmaceut. Res. 12 (1995) 1184–1188.
58. L. M. Lieb, G. Flynn, and N. Weiner, Follicular deposition and pharmacological behavior of cetmetidine as a function of formulation. Pharmaceut. Res. 11 (1994) 1419–1423.
59. D. B. Yarosh, J. Tsimis, and V. Yee, Enhancement of DNA repair of UV damage

in mouse and human skin by liposomes containing DNA repair enzyme. J. Soc. Cosmet. Chem. 41 (1990) 85–92.

60. D. Yarosh, C. Bucana, P. Cox, L. Alas, J. Kibittel, and M. Kripke, Localisation of liposomes containing a DNA repair enzyme in murine skin. J. Invest. Dermatol. 103 (1994) 461–468.
61. E. Bernard, J-L. Dubois, and J. Wepierre, Importance of sebacious glands in cutaneous penetration of an antiandrogen: target effect of liposomes. J. Pharm. Sci., 86 (1997) 573–578.
62. L. Li, V. K. Lishko, and R. M. Hoffman, Liposomes can specifically target entrapped melanin to the hair follicles in histocultured skin. In Vitro Cell. Dev. Biol., 29A (1993) 192–194.
63. L. Li, V. K. Lishko, and R. M. Hoffman. Product delivering liposomes target hair follicles in in histocultured intact skin. In Vitro Cell. Dev. Biol., 28A (1993) 679–681.
64. L. Li, V. K. Lishko, and R. M. Hoffman, Liposomes can specifically targeting of high molecular weight DNA entrapped to the hair follicles in histocultured skin: a model for gene therapy of the hair growth process. In Vitro Cell. Dev. Biol., 29A (1993) 258–260.
65. V. Gabrijelcic, M. Sentjurc, and J. Kristl., Evaluation of liposomes as drug carriers into the skin by one-dimensional EPR imaging. Int. J. Pharmaceut. 62 (1990) 75–79.
66. J. Lasch, R. Laub, and W. Wohlrab, How deep do intact liposomes penetrate into human skin? J. Controlled Rel. 18 (1991) 55–58.
67. M. Kirjaivainen, A. Urtti, I. Jääskeläinen, T. M. Suhonen, P. Paronen, R. Valjakka-Koskela, J. Kiesvaara, and J. Mönkkönen, Interactions of liposomes with human skin in vitro- the influence of lipid composition and structure. Biochim. Biophys. Acta, 1304 (1996) 179–189.
68. H. E. J. Hofland, J. A. Bouwstra, H. E. Bodde, F. Spies, and H. E. Junginger, Interactions between liposomes and human stratum corneum in vitro: freeze fracture electron microscopical visualization and small-angle x-ray scattering studies. Br. J. Dermatol. 132 (1995) 853–856.
69. H. E. Bodde, L. A. R. M. Pechtold, M. T. A. Subnel, and F. H. N. de Haan, Monitoring in vivo skin hydration by liposomes using infrared spectroscopy in conjunction with tape stripping. In: Liposome Dermatics, O. Braun Falco, H. C. Korting, eds. Springer-Verlag, Wien (1992) 137–149.
70. H. C. Korting, W. Stolz, M. H. Schmid, and G. Maierhofen, Interactions of liposomes with human epidermis reconstructed in vitro. Br. J. Dermatol. 132 (1995) 571–579.
71. M. Foldvari, A. Gesztes, and M. Mezei, Dermal drug delivery by liposome encapsulation: clinical and electron microscopic studies. J. Microscopic Encaps. 7 (1990) 479–489.
72. H. E. J. Hofland, J. A. Bouwstra, F. Spies, H. E. Bodde, J. Nagelkerke, C. Cullander, and H. E. Junginger, Interactions between non-ionic surfactant vesicles and human stratum corneum in vitro, J. Liposome Res. 5 (1995) 241–263.
73. W. Abraham and D. T. Downing, Interaction between corneotyes and stratum corneum lipid liposomes in vitro, Biochim. Biophys. Acta, 1021 (1990) 119–125.

74. D. A. van Hal, Nonionic surfactant vesicles for dermal and transdermal drug delivery, thesis, Leiden University (1994) pp 149–176.
75. A. Schätzlein and G. Cevc, Skin penetration by phospholipid vesicles, transfersomes, as visualized by means of confocal laser scanning microscopy. In: Phospholipids: Characterization, Metabolism, and Novel Biological Applications, G. Cevc, F. Paltauf, eds. AOCS Press, Champaign, IL (1993) pp 189–207.
76. G. Cevc, Transfersomes, liposomes and other lipid suspensions on the skin: permeation enhancement, vesicle penetration, and transdermal drug delivery, Crit. Res. Ther. Drug. Carrier Syst. 13 (1996) 257–388.
77. H. M. Reinl, A. Hartingen, P. Dettman, and T. M. Bayerl, Time-resolved infrared ATR Measurements of liposme transport kinetics in human keratinocyte cultures and skin reveals a dependence on liposome size and phase state, J. Invest. Dermatol. 105 (1995) 291–295.

9

Ophthalmic Emulsions and Suspensions

M. Fabrizio Saettone, Boris Giannaccini and Daniela Monti
University of Pisa, Pisa, Italy

I. INTRODUCTION

The most common ophthalmic dosage forms are *solutions, ointments,* and *suspensions*. According to a recent article (1), these account for nearly 90% of currently available ophthalmic formulations in the United States, and a similar percentage is presumably valid for the global market. The individual shares are 62.4% (solutions), 17.4% (ointments), and 8.7% (suspensions). Solutions, in spite of their drawbacks and limitations (first of all, a quick elimination from the precorneal area, resulting in poor bioavailability), are given top priority by formulators, since they are relatively simple to prepare, filter, and sterilize.

Substantial formulative difficulties arise when the drugs are water-insoluble or only lipid-soluble. Petrolatum-based ointments, in which such drugs can be dissolved or suspended, represent one traditional solution to this problem. However, these vehicles are also not devoid of drawbacks on account of their poor tolerance, greasiness, vision blurring effect, etc. Another possibility consists of preparing an appropriate aqueous suspension of the drug. Ophthalmic suspensions are nowadays mainly used for formulations involving steroids and came into general use with the post–World War II availability of these drugs for treatment of ocular inflammatory diseases. The early resistance of ophthalmologists and patients to these forms, based on the emotional fear that ''particles in the eye are bad,'' have largely been overcome as particle sizes have been reduced and as the benefits of steroidal antiinflammatory therapy has been demonstrated. Emulsions (particularly, microemulsions) might constitute a valid alternative to lipophilic ointments as vehicles for lipid-soluble drugs.

This chapter will deal essentially, as specified in the title, with emulsions and suspensions for ophthalmic use. Some of its limitations and features will be anticipated for the reader's information.

Traditional emulsions or macroemulsions, defined as systems of two immiscible liquid phases one of which is dispersed as fine globules (0.5–100 μm) throughout the other, have only rarely been used as ophthalmic drug vehicles. This is clearly understandable when the sterility requirements of ophthalmic preparations and the irritancy and overall toxicity for the corneal and conjunctival tissues of most emulsifying agents are considered. In recent years, however, there has been a surge of interest in the literature for *microemulsions* and *multiple emulsions* as ocular vehicles. Therefore, particular attention will be given to these particular systems, even if none of them has so far found commercial applications.

Dispersions of *solid* particulates, depending on the range of particle sizes, can be classified as molecular, colloidal, and coarse dispersions (2). Only ''traditional'' ophthalmic suspensions, belonging to the latter type (indicative size range, 1 to 20–30 μm), will be dealt with here. These can be defined as preparations containing finely divided *drugs* in aqueous or nonaqueous media, plus suitable adjuvants. This stipulation will therefore exclude polymeric, drug-loaded nanoparticles (which are in the colloidal size range, 1.0 nm to 0.5 μm) and the so-called ''microparticulates'' (microcapsules and microspheres) whose size is two to three orders of magnitudes larger. The interested reader will find in the recent literature many excellent reviews on nano- and microparticulates for ophthalmic drug delivery; samples are given in Refs. 3–5.

Finally, it will be assumed throughout the text that the reader has a sufficient knowledge of the problems and intricacies of ocular drug delivery, including ocular anatomy and physiology, pharmacokinetics, and bioavailability. Thus, this background information will be dispensed with; the reader willing to expand his or her knowledge will find adequate information in the reviews mentioned in the references (6–10).

II. EMULSION SYSTEMS FOR OPHTHALMIC USE

A synopsis of emulsion systems (standard or macroemulsions, microemulsions, and multiple emulsions) investigated as ophthalmic vehicles is presented in Table 1. The well known sensitivity of corneal/conjunctival tissues to surfactants has severely limited the use of oil-in-water (O/W) or water-in-oil (W/O) macroemulsions as vehicles for ophthalmic drugs. These, due to a relatively high volume fraction of dispersed phase, must contain correspondingly high amounts of surfactants, so that their ocular application may be irritating (everybody is familiar with the stinging sensation produced by accidental introduction into the eye of even minute amounts of alkaline soap). When not immediately irritant, emulsions containing

Table 1 Main Studies on Macro-/Microemulsion Systems Investigated as Prospective Vehicles for Ophthalmic Drugs

Type of vehicle	Drug	Type of study	Ref.
W/O emulsion	Pilocarpine	In vivo (rabbits)	11
		In vitro/in vivo	12
O/W emulsion W/O emulsion	4-Biphenylacetic acid	In vitro/in vivo (rabbits)	13
Multiple emulsions	Prednisolone	In vivo (rabbits)	16, 17
Multiple emulsions	Hydrocortisone	In vivo (rabbits)	18
Microemulsion	Timolol	In vitro	26
		In vivo (rabbits)	27
Microemulsion	Levobunolol	In vitro	29
Microemulsion	Δ-8-THC,	In vivo (rabbits)	30
	HU-211	Stability	35
Microemulsion	Pilocarpine	In vivo (rabbits)	34
	Pilocarpine		35
Microemulsion	—	Ocular toxicity (rabbits)	36
Microemulsion	Indomethacin	In vivo (rabbits)	32
		Ex vivo (rabbits)	25
Microemulsion	Piroxicam	Stability	37
Microemulsion	Miconazole	Optimization	38
Microemulsion	Amphotericin B	In vivo (rabbits)	39
Microemulsion	Atropine, chloramphenicol, indomethacin, diclofenac	Characterization, release in vitro, mucosal tolerance	40–42

standard surfactants may produce corneal lesions on long-term ocular application. Furthermore, emulsions are difficult to obtain in sterile form. However, studies on emulsions, particularly of the W/O type, have occasionally been published.

Investigations on pilocarpine release from W/O emulsions were reported some 20 years ago by Sieg and Robinson (11,12). These vehicles were prepared by incorporating pilocarpine aqueous solutions into commercially available, petrolatum-based ophthalmic ointments. The amount of incorporated water phase was 5% or 25%. Pilocarpine release was studied in vitro, as well as in vivo using albino rabbits. Administration of some of these vehicles resulted in aqueous humor pilocarpine levels increased above those provided by an equivalent dose of aqueous solution. The mechanism of this increase was attributed to a higher effective concentration of drug in the ointment vehicle, to an increased time of contact, and to prevention of uniform mixing of the dose with the lacrimal fluid. The authors also found that release of the drug to the ocular fluids was dependent on shear, i.e., blinking, and on the emulsifying efficiency of the dosing system. Since the system showing the best results in vivo (one ointment base containing 5% water) possessed the lowest emulsifying efficiency, shear-facilitated emulsion rupture was proposed as the most likely mechanism for drug release. In any case, these W/O "ointment" vehicles had no follow-up in the literature, possibly on account of the drawbacks indicated in the previous paragraph and of their greasy nature.

Two macroemulsions, one O/W and another W/O, were tested, together with other classical formulations, as vehicles for 4-biphenylacetic acid, a nonsteroidal anti inflammatory agent (13). According to the authors, the O/W emulsion, prepared with Polawax, was better tolerated than the W/O emulsion (containing Dehymuls E, a mixture of high molecular weight esters) and displayed significant antiinflammatory activity in a rabbit model of ocular inflammation.

Multiple emulsions have been occasionally investigated as vehicles for ophthalmic drugs. These can be of the W/O/W or O/W/O type, and can be defined as systems where the globules of the dispersed phase contain in turn smaller globules of the continuous phase. Multiple emulsions can be considered as liquid membrane systems: applications to drug localization in the body and to prolongation of drug action have been reported (14,15). Safwat et al. (16) and Kassem et al. (17) investigated W/O/W and O/W/O multiple emulsions containing prednisolone. The preparations were tested in normotensive rabbits for their effect on intraocular pressure (IOP): decreasing the hydrophile–lipophile balance (HLB) of the surfactant mixture in a W/O/W emulsion was found to favor the drug's bioavailability, along with its intensity and duration of action. Another report by the same group (18) dealt with the effect on IOP of rabbits of different emulsions containing hydrocortisone. The bioavailability of the drug was found to increase in the order solution, O/W, W/O/W, and O/W/O emulsion. According to the authors, O/W/O multiple emulsions may represent an optimal delivery system for hydrocortisone, maximizing drug bioavailability and minimizing individual variations.

Recently, the attention of some researchers was focused on the potentiality of *microemulsions* as carriers for ophthalmic drugs (19). Investigations on possible ocular applications of microemulsions were stimulated by the discovery of their potentiality as injectable vehicles for lipophilic drugs such as physostigmine or diazepam (20–23).

As indicated by Gasco (19), ''microemulsions are transparent, thermodynamically stable dispersions of water and oil, usually stabilized by a surfactant and a cosurfactant. They contain particles smaller than 0.1 μm. Microemulsions are often defined as thermodynamically stable liquid solutions: this includes normal micellar solutions, reverse micelles, cores or droplets of water or oil, and, for some systems, even bicontinuous structures, in which neither oil nor water surrounds the other. Analysis of the thermodynamic stability indicates that microemulsions, being liquid–liquid dispersed systems, consist of two bulk phases separated by an interface region. In this respect, microemulsions differ from micellar solutions; there are no direct means, however, of distinguishing between the two states. The transparency of these systems arises from their small droplets diameter, typically less than 140 nm; such small droplets produce only weak scattering of visible light when compared with that from the coarse droplets (1–100 μm) of macroemulsions (24). For an updated, exact definition of all emulsion systems, the reader is referred to Sec. I of Chapter 2 of this book by J.-L. Salager. As pointed out by Salager, ''packed swollen micelle structures that could solubilize large amounts of both oil and water have been called microemulsions, because they were first thought to be extremely small droplets emulsions.'' However, the term microemulsion is a misnomer because (a) contrary to emulsions, microemulsions are above all single-phase, thermodynamically stable systems, and (b) many microemulsions cannot be considered as dispersions of very small droplets, but rather as percolated, bicontinuous structures in which there is no internal nor external phase.

It should be pointed out that the term microemulsion, or *submicrometer emulsion*, has also been applied to thermodynamically unstable, two-phase systems in which the drop size was made extremely small by physical methods, such as high-pressure hydraulic shear (cf. Ref. 22). Such systems should be more appropriately called *mini-* or *nanoemulsions*.

Due to the rather confusing situation in the pharmaceutical literature, where the term ''microemulsion'' is indifferently used to indicate systems of presumably unlike structure (''true'' microemulsions and miniemulsions), in the present chapter it will be used without actual concern for the effective structure of the reported formulations.

Historically, the study of ocular applications of microemulsions seems to have been initiated by the Italian research group led by Maria Rosa Gasco at the University of Turin (25–29). These workers first investigated in 1988 the permeability of timolol, incorporated in O/W microemulsions containing 28.7% w/w egg lecithin as surfactant, through a lipoid artificial membrane (26). The permeability constant of timolol was found to depend on the concentration of octanoic acid (which increased the lipophilicity of the drug by ion pair formation) in the microemulsion. Ion pair formation transformed the dispersed phase into a reservoir of the drug, whose concentration in the reservoir could be modulated by varying the counterion concentration. In a subsequent study (27), the same microemulsions, containing 2.6% w/v timolol as ion pair with octanoic acid, were tested in vivo in rabbits in comparison with aqueous solutions containing timolol as ion pair (0.33% w/w) or timolol alone (1.4% w/v). The area under the curve (AUC) values for timolol in the aqueous humor after administration of the microemulsion and the ion pair solution were 3.5 and 4.2 times higher, respectively, than the AUC value observed for timolol alone. The microemulsion proved to be stable and was perfectly tolerated by the animals. These experiments proved the suitability of microemulsions as ophthalmic vehicles and the good corneal penetration properties of timolol as ion pair.

In a further investigation (29), microemulsions containing another β-blocking agent, levobunolol, as ion pair were tested for permeation through different membranes (one artificial lipid membrane and hairless mouse skin) and for eye irritation in rabbits. The emulsions contained soya phosphatidylcholine (SPC) and sodium octyl phosphate (or sodium cholate) as surfactants and sodium hexanoate (or monobutyrylglycerol) as cosurfactant. The study confirmed the validity of the ion pair approach to increase the permeability of drugs through membranes simulating the corneal epithelium. Incorporation of the ion pair into a microemulsion provided a sustained release (reservoir) effect, potentially useful in ocular delivery. One microemulsion, prepared with SPC and sodium cholate proved nonirritating in rabbit eyes.

Significant contributions in the field of microemulsions (for parenteral and ocular applications) have been provided by the research group led by Simon Benita at Hebrew University in Jerusalem. In a first ophthalmic investigation (30), Δ-8-tetrahydrocannabinol (Δ-8-THC), a lipophilic compound extracted from *Cannabis sativa* that is pharmacologically active against glaucoma, was incorporated (0.4% w/w) into a submicrometer emulsion. The microemulsion, described in a previous paper by the same group (22), contained a nonionic emulsifier (poloxamer 188, Pluronic F-68) as surfactant and egg yolk phospholipids as co-surfactant; its mean droplet size was 130 ± 41 nm, and no droplet was larger than 400 nm. It was not altered after steam autoclaving or long-term storage. An intense and long-lasting reduction of IOP was observed after instillation in ocular hypertensive rabbits (chymotrypsin model), while inferior effects were observed in normotensive rabbits. Both the Δ-8-THC emulsion and the drug-free emulsion vehicle were devoid of irritant effects on rabbit eyes.

A microemulsion of identical composition, containing 0.125% w/w indomethacin, was submitted to an ocular irritation study in rabbits and to permeation tests through freshly excised rabbit corneas (31). Again, no irritant, inflammatory, or toxic effects were detected after hourly administration of the blank vehicle during 5 days. The apparent corneal permeability coefficient of the drug incorporated in the emulsion was 3.8 times greater than that of indomethacin in a commercial aqueous solution. The authors advanced the hypothesis that a close adhesion of the microemulsion oil droplets to the lipophilic surface of the corneal epithelium could promote the permeation rate of the drug. A lower ionization of the drug in the emulsion (pH 3.8) than in the solution (pH 6.8) was also indicated as a possible reason for the increased permeability coefficient.

The topic of ocular bioavailability of indomethacin administered in a submicrometer emulsion was also examined by the research group led by M. J. Alonso at the University of Santiago de Compostela, Spain (32). These workers tested several carriers in vivo in rabbits: nanoparticles, nanocapsules, microparticles, and a submicrometer emulsion. The latter, prepared by the method of Yu et al. (33) with poloxamer 188 and soybean lecithin and containing 0.1% w/v indomethacin, had particles of diameter 0.21 ± 0.02 μm. The submicrometer systems (nanoparticles, nanocapsules, and emulsion) increased by more than three-fold the indomethacin concentration in the cornea, aqueous humor, and iris-ciliary body at 0.5 and 1.0 h post instillation. Furthermore, an increased indomethacin bioavailability of 300% was observed in comparison with the value obtained for an aqueous commercial solution. Confocal laser scanning microscopy studies indicated that the submicron particles penetrated into the corneal epithelium cells by endocytosis; the authors further suggested that the vehicle components (lecithin in the case of emulsions) may act as penetration enhancers or as endocytotic stimulators.

In another investigation by the Benita group (34), pilocarpine base was incorporated (1.7% w/w) in a specially developed microemulsion, stable to steam autoclaving and long-term storage, and complying with all requirements for ocular application. The submicrometer emulsion was prepared with an amphoteric surfactant (Miranol MHT) and phospholipids from egg yolk (Lipoid E-80). A single-dose application of the preparation in normotensive rabbits induced a prolonged and progressive decrease in IOP, starting at 11 h post instillation and reaching its maximal value at 29 h. The effect was more long-lasting and intense than that induced by a corresponding dose of standard pilocarpine-HCl eyedrops (Fig. 1).

Contrary to instillation of standard eyedrops, unilateral application of emulsified pilocarpine induced a biphasic effect in the contralateral eye, presumably due to systemic absorption. The authors concluded that, in spite of this drawback, the pilocarpine submicrometer emulsion might serve as a long acting antiglaucoma preparation, requiring a single daily instillation. In another

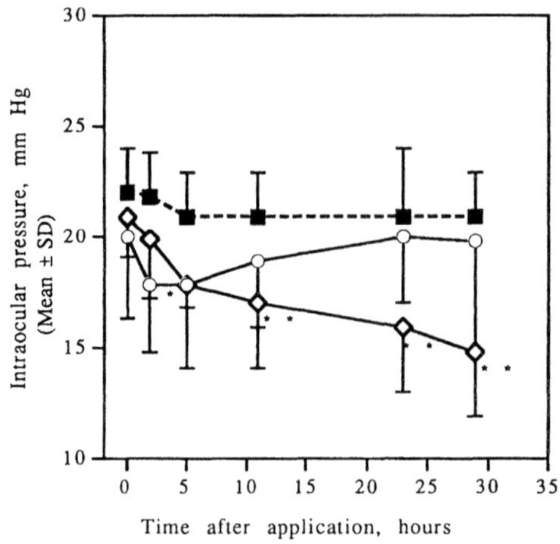

Figure 1 Effect of the intraocular pressure of rabbits of two preparations containing 1.7% pilocarpine base. ■, Baseline tonometry; ○, aqueous solution; ◇, microemulsion. (Redrawn from Ref. 34.)

investigation (35), the same microemulsion vehicle containing two drugs for the treatment of glaucoma, pilocarpine and HU-211 (an isomer of Δ-9-tetrahydrocannabinol), was characterized physicochemically and evaluated for long-term stability.

The issue of the toxicity of a novel, stable, positively charged microemulsion designed as a vehicle for intravenous or ocular drug administration was examined in another paper by the Benita group (36). This vehicle was formulated with a combination of emulsifiers comprising egg yolk phospholipids, poloxamer 188, and stearylamine. The results of an ocular tolerance study in rabbit eyes, as well as scanning electron microscopy (SEM) studies of the treated corneas, indicated that the vehicle was well tolerated and devoid of toxic corneal effects. After intravenous injections in mice and rats, no acute toxicity, hepatotoxic or nephrotoxic effects were observed. The overall results pointed, according to the authors, to the suitability of the positively charged submicrometer emulsion for parenteral and ocular applications. Physicochemical studies on the same emulsion containing piroxicam (37) and miconazole (38) were the subject of two papers. The investigations were carried out in an attempt to identify the formulation and process parameters providing optimal experimental conditions for fabrication.

The development of an amphotericin B microemulsion for the treatment of ocular mycosis was reported in 1996 by Cohen et al. (39). The vehicle was prepared with Intralipid 20%, an intravenous fat emulsion marketed by Pharmacia and Upjohn, containing soybean oil and egg yolk phospholipids. The emulsion, containing 0.5% w/v drug, was better tolerated in rabbits eyes in comparison with reference eyedrops prepared extemporaneously from Fungizone injectable. However, the ocular penetration of amphotericin B from the emulsion was not improved. The authors attributed this effect to the presence of sodium deoxycholate, acting as absorption promoter, in the reference eyedrops.

Microemulsions as prospective ophthalmic vehicles have also been investigated by Siebenbrodt and Keipert (40). These workers solubilized in microemulsions based on lecithin and Tween-80 three drugs often used in topical ocular therapy: atropine, chloramphenicol, and indomethacin. The emulsions were optimized and characterized physicochemically (40). Microemulsions based on lecithin and different poloxamers (41) or lecithin and poloxamer 184 (Synperonic L64) (42) were also evaluated as vehicles for indomethacin, diclofenac sodium, and chloramphenicol. These preparations, when tested for drug release through a hydrophilic membrane in comparison with aqueous or oily standard vehicles, and for physiological tolerance, showed prolonged release properties, produced low irritation, and, according to the authors, appeared suitable as potential carriers for the selected drugs.

III. OPHTHALMIC SUSPENSIONS

A discussion encompassing all complex aspects of ophthalmic suspensions would exceed the scope of this chapter. Basic, comprehensive chapters on suspension-type collyria and ointments, with literature references up to the early 1980s, have been presented by Skinner in the book *Ophthalmika* (43,44). In this section, some essential points concerning these medications will be briefly reviewed, and some recent reports on the subject will be presented. The main points to be considered when dealing with ophthalmic suspensions are listed in Table 2.

Table 2 Main Points to Ponder Concerning Ophthalmic Suspensions

Type of vehicle: eyedrops, petrolatum ointment, or hydrogel
Particle size and size distribution of dispersed phase
Particle dissolution rate
Presence and choice of adjuvants (wetting agents, viscosity-increasing polymers, electrolytes, preservatives, etc.)
Formation and easy redispersion of sediment; patient compliance
Sterilization and preservation
Crystal growth

A. Suspension Eyedrops vs. Semisolid Suspension Vehicles

Liquid and semisolid vehicles containing *suspended* drugs bear the same reciprocal relationship as standard eyedrops and ointments/hydrogels containing *dissolved* drugs, and have the same basic requirements and indications. Semisolid vehicles have the advantage of a longer time of residence in the conjunctival sac; therefore, they can be considered as sustained/prolonged release medications. This point is adequately demonstrated in Fig. 2 [from a paper by Sieg (45)], where the aqueous humor concentrations of fluorometholone in rabbits after application of a solution, an aqueous suspension, and a suspension ointment are compared.

The vast majority of ophthalmic ointments are essentially lipophilic and consist of mixtures of hydrocarbons. To improve spreading over the corneal surface and mixing with the precorneal film, small amounts of lanolin and/or lanolin alcohols, imparting to the ointment base W/O emulsifying properties, can be added. Suspension-type hydrocarbon ointments share with their ''solution-type'' homologs a poor patient acceptance, mostly due to their greasiness and vision-blurring effect; therefore, they are preferentially applied as nocturnal medications.

The disadvantages of petrolatum ointments have stimulated research on alternative, more acceptable semisolid vehicles. Polymeric hydrogels containing *dissolved* drugs have been developed and now account for 0.7% of the ocular

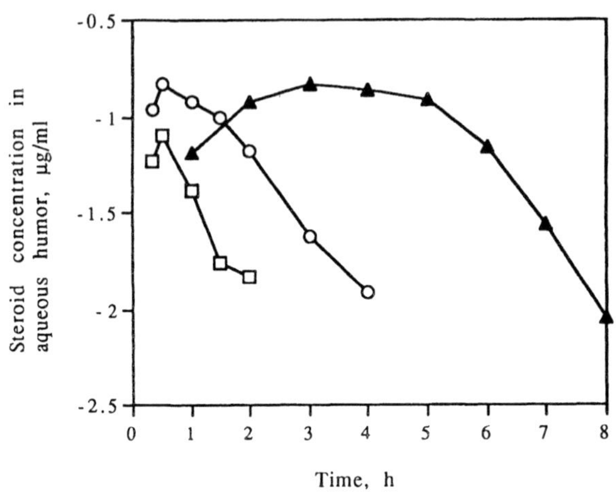

Figure 2 Aqueous humor concentration profiles of fluorometholone in rabbits after dosing with 50 µL of saturated solution (□); 50 µL of 0.1% aqueous suspension (○); 50 mg of suspension-ointment (▲). (Redrawn from Ref. 45.)

medications available on the U.S. market (1). Investigations on suspension-type *aqueous gels* have also been reported in the literature (46–49). The latter vehicles, when made up of carbomer (carboxypolymethylene) or of other mucoadhesive polymers (50), as in the mentioned references, have the added advantage of adhering to the external ocular tissues, thus providing longer times of residence in the precorneal area. For example, Burgalassi et al. (48) tested in rabbits some ophthalmic hydrogels based on different polymers and containing dissolved pilocarpine nitrate ($PiNO_3$) or suspended pilocarpine tannate (Pi-TA). As indicated in Table 3, administration of a xyloglucan-based mucoadhesive hydrogel containing suspended Pi-TA resulted in more favorable pharmacokinetic parameters (apparent rate of elimination and half-life of pilocarpine in tear fluid, AUC for miotic activity) with respect to administration of (a) the same hydrogel; (b) a less mucoadhesive polyvinyl alcohol hydrogel; and (c) an aqueous solution, all containing dissolved $PiNO_3$.

Drug release from suspension-type semisolid vehicles has been investigated by T. Higuchi (51) and is described by the now classical equation:

$$Q = \sqrt{2ADC_s t}$$

where Q is the amount of drug released per unit area at time t, A and C_s are the drug concentration and solubility in ointment, respectively, expressed in U/mL, and D is the diffusion coefficient of the drug in the vehicle. The basic assumptions for validity of the equation are that (a) the suspended drug is in a fine state such that the particles are much smaller than the thickness of the applied layer; (b) the amount of drug, A, present per unit volume of ointment, is substantially greater than C_s, the drug solubility per unit volume of vehicle; (c) the surface to which the ointment is applied is immiscible with respect to the ointment and constitutes a perfect sink for the drug. According to the equation, the size of

Table 3 Pharmacokinetic Parameters of Different Vehicles Containing Pilocarpine

Formulation	Ke_{tf}[a] ($min^{-1} \cdot 10^2$)	$t_{1/2tf}$[b] (min)	AUC[c] (min · mm)	Relative AUC
Reference solution	22.0	3.15	400.4	1
PVA gel $PiNO_3$	17.6	3.94	504.1	1.26
XYL gel $PiNO_3$	11.4	6.08	559.9	1.38
XYL gel Pi-TA	2.4	28.87	713.6	1.78

[a] Apparent rate constant for elimination of pilocarpine from tear fluid.
[b] Apparent half-time for elimination of pilocarpine from tear fluid.
[c] Area under the miotic activity vs. time curve.
Source: Data from Ref. 48.

particles should not influence release. However, this parameter has often been found to play an important role in drug bioavailability from suspension ointments and eyedrops, as will be discussed in the next section.

B. Particle Size and Dissolution of Dispersed Phase; Presence of Adjuvants

This point has been the subject of thorough discussion and much research during the last 30–40 years (44). Particles whose diameter exceeds 30–40 μm, while better retained, may cause corneal abrasions and can be irritating (52). Sieg and Robinson (45) suggested that the size should be less than 10 μm to minimize irritation. Smaller particles have the advantage of an increased surface area, hence of a faster dissolution rate, in agreement with the Noyes-Whitney law. Schoenwald and Stewart (53) administered to rabbits 0.1% dexamethasone suspensions with mean particle size of 5.75, 11.5, and 22.0 μm, and found significantly greater drug levels in the cornea and aqueous humor with decreasing particle size. They suggested that as the particles increase in size the dissolution rate in vivo decreases such that the particles are removed from the conjunctival sac before dissolution is completed. Therefore, both the rate and extent of penetration into the aqueous humor are decreased. On the other hand, reducing the diameter to a few micrometers may result in very quick elimination of the suspended particles from the conjunctival sac. This was nicely demonstrated by Sieg and Triplett (54), who tested the precorneal retention in rabbits of homogeneous, radiolabeled microspheres with 3 and 25 μm diameter. They found that the smaller spheres underwent a much faster elimination and that the elimination rate increased with increasing the dosing volume. They concluded that ''a lower limit probably exists for particle-size retention in the eye, in addition to the previously recognized upper limit for irritation.''

A kinetic model predicting ocular tissue drug levels for suspensions was recently developed by Hui and Robinson (55). According to the authors, the model, which was tested in rabbits with fluorometholone suspensions, was able to predict the effects of particle size, concentration, and changes in drainage rate. The results indicated that decreasing the particle size from 20 to 10 μm produced a 45% increase in AUC (area under the aqueous humor concentration vs. time curve). However, decreasing the particle size below 2 μm did not produce, on a relative basis, further significant AUC increases (e.g., the AUC increase was only 8% when the particle size was decreased from 2 μm to 1 μm). The effects of particle size and dissolution rate on permeability through excised pig corneas were also investigated by Bisrat et al. (56) using $\ll$5 and 10- to 15-μm sieve fractions of two sparingly soluble drugs: griseofulvin and prednisolone acetate (a prodrug). These authors found no significant differences in the permeability coefficients of the two fractions: in the tested perfusion model, the dissolution rates of both fractions were greater than the permeability rates and thus did not represent a rate-limiting step. The authors acknowledged, however, the limitations of their in vitro model. In their opinion, the differences observed in vivo with the different particle sizes of suspensions of poorly soluble drugs are obviously not linked to dissolution rate alone. Other factors, such as clearance rate and tear flow, might play an important role in governing bioavailability.

The effects of size distribution and of the presence of suspending agents on dissolution of different lots of prednisolone acetate (both commercial and formulated) were investigated by Howard et al. (57,58). The observed lot-to-lot variations in the dissolution rate were related to the percentage of fine particles in the distribution; furthermore, the presence of hydroxypropylmethylcellulose and sodium carboxymethylcellulose was found to reduce the dissolution rate in some lots.

These and other published data suggest that a proper particle size and a narrow size range, ensuring low irritation and adequate bioavailability, should be sought for every suspended drug. Most pharmacopoeias today require, both for suspension-type collyria and ointments, that "on 10 μg solid phase no more than 20 particles have a maximum diameter greater than 25 μm, not more than 2 particles have a maximum diameter greater than 50 μm, and no particle has a maximum diameter greater than 90 μm." Other formulation factors, i.e., the use of correct wetting, suspending, and buffering agents, protective colloids, preservatives, and so forth, should also be attentively considered (59).

C. Formation and Easy Redispersion of Sediment; Patient Compliance

As properly stated by Patel et al. (59):

> The basic concern involves the fact that suspensions settle, and it is necessary to redistribute them before using or dispensing them as products. A desirable suspension should be easily redispersed by shaking, should remain suspended long enough to withdraw an accurate dose, and should have the desired flow properties. . . . The particles . . . have a tendency to form a sediment or cake that is difficult to redisperse.

In all types of pharmaceutical suspensions, if a full and durable homogeneity of dispersion cannot be obtained, formation of a visible, evident sediment is to be preferred to warn the patient that shaking is required. However, the poor compliance of ophthalmic patients is well known (60–62). The issue of patient proper use of topical corticosteroid suspensions was examined by Apt et al. (63). These investigators, wondering if the irregular response of ocular inflammation in some of their patients might be related to inadequate dosage resulting from insufficient shaking of the formulation, studied the compliance of 100 adult patients who had been given appropriate instructions for use of a 1.0% prednisolone acetate test suspension. Of the patients 63% did not shake the bottle at all even after reading the instructions, which were printed clearly on the container. A comparison with four commercially available corticosteroid suspensions revealed that these patients would have delivered to the eye less than one third of the maximum concentrations of the four preparations. Almost half of the patients who did shake the bottle (1–10 times) would have delivered half or less of the required dose. The fact that even the most compliant and carefully instructed patients are unable to resuspend the best corticosteroid formulations was also reported by MacKeen (64). A reproducible test by which the dose uniformity of suspension eyedrops can be assessed under therapeutically relevant conditions was recently reported (65).

Advanced formulations, aimed at obviating totally or in part the aforementioned problems, have been developed. A novel suspension (Betoptic S, Alcon Laboratories, Inc.) containing 0.25% betaxolol bound to a cationic ion-exchange resin (5 μm in diameter), incorporated in a structured polymeric vehicle, was claimed, for example, to provide uniform dosing of betaxolol for up to 4 weeks without resuspending the product (66).

D. Sterilization and Preservation

The preparation of sterile ophthalmic suspensions, as required by all official compendia, presents some difficulties since filtration as a sterilization method cannot be used. Even if heat sterilization is possible in some cases, it is to be avoided since it might cause partial dissolution of the drug at high temperature and separation of larger crystals on cooling. Therefore, aseptic methods are preferable. In some cases, final sterilization by γ irradiation is possible. A detailed picture of different techniques for aseptic preparation of suspensions is given in Refs. 43 and 44.

In the case of multidose suspensions, preservation is also required. Compounds suitable for preservation include, in analogy with multidose ophthalmic solutions, quaternary ammonium compounds, chlorobutanol, chlorhexidine, organic mercurials, *p*-hydroxybenzoates, etc. The possibility of partial inactivation of the preservative by adsorption onto the suspended solid has been anticipated, even if, depending on the bactericidal agent and on the type of adsorbing solid, adsorption does not necessarily result in inactivation. Bean and Dempsey (67) actually found that the adsorbed phase of benzalkonium chloride on the suspended solid acted as a reservoir for the preservative. Since adsorption was in part reversible, the concentration of the agent in the aqueous phase was partially restored from the solid following uptake by the bacteria. Vulovic et al. (68) also found that indomethacin suspensions could be satisfactorily preserved by 0.02% w/v phenylmercuric nitrate despite the fact that 90% of the added preservative was adsorbed to the drug particles. Some collateral effects of preservatives should also be remembered: Camber and Edman (69) found in experiments on isolated corneas in vitro that chlorobutanol and benzalkonium chloride significantly increased the corneal permeability of pilocarpine and dexamethasone. This might result in lack of bioequivalence, in the case of reformulation of a multidose preserved suspension to an unpreserved single-dose formulation.

E. Crystal Growth

The issue of crystal growth and polymorphic changes in suspensions has always been a source of worry to formulators. In the words of K. J. Frederick (70):

> A much more difficult area, where black magic so often appears to enter the scene, occurs when changes in crystal size and shape take place either soon after the suspension is put together or slowly and unsuspectingly. Obviously, it will do little good to tailor a dispersed phase to exact specifications, if elements of growth and transformation subsequently upset everything.

Indeed, changes in crystal structure may occur during storage, resulting in an increase (or decrease) of crystal size and alteration of the suspension characteristics: the resulting solubility changes may be reflected in increased or decreased bioavailability (71). The property of some drugs to form polymorphs or pseudopolymorphs, and to undergo, when suspended in aqueous media, a more or less rapid transformation with formation of new crystals is well known. Kuhnert-Brandstätter (72) reported that about 60% of about 100 reviewed steroids are polymorphic and form crystal solvates, in which the hydrate always represents the most stable form. If an unstable, micronized drug polymorph is used to prepare a suspension, conversion into the stabler form will invariably occur, and this will result in crystal growth and potential formation of cakes and aggregates. As also pointed out by Carless et al. (73) in experiments with cortisone acetate, crystal growth is mainly initiated by polymorphic transformation: lattice energies, heats of wetting, and solution of the different crystal forms were indicated as the rate-controlling factors. The factors controlling the sensitivity of crystalline suspensions to fluctuating temperatures were also investigated by Varney using a temperature-cycling apparatus (74). Some practical implications of his work are that (a) particle growth in suspension is influenced by the addition of excipients which increase the particle solubility (e.g., wetting agents); (b) the growth rate is reduced under quiescent conditions; and (c) where sedimentation occurs, the local increase in particle concentration will increase particle growth due to the shortened interparticle diffusion distances. For example, List and Grönig (75) in 1976 examined some 65 suspension-type ophthalmic ointments, and found that 6 of them contained particle agglomerates larger than 50 μm, while 8, 12, and 25 ointments contained particles larger than 200, 150, and 100 μm, respectively. Analogous findings were reported by Lippman in 1957 (52).

It is to be assumed that progress in the formulation technique and a careful choice of adjuvants has obviated, totally or in great part, the problems associated with crystal growth in pharmaceutical suspensions. Prolonged wet milling of steroids in a ball mill with part of the final dispersion medium may cause the transformation of the crystals to occur during the milling process: at the end the crystals are not only in a finely divided state but also in the stabler form (76).

IV. CONCLUSIONS

The following conclusions can be offered at the end of this brief review. Emulsion systems, particularly, microemulsions, seem to be developing rapidly toward commercial applications and possibly will become the vehicles of choice for lipid-soluble ophthalmic drugs. Traditional suspensions, used for ocular administration of drugs both water- and lipid-insoluble, after some initial uncertainties have apparently reached a state of technical perfection. However, both emulsions and suspensions are not exempt of drawbacks and problems such as stability, sterilization, bioavailability, etc., which may limit their practical usefulness.

Recent years have witnessed an unceasing interest and substantial progress in drug solubilization by different techniques, such as complexation, use of surfactants and cosolvents, etc. (77). Thus, novel and more efficient ocular delivery systems for scarcely soluble drugs might be in store for the future.

REFERENCES

1. JC Lang. Ocular drug delivery conventional ocular formulations. Adv Drug Deliv Rev 16:39–43, 1995.
2. A Martin, J Swarbrick, A Cammarata, AHC Chun. Physical Pharmacy, 3rd ed. Philadelphia: Lea & Febiger, 1983, p. 470.
3. J Kreuter. Particulates (nanoparticles and microparticles). In: AK Mitra, ed. Ophthalmic Drug Delivery Systems. New York: Marcel Dekker, 1993, pp. 275–287.
4. A Zimmer, J Kreuter. Microspheres and nanoparticles used in ocular delivery systems. Adv Drug Deliv Rev 16:61–73, 1995.
5. A Joshi. Microparticulates for ophthalmic drug delivery. J Ocul Pharmacol 10:29–45, 1994.
6. J Stjernschantz, M Astin. Anatomy and physiology of the eye, Physiological aspects of ocular drug therapy. In: P Edman, ed. Biopharmaceutics of Ocular Drug Delivery. Boca Raton: CRC Press, 1993, pp. 1–25.

7. VHL Lee. Precorneal, corneal and postcorneal factors. In: AK Mitra, ed. Ophthalmic Drug Delivery Systems. New York: Marcel Dekker, 1993, pp. 59–81.
8. RD Schoenwald. Ocular pharmacokinetics/pharmacodynamics. In: AK Mitra, ed. Ophthalmic Drug Delivery Systems. New York: Marcel Dekker, 1993, pp. 83–110.
9. K Järvinen, T Järvinen, A Urtti. Ocular absorption following topical delivery. Adv Drug Deliv Rev 16:3–19, 1995.
10. M Van Ooteghem. Formulation of ophthalmic solutions and suspensions: problems and advantages. In: P Edman, ed. Biopharmaceutics of Ocular Drug Delivery. Boca Raton: CRC Press, 1993, pp. 27–42.
11. JW Sieg, JR Robinson. Vehicle effects on ocular drug bioavailability II: evaluation of pilocarpine. J Pharm Sci 66:1222–1228, 1977
12. JW Sieg, JR Robinson. Vehicle effects on ocular drug bioavailability III: shear-facilitated pilocarpine release from ointments. J Pharm Sci 68:724–728, 1979.
13. G Puglisi, S Spampinato, G Giammona, R Pignatello, T Bachetti, S Mangiafico, A Marino, G Mazzone. Effect of different ointment bases on ocular anti-inflammatory activity of 4-biphenylacetic acid in the rabbit. Drug Design Delivery 5:341–352, 1989.
14. A Martin, J Swarbrick, A Cammarata AHC Chun. Physical Pharmacy, 3rd ed. Philadelphia: Lea & Febiger, 1983, p. 565.
15. M De Luca, C Vaution, A Rabarron, M Seiller. Classification et obtention des émulsions multiples. STP Pharma 4:679–687, 1988.
16. SM Safwat, MA Kassem, MA Attia, M El-Mahdy. The formulation-performance relationship of multiple emulsions and ocular activity. J Controlled Rel 32:259–268 (1994).
17. MA Kassem, SM Safwat, MA Attia, M El-Mahdy. Influence of the phase volume ratio of multiple emulsions on the ocular activity of prednisolone. STP Pharma 5: 309–315 (1995).
18. MA Kassem, MA Attia, SM Safwat, M El-Mahdy. Preparation and evaluation of hydrocortisone multiple emulsions in rabbit's eye. Pharm Ind 56:584–588 (1994).
19. MR Gasco. Microemulsions in the pharmaceutical field: perspectives and applications. In: C Solans, H Kunieda, eds. Industrial applications of Microemulsions. New York: Marcel Dekker, 1997, pp. 97–122.
20. O Von Dardel, C Mebius, T Mossberg, B Svensson. Fat emulsions as a vehicle for diazepam. A study of 9492 patients. Br J Anaest 55:41–47, 1983.
21. S Benita, D Friedman, M Weinstock. Physostigmine emulsion: a new injectable controlled release delivery system. Int J Pharm 30:47–55, 1986.
22. MY Levy, S Benita. Design and characterisation of a submicronized o/w emulsion of diazepam for parenteral use. Int J Pharm 54:103–112, 1989.
23. S Benita, MY Levy. Submicron emulsions as colloidal drug carriers for intravenous administration: comprehensive physicochemical characterization. J Pharm Sci 82: 1069–1079, 1993.
24. D Attwood. Microemulsions. In: J Kreuter, ed. Colloidal drug delivery systems. New York: Marcel Dekker, 1994, pp. 31–71.
25. MR Gasco, M Gallarate, F Pattarino. On the release of prednisone from oil in water microemulsions. Pr Ed (Il Farmaco Ed) 10:325–330, 1988.
26. M Gallarate, MR Gasco, M Trotta. Influence of octanoic acid on membrane perme-

ability of timolol from solutions and from microemulsions. Acta Pharm Technol 34: 102–105, 1988.
27. MR Gasco, M Gallarate, M Trotta, L Bauchiero, E Gremmo, O Chiappero. Microemulsions as topical delivery vehicles: ocular administration of timolol. J Pharm Biomed Anal 7:433–439, 1989.
28. M Trotta, MR Gasco, S Morel. Release of drugs from oil-water microemulsions. J Controlled Rel 10:237–243, 1989.
29. M Gallarate, MR Gasco, M Trotta, P Chetoni, MF Saettone. Preparation and evaluation in vitro of solutions and o/w microemulsions containing levobunolol as ion pair. Int J Pharm 100:219–225, 1993.
30. S Muchtar, S Almog, MT Torracca, MF Saettone, S Benita. A submicron emulsion as ocular vehicle for delta-8-tetrahydrocannabinol: effect on intraocular pressure in rabbits. Ophthalmic Res 24:142–149, 1992.
31. S Muchtar, M Abdulrazik, J Frucht-Pery, S Benita. Ex-vivo permeation study of indomethacin from a submicron emulsion through albino rabbit cornea. J Controlled Rel 44:55–64, 1997.
32. P Calvo, MJ Alonso, J Vila-Jato, JR Robinson. Improved ocular bioavailability of indomethacin by novel ocular drug carriers. J Pharm Pharmacol 48:1147–1152, 1996.
33. W Yu, ES Tabosa do Egito, G Barratt, H Fessi, JP Devissaguet, F Puisieux. A novel approach to the preparation of injectable emulsions by a spontaneous emulsification process. Int J Pharm 89:139–146, 1993.
34. N Naveh, S Muchtar, S Benita. Pilocarpine incorporated into a submicron emulsion vehicle causes an unexpectedly prolonged ocular hypotensive effect in rabbits. J Ocul Pharmacol 10:509–520, 1994.
35. S Muchtar, S Benita. Emulsions as drug carriers for ophthalmic use. Colloids Surf 91:181–90, 1994.
36. SH Klang, J Frucht-Pery, A Hoffman, S Benita. Physicochemical characterization and acute toxicity evaluation of a positively charged submicron emulsion vehicle. J Pharm Pharmacol 46:986–993, 1994.
37. SH Klang, A Baszkin, S Benita. The stability of piroxicam incorporated in a positively charged submicron emulsion for ocular administration. Int J Pharm 132:33–44, 1996.
38. P Wehrle, D Korner, S Benita. Sequential statistical optimisation of a positively-charged submicron emulsion of miconazole. Pharm Dev Technol 1:97–111, 1996.
39. T Cohen, H Sauvageon-Martre, D Brossard, F D'Hermies, C Bardin, F Chast, JC Chaumeil. Amphotericin B eye drops as a lipidic emulsion. Int J Pharm 127:249–254, 1996.
40. I Siebenbrodt, S Keipert. Versuche zur entwicklung und charakterisierung ophthalmologisch verwendbarer tensidhaltiger mehrkomponentensysteme. Pharmazie 46: 435–438, 1991.
41. I Siebenbrodt, S Keipert. Poloxamer-systeme als potentielle ophthalmika. Teil 1: Viskose polymertensidlosungen. PZ Wiss 137:135–141, 1992. Chem Abstr 117: 178192m, 1992.
42. I Siebenbrodt, S Keipert. Poloxamer-systems as potential ophthalmics II. Microemulsions. Eur J Pharm Biopharm 39:25–30, 1993.

43. FS Skinner. Suspensions-Augentropfen. In: R Dolder, FS Skinner, eds. Ophthalmika: Pharmakologie, Biopharmazie und Galenik der Augenarzneimittel, 4th ed., Stuttgart: WVG mbH, 1990, pp. 441–447.
44. FS Skinner. Die Teilchenfeinheit in augensalben. In: R Dolder, FS Skinner, eds. Ophthalmika: Pharmakologie, Biopharmazie und Galenik der Augenarzneimittel, 4th ed., Stuttgart: WVG mbH, 1990, pp. 469–483.
45. JW Sieg, JR Robinson. Vehicle effects on ocular drug bioavailability I: evaluation of fluorometholone. J Pharm Sci 64:931–936, 1975.
46. RD Schoenwald, JJ Boltralik. A bioavailability comparison in rabbits of two steroids formulated as high viscosity gels and reference aqueous preparations. Invest Ophthalmol Vis Sci 18:61–66, 1979
47. A Kupferman, WJ Ryan, HM Leibowitz. Prolongation of anti-inflammatory effect of prednisolone acetate. Arch Ophthalmol 99:2028–2029, 1981.
48. S Burgalassi, P Chetoni, MF Saettone. Hydrogels for ocular delivery of pilocarpine: preliminary evaluation in rabbits of the influence of viscosity and of drug solubility. Eur J Pharm Biopharm 42:385–392, 1996.
49. S Johansen, E Rask-Pedersen, JU Prause. A bioavailability comparison in rabbits after a single topical ocular application of prednisolone acetate formulated as high-viscosity gel and as an aqueous suspension. Acta Ophthalmol Scand 74:253–258, 1996.
50. EM Slovin, JR Robinson. Bioadhesives in ocular drug delivery. In: P Edman, ed. Biopharmaceutics of Ocular Drug Delivery. Boca Raton: CRC Press, 1993, pp. 145–157.
51. T Higuchi. Rate of release of medicaments form ointment bases containing drugs in suspension. J Pharm Sci 50:874–875, 1961.
52. O Lippmann. Local irritating effect caused by topical use of steroids in the eye. AMA Arch Ophthalmol 57:339–344, 1957.
53. RD Schoenwald, P Stewart. Effect of particle size on ophthalmic bioavailability of dexamethasone suspensions in rabbits. J Pharm Sci 69:391–394, 1980.
54. JW Sieg, JW Triplett. Precorneal retention of topically instilled micronized particles. J Pharm Sci 69:863–864, 1980.
55. H-W Hui, JR Robinson. Effect of particle dissolution rate on ocular drug bioavailability. J Pharm Sci 75:280–287, 1986.
56. M Bisrat, C Nyström, P Edman. Influence of dissolution rate of sparingly soluble drugs on corneal permeability in vitro. Int J Pharm 63:49–56, 1990.
57. SA Howard, JW Mauger, L Phusanti. Dissolution profiles for multisized prednisolone acetate suspensions. J Pharm Sci 66:557–559, 1977.
58. SA Howard, JW Mauger, JW Hsieh, K Amin. Suspending agent effects on steroid suspension dissolution profiles. J Pharm Sci 68:1475–1479, 1979.
59. NK Patel, L Kennon, RS Levinson. Pharmaceutical Suspensions. In: L Lachman, HA Lieberman, JL Kanig, eds. The Theory and Practice of Industrial Pharmacy. 3rd ed. Philadelphia: Lea & Febiger, 1986, pp. 479–501.
60. JD Bartlett. Ophthalmic drug delivery. In: JD Bartlett, SD Jaanus, eds. Clinical Ocular Pharmacology, 3rd ed. Boston: Butterworth-Heinemann, 1995, pp 47–74.
61. SE Norell, PA Granström. Self-medication with pilocarpine among outpatients in a glaucoma clinic. Br J Ophthalmol 64:137–141, 1980.

62. MA Kass, E Hodapp, M Gordon, AE Kolker, I Goldberg. Patient administration of eyedrops: observation, Part II. Ann Ophthalmol 14:889–893, 1982.
63. L Apt, A Henrick, LM Silverman. Patient compliance with use of topical ophthalmic corticosteroid suspensions. Am J Ophthalmol 87:210–214, 1979.
64. DL MacKeen. Aqueous formulations and ointments. Int Ophthalmol Clin 20:79–92, 1980.
65. KA Kwon, M Diestelhorst, R Säuverkräup. Dosage problems in suspension eyedrops. Klin Monatsbl Augenheilkd 209:144–149, 1996.
66. R Jani, O Gan, Y Ali, R Rodstrom, S Hancock. Ion exchange resins for ophthalmic delivery. J Ocul Pharmacol 10:57–67, 1994.
67. HS Bean, G Dempsey. The effect of suspensions on the bactericidal activity of m-cresol and benzalkonium chloride. J Pharm Pharmacol 23:609–704, 1971.
68. N Vulovic, M Primorac, M Stupar, MW Brown, JL Ford. Some studies on the preservation of indomethacin suspensions intended for ophthalmic use. Pharmazie 45:678–679, 1990.
69. O Camber, P Edman. Influence of some preservatives on the corneal permeability of pilocarpine and dexamethasone, in vitro. Int J Pharm 39:229–234, 1987.
70. KJ Frederick. Performance and problems of pharmaceutical suspensions. J Pharm Sci 50:531–535, 1961.
71. G Hecht. Ophthalmic preparations. In: AR Gennaro, ed. Remington: The Science and Practice of Pharmacy, 19th Ed. Easton, PA: Mack Publ. Co., 1995, pp. 1563–1568.
72. M Kuhnert-Brandstätter. Polymorphe and pseudopolymorphe kristallformen von steroidhormonen. Pharm Ind 39:377–383, 1977.
73. JE Carless, MA Moustafa, DC Rapson. Dissolution and crystal growth in aqueous suspensions of cortisone acetate. J Pharm Pharmacol 20:630–638, 1968.
74. G Varney. Crystal growth in aqueous suspensions. J Pharm Pharmacol 19:19S–23S, 1967
75. PH List and JM Grönig. Teilchengrössenbestimmung in augensalben. Pharm Ind 38: 1159–1171, 1976.
76. RE Collard. Formulation and manufacture of corticosteroid preparations. Pharm J 186:113–117, 1961.
77. SH Yalkowsky, ed. Techniques of Solubilization of Drugs. New York: Marcel Dekker, 1981.

10

Administration of Emulsions to the Gastrointestinal Tract

Duncan Q. M. Craig
Queen's University of Belfast, Northern Ireland, United Kingdom

Manisha J. Patel
University of London, London, United Kingdom

Marianne Ashford
AstraZeneca, Macclesfield, Cheshire, United Kingdom

I. INTRODUCTION

Oils and emulsions may be given via the oral route for two principle reasons. First, the oil may itself have an effect on the functioning of the gastrointestinal (GI) tract, a simple example being the treatment of constipation. Second, the oil or emulsion may be used as a vehicle for the administration of drugs. It is this consideration that has received particular attention, especially due to the possibility of enhancing the oral bioavailability of poorly absorbed drug substances. This second consideration will form the primary focus of this chapter. More specifically, drug absorption from oils, emulsions, and self-emulsifying drug delivery systems (SEDDS) will be considered. While these are by no means the only lipid-based oral dosage forms available [e.g., solid systems such as lecithin (1), glyceryl tristearate (2), and gelucire (3) solid dispersions have been widely studied], primary discussion of liquid systems will allow appreciation of most of the pertinent considerations relating to lipid delivery to the GI tract in general. Further details of the current knowledge base regarding lipid-based oral drug delivery are available in an excellent recent special issue of *Advances in Drug Delivery* (4) which provides a detailed and comprehensive discussion of most of the topics outlined in this chapter.

Before proceeding to discuss the systems themselves, it is helpful to emphasize the nature and scale of the problem of poor absorption from the GI tract. While some drugs are rapidly and completely absorbed across the gut, difficulties in oral bioavailability are commonplace. The essential problem lies in the necessity for a drug to dissolve in the aqueous fluid of the GI tract prior to absorption while also being sufficiently lipophilic to allow partitioning across the lipid membrane of the gut. The balance between considerations pertinent to dissolution and absorption has led to the development of the recent classification system suggested by Amidon et al. (5) whereby drugs are considered to belong to one of four categories depending on their dissolution and absorption characteristics. There is now virtually universal acceptance of the need to consider formulation strategies that will enhance the oral bioavailability of a range of drugs, particularly those with a solubility of less than 100 μg/mL. Drug delivery via lipid vehicles undoubtedly provides opportunities to enhance the absorption of many such drugs, although there are still numerous challenges associated with this delivery approach, many of which will be discussed in this chapter.

II. SELF-EMULSIFYING DRUG DELIVERY SYSTEMS

A. Definition and Development

Self-emulsifying drug delivery systems (SEDDS) are a class of emulsion that have received particular attention as a means of enhancing oral bioavailability of poorly absorbed drugs. These systems are essentially mixes of oil and surfactant (sometimes with added cosurfactant) that form emulsions on mixing with water with little or no energy input (6). The question as to whether these emulsions form truly spontaneously in the thermodynamic sense is a matter of some conjecture. Similarly, the issue as to whether self-emulsifying systems are identical to microemulsions or whether they should be considered as a separate category of emulsifying system is still under debate. In general, while microemulsions are by definition clear isotropic systems that are thermodynamically stable, the term SEDDS has been applied to a wide range of systems that may not necessarily be either clear or stable. Indeed, the term is used somewhat loosely to describe systems for oral delivery that mix readily with water and show reasonable short-term stability. It should also be stated that many solid or semisolid systems such as Gelucire 44/14 dispersions show behavior in water that bears many resemblances to SEDDS, although detailed consideration of these systems is beyond the scope of this chapter.

The development of SEDDS may be traced back to the pesticide industry whereby many herbicides are highly lipophilic and may not be prepared as concentrated aqueous solutions. As concentrates are necessary for ease of transportation, it was necessary to develop systems that could be reconstituted prior to spraying without the need for elaborate mixing equipment. The herbicides were therefore dissolved in organic solvents containing surfactants that could then be emulsified by simple mixing with water. The concept of using such systems for pharmaceutical purposes was initially developed by the group of Groves (7–9). Early studies (7) involved the measurement of the spontaneity of emulsion formation using turbidity measurements with a particular view to relating the emulsification properties of the oil–surfactant mixes to liquid crystal formation behavior in the presence of water. From this early work, further studies have addressed the key issues relating to this technology, which are arguably the mechanisms by which emulsification takes place, the formulation of self-emulsifying systems, and the mechanism by which bioavailability enhancement takes place. The mechanistic and formulation aspects of SEDDS usage will be discussed in this section, whereas the bioavailability issue will be discussed in the context of the absorption of drugs from liquid systems in general in a later section.

B. The Mechanism of Self-emulsification

The free energy associated with the emulsification process (ΔG) may be given by

$$\Delta G = \sum_i N_i \pi r_i^2 \sigma \qquad [1]$$

where N is the number of droplets with radius r and σ is the interfacial energy (10). It is apparent from Eq. [1] that the spontaneous formation of the interface between the oil and water phases is energetically unfavored; hence the concept of self-emulsification raises immediate conceptual difficulties. In addressing the question of the mechanism of self-emulsification, the definition of these systems again becomes relevant, particularly in terms of the differentiation between these systems and microemulsions. The systems commonly classified as SEDDS have not yet been shown to emulsify spontaneously in the thermodynamic sense. Therefore, for the purposes of the present discussion it is helpful to consider the process to occur with minimal agitation rather than to necessarily involve a negative free energy.

Early work considered the association between the spontaneity of emulsification and the tendency of the oil–surfactant systems to form liquid crystals. Groves and Mustafa (7) developed a method of quantitatively assessing the ease of emulsification by monitoring the turbidity of the oil–surfactant systems in a water stream, using phosphated nonylphenoloxylate (PNE) and phosphated fatty alcohol ethoxylate (PFE) in *n*-hexane. The authors then related the emulsification (expressed as the equilibration time) to the phase behavior, as shown in Fig. 1. Previous studies (8) had indicated that these mixes existed in liquid crystal and gel phases in water, depending on the composition. Consequently, the authors were able to relate the phase behavior to the spontaneity of emulsification, with liquid crystal–forming compositions tending to form emulsions more readily, as indicated by the lower equilibration times. The authors (7) suggested that the emulsification process may be associated with the ease with which water penetrates the oil–water interface, with the formation of liquid crystalline phases resulting in swelling at the interface, thereby resulting in greater ease of emulsification.

Further investigations (9) involved a more extensive study into the correlation between the ease of emulsification and the phase behavior of oil–surfactant–water systems, this time using a range of pharmaceutically relevant mixes. The correlation was found to hold for some but not all systems. For example, liquid

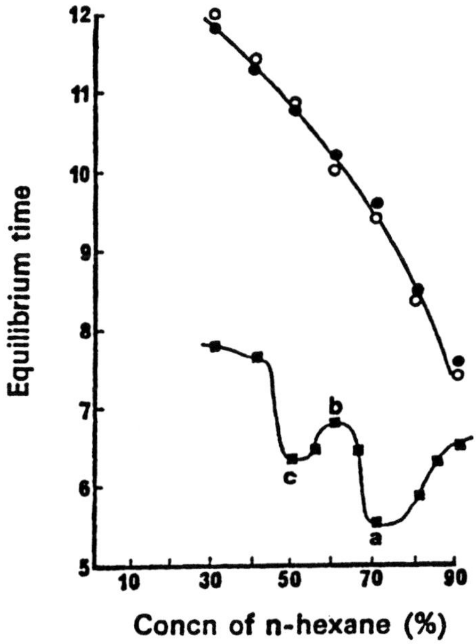

Figure 1 Effect of changing the *n*-hexane concentration of the emulsion equilibration time for systems containing different proportions of phosphated nonylphenol ethoxylate (PNE) and phosphated fatty alcohol ethoxylate (PFE). ●, System containing 9:1 PNE/PFE; ○, system containing 1:9 PNE/PFE; ■, system containing 1:1 PNE/PFE. (a) Mixture consists of middle phase liquid crystal. (b) Mixture consists of gel phase. (c) Mixture consists of neat phase liquid crystal. (From Ref. 7.)

paraffin, Tween-80, and Span-80 mixes showed reasonable correlations between liquid crystal formation and emulsification, as shown in Fig. 2, while benzyl alcohol mixes with the same surfactants showed little relationship between the two parameters. This may have been at least partially a reflection of the differences between phase equilibrium studies involving the addition of limited quantities of water compared to the kinetic studies involving emulsification in excess aqueous phase.

Later studies included those of Pouton (6) who examined the relationship between particle size and composition of Miglyol-812/Tween-85 mixes, demonstrating marked composition dependence of the emulsification process. The correlation between liquid crystal formation and self-emulsification has been further supported by Iranloye (11), who examined a range of pure hydrocarbon oils. The author reported a correlation between the extent of liquid crystal formation, the oil–water interfacial tension, and the ease of emulsification. Pouton (12) has also

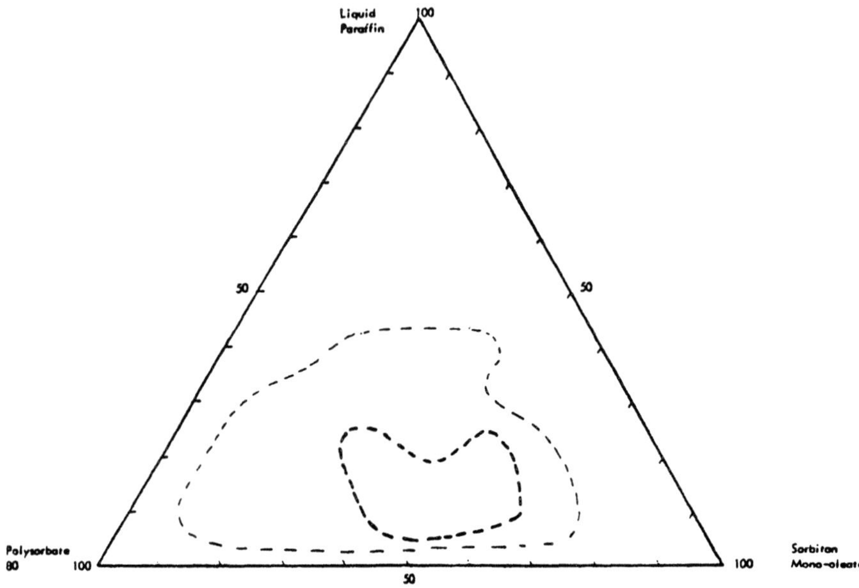
Liquid
Paraffin
100
50
50
Polysorbate
80
100
50
100
Sorbitan
Mono-oleate

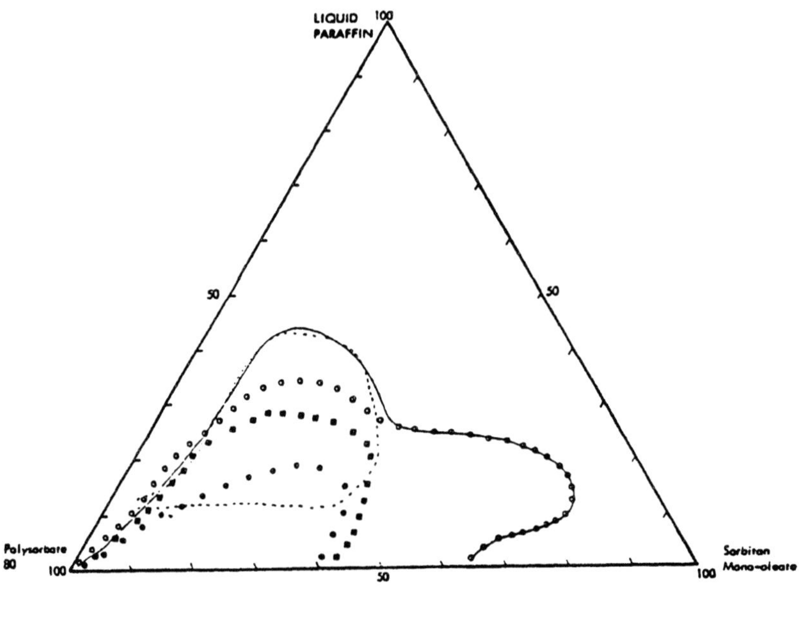
LIQUID
PARAFFIN
100
50
50
Polysorbate
80
100
50
100
Sorbitan
Mono-oleate

argued that the emulsification properties of the surfactant may be related to the phase inversion behavior of the system. If, for example, one increases the temperature of an oil-in-water (O/W) system stabilized using nonionic surfactants, the cloud point of the surfactant will be reached followed by phase inversion (13). The surfactant is highly mobile at the phase inversion temperature; hence the O/W interfacial energy is minimized, leading to a reduction in energy required to cause emulsification. Pouton (12) has suggested that the specificity of surfactant combinations required to allow spontaneous emulsification may be associated with a minimization of the phase inversion temperature, thereby increasing the ease of emulsification. This may also explain the empirical observation by Lin et al. (14) that there is an inverse relationship between the ability of Span-20/Tween-20 mixes of varying ratio to solubilize water and the particle size of emulsions formed containing mineral oil, i.e., those combinations of surfactants that showed the greatest water solubilization also resulted in emulsions with the lowest particle size.

Craig et al. (15) used dielectric analysis as a means of examining the relationship between the composition of Imwitor-742/Tween-80 mixes and the emulsification proeprties. This technique involves the application of an alternating electric field to a sample and the measurement of the response over a range of frequencies, thereby generating a spectrum from which information on the structure and behavior of the sample may be obtained (16). The spectra of oil–surfactant mixes containing equivalent quantities of water showed marked differences at two compositions (30:70 and 70:30 Imwitor 742/Tween 80). These compositions were found to show the greatest ease of emulsification. The authors suggested that the dielectric method was detecting structure formation in these particular systems, thereby supporting the hypothesis of there being a relationship between the tendency for self-assembly and the ease of emulsification. However, Gershanik et al. (17) recently questioned the necessity of oil–surfactant–water liquid crystal formation for satisfactory emulsification. These authors have suggested that the formation is associated first with the penetration of water into the oil–surfactant mix, with secondary water penetration taking place due to the

Figure 2 (top) Equilibrium phase diagram at 25°C for liquid paraffin, Tween-80, and Span-80. Liquid crystalline areas indicated for changing water concentrations: 15% w/w water (lamellar phase); 20% w/w (lamellar phase); ○ 25% w/w (lamellar phase); ■ 30% w/w (hexagonal phase); ● 35% w/w (hexagonal phase). (bottom) The effect of composition on the ease of emulsion formation for liquid paraffin, Tween-80, and Span-80 systems. The area enclosed between the heavy broken line has "good" spontaneity and the light broken line "moderate" spontaneity. The remainder is "poor". (From Ref. 9.)

establishment of concentrated surfactant structures. In terms of the systems under study in that particular investigation (Tween-80, benzyl alcohol, ethyl oleate and oleylamine mixes), the authors also suggested that the emulsification process may be dependent on the presence of the benzyl alcohol, with rapid diffusion of this component from the organic to the aqueous phase inducing convection currents and interfacial turbulence (the Marangoni effect). This in turn results in disruption of the oil–water interface and the spreading of the emulgent on the droplet surface.

As suggested, more global explanation of SEDDS formation has been put forward by Pouton (12). The author argues that for systems containing Tween-85 or Tagat TO, liquid crystal formation at the oil–water interface allows penetration of water into the oil–surfactant interface, thereby causing an increase in surface pressure and hence interfacial disruption. However, a second scenario is suggested for systems containing cosurfactants. In this case, significant partitioning of components between the oil and aqueous phases may take place, leading to a mechanism described as ''diffusion and stranding'' whereby the oil is solubilized, leading to migration into the aqueous phase. The concomitant dilution of the cosurfactant in the bulk water phase causes the oil to precipitate out as fine droplets. This mechanism is still speculative but provides a useful basis for further study and debate.

C. Formulation of SEDDS

The formulation of SEDDS is, in theory, comparatively simple, as all that is required is to incorporate the drug into a suitable oil–surfactant mix. The solution or suspension may then be presented as a dosage form such as a liquid-filled soft or hard gelatin capsule. In practice, there are a number of ''political'' and practical considerations. The pharmaceutical industry will, in general, only use nonconventional dosage forms when tableting or capsule filling processes are found to be unsatisfactory, largely due to the considerable prior financial commitment to conventional equipment. The SEDDS approach also suffers from the disadvantage shared by many nonconventional dosage forms, namely, the comfort factor problem whereby the current gap in the knowledge base regarding the formulation and bioavailability aspects of these systems leads to a poor predictability of which factors are important in designing such dosage forms. For example, questions regarding the importance of emulsion particle size to bioavailability and the necessity of presenting the drug as a solution rather than a fine suspension in the oil–surfactant mix still have not been fully answered. However, the possibility of being able to formulate and deliver a drug with favorable intrinsic biological activity but poor oral bioavailability will often supersede the above difficulties.

The use and availability of the oil is also important, with interbatch variation, stability, and the possibility of being forced to use a single supplier all being pertinent considerations. The toxicity of the surfactants should also be considered, as must the solubility of the drug in the oil–surfactant mixture. A wide range of oils has been studied either as model systems or as potential vehicles for dosage forms, and the reader is referred to a comprehensive survey by Constantinides (18), with many of the oils under study being medium chain triglycerides. A number of investigators have suggested an association between the polarity of the oil and the emulsification properties. Groves and De Galindez (9) suggested that the oil polarity determines the tendency of the oil to form liquid crystals, which would then influence the emulsification process, whereas Pouton (6) showed that very polar or nonpolar oils tend to form poor emulsions, with Miglyol-812 and 840, both of which may be considered to have intermediate polarity, showing favorable emulsification properties with Tween-85. Craig et al. (19) studied a range of Labrafils (poloxyethylated oleic glycerides) and found that, using Tweens as surfactants, the more hydrophilic oils tended to form smaller emulsion droplets, while hydrophobic oils tended to form coarser emulsions; in addition, these hydrophobic oils demonstrated a tendency to form multiple emulsions.

Primarily for toxicity reasons, work has focused on the use of nonionic surfactants, particularly Tweens and Spans. Pouton (20) and Wakerley et al. (21) have screened a range of surfactants, finding that in general molecules with unsaturated acyl chains were most efficient emulsifiers, particularly the oleates with an HLB value of approximately 11. The authors also reported that the sorbitan esters and ethoxylated triglycerides such as Tagat TO were more efficient than the fatty acid ethoxylates, possibly due to the polydispersity of the latter.

It is now well established that one of the most important considerations relating to the formulation is the oil/surfactant ratio. Indeed, numerous studies have demonstrated that the emulsification properties may be highly sensitive to relatively small changes in this ratio. For example, Pouton (6) showed that the time taken for emulsification of Tween-85/Miglyol-840 mixes was highly dependent on the oil/surfactant ratio, as shown in Fig. 3. Studies have also been conducted into the effects of the emulsification environment. In particular, Wakerley et al. (22) have studied the particle size of emulsions produced using Tagat TO and Miglyol-812, finding a minimum in size at approximately 50% w/w surfactant. The position of the minimum and the size of the particles were both found to be temperature-dependent, with a lower emulsification efficiency above 60°C. This effect was ascribed to dehydration of the oxyethylene units of the surfactant, resulting in an effective lowering of the HLB.

An additional, highly important consideration in the formulation of SEDDS is the effect of incorporating the drug, in terms of both the solubilization within the oil–surfactant mix in order to allow a suitable dose to be given and the effect that the drug may have on the emulsification properties. In terms of the solubility of the drug within the vehicle, these systems are generally used for drugs with

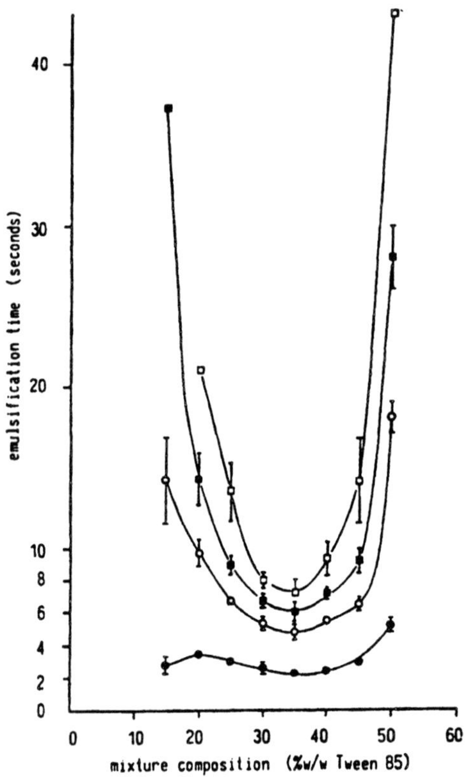

Figure 3 Self-emulsification times of Miglyol-840/Tween-85 mixtures in distilled water at 25°C (mean ± SD, $n = 5$). ● = t_0; ○ = t_{50}; ■ = t_{75}; □ = t_{90}. (From Ref. 6.)

poor aqueous solubility; hence theoretically their solubility in oil–surfactant mixes should be reasonably favorable. In addition, the use of SEDDS may reduce the required oral dose if an extensive improvement in oral bioavailability is established; hence, the quantity of drug that must be incorporated may be minimized. In practice, however, anecdotal evidence strongly suggests that many of the drugs investigated by the pharmaceutical industry show poor solubility profiles in most pharmaceutically acceptable solvents. This effectively means that, in practice, oil–surfactant mixes with suboptimal emulsification properties may have to be used in order to allow satisfactory dosing. Therefore it is perhaps surprising that few studies have been conducted into the solubility of drugs in these mixes. In a limited study, Barker et al. (23) examined the solubility of L701,324 and L365,260 in a range of oils and oil–surfactant mixes, finding a relationship be-

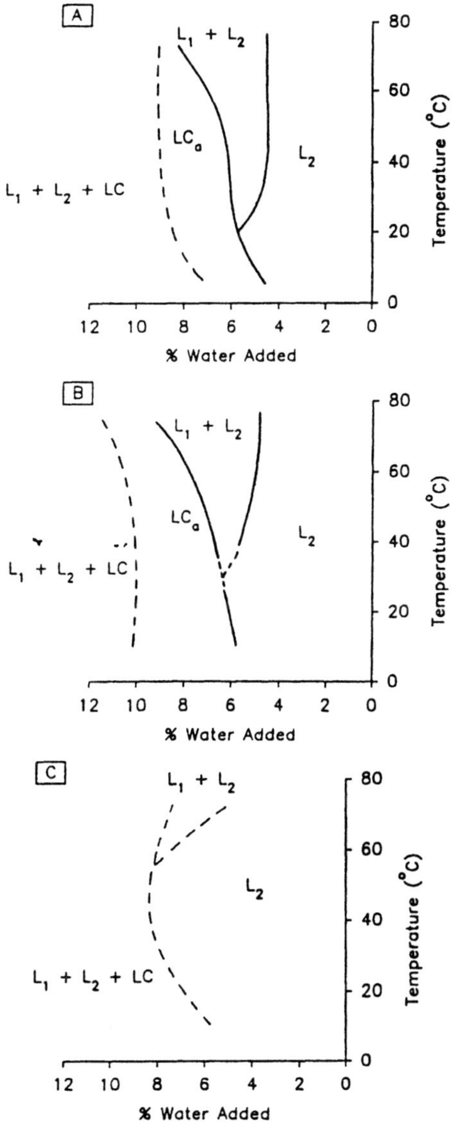

Figure 4 Partial equilibrium phase diagram for mixtures of Tagat TO, Neobee M5 and WIN-5495 diluted with distilled water. (A) 30% w/w Tagat TO, 70% w/w Neobee M5. (B) 5% w/w WIN-54954, 30% w/w tagat TO, and 65% w/w neobee M5. (C) 30% w/w WIN-54954, 30% w/w tagat TO, and 40% w/w Neobee M5. Aqueous-based liquids are designated L_1, oil-based liquids L_2, liquid crystal phases (''white'' birefringence) LC, liquid crystal phases (''multicolored'' birefringence), LC_a. (From Ref. 25.)

tween the dielectric response of the vehicle and the ability to incorporate the drugs. A further point that should be emphasized is that the measurement of the drug solubility in these vehicles may not in itself be a trivial task; the viscosity of the solvent renders equilibration times relatively long, while difficulties in establishing assay methods for nonaqueous solvents (particularly if one is trying to screen a range of solvents, which is often the case) has resulted in most workers initially using visual methods. In fact, our own experience suggests that several systems which appear to be clear when inspected visually may in fact be seen to be fine suspensions of the drug when viewed under a microscope.

More extensive studies have been conducted on the effects of the incorporated drug on the emulsification properties. Wakerley et al. (22) found an increase in particle size of Tagat TO/Miglyol-812 emulsions on addition of probucol, with a decrease in the proportion of surfactant required to produce a minimum in size. Pouton (24) examined the effects of a model drug, benzoic acid, on Tween-85/Miglyol-812 systems, noting an increased propensity for liquid crystal formation and a decrease in the critical solubilization temperature on addition of the drug, this being ascribed to the drug's competing with water for binding to the ethoxy chains of the surfactant. Charman et al. (25) studied the effects of the incorporation of a lipophilic drug (WIN-54954) on the phase behavior of Neobee M5 (a medium chain triglyceride) and Tagat TO (Fig. 4). The authors noted that at 30% drug, the LC_a phase was abolished. Craig et al. (26) studied the effects of adding the drug L365,260 to Labrafil M2125 CS using dielectric analysis, noting a decrease in conductivity on adding the drug. This is a surprising result, given that such additions may be expected either to have no effect or, if the drug is acting as a charge carrier, to increase the response. The authors suggested that the drug may be associating with low molecular weight molecules in the system, although more work is required to clarify the mechanisms involved.

III. THE FATE OF EMULSIONS IN THE GASTROINTESTINAL TRACT

The use of emulsions (either SEDDS or conventional emulsions) as vehicles for oral delivery of oil-soluble drugs will theoreticallly result in that drug being presented to the GI tract as a molecular dispersion in a system with a comparatively large surface area from which drug dissolution into the aqueous phase may take place. However, there is now overwhelming evidence that the metabolism and absorption of the vehicle itself may play a central role in the drug uptake process. In this section, the processes associated with fat digestion will be outlined in order to allow consideration of the relevance of these mechanisms to drug absorption from emulsion systems.

A. Absorption of Lipids from the Gastrointestinal Tract

The major component of dietary lipids are triglycerides, which are fatty acids esters of glycerol with the general structure:

$$
\begin{array}{l}
CH_2\text{—O—CO—}R^1 \\
| \\
CH\text{—O—CO—}R^2 \\
| \\
CH_2\text{—O—CO—}R^3
\end{array}
$$

where R^{1-3} represents saturated or unsaturated alkyl chains from the constituent fatty acids, which are, for natural products, usually long chain unsaturated (palmitic and stearic acids, C16, C18) and long chain saturated molecules (oleic and linoleic, C18, one and two double bonds respectively). Medium chain (C8–12) triglycerides may also be present in smaller quantities.

The processes involved in the digestion of fats, with particular emphasis on the role of lipases, has been discussed in detail by Embleton and Pouton (27); only a brief summary will be given here. On ingestion of the triglycerides, a coarse emulsion is believed to be formed in the stomach via mechanical agitation, with dietary phospholipids, proteins, and polysaccharides acting as emulsifying agents (28). In particular, dietary phospholipids are believed to be potent emulsifiers, forming a monolayer around the triglyceride droplets. Ten to forty percent of normal fat digestion takes place in the stomach, involving hydrolysis to diglycerides and fatty acids. This process takes place via human gastric lipase (HGL), a glycoprotein that is secreted from the chief cells in the fundic region of the stomach (29,30). This enzyme comprises a 379-amino-acid polypeptide (calculated molecular weight 43,162) and a carbohydrate moiety with a molecular weight of 8000–10,000 (31). Human gastric lipase acts primarily at the 1 and 3 position of the triglyceride, although the enzyme has been reported to have limited activity at the 2 position (32). In addition, the length of the fatty acid chain may influence the digestion process, although the relationship appears to be complex. In particular, studies using rat lingual lipase indicated that C14 and C16 chains were not cleaved from the 3 position of the glycerol backbone, whereas shorter (C10 and C12) and longer chains (C18) chains were cleaved irrespective of the degree of saturation (33).

Short chain fatty acids may dissolve into the aqueous phase, followed by absorption across the stomach mucosa or in conjugation with albumin (34,35), while longer chain fatty acids, which constitute approximately 95% of the ingested material, may remain incorporated in the emulsion droplet core, with a small percentage adsorbing on the surface. There is some evidence that this surface fraction may play a role in the inhibition of gastric lipase activity, particularly in the acidic environment of the stomach, although the evidence for this has arisen largely from in vitro studies (36) and has not yet been corroborated in vivo. Overall, the primary role of HGL is to facilitate efficient pancreatic lipolysis by generating long chain fatty acids that promote emulsification and activate pancreatic lipase (27).

The emulsion passes to the upper section of the large intestine, where particle size reduction of the droplets takes place due to the presence of a range of emulsifying agents including bile salts, monoglycerides, cholesterol, lecithin, and lysolecithin, yielding an approximate size range of 0.5–1 μm. Bile is released from the gall bladder via the influence of cholecystokinin (CCK) and contains bile salts, bile pigments, phospholipids, and cholesterol. The bile salts are classified as primary salts (sodium and potassium salts of cholic and chenodeoxycholic acid), which are synthesized in the liver, and secondary salts (deoxycholic acid and lithocholic acid), which are reabsorbed metabolites of the primary salts produced by cholonic bacteria.

The mechanisms responsible for such efficient emulsification of ingested triglycerides are not yet clear, although in vitro studies have shown that monoolein, oleic acid, and monomeric bile salts may, at intestinal pH, significantly lower the interfacial tension of triolein droplets (37), thereby allowing emulsification to take place under conditions of comparatively low shear. Indeed, Embleton and Pouton (27) have argued that while individual amphipathic compounds are relatively poor emulsifiers, the integrated action of these compounds allows highly efficient particle size reduction to take place.

Lipolysis takes place via triacylglycerol acyl hydrolase, typically referred to as pancreatic lipase. This enzyme acts specifically at the surface of the emulsion droplets and causes hydrolysis of the triglyceride at the 1 and 3 positions to produce fatty acids, diglycerides (which are further metabolized), and monoglycerides. An interesting aspect of the activity of this enzyme is that while it is capable of catalyzing the hydrolysis of triglycerides presented as molecular dispersions (38), the activity of the enzyme is greatly enhanced in the presence of O/W emulsions. This is due to pancreatic lipase being activated by interfacial binding (39), with the molecule possessing distinct sites for binding and hydrolysis.

In order to prevent desorption by bile salts, the enzyme complexes via a third site with human colipase, a cofactor with no intrinsic enzymatic activity that binds extensively to the oil–water interface. This molecule is a single nonglycosylated polypeptide containing 86 amino acids and is secreted from the pancreas in the form of a proenzyme, procolipase, which has a similar activity to colipase in terms of augmenting the hydrolysis of triglycerides at normal duodenal pH but is believed to be involved in regulating the secretion of enterostatin into the intestinal lumen. Colipase is believed to bind to pancreatic lipase via both hydrophobic and ionic interactions (40). Furthermore, on binding to an interface the cofactor undergoes a conformational change that further promotes the affinity for pancreatic lipase. This interfacial binding may take place prior to the interaction with the enzyme, with the unbound cofactor having greater lateral mobility along the interface, hence facilitating spreading across the surface of the oil droplet (27).

The oil–water interface will be a complex mix of amphipaths, one of the major constituents being phospholipids (especially lecithin); hence the influence of lecithin on the activity of the enzyme–cofactor complex is of some importance. It is now established that the presence of phospholipids inhibits enzyme activity in the presence of bile salts (27). The mechanisms associated with this inhibition are still not fully understood, with some workers suggesting that the phospholipid simply blocks access of the enzyme to the surface (28). Alternatively, Patton and Carey (44) have suggested that colipase binds to bile salt–phospholipid associations; hence in the presence of excess bile salts and phospholipids, whereby mixed micelles are formed, the colipase is effectively removed from the proximity of the surface of the oil droplet. The inhibition of pancreatic lipase activity by phospholipids may also shed light on the physiological significance of the necessity for both gastric and pancreatic lipase. Bernback et al. (36) demonstrated that human milk triglyceride was not appreciably hydrolyzed following incubation with human pancreatic juice due to the presence of a phospholipid layer on the surface of the emulsion particles. However, hydrolysis commenced immediately in the presence of bovine gastric lipase, as this enzyme does not require a cofactor and hence is not inhibited by phospholipids. The hydrolysis process ceased after 30 min due to enzymatic inhibition as a result of the liberation of free fatty acids. Exposure of the same triglyceride system to both enzymes showed a similar hydrolysis rate to that found for the gastric lipase alone. However, after 30 min a marked increase in rate was observed, which was ascribed to the fatty acid metabolites having an activation effect on the pancreatic lipase.

The physical form of the lipid material within the intestine has been the subject of some controversy (27). Early work (42,43) suggested that the lipids partition between oil-rich phases containing tri- and diglycerides and an aqueous phase containing mixed micelles of bile salts and monoglycerides and partially ionized fatty acids, although this model has now been effectively superseded. In particular, Patton and Carey (44) suggested the presence of liquid crystalline phases on the basis of microscopy studies, while Holt et al. (45) reported a viscous birefringent phase between oil and aqueous micellar phases on ultracentrifugation of human upper intestinal contents. Rigler et al. (46) later confirmed the presence of lamellar structure of this phase using electron microscopy. In addition, the authors reported vesicular structures that appeared to emanate from the liquid crystalline phase which, depending on the proportions of bile salt and lipolytic products, ranged from unilamellar vesicles to large multilamellar structures up to 3 μm in size.

The processes involved in the digestion process may be summarized as follows (27,46,47). The oil droplets, covered with monoglycerides, fatty acids, phospholipids, and bile salts, are hydrolyzed in the duodenum via the action of the pancreatic lipase-colipase. As digestion proceeds, the droplets become proportionately richer in surface amphiphiles (particularly monoglycerides and fatty acids), which form liquid crystalline phases at the oil–water interface. In the presence of bile salts, the lamellar phases form multilamellar vesicles that subsequently break down to unilamellar vesicles in the presence of the bile salts. These vesicles are then solubilized by the bile salt micelles, thus forming intestinal mixed micelles. It is generally believed that such micelles migrate to the intestinal mucosa, across which diffusion of lipophilic materials takes place. Once in the intestinal epithelial cells, monoglycerides and fatty acids may be resynthesized into triglycerides which, in combination with lipoprotein, cholesterol, and phospholipids, form chylomicrons which are subsequently taken up into the lymphatic system. In addition, a smaller quantity of fatty acids may also be incorporated into very low density lipoproteins (VLDLs), while lower chain length fatty acids may be absorbed directly into the portal system. The main features of this pathway are shown in Fig. 5 (47).

B. Effect of Oils on Gastric Function

In addition to considerations relating to the fate of oils in the GI tract, it is also worth considering the effects that the oils themselves may have on gastric function and hence drug absorption. Hunt and Knox (48) noted that lipid ingestion decreased gastric emptying time; this consideration has become an important feature in considering the effect of food on drug absorption. Indeed, Bates et al. (49) reported an enhancement in the absorption of aqueous formulations of griseofulvin following administration of the gastric motility inhibitor propantheline. This was ascribed to the increased time span available for drug dissolution and absorption. However, this effect is far from universally beneficial; Veda et al. (50) reported a decrease in cyclosporine A absorption when given in an emulsion, with this being ascribed to the increased residence time in the stomach leading to more extensive drug degradation. Similarly, other authors (51) have reported a delay in time to C_{max} due to the prolonged gastric emptying, although again this effect is not ubiquitous (52).

A related effect is the effect of lipid volume on gastric function and drug absorption. There is still some debate on this issue, with conflicting results being reported. For example, Fischler et al. (53) reported that doubling the oil volume administered increased plasma C_{max} values for chlomethiazole to the extent of producing a profile similar to that of a dosage form containing approximately twice the drug dose. Other studies, however, have found no effects of increased oil loading on drug absorption. For example, Charman and Stella (54) reported

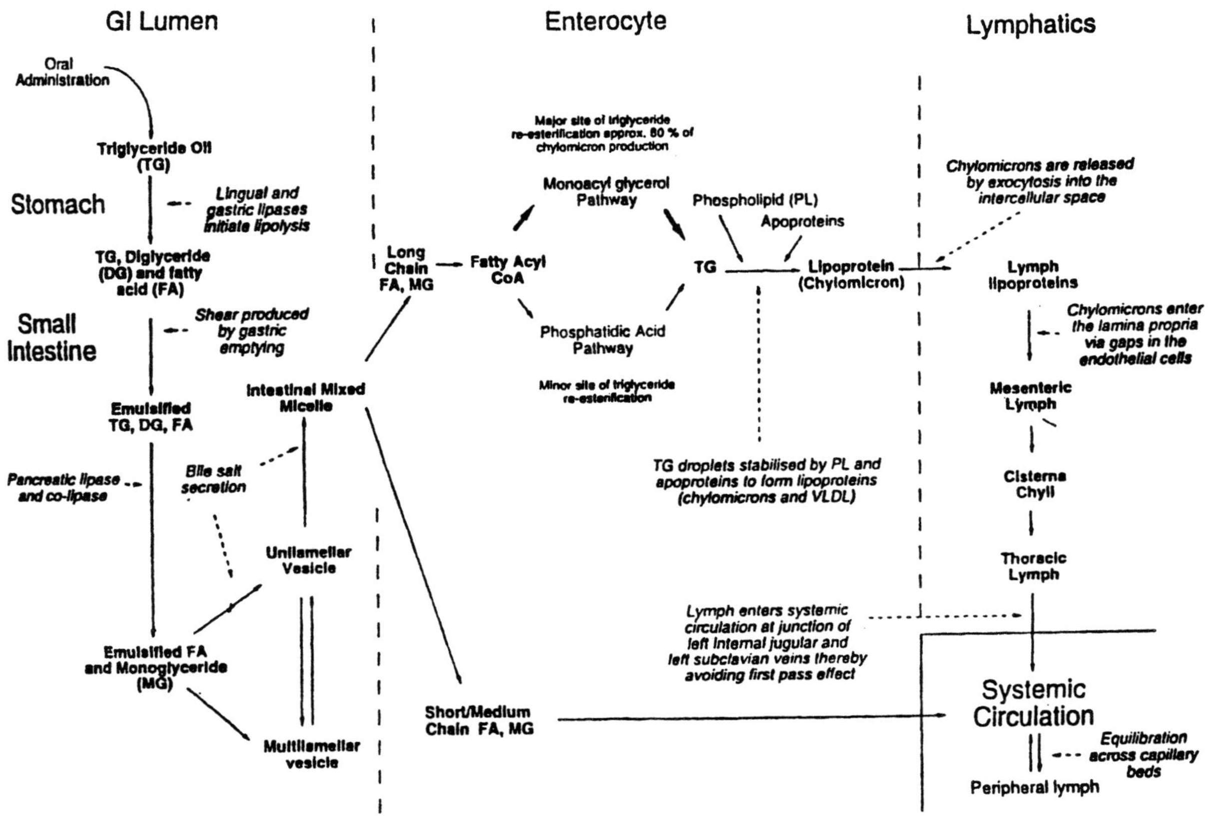

Figure 5 Schematic diagram showing the sequential steps in the digestion of lipids and subsequent absorption via the portal blood and intestinal lymphatics. (From Ref. 47.)

no change in lymphatic uptake of DDT when administered to anesthetised rats in lipid volumes of 200 μL and 50 μL. Other authors have reported the volume of coadministered lipid to have no discernible effect on plasma profiles (55–57). These conflicting reports may reflect differences in the nature of the lipid digestion and drug absorption profiles, but may also be a function of the drug-specific effects of prolonging gastric residence time on absorption.

IV. BIOAVAILABILITY OF DRUGS FROM OILY DELIVERY SYSTEMS

A. Effect of Experimental Methodology

The possibility of enhancing the absorption of drugs from the GI tract by the use of emulsion systems is clearly of great interest. However, before discussing this topic it is worth briefly considering the models used to study drug absorption, particularly for in vivo investigations. However this is a major subject in its own right and only a few of the many issues involved will be outlined here.

The choice of animal model is the subject of continuous debate. For example, in rats a continuous production and secretion of bile occurs even in the absence of food, which is not the case in humans. The use of conscious animals avoids problems associated with anesthetics and facilitates oral dosing, but over and above ethical considerations the stress caused to the animal may significantly alter gastric motility. The site of administration may also be important; Charman et al. (58) investigated the effects of administering oleic acid and peanut oil formulations containing DDT perorally and intraduodenally to the rat, finding no significant difference between the drug absorption profiles for peanut oil but a greater lymphatic uptake for the oleic acid systems when given by the intraduodenal route. The use of fasted and nonfasted models is also of importance. Charman et al. (59) noted greater lymphatic transport of DDT in the fasted state, although De Marco and Levine (60) have argued that as sugars such as glucose are essential for chylomicron formation, one would expect a reduction in lymphatic transport in the fasted state. Similarly, Shaiu et al. (61) have suggested that VLDL is the major transport lipoprotein in the fasted state compared to chylomicrons in the fed state.

A number of methods are available for studying intestinal lymphatic drug transport (62,63). Lymphatic transport may be affected by factors such as the site of cannulation and the period of fasting prior to dosing (54,58,62,63). The triple-cannulated rat model, involving the mesenteric lymph duct, jugular vein, and duodenum, have been widely used for both anesthetized and unanesthetized studies. The use of lymph cannulation as a technique has also attracted some debate. For example, Sieber (64,65) has argued that care is required in the interpretation of data obtained using radiotracers to determine whether lymph uptake has taken place due to the possibility of indirect uptake via the portal system rather than directly into the lacteals. The cannulation process itself may also influence the outcome of the study. Noguchi et al. (66) have suggested that thoracic cannulation may lead to overestimates of intestinal lymphatic uptake due to the thoracic duct draining not only the mesenteric but also the hepatic and other peripheral lymph systems.

B. Drug Absorption from Oils

The use of oily vehicles as a means by which to administer drugs to the GI tract is well recognized. Similarly, the possibility of enhancing drug absorption in this manner has been frequently described, even if the mechanisms involved are not year clear. For example, Bates and Sequeira (49) compared the absorption of griseofulvin from a commercial tablet, a corn oil emulsion, and an aqueous suspension in humans, with more rapid excretion of the metabolite 6-desmethylgriseofulvin being noted for the emulsion. Interestingly, the authors also reported that the enhancement was greater in emulsified corn oil than for comparable doses of an unemulsified system. This touches on one of the as-yet-unanswered questions regarding this approach to drug delivery, namely, does the physical form of the oil in the dosage form make any difference to the bioavailability? If the most important factor in the absorption enhancement process is the inclusion and subsequent digestion of the oil, the nature of the administered formulation may not be critical as the oil will emulsify in the GI tract in any case. This study at least suggests that the formulation may indeed be important, although clearly the generalizability of this conclusion is uncertain.

A number of mechanisms and influencing factors have been suggested in association with drug absorption from oils. Armstrong and James (67) have reviewed the available knowledge base up to 1980, discussing several examples such as the one given above whereby the absorption of drugs may be considerably enhanced by incorporation into an oily base. However, the authors also point out that the literature is contradictory in this respect, with other drugs showing little or no enhancement compared to more conventional formulations. The authors suggest that this may be a function of the lipophilicity of the drug and the route of uptake. In essence, the authors suggested that lipophilic drugs may be incorporated into the mixed micelles formed on digestion of the oil and are thereby incorporated into the fatty acid uptake mechanism; such drugs would therefore be taken up into the lymphatic system, while more hydrophilic molecules would be absorbed via the hepatic portal vein. Evidence for this include studies by Horst et al. (68) who demonstrated that the activity of the highly lipophilic undecanoate ester of testosterone in humans is greater than testosterone itself, with a further increase in activity seen on incorporating the ester in oil. Absorption studies showed that testoterone undecanoate was found exclusively in the lymphatics while the free base was found in the hepatic portal circulation. Clearly, therefore, consideration of drug uptake from oil bases must be considered in the context of the absorption route involved (69); hence the pertinent issues relating to drug uptake into the lymphatics will be considered briefly here. The interested reader is referred to a recent review by Porter and Charman (47) for more details of drug absorption via the lymphatics, while Humberstone and Charman (70) have discussed drug uptake from lipid vehicles, including numerous recent examples of relevant studies.

One factor that almost certainly merits further consideration is the relevance of the physical form of the drug in the lipid to bioavailability. It was previously argued that there is a paucity of knowledge regarding the solubility of drugs in oils and SEDDS from a formulation viewpoint. However, there is also evidence that the physical form of the drug (more specifically, whether the drug is presented to the body as a solution or a suspension) may influence the uptake through the GI tract. For example, Hargrove et al. (71) describe a study whereby micronized progesterone was administered as an oral suspension in a long chain fatty acid vehicle, producing peak progesterone concentrations of 30.3 ($\pm$7.0) ng/mL compared to 13.24 ($\pm$2.4) ng/mL when administered as a micronized powder. However, this improvement in bioavailability was lost when the drug was administered in a nonmicronized form as a lipid suspension. It may therefore be concluded that, in this cast at least, the improvement may be at least partially ascribed to a reduction in particle aggregation, the high surface energy of the micronized powder may be expected to lead to clumping, thereby reducing the effective surface area of the drug. Dispersion in a lipid base may improve particle dispersion, possibly via a wetting effect.

Abrams et al. (72) studied the oral bioavailability of a lipophilic steroid (17β-acetoxy-2α-chloro-3-(*p*-nitrophenoxy)imino-5α-androstane) in sesame oil, finding a threefold increase in bioavailability compared to an aqueous suspension. The authors also noted that the extent of absorption increased in a nonlinear manner with increasing concentration in oil, while the relationship was linear over the concentration range studied for the aqueous suspension. This effect was ascribed to the drug reaching saturation concentration within the oil; hence any increase in drug loading led to an increase in quantity of solid drug present rather than drug in lipid solution. The implication here, once again, is that the physical form of the drug within the lipid solvent may have a profound effect on bioavailability.

It is, of course, well recognized that the physical form of a drug in any formulation may have a profound effect on bioavailability. However, understanding the influence of the drug's physical form on absorption from lipid vehicles is complicated by a lack of knowledge regarding how the form of the drug will influence drug uptake with the lipid digestion products (assuming that this process is of relevance). It is reasonable to assume that the drug will remain in the same form within the oil droplets until digestion takes place, as the microenvironment within the droplets may not change markedly up to this point. However, if the drug is suspended within the oil droplets it is interesting to speculate at what stage and into what medium it will dissolve prior to absorption. Studies by Tokumura et al. (73) demonstrated that cinnarizine showed enhanced absorption in beagle dogs from an oleic acid vehicle compared to a tablet formulation, with an approximately fourfold increase in AUC seen for the former (Fig. 6a). However, the authors also conducted dissolution studies into bile salt solutions showing enhanced dissolution rate and extent for the oleic acid systems (Fig. 6b), which was mirrored by enhanced solubility of the drug in the bile salt medium in the presence of oleic acid. The authors attributed these effects to the favorable solubility of cinnarizine in mixed oleic acid/bile salts micelles; in the context of the above arguments, the study also demonstrates the importance of the ease of incorporation of the drug into the mixed micelles in the absorption process. Several other studies have indicated that bile salt mixed micelles are good solvents for lipophilic drugs (74–78). Clearly, the solubility of the drug in the lipid vehicle and/or the bile salt mixed micelles must also be considered with respect to the solubility of the drug in the triglyceride core of the chylomicrons (discussed below). However, these reports indicate that the basic physicochemical behavior of the drugs in the lipid/micellar environment may be a determining factor in the uptake process.

V. FACTORS INFLUENCING DRUG UPTAKE FROM OILS

A. Nature of the Drug

The influence of drug solubility in the lipid vehicle has been discussed in the previous section. Further considerations include the relationship between the molecular structure and physicochemical properties of the drug and the route of uptake. The subject of drug uptake into the lymphatics has been discussed in detail by Porter and Charman (47), and the interested reader is referred to their text for more information. The lymphatic system is believed to provide a pathway to the systemic circulation for highly lipophilic drugs such as cyclosporine (79), naftifine (80), probucol (81), polychlorinated biphenyls (82), and lipophilic vitamins and vitamin derivatives (83). Uptake into the intestinal lymphatic system leads to access to the left internal jugular vein via the thoracic lymph duct. However, it is estimated that the rate of fluid flow through the intestinal lymphatics is approximately 500 times lower than through the portal system (84,85). Furthermore, the lipid content of lymph is only 1–2% w/v during peak lipid transport; hence, as Porter and Charman (47) have pointed out, the mass ratio of lipid lymph to portal blood is in the region of 1 : 50,000 (although it is chylomicron flux rather than absolute flow that is the important factor). Consequently, lymphatic transport

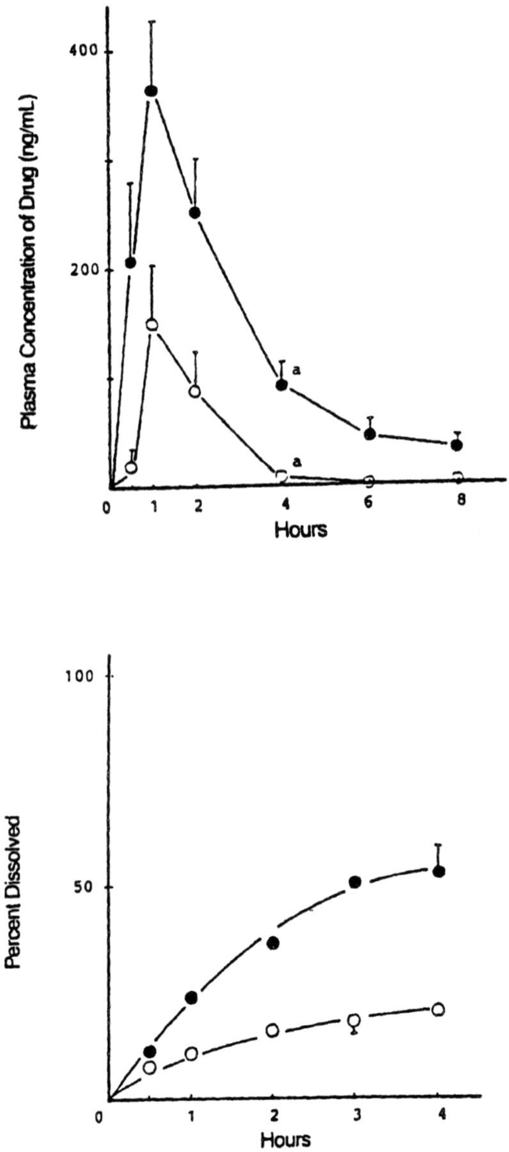

Figure 6 (a) Plasma concentration of cinnarizine after the oral administration of (○) a cinnarizine tablet and (●) the oleic acid solution of cinnarizine. Each point represents the mean ± SEM of 3–4 dogs; points marked have a $p < 0.05$. (b) Dissolution of cinnarizine from both tablet and capsule (containing the oleic acid solution of cinnarizine) dosage forms into a medium containing glycochenodeoxycholate. ○, tablet; ●, capsule. Each point represent the mean ± SEM of three determinations. (From Ref. 73.)

is only likely to be of relevance to molecules with a log octanol/water partition coefficient of at least 5 (86). In addition, it is necessary to consider the lipid solubility of the drug or, more specifically, the solubility of the drug in triglycerides as the lipid digestion products are resynthesized to triglycerides that form the chylomicron core; indeed, Charman and Stella (86) have suggested that the drug loading in chylomicrons is approximately 25% of the saturated lipid solubility in the core. For example, comparison of DDT and hexachlorobenzene, both of which have similar log P values, indicates that DDT showed a 14-fold increase in lymphatic transport using an anesthetised rat model, which corresponds well to a 13-fold greater triglyceride solubility (86). Clearly, such physicochemical measurements alone will not reliably predict the lymphatic uptake of drug compounds but do demonstrate the existence of a relationship between these parameters and the route of absorption. Charman and Stella (86) have suggested that a triglyceride solubility of at least 50 mg/mL is required for effective uptake. Similarly, Porter et al. (87) have compared the lymphatic transport of halofantrine using the free base and the HCl salt, the former having an approximately 200-fold greater solubility in triglycerides. In these studies, the lymphatic transport of the free base was 20% of the administered dose when coadministered with lipid compared to 5% after administration as the HCl salt in the presence or absence of lipid. This was ascribed to the increased lipophilicity of the free base, thereby facilitating greater association of halofantrine base with the products of luminal lipid digestion and the subsequent interaction with the intestinally derived chylomicrons responsible for lymphatic drug transport.

B. Nature of the Oil

A number of studies have examined the influence of the nature of the oil on drug uptake from lipid vehicles. For example, Palin et al. (88) compared plasma concentrations of DDT after oral administration in arachis oil, Miglyol-812, liquid paraffin, and as a suspension in water. The authors reported marked differences in absorption profile, as shown in Fig. 7. The greatest absorption was found to take place from arachis oil, followed by Miglyol-812 and the aqueous suspension; liquid paraffin was found to decrease the absorption of the drug. In addition, high concentrations of DDT were found in lymph, particularly when administered in arachis oil, indicating that lymphatic uptake was being enhanced. The authors suggested that the observed effects may be associated with the chain length of the oil, with Miglyol-812 comprising shorter chain length fatty chains which result in metabolites that are then absorbed via the hepatic portal vein. Liquid paraffin is composed of nondigestible hydrocarbons; hence drug absorption may be diminished due to entrapment of the DDT within the oil droplets that pass through the GI tract essentially unchanged. The authors suggest that arachis oil may promote uptake due to a combination of solubilization of DDT in mixed

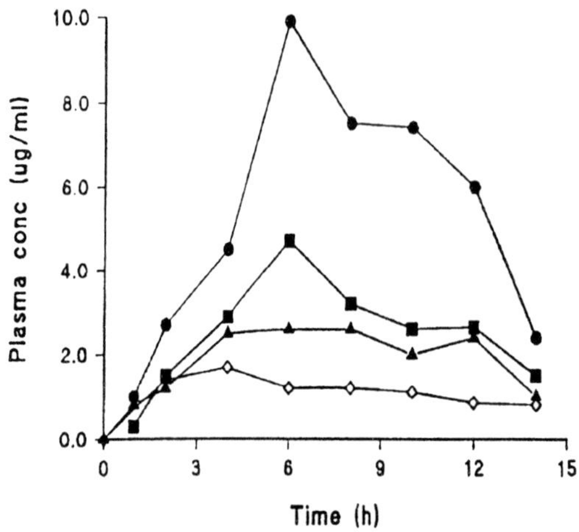

Figure 7 Effect of different vehicles (1-mL volumes) on the absorption of orally administered DDT (100 mg/kg) in unanesthetised rats. ●, Arachis oil; ■, Miglyol-812; ◇, liquid paraffin; ▲, water containing 6% Tween-80. Data are given as means ± SEM of 4 values). (From Ref. 88.)

micelles, enhanced membrane permeability due to the presence of the fatty acid metabolites, and the long chain unsaturated fatty acids promoting synthesis of chylomicrons.

There are clearly a number of issues associated with the choice of oil, particularly in terms of the relationship between the basic chemical structure of the major oil components (bearing in mind that all natural oils are mixes of several components) and the interaction of those components with the various enzymes, surfactants, and proteins associated with the digestion and absorption processes. For example, fatty acid chain length has been widely recognized recognized as being an important factor in chylomicron formation. Short and medium chain acids are predominantly absorbed by the portal blood system, while longer chain fatty acids may be reesterified in the cells lining the small intestine and absorbed via the lymphatics, as discussed in an earlier section. Short and medium chain fatty acids are also not substrates for chylomicron formation; this may be associated with the weak affinity of one of the essential proteins associated with lipoprotein synthesis, fatty acid binding protein, for short or medium chain fatty acids (89,90). In addition, increasing the chain length of the fatty acid tends to decrease both the hydrolysis and absorption of the oil (90).

A number of workers have also reported that absorption enhancement is greater when using unsaturated fatty acids. Cheema et al. (91) reported that a greater degree of unsaturation led to a more rapid onset of lipoprotein synthesis, possibly as a result of faster absorption or a greater affinity of fatty acid binding protein for unsaturated fatty acids. In addition, unsaturated fatty acids have lower melting points and increased fluidity compared to saturated, which may be relevant when considering drug solubilization. Muranushi et al. (92) investigated the effects of bile salt mixed micelle composition on the absorption of streptomycin. Micelles composed of unsaturated fatty acids (e.g., oleic or linoleic acids) markedly improved streptomycin absorption in the large intestine, with saturated fatty acids (e.g., lauric and myristic acids) having only marginal effects. Triolein, diolein, and oleyl alcohol had no apparent enhancement effect. The authors suggested an association between the choice of lipid and the enhancement of mucosal permeability, with mixed micelles containing unsaturated fatty acids leading to greater permeability enhancement than saturated. However, these studies were conducted on the large intestine; hence the direct relevance to the small intestine is uncertain. The degree of unsaturation may also influence chylomicron formation; Ockner et al. (93) investigated the uptake of two long chain fatty acids, palmitic and linoleic acid, suggesting that chylomicrons were formed when unsaturated linoleic acid was absorbed while VLDLs were formed when the saturated palmitic acid was administered.

The issue of oil digestibility is also highly important, although the relationship between the ease of digestibility and absorption does not appear to be straightforward; a full discussion of the issues relating to this topic has been given by Humberstone and Charman (70). Bloedow and Hayton (56) studied the effects of lipid digestibility on the bioavailability of three poorly soluble drugs (acetyl sulfisoxazole, dicumerol, and griseofulvin) in the rat, finding that the effect of the lipid was drug-specific. For example, griseofulvin was found to show decreased bioavailability in all of the oils studied, while acetyl sulfisoxazole absorption was increased when incorporated into digestible lipids and decreased for nondigestible lipids. However, Humberstone and Charman (70) have pointed out that the lipid loadings used were very large (5 mL/kg), which may have had an influence on the absorption process.

Yamahira et al. (94) compared oral absorption of the drug SL-512 from an easily digested medium chain triglyceride and a synthetic, poorly digested lipid, MBLA (*N*-α-methylbenzyllinoleamide); the drug partition coefficients with respect to the two lipids were similar. The medium chain triglyceride formulation showed a fourfold increase in maximum serum blood levels compared to MBLA, indicating that lipid digestibility was a primary factor determining drug uptake. Similar results were obtained when the formulations were given intraduodenally, indicating that alterations in gastric emptying were not responsible for the effect. The authors also showed using a recirculating intestinal perfusion experiment that absorption did not take place directly from the lipid formulations but that lipid digestion was a prerequisite to drug absorption. Similarly, Ritschel et al. (95,96) demonstrated that bile duct ligation decreased absorption of cyclosporine from long chain fatty acid ester microemulsions, thereby indicating the necessity for lipid digestion and mixed micelle formation for the absorption of this drug.

Myers and Stella (98) used digestible and nondigestible lipid formulations of penclomedine, showing that bioavailability was greatest from medium chain triglycerides, followed by long chain triglycerides and then mineral oil, with this rank order corresponding to the digestibility of the lipids. Interestingly, the authors also found that short chain triglycerides gave poor bioavailability profiles despite the rapid digestion of these oils, this being ascribed to rapid dispersion and dissolution of the digestion products, resulting in precipitation of the drug in the lumen of the GI tract. This observation demonstrates the importance of considering not only the extent of digestion but also the kinetics of the process. This feature of drug absorption is exemplified by a study by Charman and Stella (54) who studied DDT absorption from peanut oil, oleic acid, and oleic acid/monoolein mixes via intraduodenal administration, finding that the fatty acid based formulations transported greater quantities of DDT at a faster rate. No differences in lymph flow or chylomicron concentration were observed; however, the amount of DDT in the cores of the chylomicrons was substantially higher for the fatty acid formulations. The authors ascribed this effect to the competing uptake processes via the hepatic portal vein and the lymphatics. Bearing in mind that the ratio of portal blood to lymph flow is in the region of 500 : 1, the additional time required for digestion of the triglyceride may result in a larger proportion of DDT being absorbed directly into the bloodstream.

VI. DRUG ABSORPTION FROM EMULSIONS AND SEDDS

The above discussion has outlined some of the studies and considerations pertinent to the uptake of drugs from oil vehicles. This then leads to question of how drug uptake will differ from an emulsion compared to an ingested oily liquid. There are essentially two differences between these delivery approaches that may be relevant. First, the oil will be presented to the stomach in a finely divided form when administered as an emulsion, as opposed to essentially a single phase. The relevance of this is still a matter of debate, as the oil will be emulsified in the GI tract in any case. The debate is further complicated by the use of SEDDS, whereby the dose is given as a single phase but emulsifies rapidly on contact with the gastric fluids. The second difference is that emulsions will include an emulsifying agent. The role of the emulsifying agent has recently become the subject of focus in the field and will be discussed in more detail below.

A. Drug Absorption from Conventional Emulsions

Numerous studies have been conducted using conventional emulsion systems and a brief summary of these studies is given below; the interested reader is referred to the text of Humberstone and Charman (70) for a more detailed summary. In general, studies have shown that administering the drug is an emulsified form leads to greater bioavailability compared to administration of the oil as a single phase. Myers and Stella (97) studied the absorption of penclomedine after intraduodenal administration to anesthetised rats in an emulsified and nonemulsified form in mineral oil and trioctanoin, showing marked increases in absorption from both emulsified systems, as shown in Fig. 8. Given the indigestibility of mineral oil, this improvement was ascribed to a particle size effect, while the improvement for trioctanoin was ascribed to improved digestibility of the emulsion.

Kimura et al. (98) demonstrated greater bioavailability of vitamin E from an emulsion in a medium chain triglylceride formulation than from a long chain triglyceride solution (soybean oil). The authors went on to demonstrate that increasing the medium chain triglyceride content of the emulsion led to decreased absorption, this was attributed, at least partially, to the increase in droplet size caused by increasing the oil loading in the lecithin-stabilized formulation. The role of emulsion droplet size has been specifically studied by Tarr and Yalkowsky (99) who prepared two olive oil–based formulations of cyclosporine but processed these oils so as to produce particles with a median size of approximately

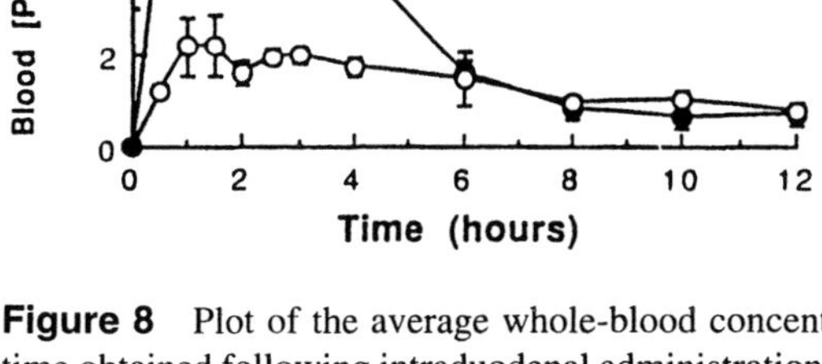

Figure 8 Plot of the average whole-blood concentration (with standard error bars) vs. time obtained following intraduodenal administration of 5 mg penclomedicine in (a) either 0.5 g of a 10% O/W trioctanoin emulsion (10 mg/g), □ (n = 8); or 50 μL trioctanoin oil (10 mg/g), ■ (n = 3) to rats (b) either 0.5 g of a 10% O/W mineral oil emulsion (10 mg/g), ● (n = 3); or 50 μL mineral oil, ○ (n = 3) to rats. (From Ref. 97.)

2 μm and 4 μm. Using a perfused rat intestine model, the authors demonstrated that the permeability of the cyclosporine through the gut was greater for the finer emulsion by a factor of approximately 1.7, which was reflected by higher blood levels attributable to the larger surface area of the finer emulsion.

B. Drug Absorption from SEDDS

There have been numerous reports outlining improvements in bioavailability when drugs are administered in SEDD formulations compared to more conventional dosage forms, some of which are outlined below; for more details, the reader is again referred to Humberstone and Charman (70). The flagship product, Sandimmune, which generated much of the interest in SEDDS, is composed of oil, ethanol, and Labrafil in the ratio 40:18:42 with 100 mg/mL cyclosporine and forms a crude emulsion on gentle mixing with water; the new product Sandimmun Neoral contains the drug, a surfactant, a hydrophilic cosolvent, and a lipid. This formulation forms a microemulsion on mixing with water and has been reported to result in a twofold increase in bioavaiability in humans compared to the earlier product (100). Inter- and intrasubject variability has also been reported to be reduced due to the more complete absorption of the drug (101), while the increase in cyclosporine bioavailability caused by the presence of food in the GI tract, particularly high-fat meals, has also been reduced (102).

The effects of formulating drugs in SEDD formulations are far from predictable, although strong evidence exists in the literature for the potential usefulness of the approach. Charman et al. (25) examined the lipophilic antiviral compound WIN-54954 administered in a formulation containing a medium chain triglyceride and a nonionic surfactant. The SEDDS was compared to a PEG-600 formulation. No significant difference in mean bioavailability was seen between the two systems but greater reproducibility was noted for the SEDD system. the drug L-365, 260 was prepared as a Labrafil M2125 and Tween-80 SEDD preparation (103) and showed a seven- to eightfold increase in C_{max} compared to a tablet or suspension formulation. However, the absorption profiles for the SEDDS were similar to those observed after administration of a PEG-600 solution, suggesting that the rate-limiting step to absorption may be dissolution of the drug in the GI fluid rather than lipid digestion. In contrast, Shah et al. (104) examined the bioavailability of Ro-150778, a highly lipophilic naphthalene derivative, demonstrating a fourfold greater bioavailability for the SEDDS formulation than a PEG-400 solution and a 20-fold greater bioavailability than from a standard tablet formulation. A further aspect (or complication) relating to the predictability of drug absorption from SEDD formulations is the particle charge. Gershanik and Benita (17) have compared progesterone absorption in female rats from a range of formulations, including SEDDS which form positively and negatively charged droplets. The positively charged systems, which include the cationic lipid oleyamine, showed the greatest bioavailability along with a solution in PEG-300, although whether this is a reflection of the surface charge or surface composition is not clear.

C. Role of Emulsifying Agent in Absorption of Drugs from SEDDS

A further aspect of SEDDS technology that is generating considerable interest is the role of the surfactant with regard to drug uptake from SEDDS, particularly in terms of the digestibility of the oil. This topic has been discussed in depth by MacGregor et al. (107) and only a brief summary will be given here.

The interest in this area arose from the observation that surfactants may inhibit the digestion of oils. For example, studies using Cremophor RH40, an ethoxylated triglyceride surfactant, inhibited in vitro lipolysis of a medium chain length triglyceride over a 90-min period, while inclusion of a lipophilic surfactant caused partial recovery of the lipolysis, with the extent of recovery depending on the nature of the second surfactant (105). Soloman et al. (106) examined a range of nonylphenylethoxylates, allowing examination of the relationship between HLB and lipolysis inhibition (Fig. 9). Lipolysis was inhibited to the greatest extent by surfactants with an HLB greater than 12, the most effective having

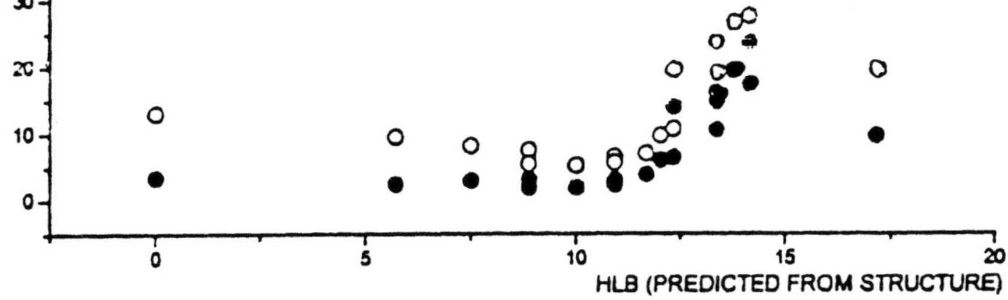

Figure 9 Effect of HLB on nonylphenol ethoxylates on inhibition of lipolysis of a medium chain triglyceride in vitro. Data is expressed as $t_{i\%}$, the time taken to reach $i\%$ of the maximum. Reaction mixtures contained 1 g triglyceride and 1 g surfactant in 100 mL. (From Ref. 106.)

an HLB between 13 and 17; further increases in HLB caused a decline in inhibition, possibly due to their weak surface activity. This has clear implications for the choice of surfactant with respect to SEDDS. Other studies referring to the role of surfactants in inhibiting lipolysis include recent work in the Japanese literature, highlighted by Porter and Charman (47), which have examined the use of milk fat globule membrane (MFGM) as an emulsifying agent for the delivery of vitamins (D_3 and A), epidermal growth factor, and insulin (107–110). MFGM is a membrane-derived material from the secretory cells of the lactating mammary gland that plays a role in the stabilization of dispersed phase in milk. Initial reports appear to suggest that this emulsifying agent may have some influence on lymphatic uptake, with a recent study (111) suggesting an increase in the lymphatic uptake of vitamin D_3 from emulsions stabilized with MFGM compared to those stabilized with polysorbate-80, although the increase was not statistically significant. Clearly, more work is required to clarify whether this emulsifying agent does indeed have a beneficial effect on lymphatic uptake, although if this proves to be the case then the implications for drug delivery are very exciting.

VII. CONCLUSIONS

This chapter has outlined some considerations regarding the delivery of drugs to the GI tract via lipid vehicles, with particular emphasis on SEDDS. Lipid delivery in general, and SEDDS in particular, offers a potentially highly useful means of enhancing the oral absorption of a number of drugs. However, a number of questions remain unanswered with respect to this approach. From the formulation viewpoint, it is necessary to consider the emulsifying properties of the lipid vehicle (simple oil, conventional emulsion, or SEDDS) and the solubility of the drug in the oil or oil–surfactant mix, both in terms of determining whether the drug will be in suspension or in solution and, if the latter is required, whether it is possible to solubilize adequate quantities of drug in the lipid vehicle. The effect of drug inclusion on the emulsification of the vehicle should also be considered.

On ingestion of the dosage form, it is then necessary to consider the particle size of the emulsion in vivo and the relationship between this parameter and the size of the ingested oil droplets or, in the case of SEDDS, the size of the droplets observed in vitro on mixing with an aqueous medium. It appears that little is known on this issue (not least due to the difficulty in measuring the droplet size in the intestine), nor is it clear how important the size of the ingested particles may be in determining the absorption rate. The study by Tarr and Yalkowsky (99) indicated that a reduction in initial size may lead to an increase in drug absorption, although interestingly the sizes used by these authors (2 and 4 μm) are of the same order of magnitude as the estimated droplet size that occurs on ingestion of oils in any case (0.5–1 μm). Given the emulsification processes that occur in the GI tract and the polydispersity of droplet sizes that may be expected when the system reaches the intestine, it is perhaps surprising that a relatively small change in initial size would lead to a significant change in absorption profile, although this is clearly the case. It is perhaps possible, given the evidence presented by MacGregor et al. (105), that the differences in absorption may be partially a function of the distribution in surfactant of the two systems; the same quantity of surfactant was used for both formulations, hence given the larger surface area of the smaller size system the surfactant density at the oil–water interface may be less than for the 4-μm systems, with a concomitant reduction in lipolysis inhibition.

In addition to considering the particle size of the emulsions, it is also necessary to consider the physical state of the drug in vivo. It is reasonable to assume that the drug will remain in the same state within the oil phase prior to lipid digestion. However, it is necessary to consider the interplay between the drug remaining dissolved in the oil, the diffusion of the drug into the aqueous phase, the solubilization of the drug in mixed micelles, the tendency for suspended drugs to simply remain in the solid state, irrespective of the state of the oil, and the possibility of dissolved drug precipitating out in the aqueous phase as the oil is digested. These considerations will be drug- and oil (or oil/surfactant mix)–dependent, but it has so far proved difficult to fully separate the influence of these effects. Furthermore, the possibility of luminal first-pass metabolism of the drug should also be considered.

The evidence for the digestion of the oil having a principal role in the absorption process is extremely strong, although the parity between absorption from SEDDS and PEG solutions reported in some studies (17,103) indicates that, at least in some cases, the key issue may be the physical form of the drug. The effects of both the rate and extent of digestion, the nature of the digestion products, and the route of drug uptake are also to be considered. The exact means by which drug uptake is enhanced is still not entirely clear. For example, does absorption enhancement occur because of the micellar transport of the drug to the absorbing membrane, thereby facilitating simple diffusion into the lymphatic system? Or is the drug associated with the lipid digestion products throughout up to the synthesis of chylomicrons? The question of the role of the surfactant, particularly in terms of lipolysis inhibition, also needs to be studied further. In addition, the relevance of the MDR (multiple drug resistance) P-glycoprotein transmembrane pump to the mechanism of drug absorption enhancement from lipid vehicles could be usefully explored. The low and variable bioavailability of certain drugs such as cyclosporine A may be partially attributed to efflux of the drug in the basolateral direction via a MDR P-glycoprotein-like pump (112). On the basis that certain agents such as chlorpromazine and progesterone are known to inhibit the pump mechanism, it is conceivable that the observed absorption enhancement of this drug may be partially due to inhibition of this pump, although at present this is purely speculation.

Clearly, there are several areas associated with lipid-based oral dosage forms in which further knowledge would be desirable. However, there is strong evidence that this delivery approach may afford significant advantages over more conventional dosage forms. Hence the use of emulsions, lipid solvents, and SEDDS is likely to continue to play an important role in the formulation of poorly soluble drugs in the future.

REFERENCES

1. S. Venkataram, J. A. Rogers. Characteristics of drug-phospholipid coprecipitates: physical properties and dissolution behaviour of griseofulvin-dimyristoylphosphatidylcholine systems. J Pharm Sci 73:757–761, 1984.
2. K. S. Chang, C. I. Jarowski. Solid dispersion of morphine-tristearin with reduced presystemic inactivation in rats. J Pharm Sci 69:466–469, 1980.
3. D. Q. M. Craig. The use of glycerides as controlled release matrices. In: D. R. Karsa, R. A. Stephenson, eds. Excipients and Delivery Systems for Pharmaceutical Formulations. London: Royal Society of Chemistry, 1996, pp. 148–173.
4. C. W. Pouton, W. N. Charman, eds. Oily formulations for administration to the gastrointestinal tract. Adv Drug Del Rev 25:1–128, 1997.
5. G. L. Amidon, H. Lennernäs, V. P. Shah, J. R. Crison. A theoretical basis for a biopharmaceutical drug classification: the correlation of *in vitro* drug product dissolution and *in vivo* bioavailability. Pharm Res 12:413–420, 1995.
6. C. W. Pouton. Self-emulsifying drug delivery systems: assessment of the efficiency of emulsification. Int J Pharm 27:335–348, 1985.
7. M. J. Groves, R. M. A. Mustafa. Measurement of the ''spontaneity'' of self-emulsifiable oils. J Pharm Pharmacol 26:671–688, 1974.
8. M. J. Groves, R. M. A. Mustafa. J. E. Carless. Phase studies of mixed phosphated surfactants n-hexane and water. J Pharm Pharmacol 26:616–623, 1974.
9. M. J. Groves, D. A. de Galindez. The self-emulsifying action of mixed surfactants in oil. Acta Pharm Suec 13:361–372, 1976.
10. H. Reiss. Entropy-induced dispersion of bulk liquids. J Colloid Interface Sci 53: 61–70, 1975.
11. T. A. Iranloye. Some factors affecting the self-emulsification of hydrocarbon oils. Prog Coll Polym Sci 70:119–126, 1985.
12. C. W. Pouton. Formulation of self-emulsifying drug delivery systems. Adv Drug Del Rev 25:47–58, 1997.
13. K. Shinoda, S. Friberg. Emulsions and Solubilization. Chichester: John Wiley and Sons, 1986.
14. T. J. Lin, H. Kurihara, H. Ohta. Prediction of optimum o/w emulsification via solubilization measurements. J Soc Cosmet Chem 28:457–479, 1977.

15. D. Q. M. Craig, S. A. Barker, D. Banning, S. W. Booth. An investigation into the mechanisms of self-emulsification using particle size analysis and low frequency dielectric spectroscopy. Int J Pharm 114:103–110, 1995.
16. D. Q. M. Craig. Dielectric Analysis of Pharmaceutical Systems. Basingstoke: Taylor and Francis Ltd., 1995.
17. T. Gershanik, S. Benita. Positively charged self-emulsifying oil formulation for improving oral bioavailability of progesterone. Pharm Dev Tech 1:147–157, 1996.
18. P. P. Constantinides. Lipid microemulsions for improving drug dissolution and oral absorption: physical and biopharmaceutical aspects. Pharm Res 12:1561–1572, 1995.
19. D. Q. M. Craig, A. Chan, N. Khan. A comparison of the self-emulsifying properties of polyglycolysed glyceride oils. Pharm Sci 1:559–561, 1996.
20. C. W. Pouton. A study of self-emulsifying oil/surfactant mixtures. PhD thesis, University of London, 1982.
21. M. G. Wakerley, C. W. Pouton, B. J. Meakin, F. S. Morton. Self-emulsification of vegetable oil–nonionic surfactant mixtures; a proposed mechanism of action. Am Chem Soc Symp Series 311:242–255, 1986.
22. M. G. Wakerley, C. W. Pouton, B. J. Meakin. Evaluation of the self-emulsifying performance of a non-ionic vegetable oil mixture. J Pharm Pharmacol 36:6P, 1987.
23. S. A. Barker, C. J. Lovegrove, D. Q. M. Craig, S. Booth, D. E. Storey. Analysis of non-aqueous solvent systems using low frequency dielectric spectroscopy. J Pharm Pharmacol 45:103P, 1993.
24. C. W. Pouton. Effects of the inclusion of a model drug on the performance of a self-emulsifying formulation. J Pharm Pharmacol 37:1P, 1985.
25. S. A. Charman, W. N. Charman, M. Rogge, T. Wilson, F. Dutko, C. Pouton. Self-emulsifying drug delivery systems: formulation and biopharmaceutic evaluation of an investigational lipophilic compound. Pharm Res 9:87–93, 1992.
26. D. Q. M. Craig, H. Lievens, K. Pitt, D. Storey. An investigation into the properties of self-emulsifying systems using low frequency dielectric spectroscopy and particle size analysis. Int J Pharm 96:147–155, 1993.
27. J. K. Embleton, C. W. Pouton. Structure and function of gastro-intestinal lipases. Adv Drug Del Rev 25:15–32, 1997.
28. M. C. Carey, D. M. Small, C. M. Bliss. Lipid digestion and absorption. Annu Rev Physiol 45:651–677, 1983.
29. H. Moreau, R. Laugier, Y. Gargouri, F. Ferrato, R. Verger, R. Human preduodenal lipase is entirely of gastric fundic origin. Biochim Biophys Acta 959:247–252, 1988.
30. C. K. Abrams, M. Hamosh, T. C. Lee, A. F. Ansher, M. J. Collen, J. H. Lewis, S. B. Benjamin, P. Hamosh Gastriclipase: localisation in the human stomach. Gastroenterology 95:1460–1464, 1988.
31. M. W. Bodmer, S. Angal, G. T. Yarranton, T. J. R. Harris, A. Lyons, D. J. King, G. Pieroni, C. Riviere, R. Verger, P. A. Lowe. Molecular cloning of a human gastric lipase and expression of the enzyme in years. Biochim. Biophys Acta 909:237–244, 1987.
32. C. Tiruppathi, K. A. Balasubramanian. Purification and properties of an acid lipase from human gastric juice. Biochim Biophys Acta 712:692–697, 1982.

33. J. S. Patton, M. W. Righler, T. H. Liao, P. Hamosh, M. Hamosh. Hydrolysis of triacylglycerol emulsions by lingual lipase. Biochim BIophys Acta 712:400–407, 1982.
34. G. L. P. Fernando-Warnakulasuriya, J. E. Staggers, S. C. Frost, M. A. Wells. Studies on fat digestion, absorption and transport in the suckling rat. 1. Fatty acid composition and concentration of major lipis components. J Lipid Res 22:668–674, 1981.
35. J. E. Staggers, G. J. P. Fernando-Warnakulasuriya, M. A. Wells. Studies on fat digestion, absorption and transport in the suckling rat. 11. Triacylglycerols; molecular species, stereospecific analysis and specificity of hydrolysis of lingual lipase. J Lipid Res 22:675–679, 1981.
36. S. Bernback, L. Blackberg, O. Hernell. Fatty acids generated by gastric lipase promote human milk triacylglycerol digestion by pancreatic colipase dependent lipase. Biochim Biophys Acta 1001:286–293, 1989.
37. J. M. Linthorst, S. B. Clark, P. R. Holt. Triglyceride emulsification by amphipaths in the intestinal lumen during digestion of fat. J Colloid Interface Sci 60:1–10, 1977.
38. J. D. de Caro, P. Rouimi, M. Rovery. Hydrolysis of p-nitrophenyl acetate by the peptide chain fragment (336–449) of porcine pancreatic lipase. Eur J Biochem 158: 601–607, 1986.
39. G. C. Ferreira, J. S. Patton. Inhibition of lipolysis by hydrocarbons and fatty alcohols. J Lipid Res 31:889–897, 1990.
40. J. S. Patton, P. A. Albertsson, C. Erlanson, B. Borstrom. Binding of porcine pancreatic lipase and colipase in the absence of substrate studied by 2-phase partition and affinity chromatography. J Biol Chem 253:4195–4202, 1978.
41. A. Larsson, C. Erlanson-Albertson. Effect of phosphatidylcholine and free fatty acids on the activity of pancreatic lipase-colipase. Biochim Biophys Acta 876:543–550, 1986.
42. A. F. Hofmann, B. Borgstrom. Physico-chemical state of lipids in intestinal content during their digestion and absorption. Fed Proc 21:43–50, 1962.
43. A. F. Hofmann, B. Borgstrom. The intraluminal phase of fat digestion in man: the lipid content of the micellar and oil phases of intestinal content during fat digestion and absorption. J Clin Invest 43:247–257, 1964.
44. J. S. Patton, M. S. Carey. Watching fat digestion. Science 204:145–148, 1979.
45. P. R. Holt, B. M. Fairchild, J. Weiss. A liquid crystalline phase in human intestinal contents during fat digestion. Lipids 21:444–446, 1986.
46. M. W. Rigler, R. E. Honkanen, J. S. Patton. Visualisation by freeze fracture, in vitro and in vivo, of the products of fat digestion. J Lipid Res 27:836–857, 1986.
47. C. J. H. Porter, W. N. Charman. Uptake of drugs into the intestinal lymphatics after oral administration. Adv Drug Deliv Rev, 25, 71–89, 1997.
48. J. N. Hunt, M. T. Knox. A relation between the chain length of fatty acids and the slowing of gastric emptying. J Physiol 194:327–336, 1968.
49. T. Bates, J. Sequeira. Bioavailability of micronised grisofulvin from corn oil in water emulsion, aqueous suspension and commercial tablet dosage forms in humans. J Pharm Sci 65:793–797, 1975.
50. C. Veda, M. Lemoure, G. Gell, P. Missin, N. Nussbaumer. Apparent dose-depen-

dent oral absorption of cyclosporin A in rats. Biopharm Drug Disp 5:141–151, 1984.

51. C. Pouton, M. Wakerley, B. Meakin. Self-emulsifying systems for delivery of drugs. Proc Int Symp Cont Rel Bioact Mater 14:113–114, 1987.
52. C. J. H. Porter, S. Charman, W. Charman. Lymphatic transport of halofantrine in the triple-cannulated anaesthetised rat model; effect of lipid vehicle dispersion. J Pharm Sci 85:351–356, 1996.
53. M. Fischler, E. Frisch, B. Ortengren. Plasma concentrations after oral administration of different pharmaceutical preparations of clomethiazole. Acta Pharm Suec 10:483–492, 1973.
54. W. Charman, V. Stella. Effect of lipid class and lipid vehicle volume of the intestinal transport of DDT. Int J Pharm 33:165–172, 1986.
55. J. M. Laher, G. A. Chernenko, J. A. Barrowman. Studies on the absorption and enerohepatic circulation of 7,12-dimethylbenz(a)anthracene. J Physiol Pharmacol 51:1368–1373, 1983.
56. D. Bloedow, W. Hayton. Effects of lipids on bioavailability of sulphisoxazole dicumarol and griseofulvin. J Pharm Sci 65:328–334.
57. J. M. Laher, M. W. Rigler, R. D. Vetter, J. A. Barrowman, J. S. Patton. Similar bioavailability and lymphatic transport of benzo(a)pyrene when administered to rats in different amounts of dietary fat. J Lipid Res 25:1337–1342, 1984.
58. W. Charman, T. Noguchi, V. Stella. Testing potential dosage form strategies for intestinal lymphatic transport: studies in the rat. Int J Pharm 33:173–179, 1986.
59. W. Charman, T. Noguchi, V. Stella. An experimental system designed to study the in situ intestinal transport of lipophilic drugs in anaesthetized rats. Int J Pharm 33: 155–164, 1986.
60. T. J. DeMarco, R. Levine. Role of lymphatics in the intestinal absorption and distribution of drugs. J Pharm Exp Ther 169:142–151, 1969.
61. Y. Shaiu, D. Popper, M. Reed, C. Umstetter, D. Capuzzi, G. Levine. Intestinal triglycerides are derived from both endogenous and exogenous sources. Am J Physiol 248:G164–169 (1985).
62. T. J. Raub, S. L. Douglas, G. W. Melchior, W. N. Charman, W. Morozowich, W. (1992). Methodologies for assessing intestinal lymphatic transport. In: W. N. Charman, V. L. Stella eds. Lymphatic Transport of Drugs. Boca Raton: CRC Press, 1992, pp. 63–111.
63. C. J. H. Porter, W. N. Charman. Model systems for intestinal lymphatic transport studies. In: R. T. Borchardt, P. L. Smith, G. Wilson, eds. Model Systems for Biopharmaceutical Assessment of Drug Absorption and Metabolism. New York: Plenum Press, 1996, pp. 85–102.
64. S. M. Sieber. The entry of foreign compounds into the thoracic lymph duct of the rat. Xenobiotica 4:265–284, 1974.
65. S. M. Sieber. The lymphatic absorption of p,p′-DDT and some structurally related compounds in the rat. Pharmacology, 14:443–454, 1976.
66. W. Charman, T. Noguchi, V. Stella. Lymphatic appearance of DDT in thoracic and mesenteric lymph duct cannulated rats. Int J Pharm 24:185–192, 1985.
67. A. Armstrong, K. C. James. Drug release from lipid-based dosage forms 2. Int J Pharm 6:195–204, 1980.

68. H. J. Horst, W. J. Holtje, M. Dennis, A. Coert, J. Geelen, K. D. Voigt. Lymphatic absorption and metabolism of orally administered testosterone undecanoate in man. Klin Wochenschr 54:875–879, 1976.
69. D. C. Bloedow. Bioavailability of lipophilic drugs in lipid vehicles. PhD thesis, Washington State University, 1974.
70. A. J. Humberstone, W. N. Charman. Lipid-based vehicles for the oral delivery of poorly water soluble drugs. Adv Drug Deliv Rev 25:103–128, 1997.
71. J. T. Hargrove, W. S. Maxson, A. C. Wentz. Absorption of oral progesterone is influenced by vehicle and particle size. Am J Obstet Gynecol 161:948–951, 1989.
72. L. S. Abrams, H. S. Weintraub, J. E. Patrick, J. L. McGuire. Comparitive bioavailability of a lipophilic steroid. J Pharm Sci 67:1287–1290, 1978.
73. T. Tokumura, Y. Tsushima, K. Tatsuishi, K. Kayano, M. Machida, T. Nagai. J Pharm Sci 76:286–288, 1987.
74. V. Bakatselou, R. Oppenheim, J. Dressman. Solubilization and wetting effects of bile salts on the dissolution of steroids. Pharm Res 8:1461–1469, 1991.
75. L. J. Naylor, V. Bakatselou, N. Rodriguez-Hornedo, N. D. Weiner, J. Dressman. Dissolution of steroids in bile salt solutions is modified by the presence of lecithin. Eur J Pharm Biopharm 41:346–353, 1995.
76. S. D. Mithani, V. Bakatselou, C. N. TenHoor, J. B. Dressman. Estimation of the increase in solsubility as a function of bile salt concentration. Pharm Res 13:163–167, 1996.
77. C.-Y. Li, C. L. Zimmerman, T. S. Wiedmann. Solubilization of retinoids by bile salt/phospholipid aggregates. Pharm Res 13:907–913, 1996.
78. L. J. Soloman, J. K. Embleton, C. W. Pouton. Solubilization od steroidal compounds by mixed bile salt/lecithin micelles. In: P. Couvrier, D. Duchene, I. Kalles, eds. Formulation of Poorly Available Drugs for Oral Administration. Paris: Editions de sante, 1996, pp 219–230.
79. C. T. Ueda, M. Lemaire, G. Gsell, K. Nussbaumer. Intestinal lymphatic absorption of cyclosporin A following oral absorption in an olive oil solution to rats. Biopharm Drug Disp 4:113–124, 1983.
80. R. C. Grimus, I. Schuster. The role of the lymphatic transport in the enteral absorption of naftifine by the rat. Xenobiotica 14:287–294, 1984.
81. K. J. Palin, C. J. Wilson. The effect of different oils on the absorption of probucil in the rat. J Pharm Pharmacol 36:641–643, 1984.
82. D. L. Busbee, J.-S. H. Yoo, J. O. Norman, C. O. Joe. Polychorinated biphenyl uptake and transport by lymph and plasma components. Proc Soc Exp Biol Med 179:116–122, 1985.
83. A. Kuksis. Absorption of fat soluble vitamins. In: A. Kuksis, ed. Fat Absorption, Vol 2. Boca Raton: CRC Press, 1987, pp 65–86.
84. J. L. Bollman, J. C. Cava, J. H. Grindley. Techniques for the collection of lymph from the liver, small intestine and thoracic duct of the rat. J Lab Clin Med 33: 1349–1352, 1948.
85. E. J. Reininger, L. A. Saperstein. Effect of digestion and distribution on blood flow in the rat. Science 126:1176, 1957.
86. W. N. Charman, V. J. Stella. Estimating the maximal potential for intestinal lymphatic transport of lipophilic drug molecules. Int J Pharm 34:175–178, 1986.

87. C. J. H. Porter, S. Charman, A. Humberstone, W. N. Charman. Lymphatic transport of halofantrin in the conscious rat when administered as either the free base or the hydrochloride salt; effect of lipid class and lipid vehicle dispersion. J Pharm Sci 85:357–361, 1996.
88. K. Palin, C. Wilson, S. S. Davis, A. Philips. The effect of oils on the lymphatic absorption of DDT. J Pharm Pharmacol 34:707–710, 1982.
89. R. K. Ockner, J. M. Manning. Fatty acid binding protein in the small intestine: identification, isolation and evidence for its role in cellular fatty acid transport. J Clin Invest 54:326–338, 1974.
90. N. J. Greenberger, J. B. Rodgers, K. J. Isselbacher. Absorption of medium and long chain triglycerides: factors influencing their hydrolysis and transport. J Clin Invest 45: 217–227, 1966.
91. M. Cheema, K. Palin, S. S. Davis. Lipid vehicles for intestinal lymphatic drug transport. J Pharm Pharmacol 39:55–56, 1987.
92. N. Muranishi, M. Kinagawa, Y. Nakajima, S. Muranishi, H. Sezaki. Mechanism for the inducement of the intestinal absorption of poorly absorbed drugs by mixed micelles. 1: Effects of various lipid–bile salt mixed micelles on the intestinal absorption of streptomycin in the rat. Int J Pharm 4:271–279, 1986.
93. R. K. Ockner, J. P. Pittman, J. L. Yager. Differences in the intestinal absorption of saturated and unsaturated long chain fatty acids. Gastroenterology 62:981–992, 1972.
94. Y. Yamahira, T. Noguchi, H. Takenaka, T. Maeda. Biopharmaceutical studies of lipid containing oral dosage forms: relationship between drug absorption rate and digestibility of vehicles. Int J Pharm 3:23–31, 1979.
95. W. A. Ritschel, S. Adolph, G. B. Ritschel, T. Schroeder. Improvement of peroral absorption of cyclosporin by microemulsions. Meth Find Exp Clin Pharmacol 12: 127–134, 1990.
96. W. A. Ritschel. Microemulsions for improved peptide absorption from the gastrointestinal tract. Exp Clin Pharmacol 13:205–220, 1991.
97. R. A. Myers, V. J. Stella. Systemic bioavailability of penclomedine (NSC-338720) from oil-in-water emulsions administered intraduodenally to rats. Int J Pharm 78: 217–226, 1992.
98. T. Kimura, E. Fukui, A. Kageyu, Y. Kurohara, T. Nakayama, Y. Morita, K. Shibusawa, S. Ohsawa, Y. Takeda. Enhancement of oral bioavailability of D-α-tocopherol acetate by lecithin-dispersed aqueous preparations containing medium chain triglycerides. Chem Pharm Bull 37:439–441, 1989.
99. B. D. Tarr, S. H. Talkowsky. Enhanced intestinal absorption of cyclosporin in rats through the reduction of emulsion droplet size. Pharm Res 6:40–43, 1989.
100. E. A. Mueller, J. M. Kovarik, J. B. Van Bree, W. Tetzloff, K. Kutz. Improved dose linearity of cyclosporin pharmacokinetics from a microemulsion formulation. Pharm Res 11:301–304, 1994.
101. J. M. Kovarik, E. A. Mueller, J. B. Van Bree, W. Tetzloff, K. Kutz. Reduced inter- and intraindividual variation in cyclosporin pharmacokinetics from a microemulsion formulation. J Pharm Sci 84:444–446, 1994.
102. E. A. Mueller, J. M. Kovarik, J. B. Van Bree, J. Grevel, K. Kutz. Influence of a fat-rich meal on the pharmacokinetics of a new oral formulation of cyclosporin in

a crossover comparison with the market formulation. Pharm Res 11:151–155, 1994.
103. J. H. Lin, I. Chen, H. Lievens. The effect of dosage forms on the oral absorption of L-365 260, a potent CCK_B receptor antagonist in dogs. Pharm Res 8:S-272, 1991.
104. N. H. Shah, M. T. Carvajal, C. J. Patel, M. H. Infeld, A. W. Malick. Self-emulsifying drug delivery systems (SEDDS) with polyglycolysed glycerides for improving in vitro dissolution and oral absorption of lipophilic drugs. Int J Pharm 106:15–23, 1994.
105. K. J. MacGregor, J. K. Embleton, J. E. Lacy, E. A. Perry, L. J. Solomon, H. Seager, C. W. Pouton. Influence of lipolysis on drug absorption from the gastrointestinal tract. Adv Drug Deliv Rev 25:33–46, 1997.
106. L. J. Solomon, J. K. Embleton, C. W. Pouton. Inhibition of lipolysis of medium chain triglycerides by non-ionic surfactants; a structure/activity study. In: P. Couvreur, D. Duchene, I. Kalles, eds. Formulation of Poorly Available Drugs for Oral Administration. Paris: Editions de sante, 1996, pp. 437–438.
107. H. Liu, I. Adachi, I. Horkoshi, M. Ueno. Promotion of intestinal drug absorption by milk fat globule membrane. Yakugaku Zasshi 111:510–514, 1991.
108. I. Adachi, H. Liu, I. Horkoshi, M. Ueno. Possibility of lymphatic absorption of epidermal growth factor from intestine. Yakugaku Zasshi 113:256–263, 1993.
109. T. Moriwaki, H. Yoshikawa, K. Takada, S. Muranishi. DDS with MFGM (milk fat globule membrane): gastrointestinal absorption of vitamin A acetate. Yakuzaikaku 50:263–267, 1990.
110. T. Honda, M. Okamoto, M. Maeda, T. Kawagoshi, H. Nihonmatsu, I. Yamashita, M. Yoshida, K. Tazawa, M. Fujimaki. DDS with MFGM (milk fat globule membrane): intestinal absorption of MFGM-insulin emulsion. Drug Deliv Syst 7:43–46, 1992.
111. H.-X. Liu, I. Adachi, I. Horikoshi, M. Ueno. Mechanism of promotion of lymphatic drug absorption by milk fat globule membrane. Int J Pharm 118:55–64, 1995.
112. L. L. Gan, D. R. Thakker. Applications of the Caco-2 model in the design and development of orally active drugs: elucidation of the biochemical and physical barriers posed by the gastrointestinal tract. Adv Drug Deliv Rev 23:77–98, 1997.

11
Dry Adsorbed Emulsions

Michel Rollet and Joëlle Bardon
Laboratoire de Pharmacie Galénique Industrielle et Biogalénique
Université Claude Bernard Lyon I, Lyon, France

Historically, the most convenient and commonly employed route of drug delivery has been by oral ingestion. According to a recent review of Ranade and Hollinger (1), the original controlled release of pharmaceuticals was through coated pills. Coating technology advanced with the discovery of gelatin and sugar coatings. The introduction of the concept of coating drugs containing beads with a combination of fats and waxes was a major development in coating technology. Since the mid-1900s, hundreds of publications and nearly a thousand patents have appeared on various oral delivery process including delayed, prolonged, sustained, and, most recently, controlled release of the active substance.

Recently, a new oral sustained drug delivery system has been proposed. It is referred to as dry adsorbed emulsion (2,3). To form a dry adsorbed emulsion, a water-in-oil (W/O) emulsion was prepared. The aqueous phase of the W/O emulsion contained a hydrophilic active drug. After the emulsion preparation, both phases of the W/O emulsion were stabilized by two inactive powders. In previous works, the first powder was a highly polar silica and could absorb the aqueous phase. The second powder was a hydrocarbon-bonded silica with a low polarity compatible with the oil phase of the W/O emulsion. The finished form was a fluid powder whose particle size diameter ranged from 300 to 2000 μm. Sodium salicylate was used as a model of hydrophilic active drug (2–6). The experiment showed the following:

1. To obtain an important sustained release effect, two silica must be used: a hydrophilic one (a silica gel) and a hydrophobic one (an alkyl-bonded silica) (2,4).
2. The role of each silica was defined as follows. The aqueous phase of the W/O liquid emulsion and the active drug were absorbed by the hydrophilic silica that produced a creamy liquid. The addition of the hydrophobic silica changed the creamy liquid into a fluid powder, i.e., the dry adsorbed emulsion (3).
3. To represent the kinetics of drug release in vitro, a theoretical model was derived from the Fick diffusion equation. This model is based on a numerical method with finite differences (5).
4. The physicochemical structure was described and observed by electron microscopy (6).
5. The dry adsorbed emulsion is stable for a year stored in darkness at room temperature in a closed flask (6).

I. PREPARATION OF A DRY ADSORBED EMULSION

A. Method Initially Proposed by the Inventors

The first step of the preparation of a dry adsorbed emulsion was to make a primary W/O emulsion by homogenizing the oil phase and the drug-containing water

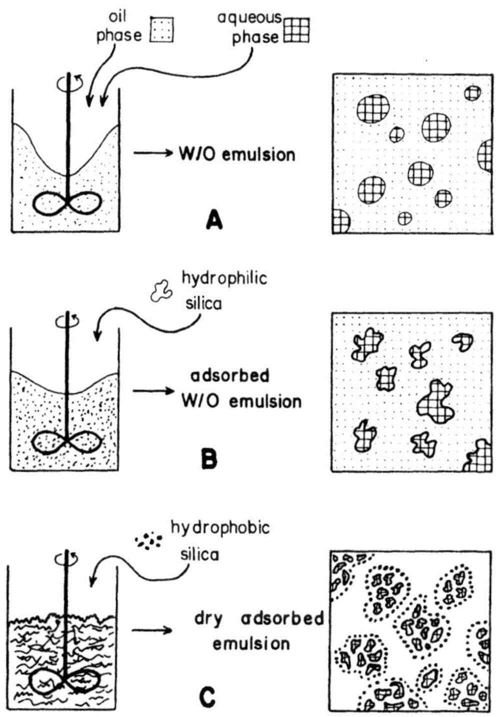

Figure 1 Steps of the dry adsorbed emulsion preparation. (Reproduced with permission of the American Pharmaceutical Association. *Source*: Ref. 6.)

phase at 60°C with a 12,000 rpm stirring rate (Fig. 1A). The composition of each phase is presented in Table 1. During this step, lime water reacted with stearic acid to form enough calcium stearate to stabilize the W/O emulsion. The melting point of pure stearic acid is 71.2°C; therefore, it was necessary to work at about 60–70°C to maintain the desired liquid W/O emulsion characteristics.

The second step was to blend the W/O emulsion with the hydrophilic silica at 60–70°C with a 12,000 rpm stirring rate. The step produced a creamy liquid (Fig. 1B).

The last step was to add with a 5000 rpm stirring rate the hydrophobic silica to the creamy liquid at 60°C. This step progressively turned the liquid into a paste and eventually into a fluid powder, the dry adsorbed emulsion (Fig. 1C). This procedure is covered by a French patent (2). The powder was sieved and three particle size classes were obtained [315–800 μm (25 ± 2% in weight of initial raw form), 800–1250 μm (27 ± 2%), and 1250–2000 μm (31 ± 2%)] for

Table 1 Dry Adsorbed Emulsion Composition

Preparation	Constituent	Compound	mass g	
Primary emulsion 1	Oil phase	Castor oil	60	
		Stearic acid	20	
	Aqueous phase	Lime water	20	
		Sodium salicylate	10	
Primary emulsion 2	Oil phase	Silicone oil	60	
		Stearic acid	20	
	Aqueous phase	Lime water	20	
		Sodium salicylate	10	
Primary emulsion 3	Oil phase	Silicone oil	60	
		Stearic acid	40	
	Aqueous phase	Lime water	20	
		Sodium salicylate	10	
Form 1	Primary emulsion 1	Tixosil 38	11	*P*[a]
	110 g	Tr. Aerosil	22	2.6×10^9
Form 2	Primary emulsion 1	Tixosil 38	11	
	110 g	Sipernat D17	22	13
Form 3	Primary emulsion 2	Tixosil 38	11	
	110 g	Tr. Aerosil	23,1	2.7×10^9
Form 4	Primary emulsion 2	Tixosil 38	11	
	110 g	Sipernat D17	27,5	16
Form 5	Primary emulsion 3	Tixosil 38	13	
	130 g	Tr. Aerosil	27,3	2.8×10^9

[a] Particle ratio parameter.
Source: Ref. 6. (Reproduced with permission of the American Pharmaceutical Association.)

the five different forms shown in Table 1. The particles > 2000 μm or < 315 μm (<17%) were disregarded.

It must be pointed out that a fluid powder has a higher surface than a paste and will disperse more easily in the release medium. A paste seems to be the best form to obtain a maximum sustained release effect. However, as a powder is easier to handle than a paste, and as the studied form is an intermediate form to obtain that will be fitted in capsules or compressed to obtain tablets or pills, the fluid powder was prepared and studied.

B. Modifications of the Initial Method

Modifications were brought to the initial method by the inventors and other authors (7) later:

1. Oil phase: castor oil, sunflower oil, silicone oil
2. Emulsifying agent: glyceryl monostearate, calcium stearate, Tween-80
3. Hydrophilic adsorbents: silica, Eudragit S, pectin, chitosan
4. The particles of dry adsorbed emulsions obtained undergo an ultimate process of drying, such as infrared radiation, freeze drying, or tray oven drying (drying in cabinets with a circulating air). This step was carried out to increase the chemcial stability of active ingredients, to change the physicochemical structure of the dry adsorbed emulsion (porosity for instance), to change the drug release kinetics, and to facilitate the manufacture process of solid oral dosage forms (such as tablets and hard gelatin capsules) in which particles of dry adsorbed emulsion will be incorporated.
5. Pharmaceutical form for administration of dry adsorbed emulsion particles: tablets or hard gelatin capsules.

Several works about these modifications have been made by the inventors, but not yet published.

II. PHYSICOCHEMICAL STRUCTURE OF THE FORM

Farah et al. (3) have showed that an important physicochemical parameter of a dry adsorbed emulsion is the particle ratio, P. This ratio P is the number of hydrophobic particles, N_{L}, divided by the number of hydrophilic particles, N_{H}.

$$P = \frac{N_{\mathrm{L}}}{N_{\mathrm{H}}} = \frac{m_{\mathrm{L}} n_{\mathrm{L}}}{m_{\mathrm{H}} n_{\mathrm{H}}} \qquad [1]$$

where m and n are the mass and number of particles per gram of adsorbent (silica), respectively; subscripts L and H refer to hydrophobic and hydrophilic silica, respectively; and P represents the mean number of hydrophobic particles surrounding one hydrophilic particle. Using the physicochemical parameters of each silica (Table 2), the particle ratio, P, can be expressed by

$$P = \left(\frac{m_{\mathrm{L}}}{m_{\mathrm{H}}}\right)\left(\frac{D_{\mathrm{H}}^3}{D_{\mathrm{L}}^3}\right)\left(\frac{\rho_{\mathrm{H}}(1 + \phi_{\mathrm{L}}\rho_{\mathrm{L}})}{\rho_{\mathrm{L}}(1 + \phi_{\mathrm{H}}\rho_{\mathrm{H}})}\right) \qquad [2]$$

where D, ρ, and ϕ are the particle diameter, density, and porosity of each silica, respectively.

Equation [2] shows that the particle diameter of the silica has more effect on the particle ratio, P, than the mass. Table 1 presents the P values obtained for each form prepared. The value of 2.2 g/cm^3 for the density of both silicas, ρ_{H} and ρ_{L}, was used for calculations: 2.2 g/cm^3 is the density of precipitated

Table 2 Physicochemical Properties of the Silica used and the Five Systems Prepared

Product	Mean particle size diameter (μm)	Specific surface area (m^2/g)	C_{BET}[a]	Porosity (cm^3/g)	%C[b]	n[c]
Tixosil 38	18	160	96	0.7	0	4×10^8
Treated Aerosil	0.012	230	29	0	3.1	5×10^{17}
Sipernat D17	9	100[d]	—	0.5	2	3×10^9
Form 1	600[f]	0.3	8	~0	50	7×10^3
Form 2	600[df]	0.3[d]	e	~0	47	7×10^3
Form 3	600[df]	0.3[d]	e	~0	27	6×10^3
Form 4	500[df]	0.6[d]	e	~0	26	6×10^3
Form 5	500[f]	0.6	38	~0	34	7×10^3

[a] Surface polarity constant produced by BET method.
[b] Carbon percentage.
[c] Mean number of particles in 1 g of product.
[d] Value given by the manufacturer (Sipernat D17) or estimated.
[e] The surface area was too small to allow a BET determination.
[f] Particle size class studied: 315–800 μm.
Forms composition are given in Table 1.
Source: Ref. 6. (Reproduced with permission of the American Pharmaceutical Association.)

amorphous silica (lechatelierite) (8). Given the low carbon content of bonded silica (<3.1%; Table 2), this value was also used for bonded silica. Forms 2 and 4 combined two silicas of similar size with about 15 hydrophobic particles per hydrophilic one. Forms 1, 3 and 5 combined a very thin hydrophobic silica with the hydrophilic silica of forms 2 and 4. This produced a huge particle size ratio parameter (about 2.7×10^9). The hydrophilic silica particles were not surrounded by the hydrophobic ones; they were almost coated (3).

It must be noted that the particle ratio, P, is theoretical. Given the particle size of the finished form, it is obvious that hydrophobic silica particles do not surround only one hydrophilic particle; they definitely surround an oily droplet of dried emulsion. This droplet may contain several hydrophilic silica particles (Fig. 1C).

III. DRUG RELEASE

Drug release studies of dry adsorbed emulsion particles were carried out as described below (Appendix I).

The characteristics of the sustained release effect of a dry adsorbed emulsion are shown in Table 3; the t_{20}, t_{50}, and t_{90} values, the time at which 20%, 50%, and 90%, respectively, of the initial amount of drug in each form was released, are presented. For each form shown in Table 1, the three particle size classes were studied in artificial gastric (pH 1.2) and intestinal (pH 7.4) media.

It was shown (3) that dry adsorbed emulsions could be assimilated to microcapsules in which the internal drug concentration decreased with time. So, as reviewed by Baker and Lonsdale (9), the drug release kinetics can be represented by an exponential relation:

$$L = L_0 + (100 - L_0)(1 - \exp^{(-kt)}) \qquad [3]$$

where L is the percentage of sodium salicylate released at time t, L_0 is the percentage of sodium salicylate released very rapidly at the beginning of the measurement (in under 5 min), and k is a kinetic constant (min^{-1}).

It is also possible to consider a dry adsorbed emulsion as a nonerodible matrix. The drug release can then be represented by a diffusion-controlled process (5). A common way to represent diffusion-controlled drug release is to use the Higuchi law (10,11)

$$L = 200\left(\phi\frac{S}{V}\right)\left(\frac{D}{\sigma}\right)^{1/2} t^{1/2} \qquad [4]$$

where ϕ is the sample porosity (cm^3/g), S is the diffusion surface (cm^2/g), V is the sample volume (cm^3/g), σ is the so-called tortuosity of the sample (cm^3/g), and D is the drug diffusion coefficient (cm^2/g). This equation is not valid for large values of time, since the highest L value is obviously 100%.

To fit our results it was possible to use a simplified form of Eq [4]:

$$L = L_0' + k't^{1/2} \qquad [5]$$

where L_0' is the percentage of drug released in under 5 min and k' is a kinetic constant (% min^{-1}) including the parameter of Eq. [4].

The k and L_0 values were obtained by using a computer fitting method (Appendix II). Plots of L vs. the square root of time (but with t less than 180 min) for Eq. [5] gave straight line with correlation coefficients greater than 0.98 in all-studied cases. Table 3 lists the values of k, k', L_0, and L_0' and the correlation coefficients for both models. Figure 2 shows the experimental points corresponding to form 1, large particles, and the fitted curves obtained with the exponential model (full lines) and the Higuchi model (dotted lines) for both pH values. For times under 20 min, the Higuchi model fits the experimental points in a better way than the exponential model. Between 20 min and 3 h, both models are suitable to

Table 3 Kinetic Parameters of the Sustained Release Effect at 37°C

pH	Form[a]	Particle size[b]	t_{20} (min)	t_{50} (min)	t_{90} (min)	Exponential[c] k (10^3) (min^{-1})	Exponential[c] L_0 (%)	Square root[d] k' (% min^{-1})	Square root[d] L'_0 (%)	r^e
1.2	1	m	5	59	280	7.5	20	5.7	6	0.996
		l	22	108	385	5.3	10	5.2	0	0.994
	2	m	5	68	312	6.7	19	5.3	6	0.992
		l	28	138	538	4.1	9	4.6	0	0.991
	3	s	6	42	176	12	17	4.6	25	0.984
		m	4	71	316	6.5	22	4.3	14	0.994
		l	23	137	495	4.4	11.5	4.3	0	0.998
	4	s	2	31	194	10	30	4.7	26	0.991
		m	6	78	298	7.1	17	4.6	9	0.998
		l	26	151	560	3.9	11.5	4.1	0	0.998
	5	s	2	43	285	6.8	31	3.9	26	0.990
		m	5	100	450	4.6	20	3.9	10.5	0.992
		l	21	281	1180	1.8	17	3.0	0	0.997
7.4	1	m	1	15	130	14	38	4.7	33	0.979
		l	1	44	233	8.5	27	6.0	10	0.987
	2	m	1	23	146	13	33	4.7	30	0.982
		l	4	83	360	5.8	19	4.6	11	0.988
	3	s	1	3	93	18	46	6.1	39	0.975
		m	2	28	145	14	24	7.6	12	0.968
		l	4	58	320	6.6	18	7.4	1	0.981
	4	s	0	1	94	17	50	5.3	38	0.985
		m	1	23	191	9.6	37	5.2	24	0.988
		l	3	78	350	6.0	20	4.7	9	0.996
	5	s	0	4	95	18	45	6.1	38	0.980
		m	2	28	145	14	24	7.6	12	0.983
		l	4	59	315	6.6	18	7.4	0	0.991
1.2	1′[f]	m	0	25	140	15	30	5.8	23	0.992
		l	1	58	270	8	22	4.9	15	0.994
7.4	1′[f]	m	0	11	100	20	39	4.2	45	0.981
		l	0	20	170	11	39	4.2	35	0.987

[a] See Table 1.
[b] Particle size: s = small, m = medium, l = large
[c] Equation [3].
[d] Equation [5].
[e] For t values lower than 180 min.
[f] Form 1 prepared using an alternative procedure (see text).
Source: Ref. 6. (Reproduced with permission of the American Pharmaceutical Association.)

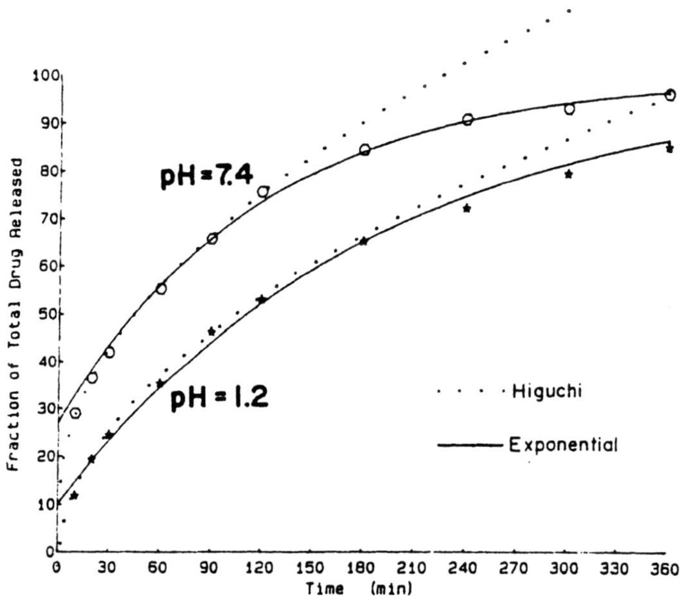

Figure 2 Sodium salicylate release by form 1 (large particles). Each experimental point is a mean of six measurements. Dotted lines: best fit with Higuchi law (Eq. [5]). Full lines: best fit with the exponential relation (Eq. [3]). Temperature 37°C. *Source*: Ref. 6. (Reproduced with permission of the American Pharmaceutical Association.)

represent the experimental results. For times greater than 3 h, only the exponential model fits the experimental points (Fig. 2).

For a given form, the k and k' values are different because the units are different. Yet the trend with both models is the same. The values of L_0 and L_0' should be identical. They are not because the computer fitting (Appendix II) was better with Eq. [3] for large values of time ($t > 30$ min Fig. 2), but better with Eq. [5] for short times ($t < 30$ min). It seems that the dry adsorbed emulsions act neither as pure matrixes (Higuchi law) nor as pure microcapsules (exponential relation). They present an intermediate behavior with at least two release processes. In one, a part of drug is released very rapidly, and this corresponds to the burst effect described by Baker and Lonsdale [9] (represented by the L_0 and L_0' values). In the second process, the remaining drug is released in a diffusion-controlled process (Higuchi law) and with first-order kinetics (9) (exponential law).

IV. PARTICLE SIZE EFFECT

The bigger the particle of dry adsorbed emulsion, the higher the sustained release effect. The half-release time t_{50} is from three times (form 3, gastric medium) up to 70 times (form 4, intestinal medium) longer with large particles than with small particles (Table 3). The hydrophilic silica and the diffusing drug were covered by a thicker hydrophobic layer in the largest particles than in the smallest ones. Figure 3 shows the particle size and pH effect on the sodium salicylate release by dry adsorbed emulsion 5.

The silica particle size effect is less important. Systems with a high particle ratio (forms 1, 3, and 5; Table 1) produced a sustained-release effect comparable to systems with a low particle ratio (forms 2 and 4; Table 1). The main role of the hydrophobic silica seems to be to separate each droplet of dry adsorbed emulsion by sticking to the oil phase surface. The hydrophobic silica does not seem to have another role in the sustained release effect.

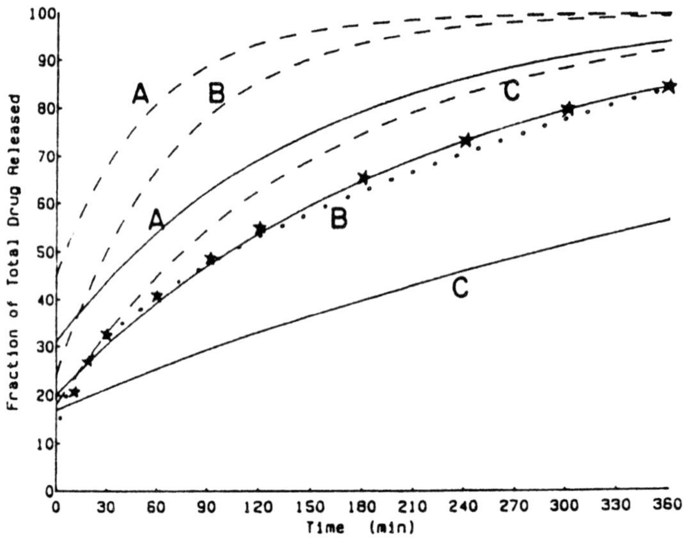

Figure 3 Effect of the particle size and pH on the sodium salicylate release by a dry adsorbed emulsion (form 5). Full lines: pH 1.2. Dashed lines: pH 7.4. Key: (A) 315–800 μm; (B) 800–1250 μm; (C) 1250–2000 μm. Curves fitted used Eq. [3] and data from Table 3, but dotted line was obtained with Eq. [5]. Stars correspond to experimental points. Temperature 37°C. *Source*: Ref. 6. (Reproduced with permission of the American Pharmaceutical Association.)

V. OIL PHASE EFFECT

The oil phase of the primary W/O emulsion is primarily responsible for the sustained release effect. It coats the hydrophilic particles that absorbed the aqueous phase and the active drug.

Drug release was faster in artificial intestinal medium than in gastric medium (Table 3; Figs. 2 and 3). At pH 7.4, stearic acid can be slowly ionized in stearic ion, which is more soluble in water; this could make the oil phase more permeable to the hydrophilic drug.

The replacement of castor oil (forms 3, 4, and 5) produced a slightly longer half-release time (Table 3) in acidic medium. The increase of the stearic content (form 3: 25%; form 5: 40%) in the oil phase produced a dramatic increase of the t_{50} and t_{90} release times in acidic medium (Fig. 4). It must be noted that the oil phase of the primary emulsion 2 (silicone oil/stearic acid, 75%:25%) was very viscous but still liquid at 37°C. The oil phase of the primary emulsion 3 (silicone oil/stearic acid, 60%:40%) was solid at 37°C. This could explain the slower diffusion of sodium salicylate in the solid stearic-rich phase. In the intesti-

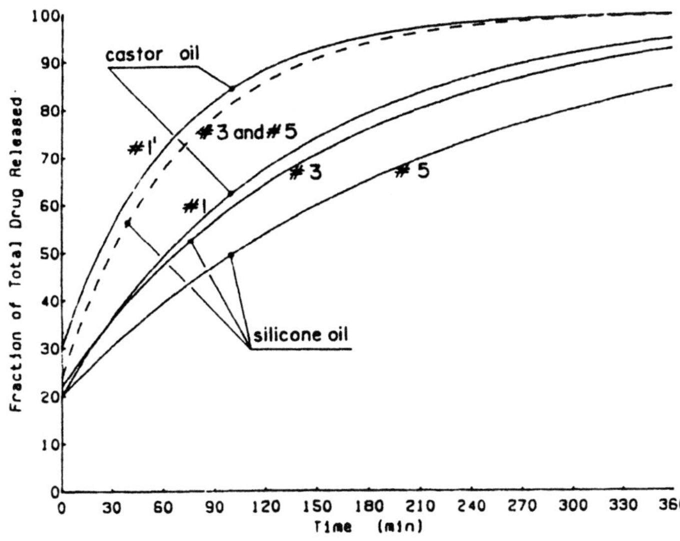

Figure 4 Effect of the oil phase and the alternative procedure (form 1′) on the sodium salicylate release. Particle size 800–1250 μm; temperature 37°C Full lines: pH 1.2. Dashed lines: pH 7.4. Curves fitted using Eq. [3] and data from Table 3. *Source*: Ref. 6. (Reproduced with permission of the American Pharmaceutical Association.)

nal medium, this effect disappeared; the stearic acid can be ionized in stearate ion, which makes the coating permeable. If the coating oil phase is richer in stearic acid, this ionization process is faster and deeper. Then the oil phase becomes more permeable to the hydrophilic drug and the release rate increases. This effect compensated for the favorable effect produced by the viscosity increase.

VI. ROLE OF HYDROPHILIC SILICA

The preparation of a dry adsorbed emulsion without hydrophilic silica (i.e., the hydrophobic silica was directly added to the primary emulsion) led to a final product that looked very similar to a genuine dry adsorbed emulsion (2). However, the sustained release effect was very poor ($t_{90} < 120$ min at pH 1.2).

To explain the hydrophilic role, the protocol of the dry adsorbed emulsion preparation was modified as follows. Instead of preparing the primary emulsion, the hydrophilic silica (Tixosil 38) and the aqueous phase (Table 1) were mixed in a beaker at room temperature. The whole aqueous solution was absorbed by the hydrophilic silica and this wet powder was added to the oil phase at 60°C. After this first step, the hydrophobic silica was mixed to the creamy liquid, following the normal procedure. Table 3 (bottom) presents the kinetic parameters obtained with the dry adsorbed emulsion prepared with the alternative procedure (form 1′). Figure 4 (curve 1′) presents the release of sodium salicylate vs. time for the medium particle size sample (800–1250 μm) at pH 1.2 compared with the results obtained with other dry adsorbed emulsions. The drug release was faster and the amounts of drug released very rapidly, L_0 and L_0' were higher.

Once the aqueous phase was absorbed by the hydrophilic silica, the wet silica was examined with an electron microscope (6) (Fig. 5) and found to have crystallites on the surface of the hydrophilic silica. Given the high concentration of the sodium salicylate in the aqueous phase (33% w/w or 2.1 mol/L), it seems that some parts of sodium salicylate were excluded from the silica interior and crystallized onto the silica surface. This could explain the observed burst effect (9), i.e., the amount of drug very rapidly released, L_0 and L_0' values. It was more difficult to coat these crystallites with the oil phase. If they remained uncoated, they could be dissolved in the release media very rapidly.

Figure 6 (6) presents an electron micrograph of a particle of form 3 (315–800 μm). At the magnification used (×1000), the hydrophobic particles of Aerosil were not visible (particle size diameter of Aerosil = 12 nm). The surface looks smooth compared with the tormented surface of the hydrophilic silica (Fig. 5). The oil phase and the hydrophobic silica coated the hydrophilic one. All of the silica pores were filled or clogged by the oil phase, producing a smooth surface (Fig. 6) and a low polarity and specific surface area (Table 2).

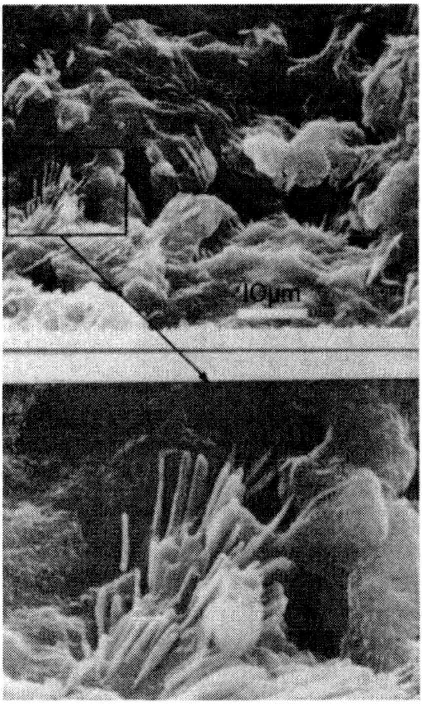

Figure 5 Electron micrograph of the hydrophilic silica after aqueous phase absorption. Crystallites of sodium salicylate can be seen. Scale bar = 10 μm. *Source*: Ref. 6. (Reproduced with permission of the American Pharmaceutical Association.)

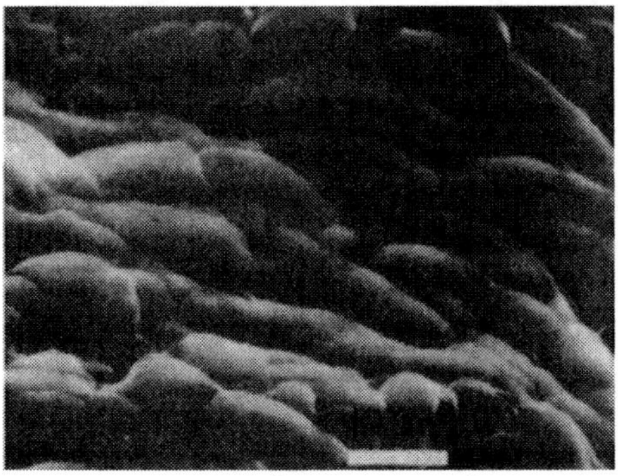

Figure 6 Electron micrograph of a particle of small size (315–800 μm) of form 3. Scale bar = 10 μm. *Source*: Ref. 6. (Reproduced with permission of the American Pharmaceutical Association.)

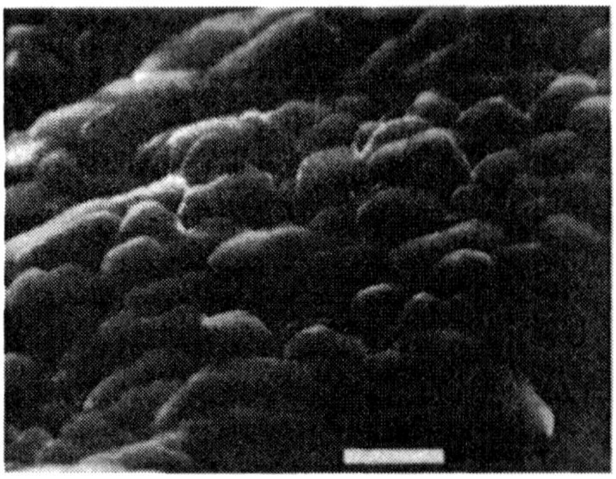

Figure 7 Electron micrograph of a particle of small size (315–800 μm) of form 3 after a release experiment. The matrix is not modified. Scale bar = 10 μm. *Source*: Ref. 6. (Reproduced with permission of the American Pharmaceutical Association.)

Figure 7 presents an electron micrograph of a particle of form 3 (315–800 μm) after the dissolution study. Figure 7 looks very similar to Fig. 6. So, the system is not modified (erosion, swelling) by the dissolution process. Indeed, the Higuchi law can be used only if the matrix is not modified by the release experiments.

VII. PREPARATION AGING

To study the stability of dry adsorbed emulsions vs. time, two samples of form 1 were put in closed flasks and stored at room temperature in a dark closet. After a year, the product aspect looked unchanged, and the drug release study gave results comparable to those obtained with a freshly prepared form. At pH 1.2 and for the 1-year-old form, the drug release was 11% and 6% faster for the medium (800–1250 μm) and large (1250–2000 μm) particle size forms, respectively. The half-release times, t_{50}, were 52 and 102 min for medium and large particle size, respectively, as compared with the values 59 and 108 min shown in Table 3 for freshly prepared forms.

At pH 7.4 and for the 1-year-old form, the drug release was 23% and 7% faster for the medium and large particle size, respectively. The t_{50} values were 11.5 and 41 min, respectively, as compared with the values of 15 and 44 min (Table 3) for freshly prepared forms. The larger the particle size, the greater the time stability.

VIII. PHARMACEUTICAL APPLICATIONS

The dry adsorbed emulsions have been formulated to be introduced into solid oral dosage forms such as tablets and hard gelatin capsules. The aim is to obtain a sustained release form of an active ingredient effective in small doses and characterized by a short half-time.

Other routes of administration are theoretically possible such as rectal route (rectal capsules and rectal suppositories) for drugs having systemic effects such as sedatives, tranquilizers, and analgesics; topical route (patchs) for drug absorption (hormones or nicotine) into the systemic circulation; parenteral route (administration of parenteral suspensions or implantation of compressed pellets).

A pharmaceutical application of dry adsorbed emulsions for oral route has been made with chlorpheniramine maleate (7), one of the most potent antihistaminic drugs. It is effective in small doses (4–6 mg) and is characterized by a long half-life (20 h). In general, only drugs with short-half lives of less than 6–8 h are considered for sustained action dosage forms. Formulations of chlorpheniramine maleate in doses greater than 4 mg require controlled release to prevent adverse effect brought about by high plasma concentration fluctuations.

A. General Formulation

Sunflower oil was selected as the oil phase. It may be associated with beeswax. The aqueous phase was distilled water or lime water containing chlorpheniramine maleate and glyceryl monostearate with or without Tween-80.

Different types of hydrophilic adsorbents with different physicochemical properties were chosen: silica is an inorganic uncharged adsorbent, Eudragit S is a synthetic anionic, whereas chitosan and pectin are cationic and anionic natural polymers, respectively.

Table 4 represents the dry adsorbed emulsion composition with chlorpheniramine maleate. Aerosil was used to cover the oil phase.

No interaction was found between the hydrophilic adsorbents studied and the drug over a 24-h period, except for the silica gel with which some adsorption ranging from 19.5% to 30% of the drug took place, according to the ratio of the adsorbent and adsorbate as well as the contact time.

Different properties of the prepared dry adsorbed emulsions were examined (yield value, determination of dry content, measurement of angle of repose, determination of flow rate, density, bulk density, percentage of porosity, particle size analysis, Karl Fisher titrimetry, drug release study, stability study) and the most appropriate selected for in vivo investigation.

B. Drug Release

Figures 8 and 9 give a representative example of the release profile of chlorpheniramine maleate from dry adsorbed emulsion using silica and Eudragit S,

Table 4 Chlorpheniramine Maleate Dry-Adsorbed Emulsion Composition

Preparation	Constituent	Compound	Weight
Primary emulsion 1	Oil phase	Sunflower oil	80 mL
	Aqueous phase	Distilled water	20 mL
Primary emulsion 2	Oil phase	Sunflower oil	40 mL
		Beeswax	38.4 g
	Aqueous phase	Distilled water	20 mL
Primary emulsion 3	Oil phase	Sunflower oil	40 mL
		Beeswax	38.4 g
	Aqueous phase	Lime water	20 mL
Form 1	Primary emulsion 1	Hydrophilic silica	20 g
		Aerosil	33 g
Form 2	Primary emulsion 1	Eudragit S	40 g
		Aerosil	33 g
Form 3	Primary emulsion 1	Chitosan	60 g
		Aerosil	33 g
Form 4	Primary emulsion 1	Pectin	50 g
		Aerosil	33 g
Form 5	Primary emulsion 2	Eudragit S	40 g
		Aerosil	33 g

Source: Ref. 7.

respectively, while Table 5 shows the time at which 20%, 50%, and 90% (t_{20}, t_{50}, and t_{90}) of the drug was released. The release of chlorpheniramine maleate from silica-based dry adsorbed emulsion (Fig. 8) is slower than that from chlorpheniramine maleate deposited on silica. The drug release from the powders was faster at pH 1.2 followed by that at pH 6.8 and lastly that at pH 7.4. This difference in drug release might be due to the pH-dependent ionization of the drug.

The dissolution rate of drug from pure chlorpheniramine maleate powder is higher than that from a dry adsorbed emulsion as well as that adsorbed on silica. The release of drug from chitosan (cationic)-based dry adsorbed emulsion or pectin (anionic) at different-pH values showed the same sequence as that using silica. This suggested that the controlling factor is still only the solubility of the drug itself. However, the release results of chlorpheniramine maleate from Eudragit S–based dry adsorbed emulsion (Fig. 9 and Table 5) indicated that the drug release rate was dependent not only on the solubility of the drug itself but on the solubility of the polymer in the dissolution media (Eudragit S is soluble in alkaline pH). Table 5 shows that, in general, the highest release rate was obtained with Eudragit S–based dry adsorbed emulsions and the lowest with silica-based dry adsorbed emulsions.

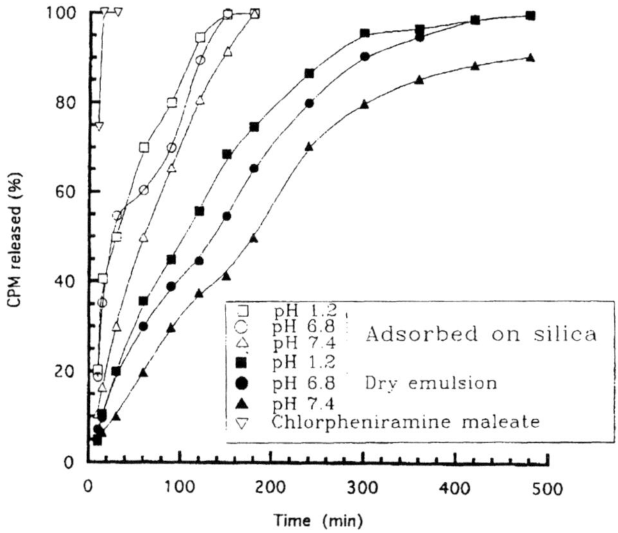

Figure 8 Release profiles of chlorpheniramine maleate from a dry adsorbed emulsion with silica as hydrophilic adsorbent and chlorpheniramine maleate powder adsorbed on silica. *Source*: Ref. 7.

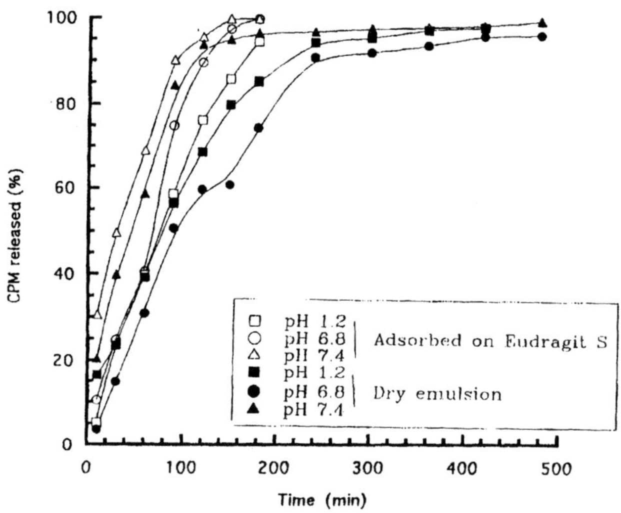

Figure 9 Release profiles of chlorpheniramine maleate from a dry adsorbed emulsion with Eudragit S as hydrophilic adsorbent and chlorpheniramine maleate powder adsorbed on Eudragit S. *Source*: Ref. 7.

Table 5 Kinetic Parameters of the Sustained Release Chlorpheniramine Maleate

pH of dissolution medium	Type of hydrophilic adsorbent	t_{20}	t_{50}	t_{90}
1.2	Silica	30	105	270
	Silica (50%)[a]	40	135	420
	Silica (50%)[b]	50	240	>480
	Eudragit S	30	73	200
	Eudragit S (50%)[a]	30	90	420
	Eudragit S (50%)[b]	46	100	>480
	Chitosan	10	90	200
	Pectin	50	100	180
6.8	Silica	30	135	300
	Silica (50%)[a]	45	180	>480
	Silica (50%)[b]	90	240	>480
	Eudragit S	15	90	240
	Eudragit S (50%)[a]	20	100	300
	Eudragit S (50%)[b]	30	135	>480
	Chitosan	20	135	360
	Pectin	15	120	240
7.4	Silica	60	180	480
	Silica (50%)[a]	75	210	>480
	Silica (50%)[b]	120	300	>480
	Eudragit S	10	50	100
	Eudragit S (50%)[a]	13	75	300
	Eudragit S (50%)[b]	20	90	>480
	Chitosan	30	180	360
	Pectin	20	140	>480

[a] Beeswax (%).
[b] Lime water as aqueous phase.
Source: Ref. 7.

The partial replacement of oil by beeswax in one formula, and of distilled water by lime water and of oil by beeswax in another (Fig. 10), retarded the drug release. This could be explained by the fact that beeswax increased the viscosity of the oil phase so that drug diffusion is slowed down.

With the use of lime water as an aqueous phase of the emulsion, the drug release rate was slower due to the in situ formation of calcium soap with the fatty acids present in the beeswax.

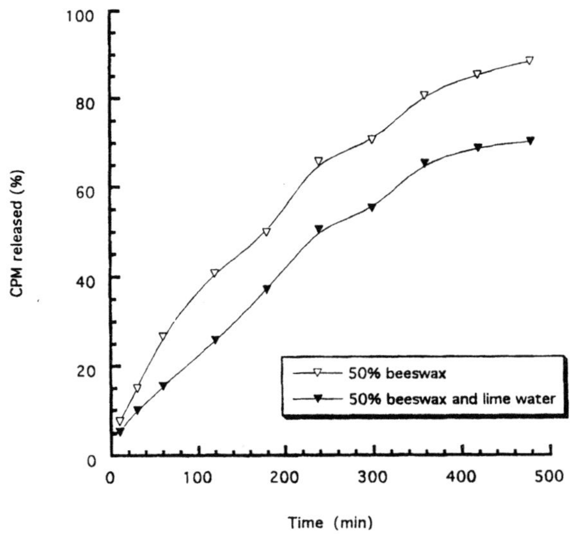

Figure 10 Release profiles of chlorpheniramine maleate from a dry adsorbed emulsion with silica as hydrophilic adsorbent in distilled water. *Source*: Ref. 7.

IX. CONCLUSIONS AND FUTURE OUTLOOK

Today the dry adsorbed emulsions are obtained from a W/O emulsion containing a hydrosoluble active ingredient, effective in small doses and characterized by a short or a long half-life. Each phase of the emulsion is fixed on a compatible polarity adsorbent so that a fluid powder with regularly shaped particles and a sustained release effect of drug is obtained.

The dry adsorbed emulsions are easy to prepare industrially (without any use of solvents) and their different particle sizes convenient to handle. Particles of dry adsorbed emulsions are incorporated in hard gelatin capsules or tablets to obtain a sustained drug delivery system for oral route that is well accepted by patients. The sustained release effect depends mainly on the particle size of the form, on the composition of the primary emulsion oil phase, and on the process of manufacturing. The adsorbent and drug nature and physicochemical properties are also important in the drug release profile.

Routes other than oral are possible for administration of the dry adsorbed emulsions such as the rectal route (soft capsules, suppositories), the parenteral route (parenteral suspensions, compressed pellets for implantation), and the topical route (patchs).

Further developments of this new oral sustained delivery system may occur in the following fields: emulsion type (O/W); emulsion phase composition (aqueous and oil phases, emulsifying agents); adsorbent nature; active ingredient solubility (water-soluble or oil-soluble); routes of administration. The dry adsorbed emulsion particle structure may also be investigated during the manufacturing process to set up as a model a structural organization of particles.

APPENDIX I: Drug Release Studies (6)

All of the drug release studies were performed at 37°C in a buffered solution of pH 1.2 [7 mL of hydrochloric acid (9 M Prolabo) and 2 g of sodium chloride in 1-L of distilled water] and in a buffered solution of pH 7.4 (6.8 g of sodium dihydrogen phosphate in 1L of distilled water ajusted at pH 7.4 using a 0.2 M NaOH solution) to simulate gastric and intestinal media, respectively. No enzyme was added. The powder under study was stirred at 60 rpm in the buffered solution thermostated at 37°C on a Dissolutest apparatus (Prolabo). The release of sodium salicylate was monitored by measuring the absorbance of the solution at 260nm with a Beckman model 25 spectrophotometer. A measurement was taken every 10 min during the first 30 min, then every 30 min up to 2 h, and then every hour up to 6 h. For a given form and a given particle size, six dissolution measurements were made, and for each time the mean value of the six measurements was used to obtain the kinetic parameters (Table 3).

APPENDIX II: Computer Fitting Method (6)

A. Higuchi Model

The $L_{(t)}$ values were plotted vs. the square root of t. Plots were linear up to t = 180 min. Table 3 presents the correlation coefficients, intercepts (L_0'), and slopes (k') obtained using a classical least-squares-method.

B. Exponential Method

As Eq. [3] was nonlinear, a computer program was designed to determine the best values of k and L_0 to fit the experimental points. For each experimental $L_{(t)}$ value, the program calculated a $k_{(t)}$ value, using:

$$k_{(t)} = \left(\frac{-1}{t}\right) \ln[1 - (L_{(t)} - L_0)/(100 - L_0)] \quad [6]$$

Equation [6] was derived from Eq. [3], and k and ν_k, the mean and variance of the $k_{(t)}$ values, respectively, were calculated. Starting with $L_0 = 0$, the program increased the L_0 value to minimize ν_k. Then, the k and L_0 values corresponding to the smallest ν_k were used to plot $L_{(t)}$ vs. t (Fig. 2–4). The K and L_0 values of three plots are given in Table 3.

REFERENCES

1. VV Ranade, MA Hollinger. Drug Delivery Systems. Boca Raton, FL: CRC Press, 1996, p 128.
2. N Farah, M Rollet. French Patent 86-02 69 4, 1986.
3. N Farah, M Rollet, A Berthod. Etude d'une préparation galénique: l'émulsion adsorbée sèche. Structure physicochimique et effet sur la libération d'un principe actif hydrosoluble. Sci Tech Pharm 22:1000–1005, 1986.
4. N Farah, C Chaumat, J Bardon, P Sébert, M Rollet. Utilisation de différentes variétés de silice dans l'obtention d'émulsions adsorbées sèches pour retarder la libération d'un principe actif très hydrosoluble. Proceedings of 4th International Conference on Pharmaceutical Technology, Paris, France, June 3–5, 1986, vol 3, pp 180–189.
5. N Farah, M Bouzon, M Rollet, JL Taverdet, JM Vergnaud. Dry emulsion–a sustained release form: modelling of drug transfers in liquids. Int J Pharm 36:81–88, 1987.
6. A Berthod, M Rollet, N Farah. Dry adsorbed emulsions: an oral sustained drug delivery system. J Pharm Sci 77:216–221, 1988.
7. MM Meshali, KE Gabr, EA El Fattah. Preparation and evaluation of different polymers based dry adsorbed emulsions of chlorpheniramine maleate as a sustained drug delivery system. STP Pharma 6:370–375, 1996.
8. RC Weast. Handbook of Chemistry and Physics. 62nd ed. Boca Raton: CRC, 1982, p B-143.
9. RW Baker, HK Lonsdale. In: Controlled Release of Biologically Active Agents. AC Tanquary, RE Lacey, eds. New York: Plenum Press, 1974, pp 15–71.
10. T Higuchi. Rate of release of medicaments from ointment bases containing drugs in suspension. J Pharm Sci 50:874–875,1961.
11. T Higuchi. Mechanism of sustained action medication. Theoretical analysis of rate of release of solid drugs dispersed in solid matrices. J Pharm Sci 52:1145–1149, 1963.

12

Nanosuspensions for the Formulation of Poorly Soluble Drugs

R. H. Müller, C. Jacobs, and O. Kayser
Free University of Berlin, Berlin, Germany

I. INTRODUCTION

Poorly soluble drugs are often a challenging problem in drug formulation, especially when the drugs are poorly soluble simultaneously in aqueous and in non-aqueous media. To achieve solubility at least in organic solvents, the drugs might be dissolved in aqueous mixtures with an organic solvent (e.g., water-ethanol). Other attempts to overcome a solubility problem are solubilization (e.g., by mixed micelles) or formation of a complex using, for example, cyclodextrins (1). However, these approaches are of limited success, as can be seen by the relatively low number of products on the pharmaceutical market based on these technologies. There is at least a few products based on solvent mixtures (water-ethanol) including injectables, such as Supertendin 2000N (dexamethasone/lidocaine), Adalat Pro Infusione (nifedipine), Alkeran (mephalan), Novodigal Injektionslösung (digoxin), Dapotum/-acutum (fluphenazine). Cyclodextrin (Cd)–based products are, for example, Brexidol (piroxicam/β-Cd complex, tablet or granulate), Stada Reise-Pastillen (β-Cd for incorporation of poorly water-soluble 8-chlortheophylline and for chemical stabilization of incorporated orange oil), and Prostavasin, introduced in 1985, for intra-arterial infusion containing prostaglandin E1 as α-Cd complex. Products based on mixed micelles are very limited. The classical example is Valium MM containing diazepam solubilized with lecithin (soya) and glycocholic acid.

Due to the limited applicability of the above approaches, one tried to improve the in vivo performance of poorly soluble drugs by reducing the particle size of the drug, thus leading to an increased surface area and an increased dissolution velocity. An established method is micronization leading to drug powders with size distributions typically in the range of about 0.1 μm up to 25 μm (2). In the case of very finely milled powders the sizes are below 10 μm. The routes of administration are oral and parenteral. However, for drugs of very low solubility, especially in combination with a desired higher blood level, micronization does not solve the problem. A further increase in dissolution rate would be possible by further increasing the surface of the drug powder, i.e., by reducing the micrometer-sized drug particles to nanoparticles. However, this is not possible using the conventional milling techniques such as jet mill or rotor stator colloid mills. The percentage of the particle population in the nanometer range of powders milled with these techniques is very low (2) and the yield therefore not acceptable.

As an alternative, Sucker produced drug nanoparticles by a precipitation step (3). The drug was dissolved in an organic solvent and then poured into an aqueous surfactant solution. In the water-solvent mixture the solubility was low and the drug precipitated. To avoid the formation of drug microparticles, growth of the nanoparticles had to be stopped using an appropriate surfactant or stabilizer mixture. To obtain products stable on long-term storage lyophilization of these so-called hydrosols was recommended (3). However, the prerequisites for the production of hydrosols are at least the solubility of the drug in an organic solvent and simultaneously the miscibility of this organic solvent with water. For many poorly soluble drugs this is not the case, therefore excluding the hydrosol technique for drug formulation. In addition, fixation of the crystal growth at a certain maximum size in the nanometer range is at least not a simple process, especially when the product is intended for intravenous injection and the content of microparticles in the nanoparticulate product needs to be very low. These aspects might have contributed to the fact that this technology is not widely used for pharmaceutical products.

An alternative to obtain drug nanoparticles is the use of more efficient milling techniques to reduce the size of drug microparticles. Pearl (ball) mills are employed by the company NanoSystems to produce a drug nanoparticulate product called NanoCrystals® (4). The drug is filled as an aqueous suspension into a pearl mill containing glass pearls or zirconium oxide pearls as milling media (Fig. 1). The pearls are moved by a stirrer, the drug microparticles are ground to nanoparticles between the moving milling pearls. Alternatively, the drug microparticulate suspension can be milled by applying a high-pressure homogenization process leading to a product referred to as a *nanosuspension*; the registered tradename is DissoCubes® (Drug Delivery Services GmbH, Krohnshagen, Germany). Piston–gap homogenizers are typically used for the production of nanosuspensions. The microparticle suspension passes a small homogenization gap at high pressure, and the cavitation forces are sufficiently high to disintegrate the microparticles to drug nanoparticles. Suspensions from 1% to 15% solid content can normally be processed. The production equipment is also used in the pharmaceutical industry for the production of emulsions for parenteral nutrition, which means that the production unit is principally acceptable by the regulatory authorities even for preparing parenterals. This chapter will focus on the production of nanosuspensions by high-pressure homogenization, the properties of drug nanoparticles, further processing for incorporation into traditional formulations (e.g., pellets, tablets), but it will also compare the major existing technologies, i.e., pearl milling and high-pressure homogenization.

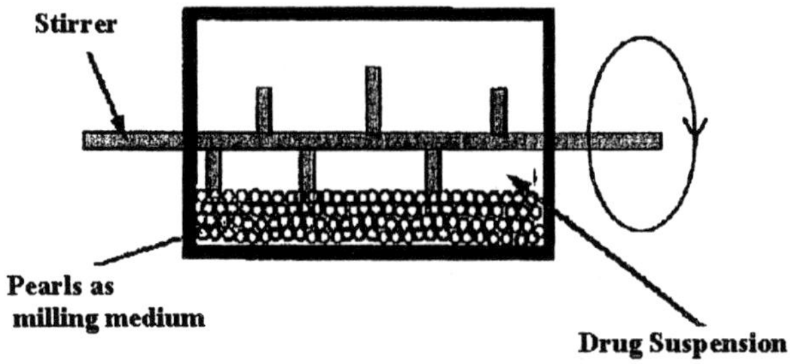

Figure 1 Cross-section through a pearl mill. (Modified from Ref. 6.)

II. PRODUCTION TECHNIQUE FOR NANOCRYSTALS®

NanoCrystals by NanoSystems are produced using a pearl mill (Fig. 1). The milling balls used in such mills range in size from approximately 0.4 mm to 3.0 mm. Current pearl materials are glass and zirconium oxide. During the milling process when using pearl/ball mills, erosion of the pearls occurs leading to contamination of the product with pearl material. Glass and zirconium oxide are insoluble in the media of the gastrointestinal tract. From the chemical point of view, eroded particles from the balls can be considered as nontoxic. The German MAK values (maximum tolerated concentration in the air) is 5 mg/m^3 for zirconium compounds. Alternatively, the pearl mills can be made from a hard polymer, especially crosslinked polystyrene. Depending on the hardness of the drug powder and the required fineness of the particle material, the milling times range from hours to days (5). The preferred size range for NanoCrystals is below 400 nm; examples are about 300 nm for naproxen NanoCrystal dispersion (diameter 50%) (6) and 100 nm for paclitaxel (7). After the milling process the drug nanoparticles must be separated from the milling balls.

III. PRODUCTION OF NANOSUSPENSIONS (DissoCubes®) BY HIGH-PRESSURE HOMOGENIZATION

A. Principle Production Technique

The drug powder is dispersed in a surfactant solution by a high-speed stirrer. The starting size of the drug powder should be as small as possible, which implies that the drug should be jet-milled. If jet-milled drug quality is not available, a special pressure program can be run on the homogenizer to achieve a first-size

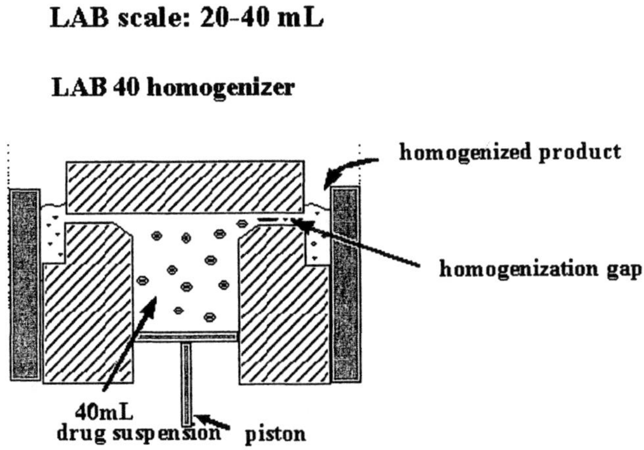

Figure 2 Cross-section of an APV Micron LAB 40 high-pressure homogenizer. (Modified from Müller, R. H.; Gohla, S.; Dingler, A.; Schneppe, T.; Large scale production of solid lipid nanoparticles (SLN™) and nanosuspensions (DissoCubes™), in *Handbook of Pharmaceuticals Controlled Release Technology*, (Wise, D., ed.) in press).

reduction of the coarse powder. Figure 2 shows the principle of a piston gap homogenizer (Micron LAB 40). Applied pressures range from about 100 bar to a maximum of 1500 bar; in some machines a maximum of 2000 bar can be reached (Rannie 56 and 118). The drug suspension is pressed through a very small gap in a size range of about 25 μm. The diameter of the cylinder, which contains the stock suspension, is 3.0 cm. That means that there is a reduction in diameter from 3.0 cm to just about 25 μm, which leads to a high streaming velocity of the suspension. According to the Bernoulli equation, the dynamic pressure increases while simultaneously the static pressure decreases. In the gap the static pressure decreases below the boiling pressure of water, which means that the water boils leading to the formation of gas bubbles, which implode when the suspension leaves the gap and normal air pressure is reached again (cavitation). The implosion (cavitation) forces are sufficiently high to disintegrate the drug microparticles to nanoparticles. Additional disintegration effects are the high shear forces in the gap and particle collision (similar to jet mill). Due to collision effects in general it is more efficient to homogenize a higher concentrated suspension, i.e., 10% instead of 1% solid content. In many cases it is not sufficient that the suspension passes the homogenizer just once; typically multiple cycles are required. Depending on the hardness of the drug, the desired mean particle size, and the required homogeneity of the product, 3, 5, or 10 homogenization cycles are run (for details, see below).

B. Aseptic Production/Processing of Highly Toxic Compounds

For a range of reasons the nanosuspensions need to be sterile or, alternatively, to be produced under aseptic conditions. In a number of cases sterilization by autoclaving is not possible because of chemical instability of the drug and/or surfactants or physical instability of the suspension. γ-Irradiation can be employed alternatively, but this is not recommended because first one must transfer the suspension in a nonaqueous system (e.g., through lyophilization) to prevent chemical stability problems, which may lead to toxic compounds. The second reason is the regulatory problems to be encountered. Therefore the possibility of aseptic production is of high importance. Aseptic production is a more tedious process when using pearl mills for particle disintegration. The APV Micron LAB 40 homogenizer has the advantage that the homogenization tower can be separated from the driving unit and placed underneath a laminar airflow cabinet (LAF)

Figure 3 Homogenization tower of an LAB 40 placed underneath a laminar airflow cabinet.

(Fig. 3). The homogenization tower can be dismantled and the parts in contact with the product can be autoclaved or heat-sterilized. The complete production process will then be performed under a LAF. The homogenization process itself supports the sterilization process because the high cavitation forces can disrupt microorganisms. In fact, high-pressure homogenization is commercially used for cell rupture process. As an additional security measure the produced nanosuspensions can be preserved (e.g., 0.005–0.02% thiomersal, 1–2% benzyl alcohol, 0.7–1.5% β-phenyl alcohol). The concentration of the preservative should be sufficiently high to guarantee no bacterial growth, or a reduction in bacteria within 2 weeks according to the USP.

This production unit can be used simultaneously for the processing of highly toxic compounds, e.g., cytotoxics, or highly active peptides and proteins, e.g., hormones. Of course the LAF containing the homogenization tower needs to be placed in a special high-security lab facility. A detailed description of such a lab facility including security measures and handling procedures is given in Ref. 8. Milling of toxic compounds such as cytotoxics with a pearl mill appears more complicated than using a high-pressure homogenizer. The electropolished metal surfaces of a homogenizer can easily be decontaminated by wiping with solvent-soaked tissues, the surfaces are smooth without too many edges and therefore easily accessible. It is more complicated with pearl mills, especially the cleaning of the mill media with possibly a less even surface structure.

IV. SIZE DISTRIBUTIONS OF NANOSUSPENSIONS

The production parameters of nanosuspensions affect the mean particle size of the bulk population and the content of particles in the micrometer range. This "contamination by microparticles" has especially to be minimized for products to be administered intravenously. The two production parameters affecting product quality are homogenization pressure and number of cycles. Examples of the effect of these parameters are displayed below using tarazepide as an example. The mean diameter of the bulk population was determined by photon correlation spectroscopy (PCS; Zetasizer, Malvern Instruments, UK). PCS yields a mean diameter (Z-average) and a polydispersity index ranging from 0 for a perfectly monodisperse particle population to 0.500 for a relatively broad size distribution. Emulsions for parenteral nutrition typically possess a polydispersity index between 0.100 to about 0.250. Monodisperse polystyrene standard latex particles have a polydispersity index in the range of 0.010 to about 0.050. PCS is limited to the size range of about 3 nm to 3 μm, i.e., particles above 3 μm are not detected. Therefore, for the analysis of microparticles laser diffraction was employed (Coulter LS 230, Coulter Electronics, Germany). Laser diffraction yields a volume distribution. As characterization parameters diameters 50%, 90%, and 99%

are currently used, i.e., 50%, 90%, and 99% of the particles are below the given size value, respectively. It should be noted that laser diffraction (LD) data are volume-based, and the PCS mean diameter is light intensity weighted size (Z-average). Therefore the PCS mean diameter and the diameter 50% from the LD are not identical; LD data are generally higher. Four different formulations of tarazepide suspensions were prepared and homogenized (Table 1). Formulations A–C have an identical drug content of 1% but vary in the composition of the stabilizing surfactant mixtures. Formulation D is identical with regard to the surfactants with A but the drug content has been increased from 1% to 10% (all weight percentages).

Table 1 Composition of Tarazepide Formulation A–D

Formulation	Composition	Wt %
A	tarazepide	1.00
	Poloxamer 188	1.00
	Tween 80	0.50
B	tarazepide	1.00
	Poloxamer 188	1.00
	sodium cholate	0.10
C	tarazepide	1.00
	Poloxamer 188	1.00
	sodium cholate	0.10
	lecithin	1.00
D	tarazepide	10.00
	Poloxamer 188	1.00
	Tween 80	0.50

A. Effect of Homogenization Pressure

The tarazepide drug powder used for the preparation of the nanosuspensions was jet-milled. Analysis by LD yielded 12.5 μm as diameter 50%, 25.0 μm as diameter 90%, and 51.0 μm as diameter 99%. The drug powder was mixed with surfactant and water using first a mortar and pestle, then an Ultra-Turrax at 1 min and 9500 rpm. Homogenization was performed at room temperature using a Micron LAB 40 (discontinuous version). As a kind of premilling a pressure profile was run, i.e., two cycles at 150 bar, two cycles at 500 bar, and then cycles with 1500 bar were run. The higher homogenization pressure lead to distinctly smaller particle sizes and lower content of micrometer particles. It confirmed the observed general rule that higher pressures lead to finer particles. This is in contrast to observations with solid lipid nanoparticles (SLNs). With SLNs it depends very much on the chemical nature of the lipids used as matrix material, if 500 bar or 1500 bar leads to smallest particles (9). The effect that nanosuspensions become smaller with increasing homogenization pressure is in agreement with the homogenization theory. A higher homogenization pressure leads to a higher power density and subsequently to an increase of the dispersitivity index. It should be noted that too high homogenization pressure can lead to particle aggregation with increasing cycle numbers (see below).

B. Effect of Cycle Number on Product Quality

After premilling, the tarazepide suspension was homogenized at 10 homogenization cycles at 1500 bar. Figure 5 shows the PCS diameter of the bulk population and the polydispersity index (PI) as a function of the number of homogenization cycles. After six homogenization cycles the mean diameter of 338 nm does not change up to cycle 10, which indicates that the maximum dipersitivity has been reached. The PI shows some fluctuations from cycle 2 to cycle 10, possibly caused by aggregation and deaggregation effects. Based on the PCS data the optimum product is obtained after four homogenization cycles with 369 nm and a polydispersity of 0.197. Application of two more homogenization cycles leads to a further small decrease in the mean PCS size to 338 nm (which is irrelevant

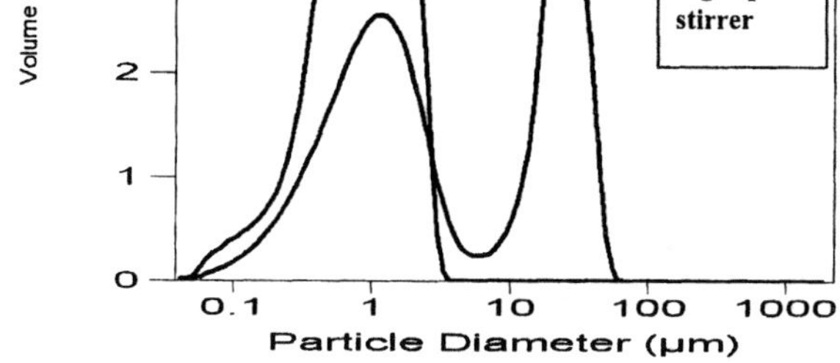

Figure 4 Laser diffraction size distribution of tarazepide nanosuspension A obtained after high-speed stirrer (Ultra Turrax) and two cycles at 1500 bar (premilling: two cycles at 150 bar, two cycles at 500 bar).

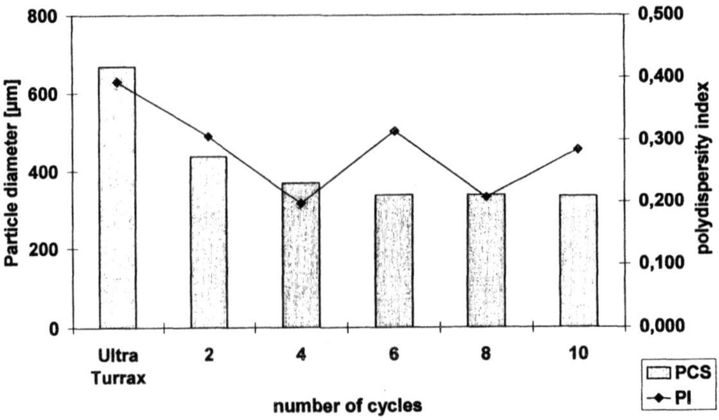

Figure 5 Mean PCS diameter and polydispersity index of tarazepide nanosuspension A as a function of homogenization cycles.

for the product quality) but broadens obviously slightly the size distribution as seen by the increase of the polydispersity index to 0.314.

The LD data show that there is little change in the bulk population (diameters 50% and 90%) but a distinct reduction in the content of particles above 2 µm. Figure 6 shows that there is little change in the diameter 50% from cycle 2 to 10, but a distinct decrease in the diameters 99% and 100% from cycle 2 to 4.

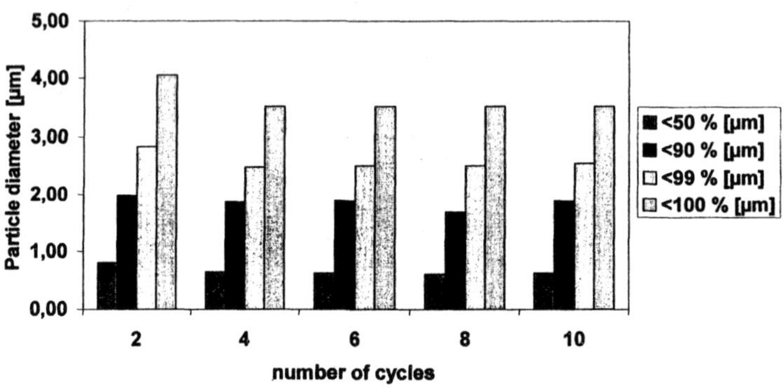

Figure 6 Diameters 50%, 90%, and 100% of tarazepide nanosuspension A from Fig. 5 as a function of the number of homogenization cycles (diameters based on volume distribution).

To summarize: If nanosuspensions are intended for oral administration, very often only two homogenization cycles are necessary to obtain a product of a sufficient quality for the oral administration route. If nanosuspensions are intended for intravenous administration, application of four or more homogenization cycles—depending on the hardness of the drug—might be required.

Figure 7 shows the LD size distribution of tarazepide nanosuspension A obtained after 4 homogenization cycles at 1500 bar. The fact that the diameter 100% is only 3.52 μm proves the high quality, i.e., the homogeneity of the nanosuspension. In most publications diameters 100% are not quoted because even a few large particles will shift this diameter to large values because it is a volume distribution. In addition, the diameters 100% fluctuate strongly in case a minor content of larger particles is present due to statistical problems of probe sampling. In the case of tarazepide nanosuspension A the diameter 100% has a low value and was highly reproducible. The diameter 100% was even found to be constantly 3.52 μm from cycle 4 to cycle 10. A nanosuspension of these characterization parameters (PCS and LD data) is suitable for intravenous injection without prior filtration through a 5-μm filter.

Table 2 shows the PCS diameters and PI values of the tarazepide nanosuspensions A–C obtained after 10 homogenization cycles at 1500 bar. Nanosuspensions A and C are relatively similar regarding diameter and PI. That means the obtained particle size is little or is not affected by the surfactant mixture used. This is in good agreement with previous results (10). In contrast, nanosuspension

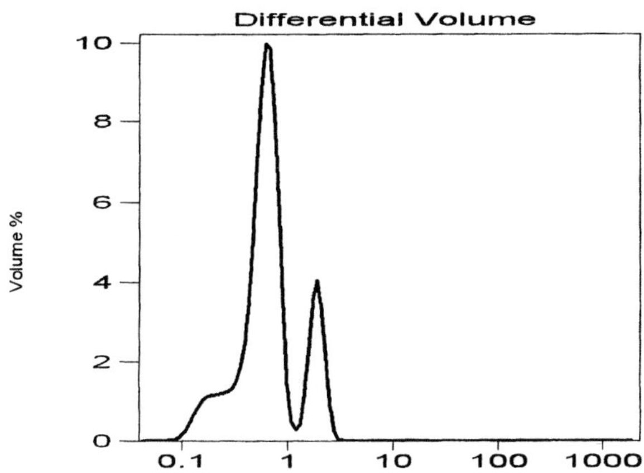

Figure 7 Laser diffraction size distribution of tarazepide nanosuspension A from Fig. 5 at four homogenization cycles and 1500 bar homogenization pressure.

Table 2 PCS Diameters and Polydispersity Index (PI) Values of Tarazepide Nanosuspensions A–C after Homogenization Cycle 10

Tarazepide nanosuspension	Mean diam. (nm)	PI values
A	334	0.284
B	902	0.316
C	384	0.256

B has almost a three times larger mean diameter of 902 nm. At first glance this seems to contradict previous results.

The large diameter of nanosuspension B can easily be explained when looking at the change of the mean diameter as a function of the number of the homogenization cycles (Fig. 8). The PCS diameter of nanosuspension B decreases until cycle 4 to 552 nm and then constantly increases. This is attributed to an overly high energy input; the surfactant mixture is unable to stabilize particles sufficiently and in consequence aggregation occurs. It is a fitting example that input of too much homogenization energy can worsen the results due to aggregation. In contrast, nanosuspension C shows a continuous decrease in the diameter reaching a minimum at 10 homogenization cycles (348 nm). This behavior is very

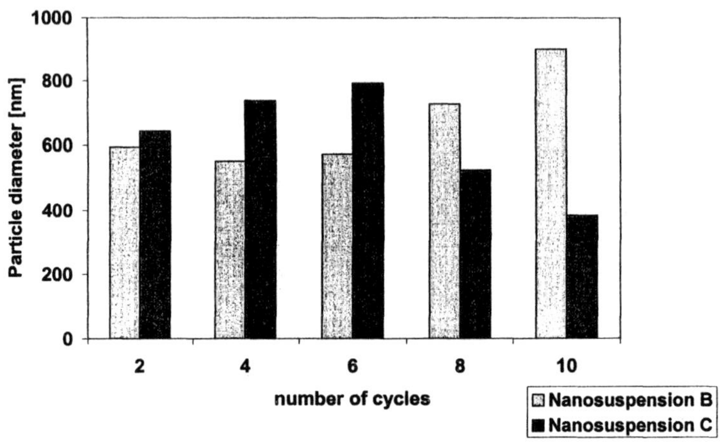

Figure 8 PCS diameters of tarazepide nanosuspension B and C as a function of the number of homogenization cycles (homogenization pressure 1500 bar).

different to nanosuspension A, reaching the minimum particle size at four to six homogenization cycles. Obviously, the surfactant mixture of C is less efficient in stabilization; aggregates form during the homogenization procedure that again need to be deaggregated in the next homogenization cycle.

To summarize: The three different surfactant mixtures show that the surfactant itself does not affect the minimum size reached; this depends solemnly on the hardness of the drug and the homogenization parameters applied. However, the surfactant mixture is the determining factor for possible aggregation of the ultrafine drug nanoparticles. Identical results were found for nanosuspensions of paclitaxel. Never was a growth of paclitaxel crystals observed, only aggregation of nanoparticles when the surfactant mixture was not optimized (see below).

V. STERILIZATION OF NANOSUSPENSIONS

Nanosuspensions can be sterilized by autoclaving or, alternatively, by γ irradiation. The physical stability of the nanosuspensions during a sterilization depends very much on the composition of the stabilizing surfactant or surfactant mixtures. RMKP-22 (4-[*N*–(2-hydroxy-2-methylpropyl) ethanolamin]2,7–bis(*cis*-2,6-dimethylmorpholin-4-yl)-6-phenyl pteridin) nanosuspensions were prepared using different surfactant mixtures (11). RMKP-22 nanosuspensions containing 9% drug and stabilized with 0.6% Phospholipon-90 showed only a negligible increase in the PCS diameter from 380 to 402 nm during autoclaving (15 min, 2 bar, European Pharmacopoeia). The LD diameter 99% increased from 3.4 μm to 4.3 μm, which also appears acceptable. In contrast, a very pronounced aggregation took place in nanosuspensions stabilized with 0.3% Tween-80. Even the LD diameter 50% increased to 7.56 μm. This indicates clearly that autoclaving is not possible. The aggregation was explained by the reduced steric stabilization efficiency of Tween-80 at increased temperatures. The increase in temperature leads to dehydration of the stabilizing polyethylene glycol chains; the thickness of the sterically stabilizing layer decreases. Some steric stabilizers might even reach their critical flocculation temperature (CFT) at the given autoclaving conditions and the concentration of electrolytes present. It is known that CFT decreases with increasing electrolyte content (12). This can be demonstrated when autoclaving sterically stabilized nanoparticle suspensions at reduced temperatures. The increase in particle size is distinctly less pronounced or even negligible (13).

To summarize: For autoclaving, nanosuspensions need preferentially to be stabilized by charged emulsifiers such as lecithin (Phospholipon). Similar to fat emulsions for parenteral nutrition, which are also stabilized by lecithin, nanosuspensions can withstand the sterilization procedure.

The Phospholipon-90 and the Tween-80–stabilized RMKP-22 nanosuspensions were also sterilized by γ irradiation (25 kGy) for reasons of comparison.

The diameters 99% before and after autoclaving were 3.78 μm and 3.70 μm and for the Tween-80 nanosuspension 4.35 μm and 4.26 μm, respectively. Stability under γ irradiation was also observed for a range of other drug nanosuspensions.

In contrast to these results, the tarazepide nanosuspension A showed a distinct increase in size during the γ-radiation process. Figure 9 shows the mean diameters measured by PCS before and after γ irradiation. What is the reason for this? Determining factors for the physical stability of suspensions are the sterically stabilizing effect and additionally repulsive charges of the particles (ζ potential). The ζ potential measurements of tarazepide nanosuspensions where taken in distilled water with conductivity adjusted to 50 mS (14). Measurements were performed using the Malvern Zetasizer at a field strength of 22 V/cm. The ζ potential before irradiation was −30.5 mV; however, it dropped to just −22.3 mV after irradiation. The drop in the ζ potential indicates clearly that there must have taken place a chemical reaction during irradiation. There is no other explanation for the change in particle charge. The ζ potential of about ±20 mV is not sufficiently high to stabilize the particles on long term. A potential above at least ±30 mV is required for a stable dispersion (15,16).

To summarize: Nanosuspensions can be autoclaved if they are sufficiently stabilized by a charged surfactant. γ-Radiation is possible with some nanosuspensions; it depends on the nature of stabilizer and drug. There seem to be exceptions when a change in the nanosuspension is induced by the irradiation process.

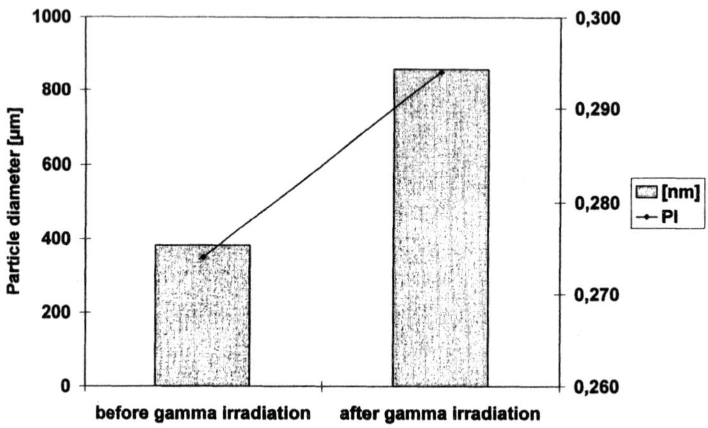

Figure 9 PCS diameters and polydispersity index of tarazepide nanosuspension D before and after γ irradiation (10 homogenization cycles with 1500 bar).

VI. PHYSICAL LONG-TERM STABILITY OF AQUEOUS NANOSUSPENSIONS

For many applications it is desirable that nanosuspensions are stable as aqueous dispersions without having the need to make a lyophilized or spray-dried product. Insufficient stability of suspensions can lead to crystal growth and/or particle aggregation. The tarazepide nanosuspensions were stored over a period of 42 days; storage was performed in the fridge because the nanosuspensions were not preserved. Figures 10 and 11 show that little or no changes in the characterization diameters occur. The surfactant mixture is optimal preventing aggregation. Despite the high dispersitivity of the suspension, no Ostwald ripening takes place. The reasons for the absence of Ostwald ripening are discussed in more detail in Sec. VIII.

Long-term storage investigations were previously performed with other drugs such as RMKP-22 and clofazemine. Storage was performed again in the refrigerator because no preservatives were added to exclude any effect of a preservative on stability. Clofazemine showed a distinct increase in the PCS diameter and the LD diameter 99% after 2 months of storage. In contrast, RMKP-22 nanosuspensions stabilized with Phospholipon-90 or Tween-80 proved to be stable for 2 years. The RMKP-22 nanosuspension stabilized with 0.6% lecithin had a PCS diameter of 564 nm after production and 575 nm after 2 years of storage; the LD diameters 99% were 4.68 μm and 4.85 μm, respectively.

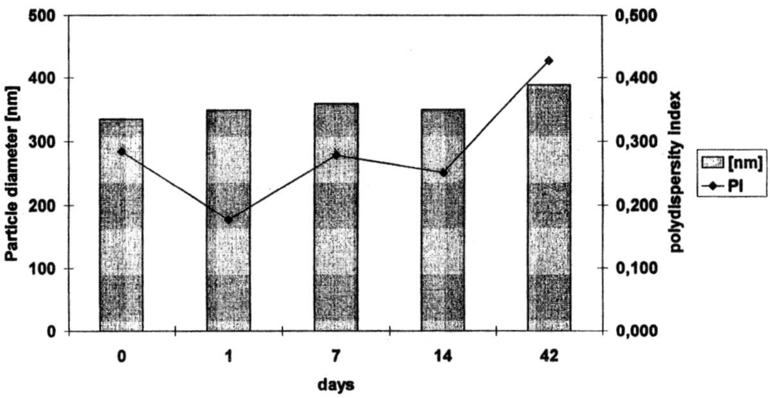

Figure 10 PCS diameter and polydispersity index of tarazepide nanosuspension A as a function of storage time (storage temperature: 2–8°C).

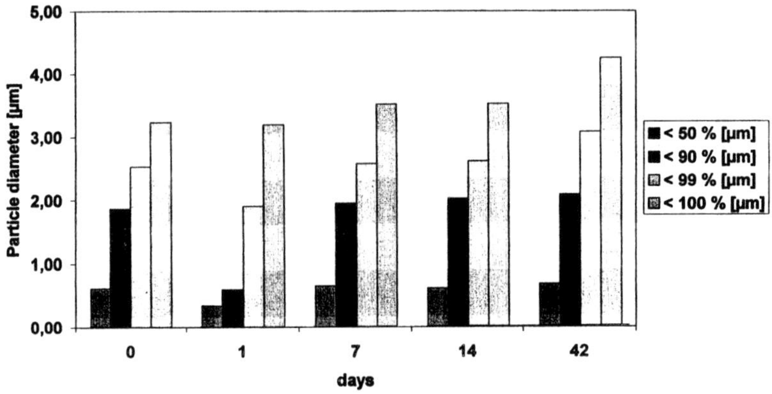

Figure 11 Laser diffraction diameters 50%, 90%, 99%, and 100% of tarazepide nanosuspension A as a function of storage time (storage temperature: 2–8°C).

To summarize: Storage of drug nanoparticles as an aqueous nanosuspension is principally possible if the stabilizing surfactant mixture is optimal. Crystal growth does not take place if the particles are relatively uniform in size.

VII. ELECTRON MICROSCOPIC CHARACTERIZATION OF NANOSUSPENSIONS

Photon correlation spectroscopy and laser diffraction do not give any information about the shape of the drug nanoparticles (or at least how to extract shape data from PCS correlation functions). Therefore nanosuspensions were intensively analyzed by electron microscopy (EM) (11). The basic result is that the form of the particles depends on the drug and its crystalline nature, not on the surfactant used. Generally, the particles are of cubic or cuboid shape, which leads to the selection of the trade name DissoCubes. Figure 12 shows the EM graph of RMBB-98 nanosuspension directly after preparation. It shows a particle population that is uniform in size and shape. This nanosuspension was stored for 3 months at 5°C and then analyzed again by laser light scattering techniques as well as EM. There was a slight increase in the diameter 99%. Possible reasons for such an increase are crystal growth or, alternatively, particle aggregation. EM analysis proved the absence of large crystals. An increase in particle size is caused by aggregation of the smaller particles, which means that there is a problem in selecting suitable stabilizers.

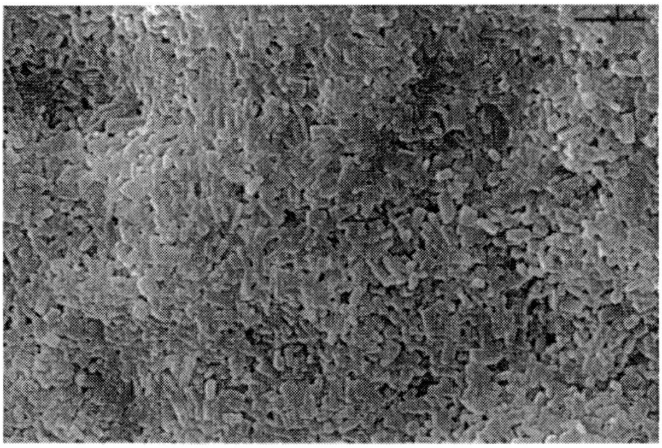

Figure 12 Electron microscopic graph of RMBB-98 nanosuspension at the day of production (bar: 5 μm, RMBB-98 is a newly developed drug. (From Ref. 2.)

VIII. PHYSICOCHEMICAL PROPERTIES OF NANOSUSPENSIONS

The reduction in particle size (e.g., by micronization) increases the surface area and consequently the dissolution velocity (*dc/dt*), according to the Noyes-Whitney equation. However, transfer of the drug microparticles to nanoparticles not only increases the solution velocity but also the saturation solubility C_s.

At first glance, the increase in saturation solubility sounds surprising because in the textbooks the saturation solubility is defined as a compound specific temperature–dependent constant. However, saturation solubility depends also on the particle size. The drug powders normally processed in pharmacy are well above the size range in which this dependency occurs. The particles need to be less than 1–2 μm to achieve this effect (17,18). A steep increase in intrinsic dissolution rate is reported for particles less than 1 μm. It can be explained by the Kelvin equation describing the vapor pressure above the surface of a liquid droplet as a function of the degree of curvature of this surface. The vapor pressure increases with increasing degree of curvature (19). This situation from the liquid–gas interphase can be transferred to a solid–liquid interphase, which means to the dissolution of a drug particle. The vapor pressure is then replaced by the dissolution pressure. Figure 13 shows this dependency.

According to the Noyes-Whitney equation, the dissolution velocity depends on the difference C_s–C_x and also on the diffusional distance h (C_x, concentration present in the bulk). According to the Prandtl equation, the diffusional distance is reduced at very small particle sizes (Fig. 13, lower). Apart from the increase

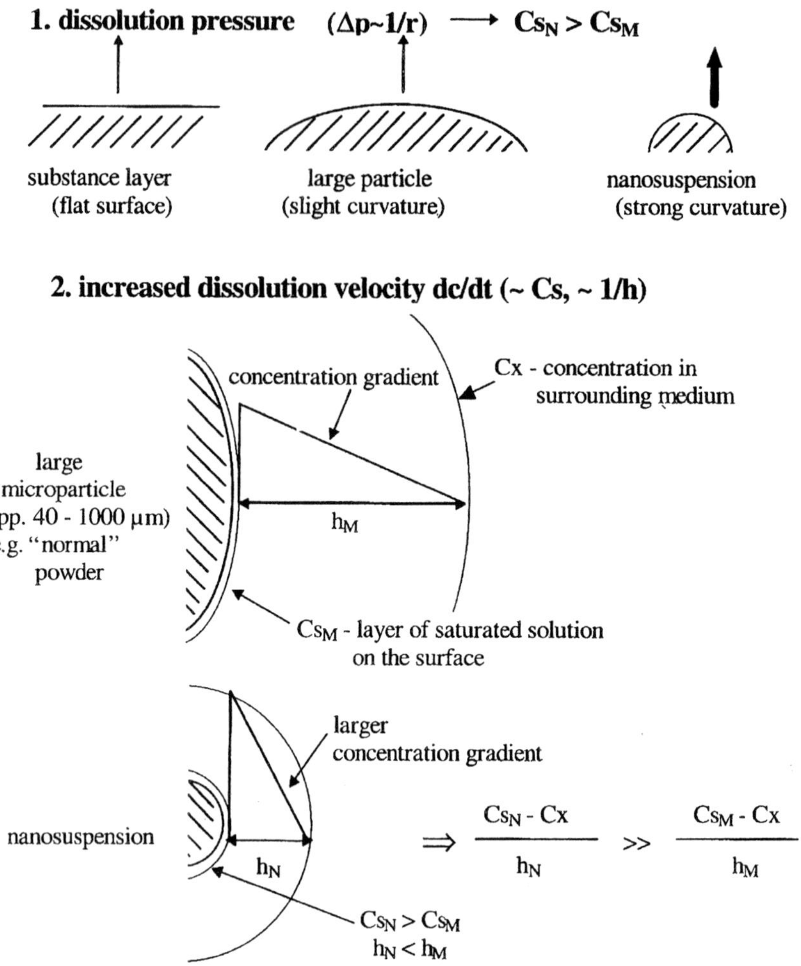

Figure 13 Increase of saturation solubility with decreasing particle size due to an increasing dissolution pressure (upper) and a reduction of the diffusional distance h as described in the Prandtl equation (lower). (From Ref. 1.)

in C_s, the decrease in h has an additional effect on the dissolution velocity. For details, the reader is referred to Ref. 20.

Nanoparticles in general possess a high adhesiveness to surfaces. After oral administration of polymeric nanoparticles they were found sticking to the mucosa of the gastrointestinal tract (21). The same is assumed for drug nanoparticles and is considered to be one of the reasons for their special in vivo performance, i.e., increased bioavailability and reduced erratic absorption.

In terms of preparation of the drug nanoparticles by high-pressure homogenization, an additional factor can contribute to a further increase in the saturation solubility. The NanoCrystals have been reported by NanoSystems to be fully crystalline (20). From tableting it is known that the energy introduced to the drug particles by high pressure during the tableting process can lead to the transformation of polymorphic forms into a higher energy form (e.g., transition from I to II). The power density in the homogenization gap is very high; the density of 10^{13} W/m^3 is similar to the power density in a nuclear power station. The main difference is the volume in which this power density is present.

High-energy input can lead to the complete transformation of the crystalline drug to an amorphous state (22). It is known that amorphous compounds possess a higher solubility than crystalline compounds. This can be explained by the Ostwald-Freundlich equation, which describes the saturation solubility as a function of particle size. An amorphous powder can be considered as particles with an indefinite small particle size, thus the saturation solubility is increased. This effect is broadly described in the literature, e.g., for griseofulvin modifications I and II compared to amorphous griseofulvin (23).

A known effect contributing to an increased saturation solubility is the formation of highly energetic surfaces by the milling process. The breaking of crystals lead to the exposure of inner parts of the crystal to the outer dispersion medium, implying that energetically less favorable surfaces are now in contact with water. The Ostwald-Freundlich equation also describes saturation solubility as a function of the interfacial tension between solid–liquid interphase. An increase in the interfacial tension γ leads to an increase in the saturation solubility.

IX. DOSAGE FORMS BASED ON NANOSUSPENSIONS

Aqueous nanosuspensions can be further processed to different traditional dosage forms. Examples are as follows:

1. *Topical formulations.* Drug nanoparticles can be incorporated into creams and water-free ointments. The nanocrystalline form leads to an increased saturation solubility of the drug in the topical dosage forms, thus enhancing the diffusion pressure into the skin.

2. *Dosage form for administration in the mouth.* Aqueous nanosuspensions—if necessary with enhancing the viscosity of the outer phase—can be used for application to areas inside the mouth, e.g., sublingual or buccal drug absorption. Potentially they could replace patches when incorporating the drug nanoparticles into a water-rejecting outer phase.

3. *Dosage forms for oral drug delivery to the gastrointestinal tract.* Aqueous nanosuspensions can be used directly as a liquid dosage form. In case of insufficient physical stability or for reasons of convenience or marketing acceptance, the aqueous nanosuspensions can be formulated as a dry dosage form, i.e., tablet or hard gelatin capsule with pellets. The aqueous nanosuspension can be used directly in the granulation process or as a wetting agent for preparing the extrusion mass of pellets. A similar process has been reported for incorporating solid lipid nanoparticles into pellets (24). Alternatively, pellets can be produced by spraying the nanosuspension onto the surface of sugar beads (spray coating). Granulates can also be produced by spray drying of nanosuspensions. For the spray drying process it is necessary to dissolve the carrier material (e.g., sugars such as mannit) in the aqueous phase of the nanosuspension. Especially spray drying leads to a very high drug loading. Benefits of using nanosuspensions for the creation of oral dosage forms are the smaller volume of these forms, an increased chemical stability of the drug, and generally a relatively simple and fast formulation development.

4. *Pulmonary delivery of nanosuspensions.* Aqueous nanosuspensions can be nebulized using mechanical or ultrasonic nebulizers for lung delivery. Compared to drug microparticles the nanosuspensions have a major advantage. Due to their small size it is very likely that in each aerosol droplet at least one drug particle is contained, leading to a more even distribution of drugs in the lungs. When using microparticles, many aerosol droplets are drug-free and others are heavily loaded with a drug microparticle. Budesonide drug nanoparticles were successfully nebulized using an Omron ultrasonic nebulizer. A good relationship was obtained between increasing drug concentrations in the formulation and the micrograms of drug delivered per 2 s actuation. Basically the nanosuspensions can be used in all nebulizers. The dispersions can be relatively high concentrated, e.g., 10% solid concentration. Due to the presence of many small particles instead of a few large microparticles, all aerosol droplets are likely to contain drug nanoparticles.

5. *Parenteral administration of nanosuspensions.* Nanosuspensions can be administered via different parenteral administration routes ranging from intraarticular via intraperitonal to intravenous injection. Especially for intravenous injection it is required that the content of microparticles be sufficiently low, and particularly that the content of particles be larger than the smallest blood capillaries (5–6 μm). There are two ways of achieving this:

(a) Preparing drug nanoparticles and using a filtration step through a 5-μm filter.

(b) Preparing the drug nanoparticles in a one-step process yielding a product of sufficient homogeneity with sufficiently low microparticle contamination.

As shown by the above data, the production of nanosuspensions by high-pressure homogenization can achieve the criteria of (b).

Nanosuspensions are an ideal tool for the activity screening of newly developed compounds that are poorly soluble. In addition, they are a tool to determine the absolute bioavailability of poorly soluble drugs that cannot be injected as a drug solution.

X. IN VIVO PERFORMANCE OF DRUG NANOPARTICLES

Some of the aspects have already been discussed in Section IX. The basic advantages of using drug nanoparticles in an oral formulation can be stressed. These advantages were nicely summarized by Liversidge at the CRS workshop in Kyoto (6). Drug nanoparticles lead to:

1. Improved bioavailability
2. Improved dose proportionality
3. Reduced fed/fasted variability
4. Reduced intersubject variability
5. Enhanced adsorption rate

The increase of oral bioavailability can be explained by the adhesiveness of the drug nanoparticles to the mucosa, the increased saturation solubility leading to an increased concentration gradient between gastrointestinal tract lumen and blood, and the increased dissolution velocity of the drug. The reduction of erratic absorption can be explained by the fact that the adhesion process of the drug nanoparticles to the mucosal surface seems to be highly reproducible and little affected by the nutritional status of the patient.

A number of in vivo studies with drug nanoparticles have been performed by NanoSystems proving the excellent in vivo performance of drug nanoparticles. For example, the oral administration of the analgesic naproxene as drug nanoparticles lead to an AUC (0–2 h) of 97.5 mg-h/L compared to just 44.7 mg-h/L for naprosyn suspension and of 32.7 mg-h/L for Anaprox tablets. The corresponding t_{max} values were 1.96 h for the drug nanoparticles, 3.33 h and 3.20 h for the two commercial products (human study, postprandial). Oral administration of the gonadotropin inhibitor danazol as a suspension of nanoparticles lead to an absolute bioavailability of 82.3% (dog study), the conventional dispersion only to 5.2% (6).

XI. LARGE-SCALE PRODUCTION OF NANOSUSPENSIONS

The possibility of large-scale production is the prerequisite for the introduction of a product to the market. NanoCrystals are produced by pearl milling. Large-volume pearl mills are available thus allowing the large-scale production of these formulations. Mills employed by NanoSystems are the LMZ-2 for 5 kg, the LMZ-4 for 20 kg, the LMZ-10 for 50 kg, and the LMZ-60 for a 450-kg batch (7). The diameter 50% is approximately 250 nm, the diameter 90% about 400 nm. Such particle sizes are problematic for a sterilization by filtration, at least excluding the use of 0.22-μm filters. Depending on the target size and the hardness of the drug, the production time required ranges from hours to several days. Problems associated with the pearl mill are erosion from the pearls as well as the microbiological quality of the product. Milling of an aqueous suspension for days carries the risk of bacterial growth in the water phase.

For the production of aqueous nanosuspension DissoCubes, various large-scale high-pressure homogenizers are available, e.g., Rannie 56 and Rannie 118 from APV Homogeniser GmbH.

For the production process, the jet-milled drug is dispersed in an aqueous surfactant solution in product container 1 using a dissolver disc. Then this microparticle suspension is passed through the homogenizer to product container 2 (= cycle 1), passed back through the homogenizer to container 1 (= cycle 2), and so on until the required number of cycles have been performed. The production time required for a 400-kg batch depends on the required homogenization pressure and the necessary number of homogenization cycles. The capacity of the Rannie 118 depends on the homogenization pressure used, i.e., 2000 L/h at 1000 bar and 1200 L/h at 2000 bar. At a low pressure 400 kg will pass the homogenizer within 12 min (= time for 1 cycle), which means that when assuming 5 cycles the product will be ready in 1 h.

When using high-pressure homogenizers, abrasion can also occur, i.e., erosion of metal from the homogenization gap. The homogenizers are made of stainless steel accepted for production of parenterals, which means it is basically an accepted material. Nevertheless, the load of the product with metal ions needs to be within the limits. To assess the contamination of the product with metal from the homogenizer, an LAB 40 (continuous version) was challenged by homogenizing a 10% nanosuspension at a maximum pressure of 1500 bar for 20 homogenization cycles. The content of iron in the product was determined by atomic absorption spectrometry. Iron (Fe) was chosen as metal for analysis because it represents the largest metal fraction in steel and will therefore be most likely to be detected in the product. The metal content was below 1 ppm (exactly 0.6 ppm).

In general, the homogenizer technology is a low-cost technology. Prices in Germany for the lab scale homogenizer LAB 40 are approximately 40,000 DM (22,000 US $), the Gaulin 5.5 (capacity 150 L/h) is about 60,000 DM (33,000 US $), and the Rannie 118 280,000 DM (155,000 US $). These prices are in the lower range for industrial equipment.

XII. NANOSUSPENSIONS (DISSOCUBES) VS. NANOCRYSTALS

The question arises regarding differences between NanoCrystals and DissoCubes. The main difference is the means of production, which might affect the quality of the products. A potential problem for NanoCrystals might be the abrasion from the pearls, at least for the parenteral route of administration. Even when preparing oral dosage forms, contamination by erosion products (glass/zirconium oxide nanoparticles and microparticles) could be a potential problem in chronic treatment. What happens when a person administers chronically insoluble nanoparticles or microparticles in the range of a few micrometers? It could be shown for polymeric nano- and microparticles that an uptake from the gastrointestinal tract takes place (25,26). The uptake is very low but detectable. Nano- and microparticles are possible systems for oral immunization via the M cells in the Peyer's patches. Has it any effect when the M cells are constantly confronted with insoluble nano- and microparticles from the pearl mills during chronic treatment? This problem can only be circumvented if the contamination is below 10 ppm (pharmaceutical inspection convention (PIC) and related state of the art, e.g., 10 ppm). Abrasion also takes place from the metal surfaces of high-pressure homogenizers. However, metals like iron represent no problem toxicologically. It should be referred to iron-based diagnostics such as magnetites for MRT diagnosis. In addition, the contamination is within the generally accepted regulations. The microbiological quality of the product might also be affected by the production methods to a different extent.

Mean particle sizes obtained by the two different production methods are considered to be in a similar range. Differences might be the contamination of the product by microparticles, especially microparticles above the critical threshold of 5 μm. To our knowledge no systematic comparison has been performed.

The increase in dissolution velocity due to the increase in surface area and the increase in saturation solubility caused by the small particle size are identical for NanoCrystals and DissoCubes. Differences might occur in the crystalline status of the product, i.e., there might be transitions of crystalline drug to the amorphous state induced by the high-power density of the homogenization process. NanoCrystals are crystalline; DissoCubes can contain an amorphous fraction or be completely amorphous (26). This amorphous status, if preserved, will lead to a further increase in saturation solubility and dissolution velocity. A further difference is that the intellectual property rights of the technologies belong to two different companies. NanoSystem and its Nanocrystal technology has recently been acquired by Elan. The nanosuspensions/DissoCubes technology belongs to the German company Medac GmbH in Hamburg. Both technologies are protected by U.S. patents.

XIII. PERSPECTIVES

The use of drug nanoparticles in the form of NanoCrystals or DissoCubes opens new perspectives in the formulation of poorly soluble drugs. For many poorly soluble drugs they might be the only solution to achieve a sufficiently high bioavailability or to remove the problem of erratic absorption from the gastrointestinal tract.

Apart from oral drug delivery, the drug nanoparticles might have a large potential in pulmonary delivery but also in parenteral dosage forms. In particular, the fact that some nanosuspensions may behave like solutions regarding their pharmacokinetics (3) appears to be a very interesting feature.

Existing formulations possessing side effects due to the excipients used (e.g., Cremophor EL) could be replaced by nanosuspensions stabilized by well-tolerated excipients (e.g., lecithin, poloxamer 188).

Apart from these aspects, the major advantage of this technology is its simplicity! Simple systems are more likely to appear on the market than highly sophisticated ones. The future number of pharmaceutical products based on drug nanoparticle technology will show if this assessment is correct.

REFERENCES

1. Frömming, K.-H. Cyclodextrine—eine vielseitig verwendbare Gruppe neuer Hilfsstoffe. In Müller R. H., Hildebrand, G. E. (eds.), Pharmazeutische Technologie: Moderne Arzneiformen, 2nd ed. WVG, Stuttgart, 1998.
2. Müller, R.H., Peters, K., Becker, R., Kruss, B. Nanosuspensions for the i.v. administration of poorly soluble drugs—stability during sterilization and long-term storage, Proc. Int. Symp. Control. Rel. Bioact. Mater., 22, 574–575, 1995.
3. Sucker, H. Hydrosole, eine Alternative für die parenterale Anwendung von schwer wasserlöslichen Wirkstoffen. In: Müller, R. H., Hildebrand, G. E. (eds.), Pharmazeutische Technologie: Moderne Arzneiformen, 2nd ed. WVG, Stuttgart, 1998.
4. Müller, R.H., Böhm, B.H.L., Grau, M.J. Nanosuspensions—Formulierungen für schwerlösliche Arzneistoffe mit geringer Bioverfügbarkeit: I Herstellung und Eigenschaften, Pharm. Ind., 74–78, 1999.
5. Liversidge, G.G. Drug Nanocrystals for Improved Drug Delivery at CRS Workshop Particulate Drug Delivery Systems, 11–12 July, 1996, Kyoto, Japan.
6. Setler, P. Identifying new oral technologies to meet your drug delivery needs for the delivery of peptides and proteins and poorly soluble molecules, IIR Limited Drug Delivery Systems, March 1999, London.
7. Liversidge, G.G., Cundy, K.C. Int. J. Pharm., 125, 91, 1995.
8. Böhm, B.H.L, Müller, R.H. Nanosuspensions of sparingly soluble cytotocix drugs: design of a lab scale production unit. Pharm. Sci. Technol. Today, in press.
9. Schwarz, C., Mehnert, W., Lucks, J.S., Müller, R.H. Solid lipid nanoparticles (SLN)

for controlled drug delivery. I. Production, characterisation and sterilisation, J. Control. Rel. 30, 83–96, 1994.
10. Müller, R.H., Peters, K., Craig, D. Electron microscopic studies of nanosuspensions–particle shapes as a function of drug and surfactant, Proc. Int. Symp. Control. Rel. Bioact. Mater., 23, 925–926, 1996.
11. Müller, R.H., Peters, K., Becker, R., Kruss, B. Nanosuspensions—a novel formulation for the i.v. administration of poorly soluble drugs, Proc. 1st World Meeting APGI/ APV, Budapest, 9–11 May, 1995.
12. Müller, R.H. Colloidal Carriers for Controlled Drug Delivery and Targeting—Modification, Characterization and in vivo Distribution. WVG, Stuttgart/CRC Press, Boca Raton, 1991.
13. Schwarz, C., Freitas, C., Mehnert, W., Müller, R.H. Sterilization and physical stability of drug-free and etomidate-loaded solid lipid nanoparticles, Proc. Int. Symp. Control. Rel. Bioact. Mater., 22, 766–767, 1995.
14. Müller, R.H., Zetapotential und Partikelladung—Kurze Theorie, praktische Meßdurchführung, Dateninterpretation, Wissenschaftliche Verlagsgesellschaft Stuttgart, 254 S., 1996.
15. Ney, P. Zetapotentiale und Flotierbarkeit von Mineralien, Springer-Verlag, Wien, 1973.
16. Riddick, T.M. Zeta-Meter Manual, Zeta-Meter Inc., New York, 1968.
17. Nyström, C., Saers, S. Solid dispersions for fast release and dissolution of drug with low aequeous solubility in: Formulation of poorly-available drugs for oral administration, European Symposium Paris, 5–6 February, APGI Swedish Academy of Pharmaceutical Sciences, 94–103, 1996.
18. Nyström, C. Dissolution properties of soluble drugs: theoretical background and possibilities to improve the dissolution behaviour. In: Müller, R.H., Benita, S., Böhm, B. (eds.), Emulsions and Nanosuspensions for the Formulation of Poorly Soluble Drugs, Medpharm Scientific Publishers, Stuttgart, 1998.
19. Simonelle, A.P., Metha, S.C., Higuchi, W.I. I. Pharm. Sci., 56, 633, 1970.
20. Müller, R.H., Benita, S., Böhm, B. (eds.), Emulsions and Nanosuspensions for the Formulation of Poorly Soluble Drugs, Medpharm Scientific Publishers, Stuttgart, 1998.
21. Durrer, C., Irache, J.M., Puisieux, F., Duchene, D., Ponchel, G. Mucoadhesion of latexes. I. Analytical methods and kinetic studies, Pharm. Res., 11, 674–679, 1994.
22. Böhm, B.H.L., Grau, M.J., Hildebrand, G.E., Thünemann, A.F., Müller, R.H. Preparation and physical properties of nanosuspensions (Dissocubes) of poorly soluble drugs, Proc. In. Symp. Control. Rel. Bioact. Mater., 25, 956–957, 1998.
23. List, P.H. Arzneiformenlehre, WVG, Stuttgart, 1976.
24. Pinto, J.F., Müller, R.H. Pellets as carriers of solid lipid nanoparticles (SLN) for oral administration of drugs, Die Pharmazie, in press.
25. Jani, P.U., Florence, A.T., McCarthy, D.E. Further histological evidence of the gastrointestinal absorption of polystyrene nanospheres in the rat, Int. J. Pharm., 84, 245–252, 1992.
26. Hillery, A.M., Florence, A.T. The effect of adsorbed poloxamer 188 and 407 surfactants on the intestinal uptake of 60 nm polystyrene particles after oral administration in the rat, Int. J. Pharm. 132, 123–130, 1996.

13

Pharmaceutical Suspensions and Their Applications

V. Gallardo, M. A. Ruiz, and A. V. Delgado
University of Granada, Granada, Spain

I. INTRODUCTION

Many textbooks on the physicochemical properties of colloidal suspensions mention the pharmaceutical field as one in which colloids find industrial and technological applications (see, e.g., Refs. 1–3). Together with, perhaps, inks, food colloids, and paints, pharmaceutical suspensions are designed to be used in most cases not by specialists but rather by people who only need to know proper directions for use, and quick and efficient therapeutic effects. This means that in their formulation there are involved factors that are not normally considered when colloid scientists prepare their suspensions. For instance, pharmaceutical suspensions

1. Should not contain any toxic or potentially hazardous components;
2. Should be stable or easily redispersed after moderately long storage periods;
3. Should be chemically stable—the therapeutic drug component cannot be altered by the composition of the dispersion medium, the presence of a normal atmosphere, or even the action of light or transportation;
4. Might have, in some cases, nice taste, and even color and fragrance.
5. No bacteria, viruses, or other infectious agents can develop during manipulation or storage.

Some of these conditions (and the list could be made longer) are very stringent, whereas others may appear as less essential, but no doubt the manufacturing and designing technologists must keep all of them in mind when sending a new formulation to the marketplace. In any case, this set of conditions makes pharmaceutical suspensions very complex colloidal systems that can display interesting and intricate physical properties, some of which are detailed in Chapter 4.

On the other hand, elucidation of the typical components and different forms of the physical systems that we call pharmaceutical suspensions requires previously to have an idea about the different types of suspensions, intimately related to their purpose and applications. Thus, most authors consider three main kinds of pharmaceutical suspensions (4–6), namely, orally administered, or simply oral, suspensions; externally applied (topical suspensions); and injectable or parenteral. Although strictly speaking, from the colloid science point of view, aerosols are simply suspensions in which the dispersion medium is a gas and the dispersed material is solid, throughout this chapter we will primarily adhere to the more classical point of view that considers only solid/liquid dispersions under the denomination of suspensions. Their most significant features are displayed in Table 1 (see also Refs. 4–6).

Figures 1 and 2 are scanning electron microscopic (SEM) pictures of typical drugs often administered in suspension, namely, phosphomicine calcium salt, and

Table 1 Main Characteristics of Pharmaceutical Suspensions (4–6)

Suspension type	% Solids conc.	Particle size (μm)	Suspension medium	Other properties
Oral	2.5–5	≈1	Water, thickeners, preservatives, sweeteners	Sometimes, must be reconstituted at the moment of use (active compound moderately unstable)
Topical	≥20	≈1	Clay or polymer dispersions, O/W or W/O emulsions, preservatives	Flow behavior is essential; normally non-Newtonian
Parenteral	0.5–5 (>20 in insoluble penicilines)	0.5–1	Solvent, preservatives, buffer	Low viscosity is essential to avoid syringe clogging; sterile vehicle

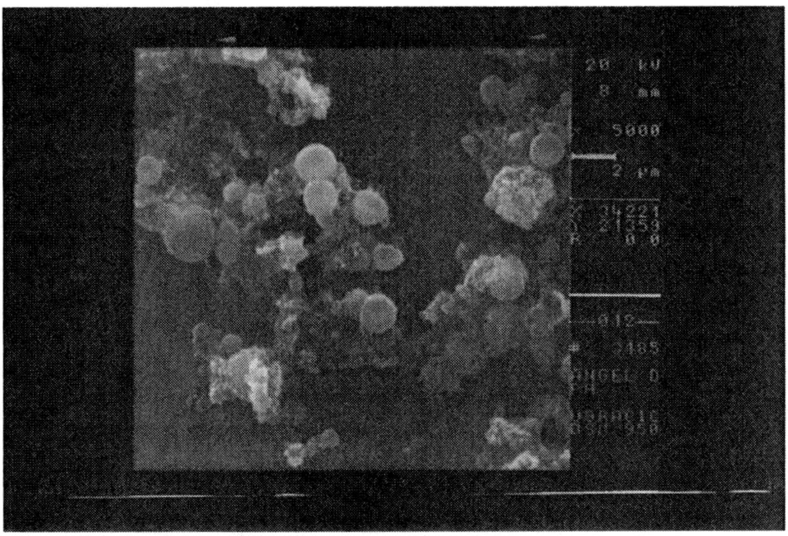

Figure 1 SEM picture of phosphomicine (calcium salt) particles.

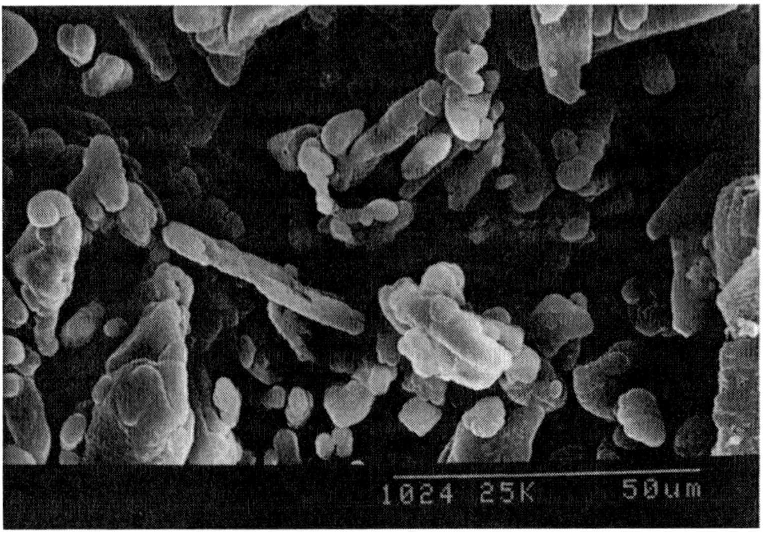

Figure 2 SEM picture of nitrofurantoin particles.

nitrofurantoin—both bactericidal antibiotics. As observed in these pictures, their average particle sizes are, respectively, ≈1.5 μm and ≈8 μm; note that in these suspensions the particle size and shape need not be controlled rigorously. Unlike other technological applications of colloidal dispersions, morphological homogeneity of the particles is not a crucial requirement.

There are a large variety of pharmaceutically useful or physiologically active compounds that are marketed as suspensions of any of the types listed in Table 1. As an example, the *General Catalog of Pharmaceutical Specialties* in Spain (7) lists over 10,000 pharmaceuticals, and more than 10% of them are available in suspension as one or the only way of administration. This gives an idea of the importance of the field. However, such a large variety also implies a comparably large amount of preparation procedures, suspending media composition, characteristics of the suspended particles, and so on. We aim here at summarizing the most general aspects of the typical composition of many of these suspensions. Hence, this chapter will be organized as follows: Some considerations are first given on the necessity or usefulness of suspensions in the pharmaceutical field. Then we will give (Sec. II) a more detailed view of the specific aspects in which such suspensions differ from others that are of interest for the general colloid scientist. In Sec. III we describe their typical components according to the therapeutic target or the desired physicochemical properties. A schematic list of applications will follow in Sec. IV. Finally, Sec. V will be devoted to the rapidly emerging field of drug transport and release by suspended particles, which are in such cases used as vehicles of the therapeutic agent.

A number of reasons can be given to justify the need or interest for selecting a suspension as a dosage form of a given therapeutic compound. The most obvious concerns active substances that are insoluble in water or aqueous solution at the necessary concentration. This is the case of, for example, hydrocortisone or neomycin (8) for ophthalmic applications: they are dispersed in sterile solutions and placed directly on the eye. In other cases, the drug can be prepared in soluble form, but the bitter taste of the solution would easily prevent the patient from using the correct dose, whereas the dispersed (i.e., nondissolved) form is essentially tasteless. Examples of suspensions designed for such organoleptic reasons are chloramphenicol or paracetamol (9–11). In other occasions, suspensions are useful when the drug must be administered to some population groups (in particular, children and elderly people), or to patients suffering from certain pathologies (in particular, those of the gastrointestinal tract) that may have troubles in swallowing, thus rendering the required dosage patterns difficult to follow (12,13).

Another set of conditions in which suspensions are recommended include those drugs that are chemically unstable when in soluble form (they can easily hydrolyze; this is the case of, for example, oxytetracycline calcium salts; see Ref. 9), or in which large specific surface areas are required. Let us mention, for instance, antacid or absorbent compounds, or radiological contrasts (8). As already discussed (see Table 1), suspending media need not be always aqueous; hence suspensions can still be used even with drugs that can degrade in aqueous solution. For example, fenoxymethylpenicillin can be dispersed in coconut oil to be administered as an oral suspension (8).

An important field of application of pharmaceutical suspensions that has developed during the last 20 years or so is that of controlled or sustained release of therapeutically useful compounds. The advantages of the method are clear (14), mainly if one takes into account the potency of many drugs presently used: (a) The therapeutic effect is maintained, essentially without peaks, for long periods; (b) there is little probability that the patient forgets its treatment; (c) the rate of drug delivery can be controlled through the characteristics of the colloidal vehicle used and the nature of the bonding between the drug and the vehicle. There are many possible routes to the formulation of controlled drug release systems, including drug entrapment in polymer particles (14–16), drug coating by a particle (mainly polymeric) film (17–19), or drug adsorption on suspended colloidal particles (20,21). As shown in Fig. 3 (18), a great control on active agent release can be achieved by changing the film thickness; the figure demonstrates that increasing the thickness of the drug-free polymer film (Eudragit NE 30D, Röhn Pharma) coating the drug–reservoir polymer matrix, up to a fivefold decrease in release rate can be obtained.

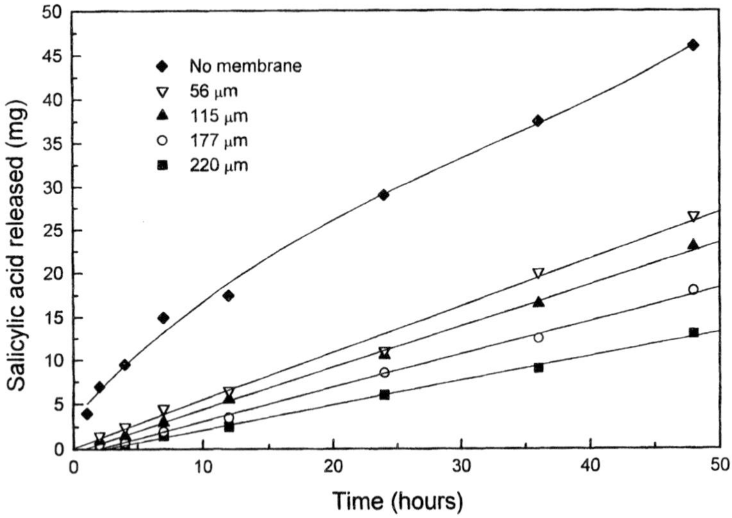

Figure 3 Effect of membrane thickness on the release of salicylic acid (10% w/w) from laminated Eudragit films, comprising a drug containing reservoir coated by a drug-free polymer. (From Ref. 18. Reproduced with permission of the American Pharmaceutical Association.)

The use of colloidal (mainly polymer) particles for drug transport and targeting is most promising in cancer chemotherapy, where high local concentrations of the therapeutic drug may be desirable without at the same time provoking unwanted toxic side effects of systemic scale (22).

II. SPECIAL FEATURES OF PHARMACEUTICAL SUSPENSIONS

In order to better understand the complexity that a pharmaceutical suspension displays as a dispersed system when it is ready for the market, it may prove useful to consider the problems associated with the formulation of a suspension (see also Chapter 4). Several authors have thoroughly reviewed these problems (5,23–26); the proper additives controlling pH, color, flavor, fragrance, etc., must be chosen. Also, care must be taken of the preservatives protecting the suspension against, e.g., bacterial growth, or decomposition or oxidation of the active drug. Complications may arise if, as is likely to occur, the presence of such a variety

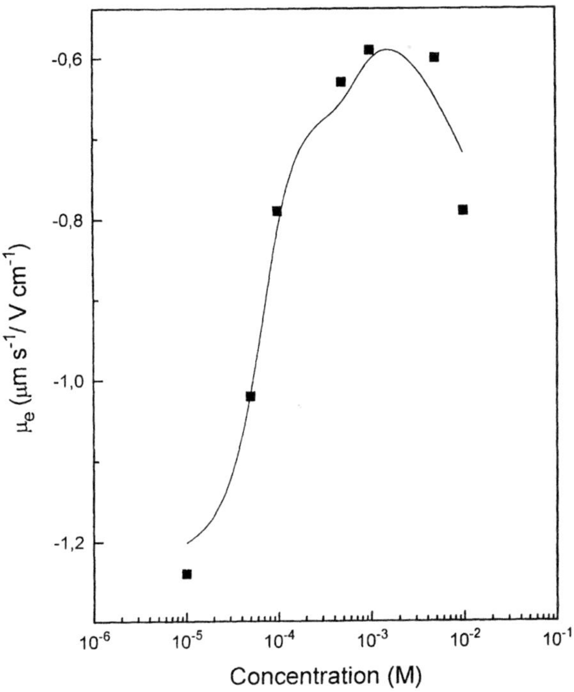

Figure 4 Electrophoretic mobility of nitrofurantoin as a function of the concentration of the photoprotector antipyrin. (Reprinted from Ref. 27, with kind permission of Elsevier Science, NL, Amsterdam.)

of additives alters some important characteristic of the system; for example, Figs. 4 and 5 (27) illustrate the effect on the electrophoretic mobility of nitrofurantoin particles in suspension of different amounts of preservatives normally used in the formulations, namely, the photoprotector antipyrin (Fig. 4) and the antimicrobial agent benzoic acid (Fig. 5). As observed, both additives show a significant effect on the electrophoretic mobility of nitrofurantoin-suspended particles. Specifically, antipyrin can decrease the mobility to almost one half its value in the absence of the photoprotector. Benzoic acid, on the other hand, adsorbs as negatively charged benzoate species on the particles, thus increasing the negative mobility of the drug (the decrease observed in Fig. 5 at higher concentrations can be related to pH changes in the suspending medium upon benzoic acid dissolution). Such effects on the ζ potential may significantly alter the stability and sedimentation behavior of the suspension because of the subsequent modification in the electrostatic repulsive interaction between the particles.

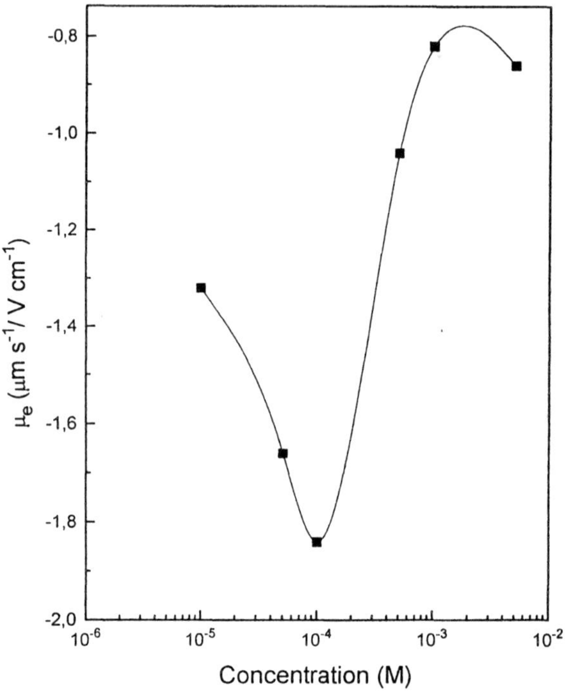

Figure 5 Electrophoretic mobility of nitrofurantoin as a function of the concentration of the antimicrobial agent benzoic acid. (From Ref. 27, with kind permission of Elsevier Science, NL, Amsterdam.)

In fact, determining the optimum stability and sedimentation behaviors is one of the main tasks in suspension formulation. Essentially, two approaches can be chosen (24): (a) the insoluble drug is allowed to settle as long as it is easy to redisperse by slight shaking; (b) conditions are such that the particles remain in suspension as almost individual entities with negligible aggregation or sedimentation. Obviously, the latter approach seems preferable: the suspension will be more pharmaceutically elegant, given its uniform appearance, with no sharp phase separation, but furthermore, the probability of a somewhat random dosage from one use to the next is greatly reduced. In addition, the so-called caking phenomenon, by virtue of which the sedimented particles are cemented together by strong short-range forces (making the suspension virtually useless), is less likely to occur in the second approach (4,28).

For these reasons, a considerable research effort has been devoted to the proper choice of the method of suspension preparation, or to the use of ''structured'' liquid vehicles (see Chapters 4 and 16). Of particular interest in this chapter is the design of structured vehicles, i.e., systems in which the viscosity is very high when the suspension is at rest, this preventing substantial particle sedimentation or coagulation by Brownian motion. For obvious syringeability reasons, this type of stabilization method is not useful for parenteral suspensions (4).

There are both natural and synthetic compounds (many of them polymeric, but also inorganic, like some types of clays) that allow one to obtain suspensions that can be considered ''ideal'' from the rheological point of view:

The system has almost infinite viscosity at zero shear stress.
It shows a yield stress below which no flow occurs (Bingham behavior).
The viscosity is greatly reduced when sufficient shear is applied (e.g., when

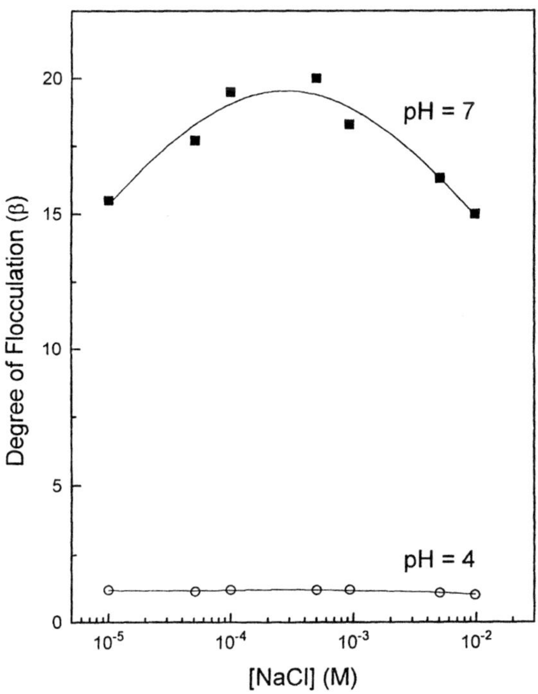

Figure 6 Extent of flocculation, β (sediment volume/sediment volume of stable suspension), of nitrofurantoin suspensions containing 0.1% Carbopol 934, as a function of the concentration of NaCl at pH 4 (○) and pH 7 (■). (From Ref. 31. Reproduced with permission of the American Pharmaceutical Association.)

Table 2 Redispersion (as Percent Optical Transmittance) of Nitrofurantoin Suspensions with 0.1% Carbopol 934 at Various Times After Preparation

	pH 4			pH 7		
Conc. (M)	A	B	C	A	B	C
10^{-5}	0	12	14	0	0	2
5×10^{-5}	0	19	11	0	0	0
10^{-4}	0	32	22	0	0	2
5×10^{-4}	0	25	17	0	0	2
10^{-3}	0	29	22	0	0	2
5×10^{-3}	0	37	34	0	0	3
10^{-2}	0	25	25	0	0	3

A: immediately after preparation; B: 24 h after preparation; C: 48 h after preparation.
From Ref. 31.

pouring the suspension for administration), but reversibly reaches again a very high value when the suspension is allowed to stand.

By way of example, Fig. 6 and Table 2 show the effects of Carbopol 934 (B.F. Goodrich) on the sedimentation and redispersibility of nitrofurantoin suspensions (29–31). As observed, the fact that the optimum conditions for using Carbopol as a stabilizer in pharmaceutical suspensions occur at neutral pH manifests in high sedimentation volumes (Fig. 6) and almost perfect redispersibility (Table 2) when the experiments are performed at pH 7.

III. TYPICAL CONSTITUENTS OF PHARMACEUTICAL SUSPENSIONS

Production of pharmaceutical suspensions on an industrial scale requires careful consideration of all of the requirements leading to the desired therapeutic and physical stability of the system, controlled bioavailability, and suitable presentation to make the suspension attractive for the patient. This means considering not only the physicochemical characteristics of the therapeutically useful particles but also those of the other components, including the solvent; wetting agents; pH buffers; antioxidant or antimicrobial agents; color, taste, and fragrance modifiers; and so on (4,5,32,33).

Concerning the dispersed insoluble drug, there are a number of variables that affect the behavior and effectiveness of the final formulation. Some of them are listed below:

Density, shape, and size distribution of the powder
Ease of wetting
Surface electric charge of the particles in suspension
Crystal structure of the drug
Chemical stability of the latter, and possible interactions and incompatibilities with other suspension constituents

The suspending medium, or vehicle, is usually distilled or deionized water, although some formulations use water or water-alcohol solutions of glycerol (always with high water content). There are relatively few examples of suspensions with nonaqueous vehicles (34); they are mostly designed for topical use and include vehicles based on water-in-oil emulsions, dermathological pastes, magmas, and so on. In some cases, soluble active principles are added when an association between several therapeutic agents is desired, although particular care must be taken to avoid possible negative effects on the stability of the powder.

Depending on the way of administration and on the required physicochemical behavior of the suspension, the final composition of the formulation will differ in general from one case to another, but generally speaking the usual components of a pharmaceutical suspension include, in addition to the aqueous solvent and the insoluble drug, one or more of the following additives:

Wetting agents
Compounds allowing control of stability and sedimentation
Additives used to regulate the flow behavior
pH regulators
Other additives (mainly protective, but also those used to ensure suitable taste, color, fragrance, and so on)

A. Wetting Agents

The primary step in the preparation of a pharmaceutical suspension is to have the drug particles wetted by the liquid vehicle because otherwise separation will occur upon the mixing of both phases. The terms *hydrophobicity* and *hydrophilicity* are often used to qualitatively distinguish particles that are difficult and easy to wet, respectively. Generally speaking, the drug particle will be wetted by the liquid if the solid–air interface is spontaneously replaced by a solid–liquid one when the drug is immersed in the vehicle. This concept is related to the thermodynamic quantity known as work of immersion, W^{i}, defined as (35):

$$W^{\mathrm{i}} = \gamma_{\mathrm{S}} - \gamma_{\mathrm{SL}} \qquad [1]$$

where γ_{S} is the surface free energy of the solid (or surface tension of the solid–air interface) and γ_{SL} is the free energy of the solid–liquid interface. The wetting process is more properly described in terms of the work of spreading, W^{S}, where by spreading we mean the process by which a thick layer of liquid spreads on a solid surface, thus giving rise to a solid–liquid and a liquid–air interface. W^S will hence be given by

$$W^S = \gamma_S - \gamma_L - \gamma_{SL} \qquad [2]$$

where γ_L is the surface tension of the liquid. A positive value of W^S ($\gamma_S - \gamma_{SL} > \gamma_L$) means that the immersion of the solid particle is spontaneous; otherwise, work must be done on the system to properly wet the solid. Alternatively, since the surface tension of the dispersion medium is often given by the requirements mentioned above (most liquid vehicles are water-based), surface treatment of the solid will be needed to change the values of the free energies of the solid–air and solid–liquid interfaces, until reaching a positive value of W^S, as desired. But still this does not mean that the solid is, strictly speaking, hydrophilic: once the particles are wetted, they can either show a tendency to form aggregates or to remain as individual entities in the liquid vehicle. This is, according to Lyklema (36), the true test of hydrophilicity: a colloidal dispersion is hydrophilic if the particles have a stronger interfacial interaction with the solvent than with each other, and hydrophobic in the opposite case.

It is hence essential to be able to measure and modify the surface free energy of the solid if criteria other than trial and error are used to predict the wettability (and the colloidal stability, for that matter) of a solid drug, since the surface tension of the liquid is readily measurable by any well-established method (Wilhelmy plate, du Nouy ring, drop shape analysis, etc.; see, e.g, Ref. 37). In the case of solids, only indirect methods are available to estimate γ_S and γ_{SL}; here we shall give details on the simplest one, contact angle measurements, and the reader is referred to other chapters of this volume for details on other methodologies. The technique is based on Young's equation (37):

$$\gamma_L \cos(\theta) = \gamma_S - \gamma_{SL} \qquad [3]$$

where θ is the contact angle (angle between the solid surface and the tangent to the liquid phase, measured on the gas side at the three-phase contact line), and the other symbols have been defined above. The interfacial free energy can be related to the free energy of adhesion, ΔG_{SL} (change of free energy associated to the process of formation of a unit surface area of the solid–liquid interface, starting from solid–gas and liquid–gas interfaces) by means of the Dupré equation:

$$\Delta G_{SL} = \gamma_{SL} - \gamma_S - \gamma_L \qquad [4]$$

Hence, Eq. [3] can be written as:

$$\gamma_L(1 + \cos\theta) = -\Delta G_{SL} \qquad [5]$$

and the problem now is to relate ΔG_{SL} to other measurable quantities. This can be done if a model of interfacial interactions is used. We will follow here the approach developed by van Oss and his group (38), which assumes that the free energy of interaction between condensed phases has two additive terms, including Lifshitz–van der Waals (LW) and acid–base (AB) interactions. The former are essentially London dispersion forces, whereas the latter are associated with the electron donor or electron acceptor character of the interfaces. ΔG_{SL} can hence be written as:

$$\Delta G_{SL} = \Delta G_{SL}^{LW} + \Delta G_{SL}^{AB} \quad [6]$$

Similarly, the surface free energy of any of the phases would also have both kinds of contributions:

$$\gamma_i = \gamma_i^{LW} + \gamma_i^{AB} \quad [7]$$

and the following relationship between the overall AB contribution and the acid–base character of the material is proposed:

$$\gamma_i^{AB} = \sqrt{\gamma_i^+ \cdot \gamma_i^-} \quad [8]$$

where γ_i^+ (γ_i^-) is the electron acceptor (electron donor) contribution to the acid–base free energy component of phase i. In terms of these components, the free energy of interaction between the solid and the liquid phases (Eq. [6]) can be written (see Ref. 38 for details):

$$\Delta G_{SL} = -2(\sqrt{\gamma_S^{LW}\gamma_L^{LW}} + \sqrt{\gamma_S^+\gamma_L^-} + \sqrt{\gamma_S^-\gamma_L^+}) \quad [9]$$

and Young's equation can be rewritten as:

$$(1 + \cos\theta)\gamma_L = 2(\sqrt{\gamma_S^{LW}\gamma_L^{LW}} + \sqrt{\gamma_S^+\gamma_L^-} + \sqrt{\gamma_S^-\gamma_L^+}) \quad [10]$$

From this equation, if contact angles are measured for three liquids of which the surface tension and its components are known, the three unknowns, γ_S^{LW}, and γ_S^+, γ_S^- can be determined, and the surface free energy of the drug material calculated from Eqs. [7] and [8]. It must be mentioned that contact angle measurements are not free of uncertainties, mainly when crystallographically smooth surfaces are not available; this is normally the case when working with drug powders, since tablets obtained by compressing the powder is our best approximation to a smooth surface. Alternative methods, and tests of validity of contact angle determinations, are discussed in Chapter 4.

Many drugs are hydrophobic, and for this reason the surface free energy of the particles must be modified by reducing the contact angle of water on them. Surfactants are very often used with that aim (39–41), their concentration in the system must be well evaluated because if it is very low the effect can be negligible, whereas high concentrations bring about easy formation of foams and other undesired organoleptic characteristics. Furthermore, their adsorption can increase the stability of the particles against aggregation, thus raising the possibility of cake formation at the bottom of the container. As a rule of thumb, the hydrophile–lipophile balance (HLB) of nonionic surfactants used to help wetting of the particles should range from 7 to 10 (see Refs. 4, 41, and Chapter 1). Typical examples of surfactants added to pharmaceutical suspensions include docusate sodium, sodium dodecyl sulfate and ammonium lauryl ether sulfate (anionic), cetyltrimethylammonium bromide (CTAB) (cationic), and propylene glycol myristate or glyceryl stearate (nonionic).

B. Stability Control

We have just mentioned the role of interfacial interactions in controlling the tendency of drug particles in suspension to remain stable or to aggregate or coagulate. But there is another, essential contribution to the overall interaction energy between dispersed particles, namely, electrostatic repulsion. The reason is that in general a solid immersed in a polar liquid acquires a net surface charge for a number of reasons (42) that depend on the nature of both the liquid and the solid, and that include differential solubility of ions in the structure of the solid, ionization of surface groups, and adsorption of charge-determining ions in solution, just to mention a few. The overall neutrality of the dispersion requires that the surface charge be compensated by ions of opposite charge in the solution close to the particle. A diffuse layer of counterions is thus formed around the particle, giving rise to a distribution of charge and potential known as electric double layer. If ψ_d is the potential at the (ideal) surface where the diffuse layer begins, and it is assumed that the particles are all identical spheres of radius a, the potential energy of electrostatic interaction for constant surface potential during particle approximation is:

$$V^{EL} = 2\pi\varepsilon_r\varepsilon_o a\psi_d^2 \ln[1 + \exp(-\kappa H)] \qquad [11]$$

where H is the distance between the surfaces of the particles, $\varepsilon_r\varepsilon_o$ is the dielectric permittivity of the liquid medium, and κ is the reciprocal Debye length:

$$\kappa = \left(\frac{10^3 N_A e^2 \sum c_i z_i^2}{\varepsilon_r \varepsilon_o kT} \right)^{1/2} \qquad [12]$$

where N_A is the Avogadro number, e is the electron charge, and c_i, z_i are the molar concentrations and valencies of each ionic species present; k stands for the Boltzmann constant, and T is the absolute temperature. The quantity κ^{-1} (Debye length) is a measure of the thickness of the double layer. Equation [11] is actually valid only for low to moderate values of the surface potential; more rigorous (and involved) expressions can be found in Ref. 43.

The total potential energy of interaction between the two spheres will include three terms: electrostatic (Eq. (12), Lifshitz–van der Waals (V^{LW}), and acid–base (V^{AB}). The H dependence of the former is (44):

$$V^{LW} = -\frac{A}{6}\left[\frac{2a^2}{H(4a + H)} + \frac{2a^2}{(2a + H)^2} + \ln\frac{H(4a + H)}{(2a + H)^2}\right] \quad [13]$$

Here the Hamaker constant, A, depends on the characteristics of the liquid and the solid; however, interestingly, it has been shown to be determined by the interfacial tension through (38):

$$A = 24\pi H_o^2\gamma_{SL}^{LW} = 24\pi H_o^2(\sqrt{\gamma_S^{LW}} - \sqrt{\gamma_L^{LW}}) \quad [14]$$

where the quantity H_o is typically 1.58 Å.

Finally, the acid–base contribution depends on the particle–particle distance H as follows (38):

$$V^{AB} = -2\gamma_{SL}^{AB}\pi a\lambda \exp\left(\frac{H_o - H}{\lambda}\right) \quad [15]$$

and the acid–base contribution to the interfacial tension is:

$$\gamma_{SL}^{AB} = 2(\sqrt{\gamma_S^+\gamma_S^-} + \sqrt{\gamma_L^+\gamma_L^-} - \sqrt{\gamma_S^+\gamma_L^-} - \sqrt{\gamma_S^-\gamma_L^+}) \quad [16]$$

In Eq. [15], λ, the correlation length of water molecules, is approximately 1 nm.

Let us finally mention that the assumption of the existence of the three contributions to the total potential energy is known as extended DLVO theory of the stability of colloidal suspensions, as opposed to the classical model, originally developed by Derjaguin and Landau, and Verwey and Overbeek, universally called DLVO theory of stability.

Having these equations in hand, one can predict the potential energy of interaction between the dispersed particles, if a complete thermodynamic characterization of the solid is performed, as described above, and if the electric potential ψ_d can be estimated. The latter is not an easy task, and in fact this potential is not directly accessible experimentally. There is, however, a relatively simple method to obtain a reasonable approximation of it, namely, measuring the electrophoretic mobility of the particles, μ_e. This quantity is related to the electrokinetic or ζ potential, which is the potential at the so-called electrokinetic or shear plane that ideally separates the mobile and immobile parts of the diffuse ionic layer (44). Although, strictly speaking, the identification of both potentials is not justified, most authors admit that ζ is a good approximation to ψ_d, as long as the surface potential is not very high (<50 mV, say).

The determination of ζ from the electrophoretic mobility of the particles requires a theoretical treatment that can be very complex for arbitrary conditions but that leads to a very simple equation for large values of the ratio particle radius/Debye length, i.e., $\kappa a \gg 1$. That is the Helmholtz-Smoluchowski equation (44):

$$\zeta = \frac{\eta}{\varepsilon_r \varepsilon_o} \mu_e \qquad [17]$$

Other theoretical approaches can be found in Refs. 44 and 45.

In order to visualize the relative weights of the different contributions in cases typically found in pharmaceutical suspensions, we have plotted in Fig. 7 each of them as a function of distance for the following choice of parameters:

$\zeta = 30$ mV
$a = 1$ μm
$A = 2 \times 10^{-20}$ J
$\gamma_{SL}^{AB} = 10$ mJ/m^2
Electrolyte: 10^{-2} M NaCl

including both the classical and extended DLVO approaches. When only LW and EL interactions are considered, the V^{TOT}–H curve shows three significant features: a deep negative minimum that if reached would mean irreversible coagulation of the particles; a potential barrier (its height and width depend on particle size, ζ potential, and electrolyte concentration) at larger distance that prevents, under the conditions chosen, the coagulation of the particles; and finally, a *secondary minimum*, rather shallow, that would give rise to the formation of loose aggregates, easy to break, sometimes called flocculi. When the acid–base interaction is also considered, the strong hydrophobic attraction provokes a rapid and irreversible particle aggregation. There is thus a wide range of possible behaviors of the system, depending on the ζ potential of the interface, the size of the particles, the ionic strength of the medium, and (something that is not always taken into consideration) the hydrophobiciy or hydrophilicity of the drug.

We have already mentioned the importance of controlling the stability (i.e., the tendency of the particles to remain unaggregated in suspension as opposed to the thermodynamic requirement of free energy reduction by decreasing the interfacial area of the system), and hence the sedimentation and redispersibility behavior, of the particles. One of the possibilities of achieving such control is to suitably modify the viscosity and rheological behavior of the vehicle; we will analyze this in the next paragraph and focus now on other methods.

It has already been noted that pharmaceutical suspensions in which a moderate particle–particle aggregation (flocculation) takes place are useful because the open and spongy structure of the flocs makes them easily redispersible: the particles form an irregular network with appreciable amounts of vehicle between them, occupying almost the whole volume of the container (Fig. 8 is an example).

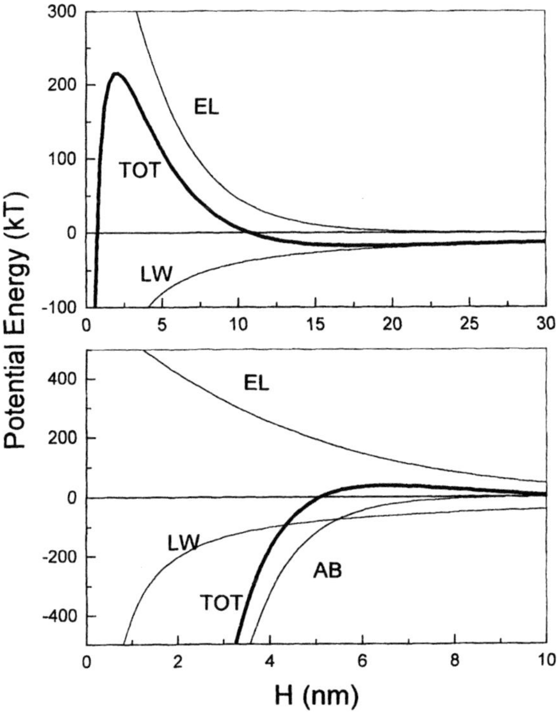

Figure 7 Plots of potential energy of interaction as a function of the distance H between the surfaces of identical spherical particles with radius $a = 1$ μm. Top: classical DLVO theory. Bottom: Extended DLVO approach. The following values were used for the calculations: $\zeta = 30$ mV; 1:1 electrolyte concentration 10^{-2} M; Hamaker constant $A = 2 \times 10^{-20}$ J; acid–base component of the solid–liquid interfacial tension $\gamma_{SL}^{AB} = 10$ mJ/m^2. See text for details.

One of the possible ways to achieve this flocculation behavior is using inorganic salts in solution. By changing their concentration, the ζ potential, as well as the double-layer thickness, will in general be modified. As a consequence, the potential energy of interaction between the particles can change from that of an essentially stable system to that of a rapidly coagulating suspension. Under certain conditions, illustrated in Fig. 7, a secondary minimum in the potential energy vs. distance curve may show up; this ideal condition would correspond, as mentioned above, to the open flocculi that can be easily redispersed and hence are suitable for oral administration.

However, this is not the best way to achieve controlled stability. The amounts of salts needed can be very high if the particle surface charge is moder-

Figure 8 Photographs of nitrofurantoin suspensions. Effect of pH on the sedimentation volume. The pH decreases from left (pH 9) to right (pH 3). The chemical instability of the drug at basic pH values is appreciable.

ate. In addition, other undesired effects cannot be ruled out, particularly when polyelectrolytes are also used, as described below. Early experiments on the effect of inorganic electrolytes on the stability of pharmaceutical suspensions were based on solutions of monovalent salts as NaCl, KCl, etc. The flocculation is more efficient with more highly charged ions, typically calcium or aluminum salts (31,46,47). Some attemps have also been made to control the stability of pharmaceutical suspensions by means of more complex electrolytes, such as amino acids (48).

In practice, polymers are very often used to control stability. They can interact with the dispersed particles, thus giving rise to polymer–particle complexes. It is known that, depending on the properties of the polymer chains and on their concentration in solution, polymeric stabilizers can be used either for stabilizing the suspension, by hindering the particle–particle contact, or for controlling the formation of aggregates (bridging or depletion flocculation), as described in Refs. 2, 5, 28, and 49–52. Figure 9 is an example: different concentrations of an ionic polysaccharide (xanthan gum) were used to control the stability of sulfamerazine suspensions; it was shown that increasing xanthan concentration brings about a logarithmic decrease in the sedimentation rate. Similar results were reported by Felmeister et al. (52) on sulfaguanidic acid suspensions.

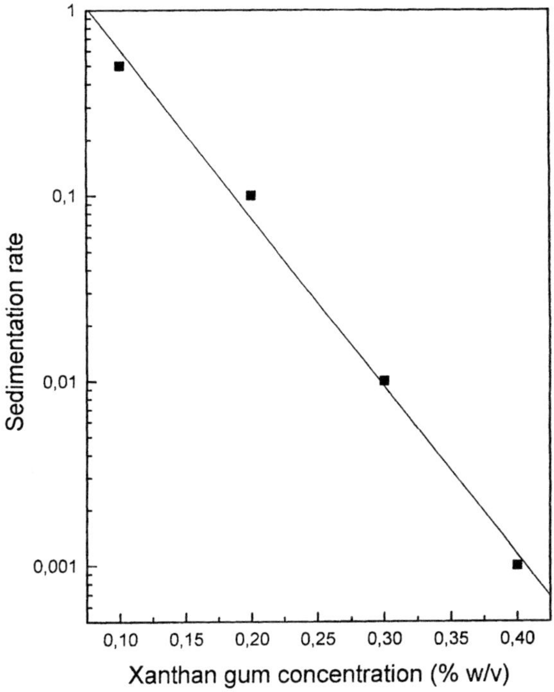

Figure 9 Sedimentation rate of 10% sulfamerazine suspensions as a function of xanthan gum concentration. (From Ref. 33. Reproduced with permission of the Society of Cosmetic Chemists.)

Nowadays there is a wide range of artificial polymers available for use as stability controllers. Carbopols (53) are among the most frequently used. They are polyacrylic acid polymers crosslinked with allyl sucrose. They are supplied as white powders, insoluble in water, that give acid pH when dispersed in the aqueous medium (pH $\approx$ 2 when Carbopol 934 concentration is 0.2% w/v), due to their content in acidic carboxyl groups; if the pH is shifted toward neutrality the dissociation of these groups gives rise to electrostatic repulsion between the anionic parts of the polymer chains, thus resulting in extended and rigid molecules and a transparent gel structure in the dispersion medium. The system obtained is very sensitive to changes in pH or electrolyte concentration; hence, the neutralizing base must be properly chosen to allow the formation of a salt that is soluble in the solvent used. Many systems can be stabilized or flocculated using Carbopol (31,54–56), but some precautions are in order, mainly concerning the possible interactions between the therapeutic drug and the polymer. For instance, cationic drugs such as promethazine or difenhydramine hydrochlorates precipitate when in contact with the negatively charged Carbopol (57). Figure 10 is a schematic representation of the overall structure of carbopol 934 before and after neutralization. As observed, when Carbopol is unsolvated, the polymer chains are folded and do not show any significant thickening efficiency; when neutralized, repulsion between the negative charges on the chains brings about an extended structure than can be quite rigid.

Colloidal silica is another example of a compound used in stability control. It has the particular advantage of forming gels also in nonpolar solvents; hence its use for the stabilization of nonaqueous pharmaceutical suspensions. In fact, while 20–30% concentration of silica is needed to form stable gels in aqueous vehicles, 3–6% SiO_2 is enough in the case of nonaqueous vehicles. The mechanisms by which silica gel is formed are shown schematically in Fig. 11, for the case of aqueous solvents (58): in basic pH, with no salt added, a monomer solution containing $\approx$1% silica will initially show the formation of aggregates, but immediately the branched polymer will condense (process C in Fig. 11) into a colloidal particle. However, at low pH, silica particles are almost uncharged (their isoelectric point, pH_{iep}, is $\approx$3.5); when two particles collide, Si—O—Si bonds form between them by condensation of silanol groups into siloxane bonds, and

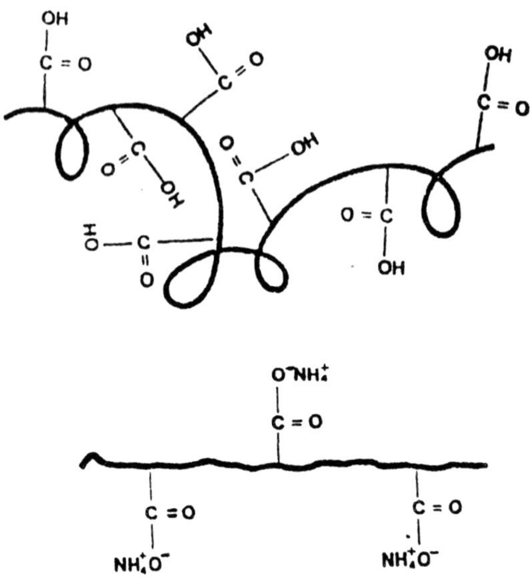

Figure 10 Schematic representation of the structure of Carbopols in acid (top) and neutral pH (bottom).

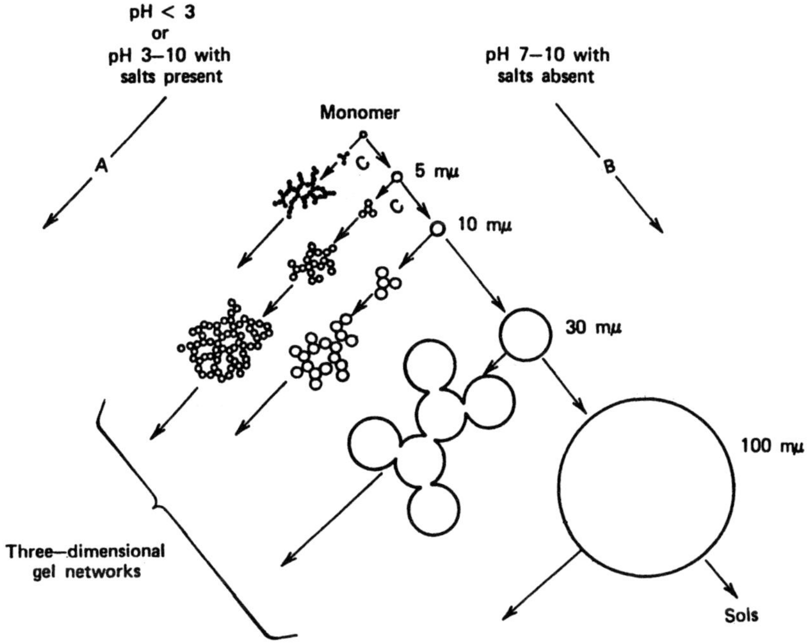

Figure 11 The formation of silica gels in aqueous media. (Taken from Ref. 50. Reprinted with permission of John Wiley and Sons, Inc.)

the particles are irreversibly aggregated (58). In the case of nonaqueous solvents, trace amounts of water suffice to cement together hydrophilic silica particles into a three-dimensional gel extending through the nonaqueous phase. There are many examples of both commercial and laboratory formulations containing silica gels in order to control the stability of pharmaceutical suspensions (see, e.g., Refs. 59–61).

Silica itself may be an example of a large variety of substances that can be named *protective colloids* (4), and the list also includes cellulose derivatives, clays, natural gums, or polymers. These are colloids that can adsorb onto the insoluble drug particles, thus increasing the strength of the hydration layer formed around them and favoring stability (62). Most of these colloids are protective, i.e., adsorb individually or in small aggregates on the drug particles when in low concentration in the medium ($\leq$1%), whereas at larger concentrations they can form gel networks and behave as viscosity builders, or thickening agents, that prevent the sedimentation of the therapeutic compound by entrapping it in the network or by sufficiently increasing the viscosity of the medium. We will focus here on their effects at low concentration and consider below their rheological action at larger concentrations.

When adsorption of any of these compounds occurs, part of the surface of the drug particles will be covered by them. The covered regions will be *protected* against flocculation with other covered regions, thus reducing the coordination number of any drug colloid in an aggregate. In the case of polymers adsorbed on the particles, thay can also produce flocculation (bridging flocculation) by simultaneously adsorbing on two particles. This possibility is conditioned by the length of the molecule and the number of charged functional groups along it.

C. Factors Affecting Flow Properties

In the simplest experimental conditions, the viscosity of a suspension depends on the concentration, shape, and even ζ potential (the ''electroviscous,'' effects) of the particles (63–65). Figure 12 (63) is an example of experimental data reported on different systems; the behavior observed is Newtonian except for very high-volume fractions, and in such concentrated suspensions the frequency of particle–particle collisions is so high that the stability of the whole system will be largely affected. It is hence difficult, if not impossible, to control the flow properties of pharmaceutical suspensions without the use of special additives, almost systematically included in practical formulations.

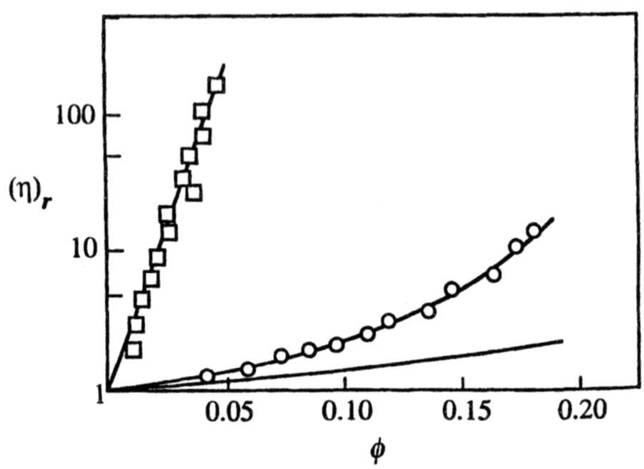

Figure 12 Relative viscosity of different suspensions as a function of the volume fraction. Circles: TiO_2 particles 0.1 μm in diameter; squares: carbon black; continuous line: ideal dilute sphere result. (Taken from Ref. 63. Reprinted with permission of John Wiley and Sons, Inc.)

It is clear that increasing the viscosity of the dispersion medium will reduce the frequency of collisions while simultaneously hindering their sedimentation—both desirable features in the design of a suspension. However, it must also be mentioned that the viscosity cannot be increased to the extent that the flow of the system out of its container is impractically low, since that circumstance would make the suspension useless from the practical standpoint. So, it is the whole rheological behavior of the system for practical shear stresses that must be controlled.

Like many other additives, the fact that the suspensions will be swallowed by or injected into human beings is determinant: the viscosizer chosen must not have therapeutic activity by itself, and must not be toxic or irritant. Concerning the physicochemical requirements of the additives chosen to control the rheological behavior of the suspension, the following could be a possible list:

They must be water-soluble or swell in aqueous media.
The rheological properties they impart to the suspension should not change significantly during the shelf life of the formulation.
They must be compatible with other components of the suspension.

In spite of these conditions, the designer has a wide range of products available for controlling the rheology of the dispersion:

1. Cellulose Derivatives

They are obtained from α-cellulose by chemical modification of the hydroxyl groups of the molecule. Such modifications yield viscosizers of well-defined, stable physical and chemical properties. Their prize is relatively low, and for this reason they have substituted natural products (such as gums) also used to control rheology.

All gels obtained from cellulose have two common properties that must be taken into consideration when using them in a formulation (33).

They can accept electrolytes at low to moderate concentrations without significantly alter their stability (microcrystalline cellulose is an exception, as it can flocculate even in the presence of small salt concentrations).
They can be easily contaminated by microbials. As a consequence of such contamination, viscosity losses by depolymerization, color changes, or chemical decomposition may occur; hence the need of using compounds inhibiting the growth of microorganisms.

The semisynthetic cellulose derivatives, such as methyl-, ethyl-, or propyl-cellulose, can display various degrees of viscosity increase in the suspension; their rheological behavior is typically pseudoplastic (see Chapter 16 in this volume), and they are compatible with most therapeutic drugs. One of the most frequently used is sodium carboxymethylcellulose, although its anionic character precludes its use with cationic drugs.

There are commercial cellulose-derived compounds in which a combination of derivatives is used: for instance, Avicel RC 591 (a registered product of FMC Corp; see Ref. 66) includes microcrystalline cellulose that provides thixotropy to the suspension, and sodium carboxymethylcellulose, with pseudoplastic nature, needed to ease the flow of suspension for administration. Its usefulness for rheology control has been checked with many drugs; for example, Córdoba et al. (67) have shown that sulfamethoxazole suspensions with 1% Avicel display excellent stability and flow behavior.

2. Clays

These inorganic crystals are able to swell and form thixotropic gels in water. This behavior, and its dependence on pH or electrolyte concentration, can be explained by the crystal structure. The basic building blocks of clay minerals are SiO_4^{4-} tetrahedra (68,69); since each oxygen of the basic silicate tetrahedron has one electron available for establishing chemical bonds, there are many possibilities for the formation of silicate structures; in the case of clays, two-dimensional arrays of tetrahedra are formed by binding of three corners in the same plane of each tetrahedron to another three units. The resulting sheets have still electrons available for binding to other structural units; for example, in the case of kaolinite $[(Si_2O_5)(OH)_4Al_5]$ a layer of silicate tetrahedra is bound to another two-dimensional arrangement of $Al_2(OH)_6$ octahedra by oxygens from the silicate tetrahedra.

When the sharing of the oxygen atoms occurs in such a way that alumina sheets have bonds with two silica sheets, the montmorillonite-type clays originate. Very often, atom substitutions take place both in the tetrahedral and octahedral lagers (68). Thus, Si is sometimes replaced by Al in the former, whereas divalent Mg may substitute Al atoms in the latter; other possibilities include substitution of Al by Fe(III), Cr,Zn, Li, and other atoms. This ''isomorphous substitution'' provides a net negative charge to the clay particles; such excess charge is compensated by the adsorption of cations on the surface of the particles. When dispersed in water, these compensating cations tend to difuse into the medium, thus giving rise to an electric double layer around the particles faces. Furthermore, the edges of the clay particles exhibit a pH-dependent charge originated by the disruption of silica or alumina sheets; both experimental and theoretical observations (68,69) suggest that the edge surface charge is positive for a wide pH range, whereas isomorphous substitutions provide a negative sign to the face surface charge. These considerations are important, since in concentrated clay suspensions clay platelet association through edge-to-edge and edge-to-face interactions may give raise to card-house-like structures that extend throughout the system, thus forming a gel; most often, such gels are found to behave as Bingham systems, as shown in the examples of Fig. 13.

These are many examples in the literature, concerning the usefulness of clays for the control of the stability of pharmaceutical suspensions; thus, Córdoba et al. (67) studied the effect of a natural sepiolite on the stability of sulfamethoxazole, Schott (70) analyzed mixtures of bismuth subnitrate and montmorillonite, and McGinity and Harris (71) analyzed how the presence of montmorillonite can affect the dissolution rates of poorly soluble drugs.

3. Natural Gums

Natural gums are also used frequently either as protective colloids (for low concentrations, <0.1%) or thickeners (for concentrations above 0.1%; see Ref. 5). They are water-soluble, and most of them negatively charged, so they can be incompatible with cationic drugs. Some drawbacks concerning the use of gums must be mentioned: they can easily get contaminated by bacteria, and furthermore the organoleptic properties of the suspensions may change with time. In many instances, the gums tend to be replaced by cellulose derivatives or by polymers.

There are, however, many gums that have been (and still are) used in the preparation of suspensions; some examples are (5,72) agar, alginates, guar gum, and xanthan gum, the latter being the most frequently found in pharmaceutical and food suspensions. Figure 14 is a schematic representation of this polysaccha-

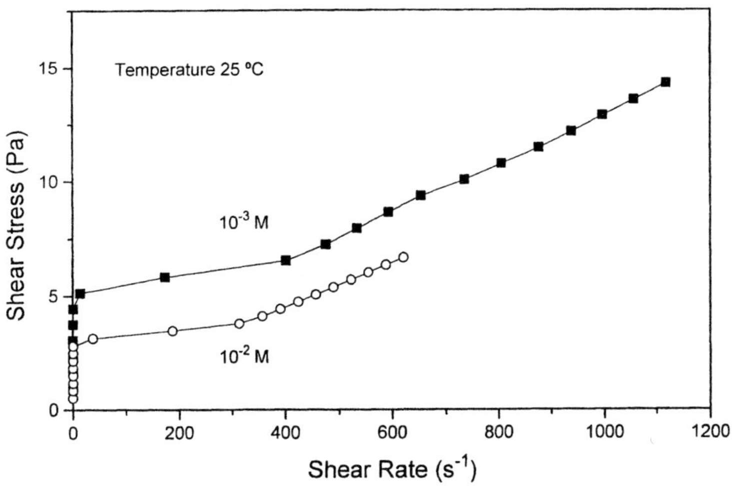

Figure 13 Rheograms of 5% montmorillonite suspensions of different NaCl concentrations; pH: 5.5–6.0.

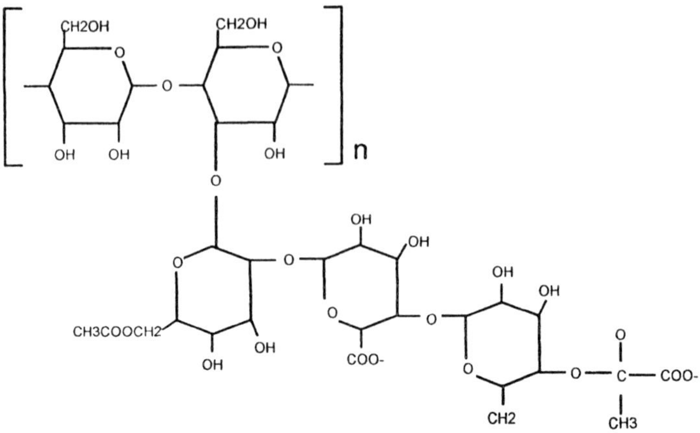

Figure 14 Structure of xanthan gum.

ride; in ordinary aqueous solution, the molecules exist as individual rodlike structures, approximately 1 μm long, stabilized by electrostatic repulsion between negatively charged trisaccharide side chains. For sufficiently high electrolyte concentrations, such repulsion is screened, and a gel network is produced by interactions between the rodlike molecules. When sufficient shear stress is applied, the gel is broken, and both the viscosity and the elastic behavior decrease.

It is interesting to mention that xanthan presents fewer drawbacks than most other gums; it has higher resistance to microbial contamination, is perfectly soluble in both cold and hot water, and the rheology of suspensions containing this gum is somewhat independent of temperature and pH in a wide range of conditions (72,73).

4. Polymers

Nowadays most of the above-mentioned natural compounds are increasingly substituted by polymers; they are mostly hydrophilic, although in general they show a tendency to form aggregates when dispersed in water. Hence it is necessary to incorporate them into the solution slowly and under constant stirring, so as to obtain an opalescent and homogeneous suspension (73,74). The most frequently used polymers are:

Carbomers (polyacrylic acids). Manufactured by B. F. Goodrich (53) under the tradenames of Carbopol 907, 934, 934P, 940, etc. As previously mentioned, initially acid dispersions of Carbopol must be neutralized with NaOH, KOH, or NH_4OH to uncoil the polymer chains by generation of negatively charged carboxylic groups along them. Typical rheograms of a 0.2% Carbopol 934 solution as a function of pH are shown in Fig. 15. As observed, the neutral solutions are very pseudoplastic and shear-thinning. This makes them very useful to produce stable creams and lotions that can be easily spread on the skin, with little shear stress. Typical Carbopol concentrations range between 0.1% and 2% in most applications.

Polyvinylpyrrolidone (PVP). It is soluble in cold water and many organic solvents. It is mainly used as a protective colloid, since PVP solutions do not show high viscosities.

Polyvinyl alcohol (PVA). It is soluble in water and water–alcohol mixtures, when completely hydrolyzed. As in the case of PVP, its effect on the viscosity of the solution is not very high, but PVA is also very useful as a protective colloid since it can easily form complexes with the dispersed therapeutic drug. However, a network can be obtained from PVA, both by chemical and physical methods (75): the chemical procedure involves the formation of bridges between the linear chains by acethalization, e.g., by addition of glutaraldehyde. Transparent hydrogels can thus be obtained. However, there is a risk that residuals of the reaction remain in the system, with possible toxic effects. These drawbacks do not exist when physical methods are used: γ radiation of suitable energy is passed through the PVA solution, and as a consequence free active radicals (H·, OH·, Cl·, and so on) are formed that in turn produce polymer radicals. The network is produced by the attachment of two adjacent macromolecular radicals. Both the physical and chemical methods described are rather involved, and hence seldom used to control the flow properties of the suspensions.

Polyethylene glycols (PEG-200, 300, etc.). They can be used between pH 3 and 10; their typical concentrations are dependent on the type used (i.e., on their molecular weight). These polymers are also used as plasticizers or lubricants.

D. pH Regulators

Ideally, a pharmaceutical suspension should be stable in a wide pH range. But in some cases (e.g., nitrofurantoin is soluble at alkaline pH and is chemically degraded at that pH range; ferrous fumarate oxidizes also under alkaline conditions), the chemical stability of the active compound requires that the pH of the suspension be maintained within specified values. Obviously, the remaining components of the suspension must be selected in such a way that they produce their intended effects (on stability, viscosity, etc.) at precisely that specified pH range.

A careful choice of suitable, pharmaceutically acceptable buffers must be carried out. Attention must be paid not only to the potential toxicity of the buffer

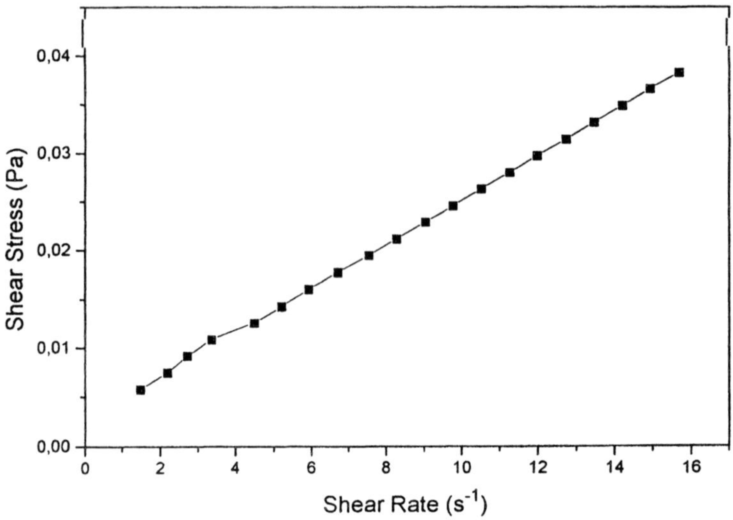

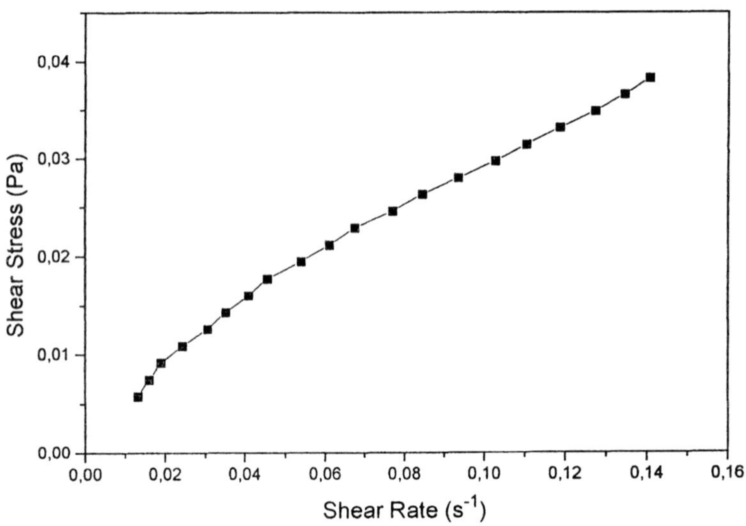

Figure 15 Rheograms of acidic (top) and neutralized (bottom) Carbopol 934 solutions for a 0.2% (w/v) concentration.

components but to their effects on the physical and chemical stability of the other suspension components: mainly in the case of highly charged cations (calcium, aluminum), small amounts can significantly alter the surface charge of the particles (Chapter 4). Precisely citrates and phosphates, the pH buffers most frequently used in pharmaceutical suspensions, include doubly charged cations and anions. It is hence suggested to perform preliminary measurements of electrophoretic mobility and stability to ensure that no undesired effects will occur.

E. Other Additives

As we have mentioned, a number of other additives are still needed (in addition to those described in the previous paragraphs) before the pharmaceutical suspension is ready to be used by the patient. Let us mention (a) *coloring agents* (must be nontoxic and soluble in the external phase, and should not react with the therapeutic drug; each country has a list of admitted colorants) and (b) *flavours* (the same restrictions and precautions mentioned above apply to flavoring substances). There are a great variety of them, and most oral pharmaceutical suspensions contain a flavoring agent (76).

However, most important is the addition of a preservative substance to prevent microbial growth in the suspension. Preservatives are needed in most cases, since the suspension components, and mainly the flavors, constitute an excellent culture medium for the growth of microorganisms. Another source of contamination is the use of natural products (like gums and clays) to control the stability of the suspension. There are many acceptable preservatives to inhibit the microbial growth in orally administered suspensions (see Ref. 5 for an exhaus-

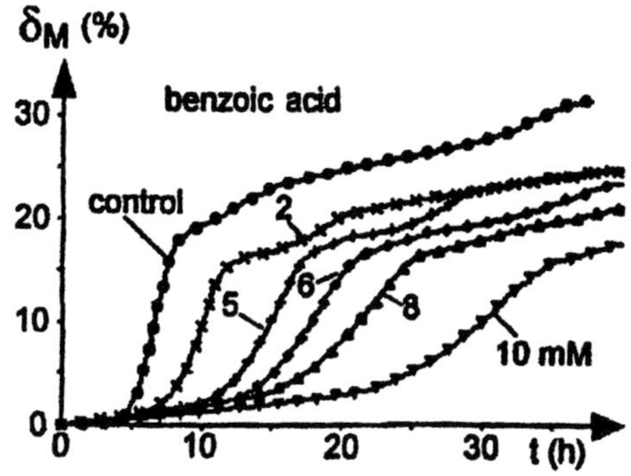

Figure 16 Growth of *E. coli* as a function of time for different concentrations of benzoic acid, as measured by the percent decrease of the impedance of the culture medium. (From Ref. 77. Reprinted with permission of Globetech Publishing Corp.)

tive list), the most frequently used being parabens, alcohol, glycerin, propylene glycol, benzoates, and sorbates. They are typically used in concentrations around 0.1–5%. Figure 16 is an example of *E. coli* growth in the presence of different concentrations of benzoic acid, as a function of time (77); the protective effect of this compound is clear. Other results can be found in Refs. 78 and 79.

IV. APPLICATIONS OF PHARMACEUTICAL SUSPENSIONS

In this part of the chapter, we give information about a number of pharmaceutical suspensions, including oral, parenteral, and topical, that are somewhat representative of different therapeutic groups. The list is not exhaustive since the number of drugs that can be administered to the patient in the form of suspension is very large and in fact increases almost every day. Throughout the chapter we have given details about the multiple requirements concerning the therapeutic agent itself, the dispersion medium, and the additives that must be included in the formulation. We shall assume that these requirements are met in all cases; hence, we will not give the details of all components of the suspensions, except in some examples, which will be discussed in more detail.

A. Oral Suspensions: General Considerations

This type of suspension is especially interesting, given the large number of drugs orally administered to the patient. As already mentioned, oral suspensions are used for a number of reasons:

- Low solubility of the active principle
- Better organoleptic characteristics
- Chemical stability of the drug
- Possibility of control of the bioavailability of the therapeutic compound through variations in the particle size, the viscosity of the vehicle, polymer adsorption and desorption on the particles, and so on.

Oral suspensions may also have drawbacks:

- Changes in the dose actually given to the patient due to particle sedimentation, formation of polymorphs of the drug in the medium, or crystal growth of the former (80–82)
- Need of microbial growth control
- Possible evaporation of solvent and changes in drug concentration
- Effect of temperature changes on the behavior of the suspensions
- Need of additives to control fragrance, taste, and color

A general scheme of the way of preparation of these suspensions can be seen in Fig. 17. Note that the usual way to proceed is as follows:

1. Disperse the desired amount of the insoluble therapeutic compound in the solvent; addition of a wetting agent to improve wettability of the drug might be necessary at this stage.
2. The suspension of particles thus obtained is added to a structured vehicle in order to obtain a deflocculated suspension; or a flocculating agent is added to the particles thus obtaining a flocculated suspension; the third possibility is to add to the already flocculated suspension a structured vehicle so as to have a final suspension that is deflocculated but that stays as it is, due to the presence of the vehicle.
3. The remaining additives are then incorporated, taking the necessary precautions to avoid the formation of agglomerates, very difficult to redisperse in these latest stages of the process.

It is particularly important to assess the stability of the final product (84) both at the laboratory and at the manufacturing scales. The chemical stability of the drug must have been checked at the initial stages of the design; its solubility in the vehicle must be low, and in fact that small but finite solubility can limit the period in which the suspension can be used. Most probably, the main part of the degradation by chemical decomposition is due to the additives of the suspension, such as thickness, protective colloids, preservatives, and so on. A rapid checklist of the possible degradation of any of the components can include:

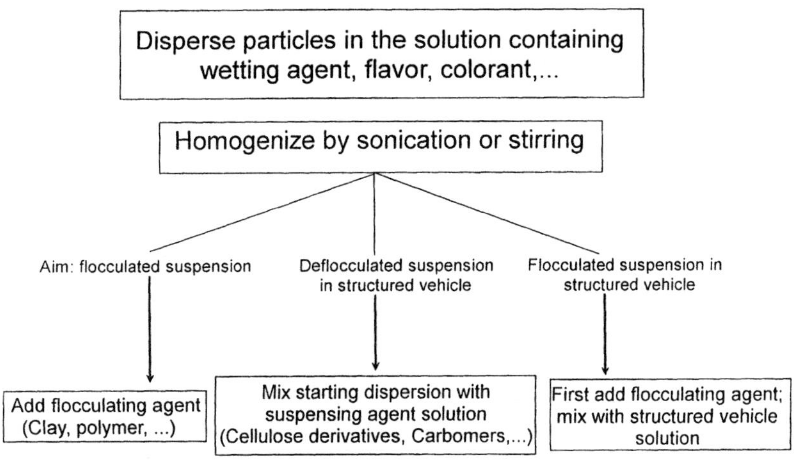

Figure 17 Schematic representation of the typical stages in the preparation of an oral suspension. (Adapted from Ref. 83, p. 252, by courtesy of Marcel Dekker, Inc.)

Estimating particle size changes by proton correlation spectroscopy (see Chapter 17 in this volume)
Measuring the electrical conductivity of the systems
Carrying out a rheological test to verify possible changes in viscosity

Unlike chemical stability, physical stability (in particular, particle aggregation and/or sedimentation) is easier to control. A list of possible tests has been given by Ofner et al. (83) and by Idson and Scheer (85). Besides considering the simple visual appearance of the suspension, it is advisable to:

Measure the sedimentation volume, if any, as a function of time. Increasing sediment may be indicative of formation of particle aggregates.
Check the electrophoretic mobility, μ_e, of the particles: changes in μ_e can suggest (unexpected) adsorption of desorption of chemical species on or from the particles. Modifications in the stability of the suspension due to those in the electric surface potential cannot be ruled out.
As before, PCS analysis of the suspension provides clear indication of the existence of aggregative phenomena in the system. Sometimes, even a simple turbidity (i.e., optical absorbance) measurement may help in deciding if a deeper stability analysis is needed.
If available, since the rheology of the system is strongly dependent on the existence of the aggregates, a rheogram is again a good check of stability.
Furthermore, since the preparation will be transported in several ocasions, stored at maybe widely different temperatures, etc., the stability of the suspension against vibrations or moderate temperature changes might be also checked, especially at the very early stages of design (83,85).

B. Oral Suspensions: Fields of Application

The therapeutic fields where oral suspensions are applicable are, as we have mentioned repeatedly throughout the chapter, very numerous. New preparations emerge continuously, either including new active compounds or modifying the components (in particular, type and concentration of suspension additives) of already existing formulations. Instead of showing an exhaustive list, we find it more useful to give a number of examples classified by therapeutic groups. Table 3 summarizes the information; we restrict ourselves to the name of the therapeutically active compound and its typical concentration in the suspension. In the next paragraph we will give more detailed examples on the actual ingredients and ways of preparation.

As observed, pharmaceutical suspensions can be therapeutically useful in almost any disease treatment. In most cases, suspensions of different compounds can be used in the same field, with similar efficiencies. Furthermore, most of the drugs cited can also be given in the form of caplets, tablets, parenteral suspensions

Table 3 Some Typical Applications of Pharmaceutical Suspensions

Therapeutic effect	Active compound	Typical concentration (mg/mL)
Antifungal	Ketoconazole	20
Antihelminthic	Pirantel pamoate	50
	Tiabenzole	60
Anxiolytic	Diazepam	0.5
Calcium antagonist	Nicardipine	20
Antacid	Almagate	130
	Aluminum hydroxide	70
	Magnesium hydroxide	200
Antianemic	Folic acid	10
	Ferrous gluceptate	30
Antibacterial	Nalidixic acid	125
	Amoxicillin	50
	Ampicillin	50
	Cefalexin	50
	Cefradoxyl	50
	Chloramphenicol palmitate	25
	Nitrofurantoin	10
Antiepileptic	Diphenylhydantoin	25
Cough relief	Codeine	6
	Dextromethorfane	0.5
Anti-inflammatory	Ibuprofen	20
Antiviral	Acyclovir	80
Nasal congestion relief	Phenylpropanolamine	3
Immunological estimulation	Palmidrole	100
Intestine motility estimulation	Cinitapride	1
Intestine motility inhibition	Albumin tannate	50

and solutions, etc. We have already discussed in which cases the suspensions are preferred.

Concerning particle concentrations, data in Table 3 show a great variability, depending on the potency and bioavailability of the each particular drug. Typically, the most concentrated suspensions contain above 100 mg/mL, whereas the lowest limit is in the range of 1 mg/mL.

C. Reconstitutable Suspensions

Some of the systems mentioned in Table 3 are in fact reconstitutable suspensions, i.e., suspensions that reach the user as a mixture of dry powders that are dispersed in the solvent (usually water), i.e., that are *reconstituted* immediately prior to their use. Once reconstituted, they have a limited shelf life, typically in the range of weeks, if properly stored. This type of suspension administration is the choice when the chemical stability of the drug is a major concern (86). For instance, amoxicillin suspended in water can be used for 15 days if kept in a refrigerator, whereas the dry powder, including all the additives, can stay on the shelf for more than 2 years.

Perhaps the main point in connection with these suspensions is that they will be finally prepared by the user, not the manufacturer. The former has not available soniccators or powerful stirrers, so the powder must be easily redispersed in water, by simple shaking. As mentioned in (86), for the same reason the number of ingredients accompaying the therapeutic compound must be kept to a minimum. In order to avoid particle aggregation, a thickener must be included in the powder mixture; again, the polymer chosen should be easily dissolved in the vehicle, and this limits the possibilities somewhat; cellulose derivatives and naturals gums are preferred, whereas Carbopols are not recommended in this case.

Another necessary ingredient in most cases is a wetting agent, used to improve the dispersibility of the drug; as mentioned in Sec. III.A of this chapter (see also Chapter 1 in this volume), surfactants are typically used with this aim. The most commonly used surfactants are polysorbate 80 and sodium dodecyl sulfate.

Finally, a list of other possible additives (pH buffers, flavors, preservatives, sweeteners) normally used can be found in Refs. 85 and 86, and an example of an actual formulation will be provided in Sec. IV.G.

D. Parenteral Suspensions

The preparation of these suspensions is carried out with consideration paid to:

Obtaining a prolonged effect of the drug by generation of a pool or depot of therapeutic agent in the place where it has been injected.

Administering a drug that is insoluble in the somewhat reduced number of injectable vehicles.

The use of injectable suspensions brings about new requirements; first of all, the agents used in their formulation must be safe, i.e., they should not be toxic, should not have pyrogenic or antigenic effects, or be irritant or hemolitic. Furthermore, wetting or stabilizing agents must be rather potent, since they must show their effect even when in very low concentrations, as well as stable, capable of maintaining their characteristics after prolonged storage or after sterilization (87). The solvents (usually water or vegetable oils) must be free of particles, and all of the operations need to be done in sterile environments with aseptic techniques, in order to avoid contamination by particles or microorganisms, and hence any pyrogenic effects. It must be noted that heat sterilization cannot be used with the final powder because its stability can be affected and, furthermore, crystal growth and protecting colloids modification are potentially possible.

The formulation of aqueous injectable suspensions follows, in general, the same lines that were established when drugs such as cortisone, penicillin, or procaine were first prepared in this form. The main ingredients are, in addition to the active compound, wetting agents and protective colloids. As in the case of reconstitutable suspensions, polysorbate 80 is most frequently used to decrease the solid–liquid interfacial tension, this promoting the sustitution of a solid–air interface by a solid–water one. Another surfactant that is often employed is sodium polysuccinate.

An efficient suspension also requires a suspending agent, normally a protective hydrophilic colloid, used in low concentration to avoid significant viscosity increases in the suspension, as compared to that on water. Very viscous vehicles provoke painful injections; the same can be said about suspensions formed by too large particles. The most frequently used colloidal protectors are sodium carboxymethylcellulose, polyvinylpyrrolidine, polyvinyl alcohol, or gelatin. These are all first choice when a new formulation is designed; however, for very insoluble particles like hydrocortisone butyl acetate or prednisolone butyl acetate, sorbitol has proven useful.

Ideally, the suspensions should be stable in a wide pH range. Nevertheless, if it is necessary to maintain a particular system in a narrow pH range, buffers can also be employed; among them, acetate, citrate and phosphate are the most often used buffers (87). They are needed mainly when the therapeutic drug has ionizable groups, most of all when pH affects solubility.

As before, some examples concerning the preparation and ingredients of parenteral suspensions will be provided in Sec. IV. G.

E. Topical and Cosmetic Suspensions

Most of the features and formulation principles discussed in connection with oral suspensions apply to topical or cosmetic preparations. There are, however, some new factors to be taken into account in the latter case. First of all, in any of these suspensions the concentration of the dispersed phase is 20% or higher in solids. It is essentially impossible to keep such a highly concentrated suspension stable without using additives with that purpose. The possibility of controlling the stability by simple electrolyte addition, i.e., by controlling the ζ potential of the particles can be discarded in most practical situations.

For these reasons, a proper choice of the additives used together with the active compound is essential. Most often, topical suspensions contain (33):

A thickener such as cellulose derivatives, Carbopols, clays, or natural gums. More than one of these components can be used, e.g., a gum combined with a silicate (33).

A wetting agent: it helps the wetting of the drug and, upon adsorbing on the surface of the particles, improves their stability by electrostatic repulsion, protective action, or both. Additives typically used include surfactants such as polyoxyethylene, sodium dodecyl sulfate, and, more recently, silicone surfactants (88), and polyols (particularly propylene glycol and PEG; see Ref. 89).

A film-forming polymer, such as Aquacoat, that if needed may help in the control of the release rate of the therapeutically active agent.

Many topical suspensions find application in dermatology. Thus, the treatment of acne through topical application of benzoyl peroxide has been particularly successful; the same can be said about local treatment of viral or fungi infection. Other drugs, while having local effect, can spread from the place of application, showing also a systemic action. This is the case, for instance, of steroidal antiinflammatory preparations.

The cosmetic applications are also very numerous, ranging from sun protectors (cinnamic acid and benzoic acid derivatives), to antiperspirants (aluminum salts), to shampoos, toothpastes, and makeup preparations.

F. Aerosol Suspensions

The use of aerosols for oral, nasal, and topical drug administration has found widespread application since the 1950s. Their ease of use and high therapeutic efficacy make them very acceptable to patients. Aerosols are mainly used to administer therapeutically active compounds suitable for topical administration, or for direct action in body cavities, most often by inhalation through the respiratory tract (90).

Aerosol suspensions have advantages over other dosage forms in that they show rapid therapeutic action, avoiding the first-pass effect, and they are a valid administration route for drugs that suffer degradation in the gastrointestinal tract. Furthermore, the doses normally used are low; hence there are few side effects. Nevertheless, they are not indicated in those cases where the active ingredient provokes irritation of the patient's mucosa (as happens with inhaled drugs that have little solubility in the respiratory fluids), or where large doses are needed to achieve the therapeutic effect (90).

The aerosol product has two necessary components: the concentrate containing the active compound(s) and the propellant gas mixture. The concentrate can also be an emulsion, solution, or semisolid paste, but here we restrict ourselves to the suspension case in which the active compound is directly dispersed in the propellant vehicle. This type of preparation is not free of difficulties, since caking, particle agglomeration, crystal growth, and so on are not as easily controllable as in the case of solid–liquid systems.

There are, however, methods to obtain efficient aerosol suspensions (90–91):

Decrease the rate of settling and the degree of flocculation by adding surfactants or dispersing agents (lecithin, oleic acid, ethyl alcohol).
Reduce the particle size below 5 μm.
Match the densities of active principle and propellant mixture.
Minimize moisture content.

With these methods, aerosol suspensions of epinephrine bitartrate, steroid compounds, and bronchodilatants (such as salbutamol, phenylephrine, disodium chromoglycate) have been successfully prepared.

G. Examples

Although it is not our aim to focus on the technological aspects of pharmaceutical suspensions preparation, it may be useful to consider in some detail how some suspensions can actually be prepared to meet the required conditions of stability, pourability, aesthetic appeal, etc., that we have mentioned. The systems described below are intended as examples of the preparation routes used in most practical situations.

Example 1: Sulfamethoxazole (Antibacterial, Ref. 59).
Oral suspension ingredients:

Sulfamethoxazole	5 g
Suspending agent (Avicel RC591)	1–2 g
Sodium methyl- and propylparaben	0.1 g
Polysorbate 80	0.05 g
Curcumin	0.08 g
Water	qs ad 100 mL

Example 2: Acetylsalicylic acid (Ref. 92)
Reconstitutable suspension ingredients

Acetylsalicylic acid	6.0 g
Sorbitol	70.0 g
Citric acid	0.1 g
Purified water	qs ad 100 mL

Example 3: Injectable betamethasone suspension (Ref. 93)
Parenteral suspension ingredients:

Betamethasone (sodium phosphate)	3.0 mg
Betamethasone acetate	3.0 mg
Sodium dibasic phosphate	7.1 mg
Sodium monobasic phosphate	3.4 mg
NaOH or HCl	qs ad pH 6.8–7.2
Water for injection	qs ad 1.0 mL

Example 4: Triamcinolone diacetate (Ref. 93)
Parenteral suspension ingredients:

Triamcinolone diacetate micronized	40 mg
PEG-4000	3.0 mg
Polysorbate 80	2.0 mg
Sodium chloride	8.5 mg
Benzyl alcohol	9.0 mg
NaOH or HCl	qs ad pH 6.0
Water for injection	qs ad 1.0 mL

Example 5: Isoprenaline aerosol (Ref. 91)
Ingredients (in %):

Isoprenaline sulfate	0.1
Atropine sulfate	0.04
Sodium dioctyl sulfosuccinate	0.02
Propellant gas mixtures	qs ad 100

Example 6: Calamine lotion (Ref. 94)
Topical suspension ingredients:

Calamine	8 g
Zinc oxide	8 g
Glycerine	2 mL
Bentonite magma	25 mL
Water	qs ad 100 mL

V. SUSPENSIONS AS DRUG DELIVERY SYSTEMS

A. Introduction

Generally speaking, the therapeutically active compounds contained in a conventional pharmaceutical formulation may become distributed throughout the body, depending on their physicochemical characteristics. Thus, a sufficient concentration in the target organ requires relatively high doses, with the possible consequence of unwanted side effects (95,97). Cancer chemotherapy is one of the most significant examples; in addition to having effect only on tumor cells, most anticancer drugs show high toxic effects on normal cells.

In fact, from a historical point of view, the importance of the dosage method increased parallel to the potency of the newly designed drugs and their associated risk of severe adverse effects. Pharmacokinetic studies soon demonstrated that it is the rate and extent of drug adsorption (its bioavailability), and not only the dose, that control the therapeutic action of the pharmaceutical.

For these reasons, a great effort has been devoted, in the last 20 years or so, to the design of drug delivery systems, i.e., systems in which the drug is associated with some other chemical, or is administered with the help of some device or process, in such a way that the rate of release, the target where it must be released, or both can be controlled (96,98). Ideally, a drug delivery system should be able to selectively supply the drug to the organ or tissue that requires it, and to do so at the required dose and time scale.

The advantages of these systems and their applications have been thoroughly studied in recent years (95, 97–99). Some of them are:

- Protection against inactivation of the therapeutic drug (by chemical, enzymatic, or inmunological factors) between the administration and the target sites.
- Better acceptance by the patient through less frequent dosage.
- Improvement in the transport of active substances to particularly inaccessible sites (in the case of bacterial or parasitic infections in intra- or extracellular regions, not reachable by simple diffusion).
- Increased safety because of the lower dosage and the lesser extent of drug delivery in nontarget tissues.
- Increased specificity of action through selective, efficient, and constant concentration of active compounds in the desired site. With lower doses, the therapeutic action is comparable to or better than that achieved with standard formulations, but with greatly reduced side effects.
- Reduction of toxic metabolites.
- Decreased undesired effects in the gastrointestinal tract.

Disadvantages, though of little significance as compared to the potential advantages above listed, can also be mentioned:

- Possibility of toxicity or insufficient biocompatibility of the drug vehicle, if used (100).
- Pain when the device is placed in the body or during its presence in it.
- High cost of the system.

The existing designs of controlled release systems are very numerous; a simple classification can be carried out attending to the mechanisms through which the active drug release is actually controlled (101–103):

1. The release takes place through a Fickian diffusion process. In this group one can find both reservoir systems, in which a membrane controls the process, and monolithic matrix systems.
2. A chemical reaction takes place at the interface between the drug carrier and the external solution; sometimes bioerosion of the polymer carrier takes place, whereas in other cases the drug is covalently bound to the carrier and is dissolved by breaking of such bonds.
3. The external solution penetrates the (normally, polymeric) carrier, dissolves the drug, and returns out as a drug solution through laser-drilled orifices.
4. Systems in which the drug release is modulated by external, e.g., magnetic, fields.

Other authors focus on the means of administration, in order to obtain a classification (96):

1. Oral drug delivery systems
2. Injectable systems
3. Implantable drug delivery systems
4. Noninvasive (transdermal, respiratory, intranasal, lymphatic, rectal, intrauterine, intravaginal, topical)

In the context of the present chapter, whatever the mechanisms of drug loading and release, or the means of administration, we must focus on suspension-based drug delivery systems, also called particulate delivery systems. They include:

1. *Liposomes and niosomes*: They are, respectively, small vehicles based on phospholipids and synthetic surfactants.
2. *Nanoparticles*: Colloidal particles in the nanometer range of sizes with drug adsorbed on them.
3. *Nanocapsules*: Colloidal hollow particles filled with the therapeutic compound.
4. *Microspheres or microcapsules*: They are similar to nanoparticles or nanocapsules but larger in size (several hundreds of nanometers in diameter) and are normally administered via intraarterial catheters.

Since liposomes are considered in detail in this volume, we will focus here on a brief account of the characteristics and applications of solid colloidal particles as drug vectors. Since most of them are polymeric, we will devote the next section to polymer-based micro- and nanoparticles.

B. Polymer Latexes as Drug Vectors

Both synthetic (biodegradable or not) and natural polymers have been proposed and tested as drug delivery systems (104,105). It was Speiser (104,106) who first prepared spherical capsules made of a polymeric material capable of being loaded with active drugs by entrapment or adsorption. The method was based on the so-called micellar polymerization of such monomers as acrylamide or methyl methacrylate. Since those first contributions, the number of monomers potentially useful in the field, as well as the polymerization routes employed have grown almost exponentially, and so have the fields of application in pharmaceutical dosage.

Polyacrylamide and polymethyl methacrylate have the important drawback of not being biodegradable. Their stability in biological fluids not only delays the release of the active drug but may induce accumulation of toxic material, particularly in liver cells. Hence, important efforts were devoted to the use of biodegradable polymers, of which polyalkylcyanoacrylate is one of the most significant examples; alkylcyanoacrylates have been used for many years in surgery as tissue adhesive and hemostatic (105). The polymer can be prepared rapidly and easily by anionic polymerization, and, furthermore, depending on the length of the alkyl chain (from one to seven carbon atoms), the degradation rate is different, allowing the design of systems with various release rates.

Due to their high specific surface area, these nanoparticles can adsorb large amounts of drug compounds, following in many cases a Langmuir adsorption isotherm (104). Loading of polyalkylcyanoacrylate (including methyl-, ethyl- isobutyl-, or hexylcyanoacrylates) with such compounds as dactinomycin, insulin, and phenoxymethylpenicillin has been demonstrated (104). Test studies with model drugs (107) using poly(methyl methacrylate/methacrylic acid) copolymer have shown that the release rate is strongly depending on the water solubility of the drug, and on the loading level (mass of drug/mass of polymer).

Polyalkylcyanoacrylate particles have shown their efficiency in the controlled release of antitumor drugs, such as dactinomycin (108); use of nanoparticles demonstrated longer survival of laboratory animals. Other applications of these particles can be mentioned; Fresta et al. (109) have shown that several antiepileptic drugs (ethosuximide, 5,5-diphenylhydantoin, and carbamazepine) entrapped in polyethylcyanoacrylate could be released at a controlled rate, diffusion through the polymer barrier being the main rate-limiting process.

Polymethyl methacrylate nanoparticles are commercially available under the trade name of Eudragit (Röhn Pharma; a comparison between the mechanical properties of its films and those of films based on cellulosic polymers can be found in Ref. 110). This has increased the number of formulations based on these particles; in the solvent evaporation method (Fig. 18) (111–113), the active compound and the polymer are first dissolved in a volatile organic solvent (such as dimethylformamide). The resulting solution is emulsified in an aqueous surfactant solution; the microspheres are then produced upon solvent evaporation. Finally, they are washed and collected.

Using this method, Yuksel et al. (112) prepared a system for the pH-dependent controlled release of nicarpidine hydrochloride. Kim et al. (114) demonstrated that treatment of hypertension with nifedipine using Eudragit (and other polymers) nanoparticles provided controlled release methods, with prolonged action and low initial burst. The kinetics and mechanism of nifedipine release by Eudragit particles have been analyzed by Chowdary and Sankar (115).

Since the early 1970s (75) there has been great interest in the use of polymers of lactic acid. Together with lactic acid/glycolic acid copolymers, they are among the most promising drug carriers because they are biodegradable in the body. They are broken into monomers by hydrolytic deesterification and finally disappear from the body (111).

Particles of poly(lactic acid) (PLA) for controlled release are prepared as injectable suspensions; most often, the solvent evaporation method described above is used for loading the particles with the drug of interest, although other methods, like phase separation (75), have also been proposed. The number of

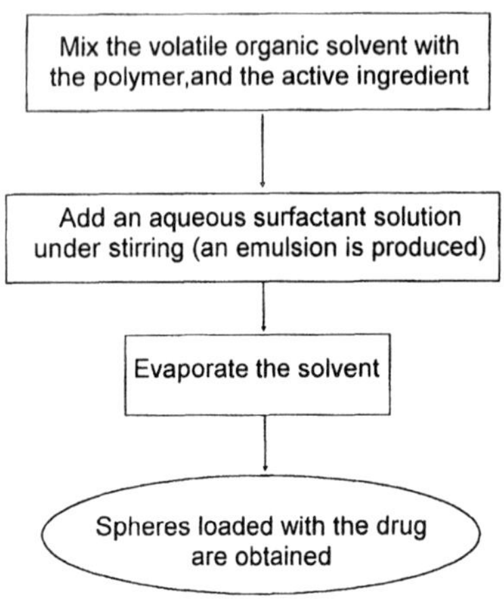

Figure 18 Scheme of the solvent evaporation method used for the obtention of polymer dispersions loaded with drug. (Adapted from Ref. 113.)

preparations for controlled release of very different drugs using PLA is very considerable. Let us mention a few examples.

Moritera et al. (111) used PLA microspheres of different molecular weights, as well as poly(glycolic-lactic acid) particles to control the release of 5-fluorouracil (used to inhibit cellular proliferation in vitreoretinopathies or after glaucoma filtering procedures) in the vitreus. Their in vitro results can be found in Fig. 19; as observed, most of the drug was released from the copolymer spheres after 3 days, whereas PLA particle released the drug at a lower rate and to a lesser extent. The process can be controlled by changing the molecular weight of the polymer. In vivo experiments demonstrated the absence of adverse side effects in the eyes of rabbits to which the suspensions was injected.

Zhang et al. (116) carefully analyzed the release mechanisms of different antibiotic (gentamicin, cefazolin) from poly(D,L-lactide) cylinders. A similar study was carried out by Niwa et al. (117) with D,L-lactide-glycolide copolymer loaded with nafarelin acetate, a compound analogous to luteinizing hormone–releasing hormone. Other drugs that can be loaded into these polymer particles to achieve controlled or sustained release are leuprorelin acetate (118), the anticancer agent Taxol (119), nonsteroidal anti-inflamatory drugs (120), tetracycline for periodontal disease therapy (121), or local anesthetics (122), to mention a few.

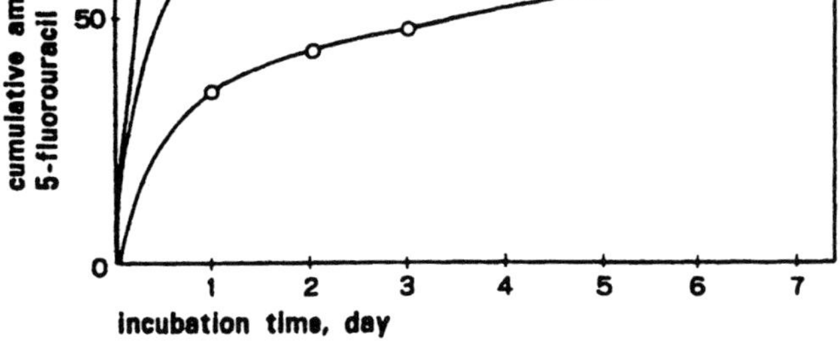

Figure 19 In vitro release of 5-fluorouracil from poly(lactic acid) microspheres (○: 4700 molecular weight; ●: 3400 molecular weight) and poly(glycolic-lactic acid) (△). (From Ref. 111. reprinted by courtesy of Lippincott-Raven Publishers.)

A related family of polymers is based on ε-caprolactone (ester of hydroxy-6-caproic acid); microspheres of poly(ε-caprolactone) homopolymer, and its copolymers with racemic lactic acid have also been used for pharmaceutical forms of controlled release (74,114). In fact, the degradation of poly(ε-caprolactone) is similar to that of poly(lactic acid); different investigations (123) have shown that the requirements of hormonal contraceptive methods and the slow biodegradation of the polymer make poly(ε-caprolactone) particularly useful in the controlled release of such drugs as progesterone, testosterone, norgestrel, or norethisterone from depot systems.

A large interest has also been reported (105) on a number of new polymer dispersions called *pseudolatexes*: as is known, a latex (or *true latex*) is made by polymerization of a monomer, usually emulsified in an aqueous medium in the presence (in some cases, not always) of a surfactant. Anionic or cationic initiators start the free radical polymerization mechanism. In the case of pseudolatexes, the preparation route starts by dissolving an already existing polymer, like ethylcellulose, in a suitable solvent and emulsifying the monomers in water using, for example, sodium lauryl sulfate and cetyl alcohol as stabilizers. For pharmaceutical use, cellulose-based pseudolatexes are preferred, since they have traditionally been approved for health use. Two of the most frequently used are Aquacoat and Aquateric (both registered trademarks of FMC Corp.); they are based, respectively, on ethylcellulose and cellulose acetate phthalate. Figure 20 shows pictures of the spherical particles of these commercially available pseudolatexes.

These polymers possess the characteristics of a true latex concerning colloidal stability, particle size uniformity, film forming properties, etc. However, the fact that they are not manufactured from monomers but by direct emulsification of already formed polymers makes them free of (potentially toxic) unreacted monomers, and they can hence be used in the body without toxicity hazards (124,125).

In our laboratory, we have checked the possibility of using Aquacoat spheres as systems for the controlled release of betamethasone phosphate (BMP) and acetate (BMA), glucocorticoids used in the treatment of rheumatological diseases, and frequently prescribed in creams for topical use. For example, Fig. 21 (126) shows that Aquacoat particles can adsorb considerable amounts of BMP, giving an isotherm very similar to that normally found in the adsorption of amphiphilic substances.

An in vitro analysis of the performance of Aquacoat to control the release of both betamethasone disodium phosphate (water-soluble) and betamethasone acetate (insoluble) from creams for topical administration was reported in Ref. 20. An example of the results found is reproduced in Fig. 22; as observed, in the case of soluble BMP, the kinetics of drug release is essentially independent of how much Aquacoat latex is added to the preparation. For all polymer concentrations, almost 90% BMP was released after 24 h, hence the only advantage of

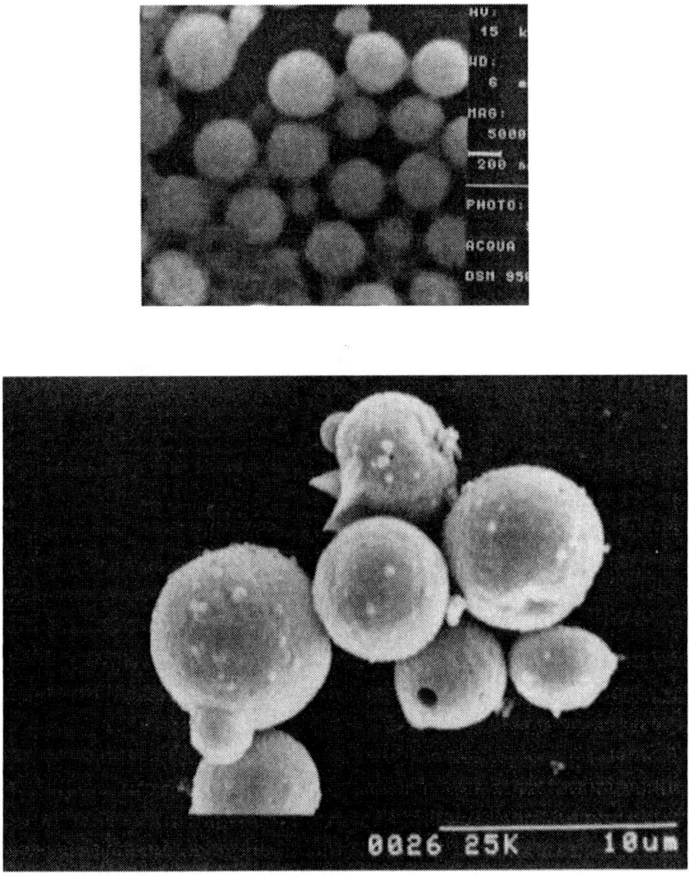

Figure 20 SEM pictures of Aquacoat (top) and Aquateric (bottom) pseudolatex particles.

using Aquacoat will be related to the pharmaceutical elegance of the formulation, not its therapeutic efficiency. On the contrary (bottom plot in Fig. 22) the release trends of insoluble BMA are quite different: the larger the latex concentration, the smaller the amount released after 5 h or longer.

Let us mention, finally, that there is also a growing number of works directed to the use of natural polymers such as albumin, gelatin, or casein (104). For instance, albumin microspheres can be prepared by denaturalization (by heat treatment, or by formaldehyde addition) of natural albumin. In addition to being useful in the exploration, for diagnostic purposes, of the reticuloendothelial sys-

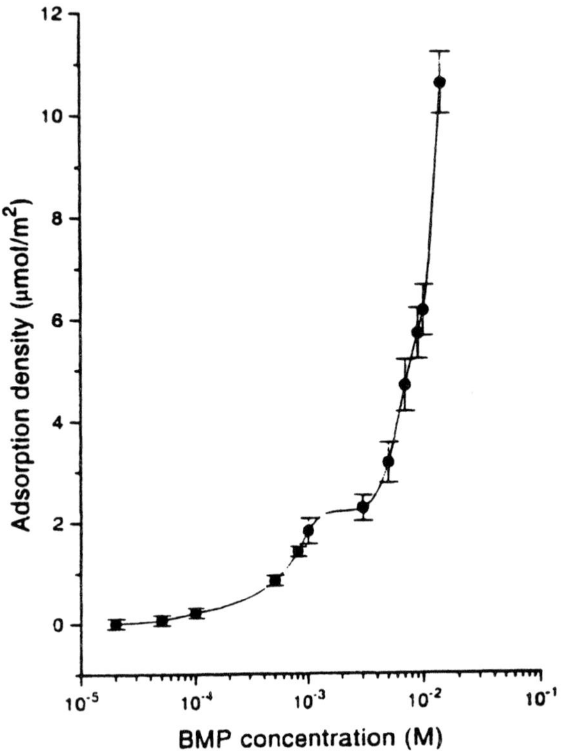

Figure 21 Adsorption of betamethasone disodium phosphate on Aquacoat as a function of initial glucocorticoid concentration. (Reprinted with permission from Ref. 126, Copyright 1996, American Pharmaceutical Association and American Chemical Society.)

tem (127), albumin particles have been tested with success as drug vehicles (128–130). Very promising results have also been reported on the use of crosslinked gelatin microspheres (130,131) in the controlled release of, for example, growth hormone (132) or solid tumor chemotherapy (133).

C. Magnetic Particles

This is an open and exciting research field: the possibility of solving the major problem of tissue specificity, in cancer treatment, by using an external field that might drive the drug-loaded particles to the specific organ affected by a tumor and, furthermore, control the drug release has been explored for many years. Most works have focused on the incorporation of magnetite (Fig. 23; see Ref.

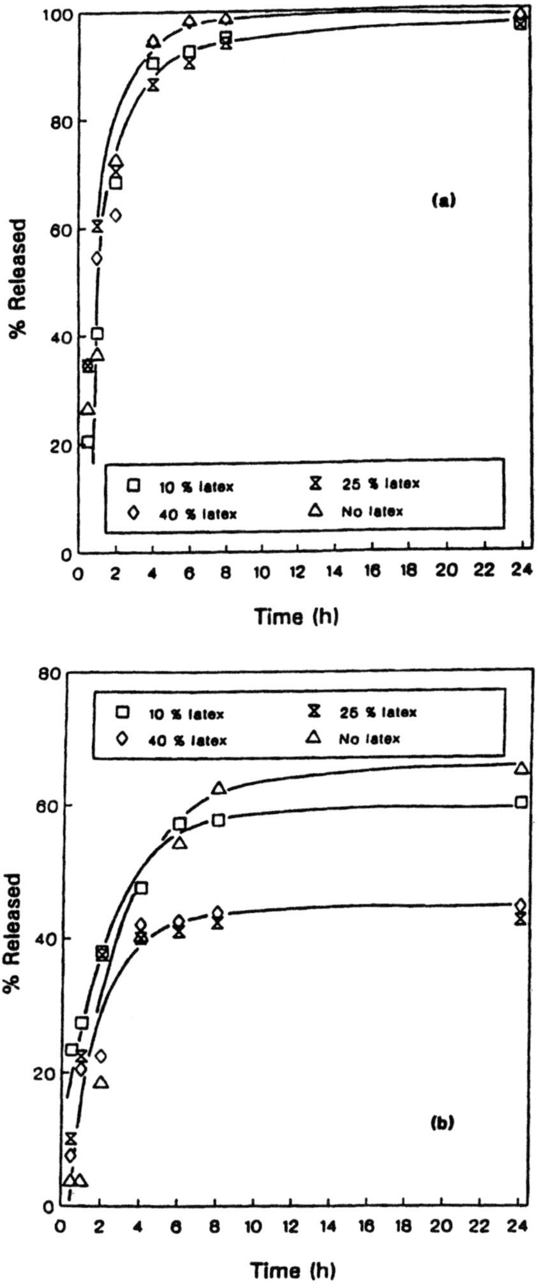

Figure 22 Amount of betamethasone disodium phosphate (a) and betamethasone acetate (b) released as a function of time for three different Aquacoat concentrations. (From Ref. 20, by courtesy of Societá Chimica Italiana.)

Figure 23 TEM picture of colloidal magnetite particles obtained as described in Ref. 134.

134) or other ferrites into natural or synthetic polymer particles (104,135). Different preparation routes have been proposed. For instance, Rembaum et al. (136) proposed emulsion polymerization of acrylate monomers using an aqueous suspension of magnetite particles as liquid medium. Albumin microspheres with entrapped magnetite have also been obtained for the controlled delivery of such drugs as mitomycin (137) or adriamycin (138). Another polymer checked with the purpose of delivering antitumor drugs has been 2–3 dicarboxycellulose (139).

Both in vitro and in vivo results are promising in the sense that, by applying a magnetic field to the part of the body where cytostatic action is needed, magnetic particles can be attracted and held in that position, thus increasing the drug concentration there and hindering the release of these potent compounds in other parts of the body (140,141).

Although the technique can be more easily applied to treatments in such organs as legs, arms, animal tails, etc., where magnetic fields are easier to apply selectively, results have been reported on the possibility of concentrating antitumor drugs in kidneys and lungs by applying magnetic fields to those organs (142).

ACKNOWLEDGMENTS

Thanks are due to the following persons who helped in the preparation of this manuscript: F. J. Arroyo, D. Mejías, J. L. Arias. Financial support by MEC, Spain, Projects #PB94-0812-CO2 and MAT98-0940 is also acknowledged.

REFERENCES

1. DF Evans, H Wennerström. The colloidal domain. Where physics, chemistry, biology and technology meet. New York: VCH, 1994, pp xxvi–xxxii.
2. RJ Hunter. Introduction to modern colloid science. Oxford: Oxford University Press, 1993, pp 300–322.
3. W Duncan-Hewitt, R Nisman. Investigation of the surface free energy of pharmaceutical materials from contact angle, sedimentation and adhesion measurements. In: K.L. Mittal, ed. Contact angle, wettability and adhesion. Utrecht: VSP, 1993, pp 791–911.
4. J Swarbrick, JC Boylan. Encyclopaedia of pharmaceutical tecnology. Vol 14. New York: Marcel Dekker, 1996, p 333.
5. RA Nash. Pharmaceutical suspensions. In: Lieberman H. A., Rieger M. M., Banker G. S. eds. Pharmaceutical dosage forms. Disperse systems. Vol. 1. New York: Marcel Dekker, 1988, pp 151–155.
6. CT Rhodes. Modern pharmaceutics. Vol. 2. New York: Marcel Dekker, 1979, p 329.
7. Spanish General Council of Pharmaceutical Associations. General catalog of pharmaceutical specialties. Madrid, 1997.
8. EM Morefield, JR Feldkamp, GE Peck, JL White, SL Hem. Preformulation information for suspensions. Int. J. Pharmaceut. 34: 263–265, 1987.
9. HC Ansel, NG Popovich. Pharmaceutical dosage forms and drug delivery systems. 5th ed. London: Lea & Febiger, 1990.
10. MR Billany. Suspensions. In: Aulton M. E. ed. Pharmaceutics: the science of dosage forms design. London: Aulton M. E. 1987, p 269.
11. CM Ofner, RL Schnaare, JB Schwartz. Oral aqueous suspensions. In: Lieberman H. A., Rieger M. M., Banker G. S. eds. Pharmaceutical dosage forms. Disperse systems. Vol 2. New York: Marcel Dekker, 1988, pp 231–264.
12. DM Collet. Suspensions. In: Collet D. M., Aulton M. E. eds. Pharmaceutical practice. London: Churchill-Livingstone, 1990, p 99.
13. American Medical Association. AMA drug evaluations, 5th ed. Philadelphia: 1983, p 60.
14. H Goodman, GS Banker. Molecular scale drug entrapment as a precise method of controlled drug release. I: Entrapment of cationic drugs by polymeric flocculation. J. Pharm. Sci. 59: 1131–1137, 1970.
15. CT Rhodes, K Wai, GS Banker. Molecular scale drug entrapment as a precise method of controlled drug release. II: Facilitated drug entrapment to polymeric colloidal dispersions. J. Pharm. Sci. 59: 1578–1581, 1970.
16. TK Chukwu, A Chukwu, OK Udeala. Hydrophilic polymers as drug release modulators from non-disintegrating tablet matrices. Acta. Pharmaceutica. 42: 181–188, 1992.
17. F. M. C. Corporation. Aquacoat: aqueous polymeric dispersion. Philadelphia: FMC, 1980.
18. R Bodmeier, O Paeratakul. Drug release from laminated polymeric films prepared from aqueous latexes. J. Pharm. Sci. 79: 32–36, 1990.
19. CR Steuernagel. Latex emulsions for controlled drug delivery. In: McGinity J. V.

ed. Aqueous polymeric coating for pharmaceutical dosage forms. New York: Marcel Dekker, 1989, pp 1–62.

20. MA Ruiz, V Gallardo, AV Delgado, P Vera. Study of in-vitro release of corticoids in topical formulations. Il. Farmaco. 49: 147–152, 1994.
21. P Vera, V Gallardo, J Salcedo, AV Delgado. Adsorption of a corticoid on colloidal hematite particles of different geometries. J. Colloid Interface. Sci. 187: 429–434, 1997.
22. CD Platsoukas. Biomedical applications of polymer particles with emphasis on cell separation. In: El-Aasser M. S., Ficth R. M. eds. Future directions in polymer colloids. Dordrecht: Martinus Nijhoff, 1987, p 321.
23. N Weiner. Introduction. In: HA Lieberman, MM Rieger, GS Banker, eds. Pharmaceutical dosage forms. Disperse systems. Vol 1. New York: Marcel Dekker, 1988, pp 1–12.
24. JC Samyn. An industrial approach to suspension formulation. J. Pharm. Sci. 50: 517–522, 1961.
25. HC Ansel, NG Popovich, LV Allen Jr. Pharmaceutical dosage forms and drug delivery systems. 6th ed. Philadelphia: Williams & Wilkins, 1995, p 254.
26. J Helman. Practical and theoretical pharmaceutical technology [in Spanish]. Vol. 2. México: Cia. Ed. Continental, 1982, p 501.
27. V Gallardo, J Salcedo, A Parera, A Delgado. Effect of the preservatives antipyrin, benzoic acid and disodium metabisulfite on properties of the nitrofurantoin/solution interface. Int. J. Pharmaceut. 71: 223–227, 1991.
28. RJ Hunter. Foundations of colloid science. Vol. I. Oxford: Clarendon Press, 1987, chapters 7, 8.
29. AV Delgado, V Gallardo, J Salcedo, F González-Caballero. A study of the electrokinetic and stability properties of nitrofurantoin suspensions. I: Electrokinetics. J. Pharm. Sci. 79: 82–86, 1990.
30. V Gallardo, AV Delgado, A Parera, J Salcedo. Electrophoretic study of nitrofurantoin in aqueous suspensions. Effect of the addition of a polymeric thickener. J. Pharm. Pharmacol. 42: 225–229, 1990.
31. AV Delgado, V Gallardo, A Parera, F Gonzalez-Caballero. A study of the electrokinetic and stability properties of nitrofurantoin suspensions. II: Flocculation and redispersion properties as compared with theoretical interaction energy curves. J. Pharm. Sci. 79: 709–715, 1990.
32. JEF Reynolds. ed. Martindale: The Extra Pharmacopeia, 29th ed. London: Pharmaceutical Press, 1989, pp 1355–1364.
33. JL Zatz. Physical stability of suspensions. J. Soc. Cosmet. Chem. 36: 393–411, 1985.
34. S Guyon, JL Abric, D Barthes. Stability study of cloxacillin in oily suspension, by high performance liquid chromatography. Il. Farmaco. 47: 1081–1094, 1992.
35. MJ Rosen. Surfactants and interfacial phenomena. New York: Wiley, 1989, chapters 2, 6.
36. J Lyklema. Fundamentals of interface and colloid science. Vol. 1. London: Academic Press, 1991, chapter 1.
37. AW Adamson. Physical chemistry of surfaces. New York: Wiley, 1990.

38. CJ van. Oss. Interfacial forces in aqueous media. New York: Marcel Dekker, 1994, chapters 2, 3, 6.
39. MM Rieger. Surfactants. In: HA Lieberman, MM Rieger, GS Banker, eds. Pharmaceutical dosage forms: Disperse systems. Vol 1. New York: Marcel Dekker, 1988, pp 285–366.
40. CF Lerk, AJM Schoonen, JJ Fell. Contact angles and wetting of pharmaceutical powders. J. Pharm. Sci. 65: 843–847, 1976.
41. WC Liao, JL Zatz. Effect of surfactant polyoxyethylene chain length on flocculation of local anesthetic suspensions. J. Soc. Cosmet. Chem. 31: 107–121, 1980.
42. J Lyklema. Fundamentals of interface and colloid science. Vol 2. London: Academic Press, 1995, chapters 3, 4.
43. H Ohshima. Interaction of electrical double layers. In: H Ohshima, K Furusawa, eds. Electrical phenomena at interfaces. Fundamentals, measurements and applications. New York: Marcel Dekker, 1998, chapter 3.
44. RJ Hunter. Zeta potential in colloid science. London: Academic Press, 1981, chapter 3.
45. H Ohshima. Interfacial electrokinetic phenomena. In: Ohshima H., Furusawa K. eds. Electrical phenomena at interfaces. Fundamentals, measurements and applications. New York: Marcel Dekker, 1998, chapter 2.
46. V Gallardo, A Parera, J Salcedo, F González-Caballero, AV Delgado. A study of the stability of aqueous phosphomycin suspensions. Comparison between two pharmaceutical forms (In Spanish). An. Real. Acad. Farm. 60: 79–90, 1994.
47. JB Kayes. Pharmaceutical suspensions: microelectrophoretic properties. J. Pharm. Pharmacol. 29: 163–168, 1977.
48. V Gallardo, MA Ruiz, P Vera, F González-Caballero, AV Delgado. Effect of amino acids on the stability properties of nitrofurantoin suspensions. Flocculation and redispersion as compared with interaction energy curves. J. Pharm. Pharmacol. 48: 450–455, 1996.
49. T Gosgrove. Adsorption of polymers at the solid/solution interface. In: Th F. Tadros, ed. Solid/liquid dispersions. London: Academic Press, 1987, chapter 6.
50. B Vincent. The stability of solid/liquid dispersions in the presence of polymers. In: Th F. Tadros, ed. Solid/liquid dispersions. London: Academic Press, 1987, chapter 7.
51. J Gregory. Flocculation by polymers and polyelectrolytes. In: Th F. Tadros, ed. Solid/liquid dispersions. London: Academic Press, 1987, chapter 8.
52. A Felmeister, GM Kuchtyak, S Koziol, CJ Felmeister. Polymer-induced flocculation of pharmaceutical suspensions. J. Pharm. Sci. 62: 2026–2027, 1973.
53. Carbopol: hydrosoluble resins. BF Goodrich, Cleveland, 1981.
54. DQM Craig, S Tamburic, G Buckton, JM Newton. An investigation on the structure and properties of Carbopol 934 gels using dielectric spectroscopy and oscillatory rheometry. J. Controlled Release 30: 213–223, 1994.
55. BM Berney, PB Deasy. Evaluation of Carbopol 934 as a suspending agent for sulphadimine suspensions. Int. J. Pharm. 3: 73–80, 1979.
56. MA Ruiz, MJ Martín, A Parera. Effect of neutralizer and of pH on aqueous dispersions of Carbopol 940. [In Spanish]. Ciencia. Ind. Farm. 6: 41–43, 1987.

57. JC Boyland, J Cooper, ZT Chowhan. Handbook of pharmaceutical excipients. Washington: American Pharmaceutical Association, 1983, p 42.
58. RK Iller. Colloidal silica. In: E Matijevic, ed. Surface and colloid science, Vol. 6, New York: Wiley, 1973, pp. 1–100.
59. JEF Reynolds. ed. Martindale. The extra pharmacopeia. London: Pharmaceutical Press, 1989, p 1437.
60. JC Boyland, J Cooper, ZT Chowhan. Handbook of pharmaceutical excipients. Washington: American Pharmaceutical Association, 1983, p 253.
61. KY Yang, R Glemza, CI Jarowski. Effect of amorphous silicon dioxides on drug dissolution. J. Pharm. Sci. 68: 560–565, 1979.
62. EN Hiestand. Physical properties of coarse suspensions. J. Pharm. Sci. 61: 268–272, 1972.
63. CW Macosko. Rheology. Principles and applications. New York: VCH, 1994, p 426.
64. AV Delgado, F González-Caballero, MA Cabrerizo, I Alados. The primary electroviscous effect in monodisperse polystyrene suspensions. Acta Polymerica 38: 66–70, 1987.
65. L Zurita, F Carrique, AV Delgado. The primary electroviscous effect in silica suspensions. Ionic strength and pH effects. Colloids Surfaces 92: 23–28, 1994.
66. U Kästner, H Hoffmann, R Dönges, R Ehrler. Hydrophobically and cationically modified hydroxyethyl cellulose and their interactions with surfactants. Colloids Surfaces 82: 279–297, 1994.
67. M Córdoba, B Carretero, JJ Torrado. Effect of the structured vehicle on the physical stability of sulfamethoxazol suspensions (In Spanish). An. Real Acad. Farm. 50: 379–388 (1984).
68. H van Olphen. An introduction to clay colloid chemistry. New York: Wiley, 1977, chapter 5.
69. WF Smith. Principles of materials science and engineering. New York: McGraw-Hill, 1990, pp 578–581.
70. H Schott. Controlled flocculation of coarse suspensions by colloidally dispersed solids. I: Interaction of bismuth subnitrate with bentonite. J. Pharm. Sci. 65: 865–871, 1976.
71. JW McGinity, MR Harris. Increasing dissolution rates of poorly-soluble drugs by adsorption to montmorillonite. Drug Dev. Ind. Pharm. 6: 35–39, 1980.
72. E Dickinson. An introduction to food colloids. Oxford: Oxford University Press, 1992, pp 66–70.
73. JL Zatz, JJ Berry, DA Alderman. Viscosity-imparting agents in disperse systems. In: Lieberman H. A., Rieger M. M., Banker G. S. eds. Pharmaceutical dosage forms. Disperse systems. Vol 2. New York: Marcel Dekker, 1988, pp 171–203.
74. JL Zatz, GP Kushla. Gels. In: HA Lieberman, MM Rieger, GS Banker. eds. Pharmaceutical dosage forms. Disperse systems. Vol 2. New York: Marcel Dekker, 1988, pp 495–510.
75. E Doelker. Polymer properties and principles of preparation of polymeric drug systems. In: P. Buri, F. Puisieux, E. Doelker, J.P. Benoit, eds. New pharmaceutical forms [in French]. Paris: Lavoisier Tec & Doc, 1985, pp 5–63.
76. M de la Luz Reyes-Vega, MC Fahara-Valdés, IC Valdés-Silva, CT Saucedo-Sala-

zar, RD Peralta-Rodríguez. Flavor selection for an aluminum and magnesium hydroxides antacid suspension. Drug Dev. Ind. Pharm. 22: 631–635, 1996.

77. G Th Kroyer, K Futschik. The impedance-splitting method for microbiologic analysis of food preservatives. Biotech. Lab. 2: 8–10, 1997.
78. J Helman. Practical and theoretical pharmaceutical technology [in Spanish]. Vol. 8 México: Cia. Ed. Continental, 1982, p 2413.
79. M Seiller, MC Martini. Pharmaceutical forms for local application [in French]. Paris: Technique Document, 1996, p 495.
80. GA Neville, HD Beckstead, HF Shurvell. Utility of Fourier transform-Raman and Fourier transform–infrared diffuse reflectance spectroscopy for differentiation of polymorphic spironolactone samples. J. Pharm. Sci. 81: 1141–1146, 1992.
81. MC Martínez-Ochárriz, C Martín, MM Goñi, C Rodríguez-Espinosa, MC Tross, M Sánchez. Polymorphism of diflunisal: isolation and solid state characteristics of a new crystal form. J. Pharm. Sci. 83: 174–177, 1994.
82. AV Gubskaya, KA Chishko, YV Lisnyak, YP Blagoy. Effect of cryogrinding on physicochemical properties of drugs. II. Cortisone acetate: particle sizes and polymorphic transition. Drug Dev. Ind. Pharm. 21: 1965–1974, 1995.
83. CM Offner III, RL Schnaare, JB Schwartz. Oral aqueous suspensions. In: HA Liebermann, MM Rieger, GS Banker, eds. Pharmaceutical dosage forms. Disperse systems. Vol 2. New York: Marcel Dekker, 1989, pp 253–256.
84. V Gallardo, MA Ruiz, A Parera, A Delgado. ADSC study of the chemical stability of nitrofurantoin in pharmaceutical suspensions. Die Pharmazie 45: 633–634, 1990.
85. BI Idson, AJ Scheer. Suspensions. In: I Wallace, ed. Problem solver. Vol I. New Jersey: FMC, 1984, pp 1–31.
86. CM Offner III, RL Schmaare, JB Schwartz. Reconstitutable suspensions. In: Liebermann H. A., Rieger M. M., Banker G. S. eds. Pharmaceutical dosage forms. Disperse systems. Vol 2. New York: Marcel Dekker, 1989, pp 317–334.
87. JB Boyett, CW Davis. Injectable emulsions and suspensions. In: Liebermann H. A., Rieger M. M., Banker G. S. eds. Pharmaceutical dosage forms. Disperse systems. Vol 2. New York: Marcel Dekker, 1989, pp 397–413.
88. M Gasperlin, J Smid-Korbar, J Kristl. Formulation and physical stability evaluation of creams containing silicone surfactants. Farm Vestn (Ljubljana) 43: 3–10, 1992.
89. JL Zatz, RY Lue. Effect of polyols on physical stability of suspensions containing nonionic surfactant. J. Soc. Cosmet. Chem. 33: 149–155, 1982.
90. JJ Sciarra, AJ Cutie. Aerosol suspensions and emulsions. In: Liebermann H. A., Rieger M. M., Banker G. S. eds. Pharmaceutical dosage forms. Disperse systems. Vol 2. New York: Marcel Dekker, 1989, pp 417–460.
91. D Rodríguez-Devesa. Aerosol formulations [in Spanish]. Madrid: Castilla SA, 1971, p 412.
92. EW Martin. Ed. Dispensing of medication. Easton: Mack, 1971, p 554.
93. JB Portnoff., RJ Harwood., EW Sunbery. The processing of small volume parenteral and related sterile products. In: KE Avis, L Lochman, HA Lieberman, eds. Pharmaceutical dosage forms. Vol I. Parenteral medications. New York: Marcel Dekker, 1984, p 246.
94. United States Pharmacopeia XX. Rockville, MD: United States Pharmacopeial Convention Inc., 1990, p 205.

95. F Puisieux. Micro- and nano-particle vectors. Introduction. In: P. Buri, F. Puisieux, E. Doelker, J.P. Benoit, eds. New pharmaceutical forms [in French]. Paris: Lavoisier Tec & Doc, 1985, pp 465–466.
96. DH Robinson, JW Manger. Drug delivery systems. Am. J. Hosp. Pharm. 48: S14–S23, 1991.
97. PS Banerjee, JR Robinson. Novel drug delivery systems. An overview of their impact on clinical and pharmacokinetic studies. Clin. Pharmacokin 20: 1–14, 1991.
98. P Goldman. Rate-controlled drug delivery. N. Eng. J. Med. 307: 286–290, 1982.
99. G Storm, DJA Crommelin. Colloidal systems for tumor targeting. Hybridoma 16: 119–125, 1997.
100. B Rihova. Biocompatibility of materials. Hemocompatibility, immunocompatibility and biocompatibility of solid polymeric materials, and soluble targetable polymeric carriers. Adv. Drug Deliv. Rev. 21: 157–176, 1996.
101. JL Vila-Jato, MJ Alonso, B Seijo, D Torres. Biodegradable systems for parenteral administration of drugs. I [in Spanish]. Ind Farm (May/June) 67–75, 1971.
102. AM Abukhalaf, MA Soliman. Drug-release from spherical particles under nonsink conditions. 1. Theoretical evaluation. Drug Dev. Ind. Pharm. 22: 465–469, 1996.
103. B Narasimhan, NA Peppas. Molecular analysis of drug-delivery systems controlled by dissolution of the polymer carrier. J. Pharm. Sci. 86: 297–304, 1997.
104. P Couvreur. Nanoparticulate drug vectors [in French]. In: Buri P., Puisieux F., Doelker E., Benoit J.P. eds. New pharmaceutical forms. Paris: Lavoisier Tec & Doc, 1985, pp 577–611.
105. JW Vanderhoff, MS El-Aasser. Theory of colloids. In: HA Liebermann, MM Rieger, GS Banker, eds. Pharmaceutical dosage forms: disperse systems. Vol. 1. New York: Marcel Dekker, 1989, pp 93–149.
106. P Speiser, G Rirrenbach. Microcapsules in the nanometric range, and a method for their production. U.S. Patent 4021364, 1977.
107. CJ Kim, PI Lee. Enhanced and retarded drug release from hydrophobic ionic beads. J. Macromol Sci. Pure Appl. Chem. A33: 1227–1238, 1996.
108. F Brasseur, P Couvreur, B Kante, L Deckers-Passau, M Roland, C Deckers, P Speiser. Actinomycin D adsorbed on polymethylcyanoacrylate nanoparticles: increased efficiency against experimental tumor. Eur. J. Cancer. 16: 1441–1445, 1980.
109. M Fresta, G Cavallaro, G Giammona, E Wehrli, G Puglisi. Preparation and characterization of polyethyl-2-cyanoacrylate nanocapsules containing antiepileptic drugs. Biomaterials 17: 751–758, 1996.
110. R Bodmeier, O Paeratakul. Mechanical properties of dry and wet cellulose and acrylic films prepared from aqueous colloidal polymer dispersion used in the coating of solid dosage forms. Pharmaceut. Res. 11: 882–888, 1994.
111. T Moritera, Y Ogura, Y Honda, R Wada, SH Hyon, Y Ykada. Microspheres of biodegradable polymers as a drug-delivery system in the vitreous. Inv. Ophthalmol. Vis. Sci. 32: 1785–1790, 1991.
112. N Yuksel, T Tincer, T Bagkara. Interaction between nicarpidine hydrochloride and polymeric microspheres for a controlled release system. Int. J. Pharm. 140: 145–154, 1996.

113. JP Benoit. Microencapsulation and chemiembolization. In: Buri P., Puisieux F., Doelker E., Benoit J. P. eds. New pharmaceutical forms [in French]. Paris: Lavoisier Tec & Doc, 1985, pp 613–656.
114. Y Kim, L Fluckiger, M Hoffman, I Lartandidjouadiene, J Atkinson, P Maincent. The antihypertensive effect of orally administered nifedpipine-loaded nanoparticles in spontaneously hypertensine rats. Br. J. Pharmacol. 120: 399–404, 1997.
115. KPR Chowdary, GG Sankar. Eudragit microcapsules and its dispersions in HPMC-MCC. Physicochemical characteristics and drug release studies. Drug Dev. Ind. Pharm. 23: 325–330, 1997.
116. XC Zhang, UP Wyss, D Pichora, MFA Goosen. A mechanistic study of antibiotic release from biodegradable poly(D,L-lactide) cylinders. J. Controlled Release 31: 129–144, 1994.
117. T Niwa, H Takeuchi, T Hino, N Kunon, Y Kawashima. In-vitro drug release behaviour of D,L-lactide/glycolide copolymer (Plga) nanospheres with nafarelin acetate prepared by a novel spontaneous emulsification, solvent diffusion method. J. Pharm. Sci. 83: 727–732, 1994.
118. H Okada, Y Doken, Y Ogawa, H Toguchi. Preparation of 3-month depot injectable microspheres of leuprorelin acetate using biodegradable polymers. Pharm. Res. 11: 1143–1147, 1994.
119. YM Wang, H Sato, I Adachi, I Horikoshi. Preparation and characterization of poly (lactic-coglycolic acid) microspheres for targeted delivery of a novel anticancer agent, Taxol. Chem. Pharm. Bull. 44: 1935–1940, 1996.
120. B Guizion, DJ Armstrong, PNC Elliot, JL Ford, C Rostron. Investigation of in vitro release characteristics of NSAID-loaded polylactic acid microspheres. J. Microencapsulation 13: 701–708, 1996.
121. E Esposito, R Cortezi, F Cervellati, E Menegatti, C Nastruzzi. Biodegradable microparticles for sustained delivery of tetracyicline to the periodontal pocket. Formulatory and drug release studies. J. Microencapsulation 14: 175–187, 1997.
122. P Lecome, JH Rytting, V Gajan, F Chevanne, R Laverge. In-vitro controlled release kinetics of local anesthetics from poly(D,L-lactide) and poly(lactide-co-glycolide) microspheres. J. Microencapsulation 14: 243–255, 1997.
123. E Doelker. Parenteral administration. In: P Buri, F Puisieux, E Doelker, JP Benoit. eds. New pharmaceutical forms [in French]. Paris: Lavoisier Tec & Doc, 1985, pp 229–299.
124. SG Banker, TG Rhodes. Modern pharmaceutics. New York: Marcel Dekker, 1990, p 327.
125. RW Croswell, CH Becker. Suspension polymerization for preparation of timed-release dosage forms. J. Pharm. Sci. 63: 440–442, 1974.
126. P Vera, J Salcedo, AV Delgado, MA Ruiz, V Gallardo. Adsorption of betamethasone disodium phosphate on ethylcellulose latex: an electrokinetic study. J. Pharm. Sci. 85: 468–472, 1996.
127. U Schieffel, BA Rhodes, TK Natarajan, J Wagner. Albumin microspheres for study of reticuloendotelial system. J. Nucl. Med. 13: 498–503, 1972.
128. P Kramer. Albumin microspheres as vehicles for achieving specificity in drug delivery. J. Pharm. Sci. 63: 1640–1647, 1974.
129. OP Rubino, R Kowlasky, J Swarbrick. Albumin microspheres as a drug-delivery

system. Relation among turbidity ratio, degree of cross-linking, and drug release. Pharm. Res. 10: 1059–1065, 1993.

130. V Dilova, V Shishkova. Albumin microspheres as a drug-delivery system for dexamethasone. Pharmaceutical and pharmacokinetic aspects. J. Pharm. 45: 987–989, 1993.
131. E Esposito, R Cortesi, C Nastruzzi. Gelatin microspheres. Influence of preparation parameters and thermal-treatment on chemicophysical and biopharmaceutical properties. Biomaterials 17: 2009–2020, 1996.
132. L Disilvio, N Gurav, MV Kayser, M Braden, S Downes. Biodegradable microspheres. A new delivery system for growth hormone. Biomaterials. 15: 931–936, 1994.
133. R Narajani, KP Rao. Solid tumor-chemotherapy using injectable gelatin microspheres containing free methotrexate and conjugated methotrexate. Int. J. Pharm. 142: 25–32, 1996.
134. E Matijević. Production of monodispersed colloidal particles. Annu. Rev. Mater. Sci. 15: 483–516, 1985.
135. DQM Craig. Dispersions as pharmaceutical dosage forms. In: RB McKay. ed. Technological applications of dispersions. New York: Marcel Dekker, pp 487–488, 1994.
136. A Rembaum, RCK Yen, RS Molday. Synthesis and reactions of hydrophilic functional microspheres for immunological studies. J. Macromol. Sci. Chem. A13: 603–632, 1979.
137. T Kato, R Nemoto, H Mori, S Sato, U Unno, M Honma, M Okada, T Minowa. Mitomycin entrapped in ferromagnetic microcapsules. IRCS. Pharmacol. 7: 621, 1979.
138. K Widder, A Senyei, D Ranney D. In-vitro release of biologically active adriamicyn by magnetically responsive albumin microspheres. Cancer Res. 40: 3512–3517, 1980.
139. MI Popa, N Aelenei, G Ionescu. Preparation of 2,3-dicarboxycellulose microspheres containing antitumor agent as a drug carrier. Cellulose Chem. Technol. 30: 33–38, 1996.
140. Y Morimoto, K Sugibayashi, M Okumura, Y Kato. Biomedical applications of magnetic fluids. I. Magnetic guidance of ferro-colloid-entrapped albumin microspheres for site specific drug delivery in-vivo. J. Pharm. Dyn. 3: 264–267, 1980.
141. Y Morimoto, M Okumura, K Sugibayashi, Y Kato. Biomedical applications of magnetic fluids. II. Preparation and magnetic guidance of magnetic albumin microspheres for site specific drug delivery in-vivo. J. Pharm. Dyn. 4: 624–631, 1981.
142. A Ilbrahim, P Couvreur, M Roland, P Speiser. New magnetic drug carrier. J. Pharm. Pharmacol. 35: 59–61, 1983.

14

Experimental Design in Emulsion and Suspension Formulations: Theoretical Aspects

Roger Phan-Tan-Luu
Laboratoire de Méthodologie de la Recherche Expérimentale, Faculté de St Jérôme, Marseille, France

Didier Mathieu
Laboratoire de Méthodologie de la Recherche Expérimentale–I.U.T Département Informatique, Aix-en-Provence, France

I. EXPERIMENTAL METHODOLOGICAL APPROACH

A. Introduction

Experiments play an important part in the research carried out in industry or university laboratories. They are essential to any scientific approach and, surprisingly, still highly thought of by experimenters. Work including a great number of experiments is considered as *serious* by nature. Moreover, over the past years the methods of study and analysis of physical, chemical, and biological phenomena have dramatically expanded, as a result from, among others:

The appearance of more and more advanced technical means and equipment (very often coupled with computers in charge of part of the analysis of the results).

The development of mathematical and statistical methods for analyzing and processing numerical data that have been used with great success in many domains (factorial analysis, pattern recognition, etc.).

For some time now, however, the tendency has been toward a decrease in the number of experiments run, a tendency that has been encouraged mostly for economic reasons. Experimenters cannot afford any longer to carry out countless experiments. And quite naturally, they have wondered why we do experiments. The answer is very simple: to obtain information. This information is used to answer questions or to confirm hypotheses. Therefore, we have first of all to state the questions to which we want answers and then, and only then, run the experiments that will provide the desired information. Because each experiment provides a specific piece of information, we must before any experimentation select the interesting experiments to run with regard to the desired information. This selection is done a priori; we thereby establish an experimental strategy, called *experimental design*. This step is an important one, still often neglected. It is indeed difficult to have people admit that *no information is contained in the result of an experiment and all of the information is contained in the experiment chosen*.

We propose a methodological approach whose principal characteristics are as follows:

The predominant role assigned to reflection rather than to experimentation. Reflection is the essential part of any experimentation. It is necessary not only during the initial phase (definition of the problem) but during each step of the research process. The new information obtained by experimentation must be analyzed, interpreted, and added to the information previously obtained to redefine the best strategy. The experimentation should always provide the maximum of *useful* information, whatever the adopted strategy.

A great flexibility of the strategy. This is made necessary and possible by *fragmentation* of the objective of the problem dealt with, which most of the time results in a succession of questions, each one depending on the resolution of the preceding question. The same fragmentation must be applied to the experimental part, which provides some of the information necessary to the resolution of these questions. Except in cases of batch experiments, it is nearly always preferable to run small series of experiments, if only to add in the subsequently built experimental designs the information as and when it is obtained.

The different steps of our methodological approach are the following:

1. First it is necessary to establish a clear formulation of the problem: objectives pursued, consequences of a wrong decision, budget (time, cost, financial means).
2. After the objectives have been clearly defined, we have to determine our state of knowledge by asking questions of colleagues and also by examining the literature in search of answers. If some necessary information is not available, it is imperative to perform some experiments, and for this we must define as exhaustively and accurately as possible the factors likely to have an influence, the responses, and the requirements. At this stage we must also define the part of the experimental domain in which the missing information is to be sought and that we call the *experimental domain of interest*.
3. Then we establish an experimental strategy (experimental design), which is to say we select the experiments to run with regard to the objectives fixed, the means at our disposal, and the information wanted. We seek a relationship between cause and effect between certain pa-

rameters of the phenomenon (*factors*), which are supposed to influence its behavior, and other parameters (*responses*) that characterize its result. The design of experiments consists of imposing special variations to the factors (entries), measuring the induced variations of the responses, and deducing the relationships between causes and consequences.

4. We run the experiments that will give the values of the responses studied.
5. We infer the answers to the questions either directly or by means of a mathematical model.

It must be noted that if the experiments are carefully planned, relatively simple data analysis techniques (variance analysis, linear regression, etc.) can be used. Conversely, if the experiments are not carefully planned, it is often necessary to use much more complex mathematical and statistical tools (factorial analyses, classifications, etc.) without even being sure of the results.

B. Terminology

Most of the specific terms that we use in *Methodology of Experimental Research* are conventional, but they may sometimes have slightly different meanings than usual. In order to make our presentation clear, let us remind you of some important definitions. We usually represent the phenomenon studied by means of the diagram shown in Fig. 1.

Figure 1 Representation of the phenomenon being studied.

1. Factors

The factors are the causes—supposed or certain—that account for the phenomenon. We can distinguish the factors that we control and that we may study, which we call *controllable factors*, from those that we cannot or do not want to control, which we call *noise factors*. We shall take into account only the factors that we can control. The respective influences of the factors can be determined by making these factors vary, which lead to changes in the behavior of the phenomenon studied. To this end, we define for each factor a domain of variation within which it can take different states or *levels*. The factors can differ in nature: they can be *qualitative* or *quantitative, continuous* or *discontinuous*.

a. Quantitative Factors

These are factors whose different levels can take numerical values within a well-defined domain of variation representing the domain of variation of the factor considered. It is theoretically possible to choose in this domain any state (or level) for the factor. Although it often occurs that only a few discrete values can be chosen in this domain of variation, we generally consider that a quantitative factor is also a continuous factor. Examples are temperature of reaction, duration of addition, concentration of a reactant, and composition of an emulsion.

b. Qualitative Factors

These are factors whose different states (or levels) cannot be arranged in ascending or descending order. By definition, these factors can take only discrete values. A qualitative factor is always discontinuous. Examples are nature of an excipient, type of machine, and nature of an emulsion.

Later on we shall represent a factor by the letter U followed by an index.

2. Coded Variables

A dimensionless variable, which we call *coded variable* and we represent by the letter X followed by an index, is associated to each factor. The passage from factors to variables and vice versa depends on the nature of the factors. While working with these coded variables, we can compare their different effects directly without taking their units into account.

a. Quantitative Factors

The most often used transformation is:

$$x_{ij} = (u_{ij} - U_j^0)/\Delta U_j \tag{1}$$

x_{ij} = value of the coded variable X_j in the experiment i
u_{ij} = value of the actual variable U_j in the experiment i
U_j^0 = value of the actual variable U_j at the center of the domain ($X_j = 0$)
ΔU_j = variation of the actual U_j corresponding to a variation of the corresponding coded variable equal to 1. ΔU_j is called step of variation.

$$\Delta U_j = (U_{j,\max} - U_{j,\min})/2 \tag{2}$$

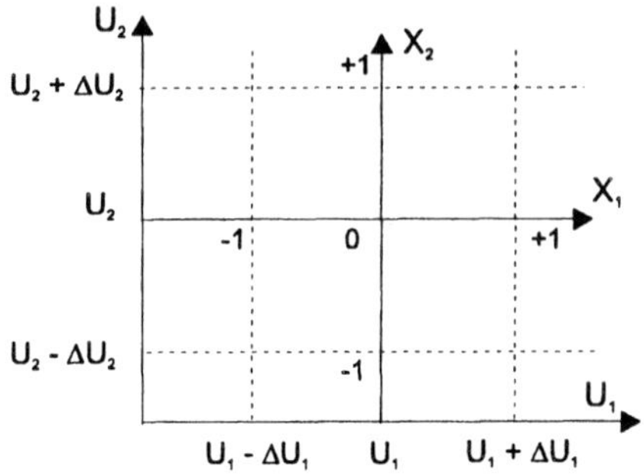

Figure 2 Passage actual variables—coded variables.

b. Qualitative Factors

The coded variable X corresponding to the qualitative factor U takes entire values corresponding to the different states taken by the factor U. When the factor U takes s levels, the coded variable X takes either the values $0, 1, \ldots, s - 1$ or the values $1, 2, \ldots, s$. In the particular case where the factor U takes only two levels, the coded variable X usually takes the value -1 and $+1$ (the level -1 is not considered as being inferior to the level $+1$).

3. Responses

An experimental response (often called *dependent variable*) is a measurable manifestation that is observed when the studied factors are made to vary. A phenomenon may be described by means of several experimental responses. All sorts of responses can be considered, such as a yield or purity in chemistry, a weight increase in biology, a gustatory quality in wine science, mechanical properties, or still the texture of a food product determined by sensory testing in food science. However, this last type of response, which can take only discrete values, is a source of problems when it comes to the interpretation of the results. It is also possible to study a function of a measured response, e.g., the logarithm of the equilibrium constant. The term *experimental response* is taken in a broad sense, as an experimental response can be the experimental result of an experiment but also the result of a numerical simulation. The studied response is represented by η and the studied response known with an experimental error by y.

$$y_i = \eta_i + e_i$$

where e_i represents the experimental error.

4. Reproducibility

Before applying the experimental strategy that we have established to solve our problem, we have to know the reproducibility of our experiments. This knowledge may be obtained by analogy with other problems of the same type, but when this is not possible we must obtain it by replicating experiments in the same conditions. In this case it is essential to run experiments independently (*reproducibility*) and not replicate only a part of the experimental procedures (*repeatability*). Even though this latter approach enables us to obtain a lower estimation of the experimental error, this estimation is distorted and does not provide us with any information.

5. Experimental Domain

We call *domain of factors* the set product of the sets of levels that each factor can take. Each element of this domain represents a set of experimental conditions. The possible experimental domain (or operability domain) is the subset of the domain of factors containing the feasible experiments. In practice, this domain is often reduced to the *experimental domain of interest*. Before each experimentation, we must define carefully its size, shape, the possibilities of extension, and the suspected discontinuities of the phenomenon.

6. Experiments

An experiment (or experimental point) is defined by a set of operating conditions. The factors that are not taken into account in the study must be set to a value that will remain constant throughout the experiment. Each experiment must be run independently of the others. When carrying out a series of experiments and in order to avoid interferences, it is advised to randomize them. This is called *randomization*.

7. Matrix of Experiments

A matrix of experiments is a mathematical object that represents, in coded or normed form, the set of experiments to run. It is a matrix of N rows, corresponding to N experiments, and of k columns, corresponding to the k-coded variables studied. The element x_{ij} of the so formed matrix corresponds to the level that the jth coded variable takes in the ith experiment. We will represent it by ξ_N.

8. Plan of Experimentation

A plan of experimentation is the *translation* of an experimental design in natural variables. It is therefore an array that contains data directly available to the experimenter. Like the matrix of experiments, this array is composed of N rows and k columns. Each of its elements u_{ij} corresponds to the level taken by the jth factor in the ith experiment.

C. Factor Screening

This phase consists of seeking, very roughly and very quickly, among a set of potentially influential factors, which ones are effectively influential in a fixed experimental domain. When beginning the study of a problem, it seems natural to make a list of the factors that could have an influence on the phenomenon studied. The number of factors considered is often thought of as being too important, simply because there is sort of empirical ''saying'' that makes the number of experiments depend closely on the number of factors: *the number of experiments increases exponentially when the number of factors increases*. Then a tendency toward simplification develops and the number of factors studied is brought down to three or four. This is an artificial reduction that is usually based on what is customary in laboratories. The experimenters perform what could be called a ''sentimental'' screening: they choose the factors they ''like'' and reject the factors they ''dislike.'' It appears quickly that this selection depends on the material available in the laboratory. The experimenters will choose to work with factors for which the technical means of control and analysis are available in the laboratory. The factors that are not chosen (*that are not liked*) are thus implicitly supposed to have no influence on the phenomenon studied.

The methodological tool contains an approach that enables to replace this sentimental screening with a scientific screening. For this approach to be accepted, the number of necessary experiments must not be too big. There are experimental strategies in which the number of experiments is even much lower than the number of factors if the latter is very big (from 50 to 10,000) [the techniques of group screening (1), and of sequential bifurcation (2), supersaturated designs (3–5), and, when the number of factors is lower, i.e., less than 50, this number of experiments is close to the number of factors, e.g., Hadamard design (6), resolution III symmetrical and asymmetrical fractional factorial designs (7)].

The Hadamard designs are certainly the most often used designs. They are also known under the name of Plackett and Bluman designs.

1. Hadamard Designs

A weighing experimental design containing only elements equal to ± 1 is called a Hadamard design if the resulting information matrix $\mathbf{X'X}$ built from the N points is such that:

$$(\mathbf{X}'\mathbf{X}) = \mathbf{N} \cdot \mathbf{I}_N$$

where **X** is the model matrix, **X′** is the matrix transposed from **X**, and **I** is the identity matrix.

Note the following exact definition of a Hadamard design:

> A Hadamard design is a square **H** matrix ($N_H \cdot N_H$) with all elements equal to $+1$ or -1, such that: $\mathbf{H}'\mathbf{H} = N_H \cdot \mathbf{I}_{NH}$. The condition of existence of such a matrix is $N_H = 2$ or $N_H \equiv 0$ (modulo 4). N_H is the Hadamard number.

Strictly speaking, it is not the experimental design that is a Hadamard design but the model matrix: **X**. But this notation is now common and we will use it. In order to build this type of experimental design, it is advised to use the mode of generation proposed by Plackett and Burman (6), which we review here: Knowing the number of factors *k*, we determine the minimum number of experiments needed to study a polynomial model of degree 1: $N = k + 1$, and we seek the Hadamard number N_H immediately superior or equal to *N*. Knowing N_H, we seek in the table of Plackett and Burman the row corresponding to that value. We give hereafter the rows proposed by Plackett and Burman (6).

- $N_H = 4$ + + -
- $N_H = 8$ + + + - + - -
- $N_H = 12$ + + - + + + - - - + -
- $N_H = 16$ + + + + - + - + + - - + - - -
- $N_H = 20$ + + - - + + + + - + - + - - - - + + -
- $N_H = 24$ + + + + + - + - + + - - + + - - + - + - - - -
- $N_H = 32$ - - - - + - + - + + + - + + - - - + + + + + - - + + - + - - +

Figure 3 Hadamard matrix construction rules.

From the selected row, we build the Hadamard design by choosing one of the methods that we present now, taking as an example the construction of the Hadamard design for three factors ($k = 3$) and $N = 4$ experiments.

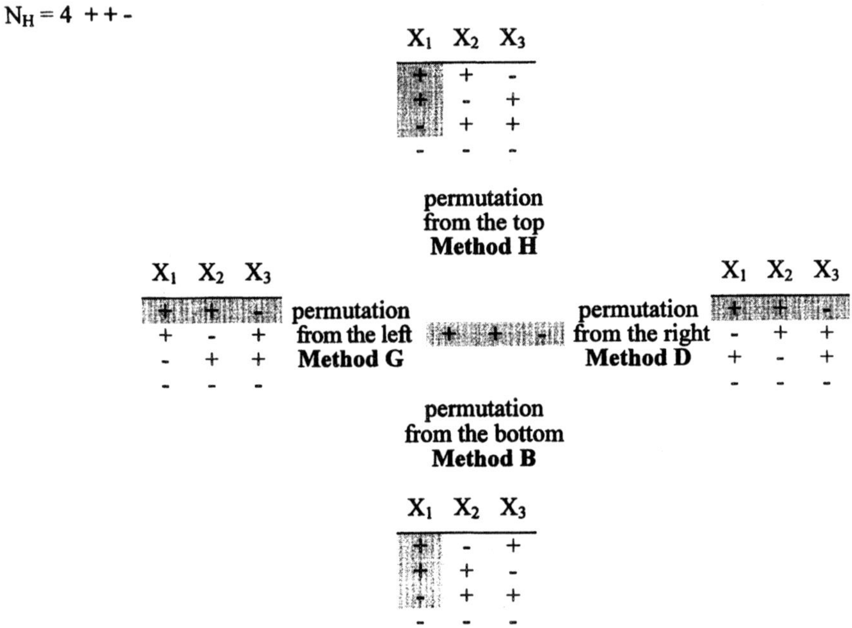

Figure 4 Hadamard design construction methods.

The Hadamard designs allow us to estimate the "weights" (β_i) of k factors in N experiments with a variance:

$$\mathrm{var}(b_j) = \sigma^2/N \tag{3}$$

2. Example: Study of the Cycle of Compression of a Composite Ceramic

To obtain a composite ceramic, one has often to resort to the compression by heat. The action of a strong pressure applied on the material is added to the action of the temperature. The research workers realize some cycles of thermic treatment with two levels of temperature and they have first carried out a preliminary study enabling them to find out, among a list of 11 factors, the ones whose variation entails a variation of the responses studied.

a. Experimental Domain

Table 1 Experimental Domain

	Factors	Level (−)	Level (+)
U_1	Temperature of the 1st plateau (°C)	1700	1800
U_2	Duration of the 1st plateau (min)	15	45
U_3	Temperature of the 2nd plateau (°C)	$U_1 + 50$	$U_1 + 150$
U_4	Duration of the 2nd plateau (min)	5	15
U_5	Rate of increase in temperature (°C/min)	10	30
U_6	Rate of decrease in temperature (°C/min)	10	30
U_7	Pressure (MPa)	20	40
U_8	Time of pressure start	Mp1	Mp2
U_9	Time of pressure release	Rp1	Rp2
U_{10}	Rate of increase in pressure (MPa/s)	0.1	1.0
U_{11}	Rate of decrease in pressure (MPa/s)	0.1	1.0

b. Experimental Responses

η_1: relative density (%)
η_2: open porosity (%)
η_3: bending strength (MPa), noted FLEX

These parameters are often used to characterize ceramic materials, especially if the simplicity of preparation of the samples is taken into account.

The compression cycle chosen will be the one that allows one to achieve the optimal properties while maintaining the temperature and duration of the plateau at the minimal values.

c. Experimental Design

The study aims at estimating the "weight" of each factor studied, which can take two distinct levels. The appropriate experimental design is a Hadamard design (or Plackett and Burman's design). This type of design exists only for numbers of experiments N multiple of 4, called Hadamard numbers. In this case, the number of factors studied is $k = 11$ and consequently at least a minimum of 12 experiments will be necessary to estimate the weight of these 11 factors.

The experimental design to construct is a Hadamard design: $N = 12$. The first row of this design is:

$$+ \quad + \quad - \quad + \quad + \quad + \quad - \quad - \quad - \quad + \quad -$$

This first row generates the following rows by circular permutation (to the right, in this case: method D) and we obtain the following experimental design with 11 columns and 12 rows:

Table 2 Hadamard Matrix $N = 12$, for the 11-Factor Study

No.	X_1	X_2	X_3	X_4	X_5	X_6	X_7	X_8	X_9	X_{10}	X_{11}
1	+	+	−	+	+	+	−	−	−	+	−
2	−	+	+	−	+	+	+	−	−	−	+
3	+	−	+	+	−	+	+	+	−	−	−
4	−	+	−	+	+	−	+	+	+	−	−
5	−	−	+	−	+	+	−	+	+	+	−
6	−	−	−	+	−	+	+	−	+	+	+
7	+	−	−	−	+	−	+	+	−	+	+
8	+	+	−	−	−	+	−	+	+	−	+
9	+	+	+	−	−	−	+	−	+	+	−
10	−	+	+	+	−	−	−	+	−	+	+
11	+	−	+	+	+	−	−	−	+	−	+
12	−	−	−	−	−	−	−	−	−	−	−

d. Plan of Experimentation

Table 3 Plan of Experimentation

No.	U_1 (°C)	U_2 (min)	U_3 (°C)	U_4 (min)	U_5 (°C/min)	U_6 (°C/min)	U_7 (MPa)	U_8	U_9	U_{10} (MPa/s)	U_{11} (MPa/s)
1	1800	45	$U_1 + 50$	15	30	30	20	Mp1	Rp1	1, 0	0, 1
2	1700	45	$U_1 + 150$	5	30	30	40	Mp1	Rp1	0, 1	1, 0
3	1800	15	$U_1 + 150$	15	10	30	40	Mp2	Rp1	0, 1	0, 1
4	1700	45	$U_1 + 50$	15	30	10	40	Mp2	Rp2	0, 1	0, 1
5	1700	15	$U_1 + 150$	5	30	30	20	Mp2	Rp2	1, 0	0, 1
6	1700	15	$U_1 + 50$	15	10	30	40	Mp1	Rp2	1, 0	1, 0
7	1800	15	$U_1 + 50$	5	30	10	40	Mp2	Rp1	1, 0	1, 0
8	1800	45	$U_1 + 50$	5	10	30	20	Mp2	Rp2	0, 1	1, 0
9	1800	45	$U_1 + 150$	5	10	10	40	Mp1	Rp2	1, 0	0, 1
10	1700	45	$U_1 + 150$	15	10	10	20	Mp2	Rp1	1, 0	1, 0
11	1800	15	$U_1 + 150$	15	30	10	20	Mp1	Rp2	0, 1	1, 0
12	1700	15	$U_1 + 50$	5	10	10	20	Mp1	Rp1	0, 1	0, 1

e. Randomization

We must carefully examine the set of experiments of the experimental design and in particular we must verify that all of the experiments are possible and that the actual operating conditions meet all requirements. If one or more experiments could not be performed, the proposed experimental design should be revised. Let us remind you of an essential principle: *The experimental design must be modified to take the problem into account*, NOT *"The problem must be modified to adapt to the design of experimentation that is known."* Theoretically, all of the nonstudied factors should been kept constant, but this is not actually possible. Even if we suppose that these variations have very little influence on the responses studied, these interferences can distort our interpretations. We must try to see to it that these distortions are randomly distributed on the set of experiments and for this we must run the experiments in a random order.

f. Calculation of Effects

In the case of a screening study, i.e., a study aiming at estimating the "weight" of each variable studied, the effects are a priori supposed to be totally additive. This implies that the relationship between the experimental responses and the variables is in the form of a first-degree polynomial model, valid only for values of $x_{ij} = \pm 1$.

$$\eta = \beta_0 + \beta_1 X_1 + \beta_2 X_2 + \beta_3 X_3 + \beta_4 X_4 + \beta_5 X_5 + \beta_6 X_6 + \beta_7 X_7 + \beta_8 X_8 + \beta_9 X_9 + \beta_{10} X_{10} + \beta_{11} X_{11} \quad (4)$$

It is demonstrated that we can construct as many independent linear combinations **L** as there are columns in the effect matrix. From these linear combinations, we can calculate an estimate (b_j) of the weight of the factors studied. The estimate of these effects b_j is obtained by multiplying the elements in column X_j by the corresponding elements of the experimental response vector Y, then calculating the algebraic sum and dividing it by the number of experiments.

After calculation, the estimates of the effects b_j for each response are:

Table 4 Experimental Responses and Calculation of Effects

Experimental responses			Calculation of the effects			
Y_1	Y_2	Y_3		Y_1	Y_2	Y_3
91.2	1.79	845	$L_0 = b_0$	90.02	1.45	902.2
92.7	0.03	1022	$L_1 = b_1$	5.20	−0.47	78.8
99.2	0.23	1067	$L_2 = b_2$	0.50	0.05	12.2
90.6	1.49	878	$L_3 = b_3$	0.20	−0.02	38.0
80.3	3.63	715	$L_4 = b_4$	0.25	0.13	8.2
90.6	1.75	911	$L_5 = b_5$	0.72	0.09	9.0
96.9	2.21	1020	$L_6 = b_6$	1.02	−0.03	6.3
92.2	1.09	891	$L_7 = b_7$	4.83	−0.42	93.5
99.1	0.47	1076	$L_8 = b_8$	−0.60	0.68	−11.3
77.3	4.13	774	$L_9 = b_9$	0.90	−0.03	7.5
92.7	0.09	987	$L_{10} = b_{10}$	−0.78	0.88	−12.0
77.4	0.49	640	$L_{11} = b_{11}$	0.38	0.10	32.0

g. Interpretation Helping Tools

There are conventional statistical tools that can help us determine which factors—passing from one state (−1) to another state (+1)—make the response vary significantly, provided that the variance of the experimental error is known.

In order to know the experimental error, it is necessary to replicate one or several experiments. If we consider that we are at the screening stage and are trying to reduce the number of experiments to the minimum, we have only rarely the possibility to perform these replicates. In the case of *saturated* (as many experiments to run as there are effects to calculate) or near-saturated experimental designs, and in the case of an unknown experimental variance, we have at our disposal interpretation helping tools that allow us to set limits of *significance* for the calculated effects. For this, we use different properties of the normal distributions that allow us to estimate a standard deviation and so fix limits characteristic of the level of significance chosen. In these tools, we use the term *active* rather than *significant* as used in the conventional quantitative statistical tools. We will suppose that the experimental error is a random variable, distributed according to a normal distribution.

The values of the effects can be represented in the form of a *bar chart*, the surface of each bar being proportional to the value of the effect.

h. Graphical Analysis of the Effects

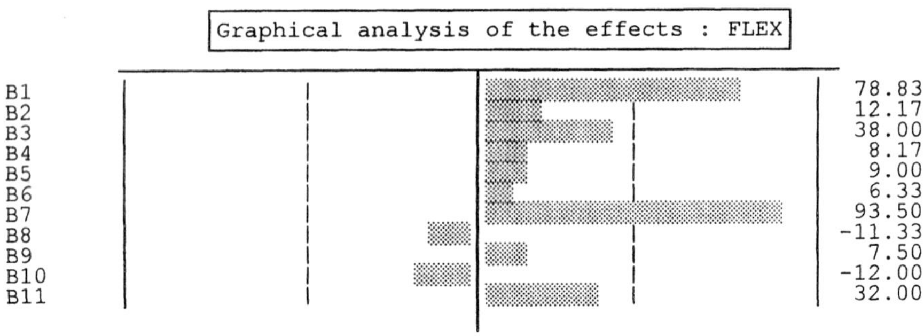

Figure 5 Graphical analysis of the effects for the response (FLEX).

i. Approach of Lenth

Lenth supposes that all the effects $b_i (i > 0)$ are null, i.e., their values are due only to the experimental errors:

If no effect is "active," i.e.:

$$H_0: \beta_i = 0$$

$$\text{so } b_i \approx f(e_i)$$

with the error e_i being a random variable, distributed according to a normal distribution. ⇒ the b_i will also be distributed according to a normal distribution around 0.

Then Lenth showed that, with this assumption of normality, there is a relation between the median of the absolute values of the b_j and the standard deviation, which takes the form:

$$S_0 = 1.5 \times \text{median}\,(|b_i|) \tag{5}$$

S_0 is rather called "pseudo–standard deviation."

Concretely, this analysis can be broken down into several stages:

1. The effects b_j are arranged in ascending order, in absolute value:

6.3, 7.5, 8.2, 9.0, 11.3, 12.0, 12.2, 32.0, 38.0, 78.0, 93.5

2. The pseudo–standard deviation is calculated:

$$S_0 = 1.5 \times 12.0 = 18.0$$

3. In this first stage, all the calculated effects have been considered but for some of them, very "remote," the null hypothesis (H_0) is probably not acceptable. In order to characterize a normal distribution, it is sufficient to know the mean and the standard deviation. In this case, we have the necessary and sufficient parameters: if we suppose that for all effects, $b_j \approx f(e_i)$, the mean is equal to 0 (since the error is a random variable centered on 0) and the estimation of the standard deviation has just been calculated (S_0).

According to this approximation, the b_j can be positioned and their distribution can be represented like this:

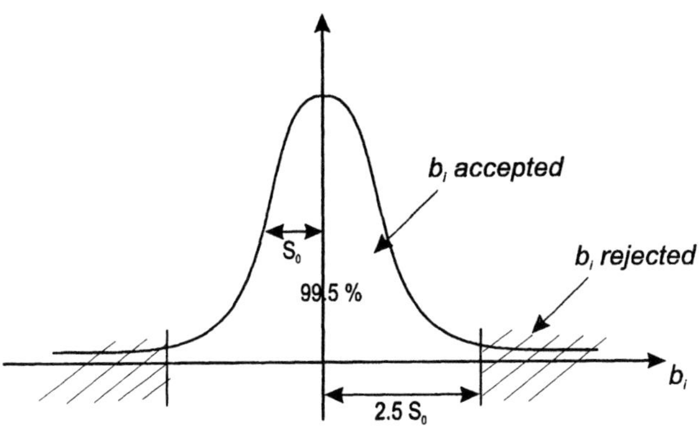

Figure 6 Normal distribution of the effects.

Lenth chooses to disregard the remote effects and considers only one part of the population. He decides to keep only 99.5% of the population: the effects located at a distance superior to $\pm 2.5 S_0$.

The disregarded effects will be such that:

$$|b_i| > 2.5 \times S_0$$

$$|b_i| > 2.5 \times 18.0 = 45.0$$

$\Rightarrow$ The two effects >45.0 ($|b_1|$ and $|b_7|$) are thus excluded.

4. Nine effects are left, for which the same approach is used:

6.3, 7.5, 8.2, 9.0, <u>11.3</u>, 12.0, 12.2, 32.0, 38.0

$$S_0 = 1.5 \times 11.3 = 16.95$$

We use the same approach as before and seek the effects superior to the limit value:

$$|b_i| > 2.5 \times 16.95 = 42.38$$

There are no effects left to exclude now. Thus we obtain an estimate of the pseudo–standard deviation:

PSE = 16.95

5. The limits of significance can be calculated by means of the formula:

$$\text{ME} = t_{\alpha/2,d} \times \text{PSE} \qquad (6)$$

where $t_{\alpha/2,d}$ is the Student-Fisher coefficient, which depends on:

$-\alpha$: level of significance, which we have chosen equal to 5%

$-d$: degree of freedom

Lenth suggests taking:

$$d = \text{number of effects left}/3$$
$$= 9/3 = 3$$

$\Rightarrow t_{0.025,3} = 3.18$

ME = 3.18 × 16.95 = **±53.9**

that is to say b_1, b_7 are *active* effects.

6. To avoid a distortion caused by interferences, Lenth suggests calculating other limits by changing the value of the level of significance.

$\Rightarrow t_{1-\gamma,d}$, with:

$$\gamma = (1 - 0.95^{1/\text{nb}\, b_j})/2 \qquad (\text{nb } b_j\text{: number of effects } b_j) \qquad (7)$$

Thus, the other limits of significance will be:

$$\textbf{SME} = \pm\textbf{113.8}$$

We can fix the limits of confidence, characteristic of the level of significance chosen, on the diagram of the graphic representation of the effects.

j. Normal Plot and Half-Normal Plot

If we represent the function of repartition of the effects (b_i or $|b_i|$) on Gausso-arithmetic paper, we obtain the following graph:

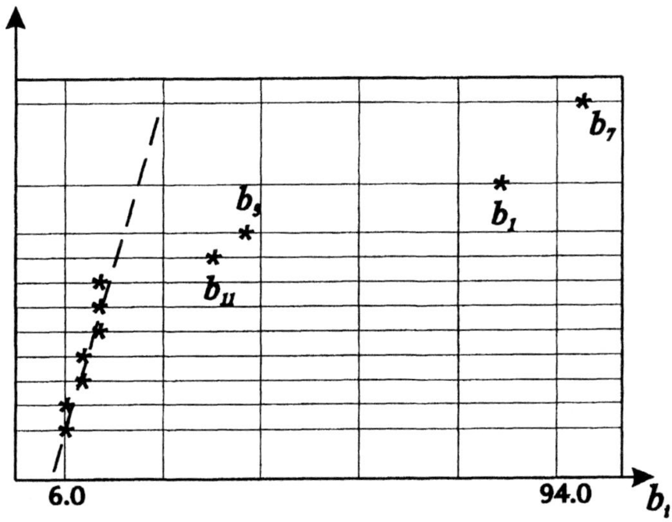

Figure 7 Half-normal plot.

The effects that are far from the straight line (below or above the line, depending on whether the effects are positive or negative) do not agree with the initial hypothesis: ⇒ these effects are thus considered as "*probably active*."

This graphic representation is done like this:

1. Classification of the k studied effects in increasing order,
2. Calculation of the ordinate that each effect would have on the combined distribution curve if all of the effects were due to experimental errors only. This value is given by:

$$p_i = 100 \cdot (i - 0.5)/k \tag{8}$$

3. Plotting each effect on Gausso-arithmetic paper, with a Gaussian scale on the y axis and an arithmetic scale on the x axis.

k. Pareto's Approach

In this approach and by analogy with the least-squares method, we consider the squares of the effects. We consider the square of each effect $b_i : b_i^2$ and we perform a normation:

$$P_i = 100\,(b_i^2/\Sigma\, b_i^2) \tag{9}$$

Table 5

L_i	b_i	b_i^2	p_i (%)	Rank
L_1	78.8	6214	34.4	2
L_2	12.2	148	0.8	5
L_3	38.0	1444	8.0	3
L_4	8.2	67	0.4	9
L_5	9.0	81	0.4	8
L_6	6.3	40	0.2	11
L_7	93.5	8742	48.3	1
L_8	−11.3	128	0.7	7
L_9	7.5	56	0.3	10
L_{10}	−12.0	144	0.8	6
L_{11}	32.0	1024	5.7	4

$[\Sigma\, b_i^2 = 18089]$

The p_i, which represent a percentage of the total dispersion, are arranged in descending order and represented graphically in the form of a bar chart.

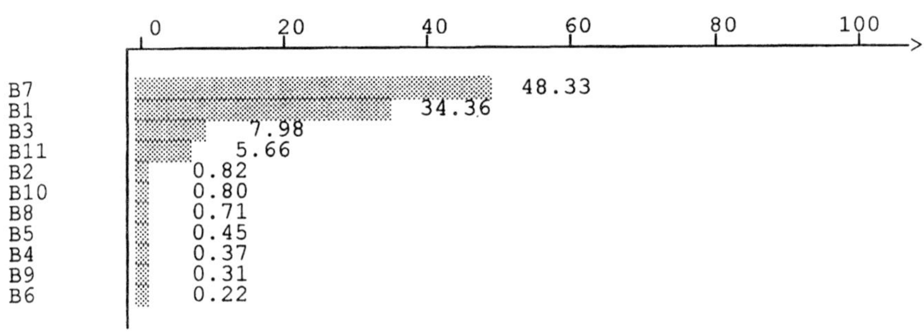

Figure 8 Graphic analysis of the effects according to Pareto: FLEX.

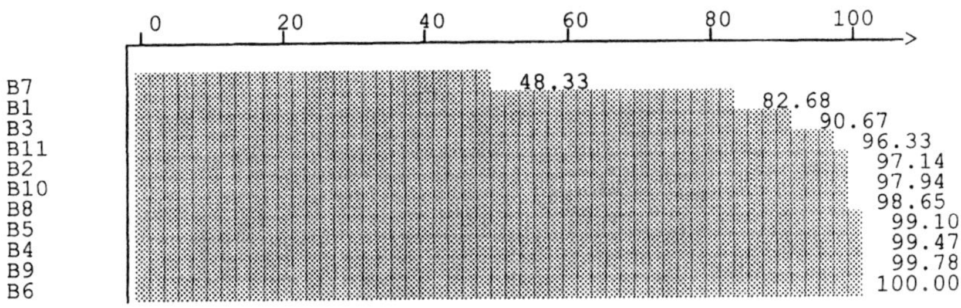

Figure 9 Cumulative function according to Pareto: FLEX.

It is very easy to determine experimentally the weight of each of the factors and to classify them, and then, in view of a more accurate study, to choose the most influential (not the most liked) factors.

II. STUDY OF THE EFFECTS OF THE FACTORS

A. Introduction

When the influential factors are known, they can be studied more accurately in a second step. We thus abandon the hypothesis of additivity that we had put forward in the screening stage by assimilating factors to objects, and we take into account the possible interactions between the different factors. The conventional strategy, which consists of studying one factor at a time while holding the others constant, still in use in some laboratories, not only represents a high cost in experiments but also ignores the interactions. The information obtained is incomplete and we will never manage to solve the problem under study. There are two sorts of possible interactions:

Those we can postulate and whose importance we want to ascertain
Those we ignore and which we would like to neglect if possible

For this type of study, the desired information is well specified and the most often used experimental designs are the full or fractional, symmetrical or asymmetrical factorial designs. These designs have all of the required qualities, in particular the quality of sequentiality. These designs are very often used [especially the two-level factorial designs (7,8)] and their applications are many.

B. Factorial Designs

We consider k factors $U_1, U_2, \ldots, U_k$, each of which can take $s_1, s_2, \ldots, s_k$ different levels. By definition, a full factorial design is composed of all of the possible combinations of the levels of each factor. The number **N** of distinct combinations is equal to the product of the numbers of levels:

$$N = s_1 \times s_2 \times \cdots \times s_k \tag{10}$$

Example: be the factors U_1, U_2, and U_3, which can take the following levels:

U_1: Temperature: 100°C, 120°C, 150°C (three quantitative levels)
U_2: Concentration of a reactant: 1%, 2% (two quantitative levels)
U_3: Nature of a catalyst: A, B, C (three qualitative levels)

The full factorial design is composed of the 18 following combinations:

Table 6 Plan of Experimentation

No.	U_1	U_2	U_3	No.	U_1	U_2	U_3
1	100	1	A	10	100	2	B
2	120	1	A	11	120	2	B
3	150	1	A	12	150	2	B
4	100	2	A	13	100	1	C
5	120	2	A	14	120	1	C
6	150	2	A	15	150	1	C
7	100	1	B	16	100	2	C
8	120	1	B	17	120	2	C
9	150	1	B	18	150	2	C

If all of the factors take the same number of levels, the number of combinations is:

$$N = \underbrace{s \times s \times \cdots \times s}_{k \text{ times}} = s^k \tag{11}$$

By extension, the design is called *full factorial design* (7,8). The factorial design is called *symmetrical.* In the opposite case, it is called *asymmetrical.* The number of experiments increases very quickly according to the value of s. It is easy to understand why the most used factorial designs are the 2^k factorial designs, i.e., the designs in which each factor can take only two distinct levels.

1. Full Factorial Design at Two Levels (2^k)

The factorial design contains k columns and N rows. The N rows correspond to the N distinct combinations. When the number of factors increases, so does the number of distinct combinations, and to facilitate the construction of the designs we usually use an order of construction called standard order. This representation is made easier by replacing -1 and $+1$ by the signs $-$ and $+$.

All columns start with the sign (−).
The first column contains an alternance of (−) and (+).
The second column contains an alternance of two (−) and two (+).
. . .
The kth column contains 2^{k-1} signs (−) followed by 2^{k-1} signs (+).

Example: 2^2 factorial design

The full factorial design is composed of the $2^2 = 4$ experiments reported in the two central columns of Table 7:

Table 7 2^2 Factorial Matrix

No.	X_1	X_2
1	−	−
2	+	−
3	−	+
4	+	+

a. Mathematical Model
We postulate a model of the form:

$$\eta_i = \beta_0 + \beta_1 x_{i1} + \beta_2 x_{i2} + \beta_{12} x_{i1} x_{i2} \qquad (x_{ij} = \pm 1; i = 1,2,\ldots,4; j = 1,2) \quad (12)$$

b. Calculation and Interpretation of Effects
We represent by y_j the experimental response obtained in the j experiment. By replacing x_{ij} and y_j by their values in the preceding equation, we obtain:

$$\begin{aligned} y_1 &= b_0 - b_1 - b_2 + b_{12} \\ y_2 &= b_0 + b_1 - b_2 - b_{12} \\ y_3 &= b_0 - b_1 + b_2 - b_{12} \\ y_4 &= b_0 + b_1 + b_2 + b_{12} \end{aligned} \qquad (13)$$

By combining these four equations, we obtain:

$$\begin{aligned} b_0 &= (+y_1 + y_2 + y_3 + y_4)/4 \\ b_1 &= (-y_1 + y_2 - y_3 + y_4)/4 \\ b_2 &= (-y_1 - y_2 + y_3 + y_4)/4 \\ b_{12} &= (+y_1 - y_2 - y_3 + y_4)/4 \end{aligned} \qquad (14)$$

We can interpret the coefficients as discussed in the following.

c. Estimate of Constant Term b_0

It is the arithmetic mean of the responses, or the *theoretical value* of the response at the center of the experimental domain (the phenomenon is supposed to be linear of the first degree in the experimental domain).

d. Estimates of Main Effects b_j

b_j is the main effect of the variable X_j.

Main Effect b_1. We can consider that b_1 represents the variation of the response obtained when the variable X_1 increases by one unit, with the variable X_2 held constant.

$$b_1 = (-y_1 + y_2 - y_3 + y_4)/4 \tag{15}$$

Main Effect b_2. We can consider that b_2 represents the variation of the response obtained when the variable X_2 increases by one unit, with the variable X_1 held constant.

$$b_2 = (-y_1 - y_2 + y_3 + y_4)/4 \tag{16}$$

e. Estimate of the Interaction Effect b_{12}

The first-order interaction effect (or interaction effect between two variables) between the variables X_1 and X_2 is represented by b_{12}. A null or low value of b_{12} means that the effect of b_1 (when $X_2 = +1$) differs only slightly from the effect of b_1 (when $X_2 = -1$); thus, the main effect of X_1 does not virtually vary as a function of the level of X_2. The effects of the variables X_1 and X_2 are said to be independent.

A great value of b_{12} means that the effect of b_1 (when $X_2 = +1$) differs markedly from the effect of b_1 (when $X_2 = -1$), and thus the main effect of X_1 varies according to the value taken by the variable X_2. The effects of the variables (and thus of the factors) are no longer independent. We say that there is an interaction effect between the two variables X_1 and X_2.

$$b_{12} = (+y_1 - y_2 - y_3 + y_4)/4 \tag{17}$$

f. Matrix Effect or Model Matrix

The model matrix (also called *effect matrix* in the case of the factorial experimental designs studied here) is composed of four rows (four experiments) and four columns (X_0, X_1, X_2, X_1X_2).

Table 8 Effect Matrix (or Model Matrix)

	X_0	X_1	X_2	X_1X_2
	1	−1	−1	+1
$X =$	1	+1	−1	−1
	1	−1	+1	−1
	1	+1	+1	+1

Generalization to the 2^k Experimental Design. The full factorial design is composed of the 2^k experiments reported in the k columns:

Table 9 2^k Full Matrix

No.	X_1	X_2	...	X_k
1	−	−	...	−
2	+	−	...	−
	−	−		−
	+	−		−
...	...	...	...	...
...	−	−	...	+
...	+	−	...	+
$N - 1$	−	+	...	+
N	+	+	...	+

a. Mathematical Model

We postulate a model of the form:
Mathematical model

$$
\begin{aligned}
\eta_i = \beta_0 &+ \beta_1 x_{i1} + \beta_2 x_{i2} + \cdots + \cdots + \beta_k x_{ik} && \text{main effects} \\
&+ \beta_{12} x_{i1} x_{i2} + \cdots + \beta_{k-1,k} x_{ik-1} x_{ik} && \text{1st-order interaction effects} \\
&+ \beta_{123} x_{i1} x_{i2} x_{i3} + \cdots + \beta_{k-2,k-1,k} x_{ik-2} x_{ik-1} x_{ik} && \text{2nd-order interaction effects} \\
&+ \cdots + \cdots + \cdots && \cdots\cdots\cdots\cdots \\
&+ \beta_{123\cdots k} x_{i1} x_{i2} x_{i3} \cdots x_{ik} && (k-1)\text{th order interaction effects} \\
(x_{ij} &= \pm 1;\ i = 1, 2, \ldots, \mathrm{N};\ j = 1, 2, \ldots, \mathrm{k})
\end{aligned} \tag{18}
$$

The number of coefficients of the model is fixed equal to p.

b. Effect Matrix or Model Matrix

The model matrix contains as many rows as there are experiments and as many columns as there are coefficients in the model. The first column (X_0) contains only (+) signs. We then add the columns of the experimental design ($X_1, \ldots,$

X_k), which we call *basic columns*. The other columns are obtained by combining the k basic columns 2×2, 3×3, $k \times k$.

Table 10 Effect Matrix (or Model Matrix)

	X_0	X_1	$\cdots$	X_k	X_1X_2	$\cdots$	$X_{k-1}X_k$	X_1X_2 X_3	$\cdots$	X_{k-2} $X_{k-1}X_k$	$\cdots$	$\cdots$	$X_1 \cdot X_{k-1}X_k$
	+	−	...	−	+	...	+	−	...	+	...	...	...
	+	+	...	−	−	...	+	+	...	+	...	...	...
	+	−		−	−		+	+		+			
	...	...	...	...	...	...	...		...		...	...	
X	...	...	...	...	...	...	...		...		...	...	...
...	...	...	...	...	...	...	...		...		...	...	...
...	+	+	...	+	−	...	+	−	...	+	...	...	...
	+	−	...	+	−	...	+	−	...	+	...	...	...
	+	+	...	+	+	...	+	+	...	+	...	...	...

c. Calculation and Interpretation of Effects

We demonstrate that it is possible to build, from the model matrix, 2^k independent linear combinations represented by the letter L followed by an index (ranging from 0 to $p - 1$). The rule of construction is very simple: from the model matrix, we combine the signs of a column and the corresponding elements of the column Y, we calculate the algebraic sum, and we divide it by the number of experiments. Each independent linear combination allows us to obtain the estimate of an effect. For example, if we go back to the 2^2 design and the corresponding model matrix, we can construct the following four independent linear combinations:

Table 11 2^2 Model Matrix for the Calculation of Effects

No.	X_0	X_1	X_2	X_1X_2	Y
1	+	−	−	+	y_1
2	+	+	−	−	y_2
3	+	−	+	−	y_3
4	+	+	+	+	y_4

Two-by-two independent linear combinations

$$\begin{aligned}
L_0 &= b_0 = (y_1 + y_2 + y_3 + y_4)/4 \\
L_1 &= b_1 = (-y_1 + y_2 - y_3 + y_4)/4 \\
L_2 &= b_2 = (-y_1 - y_2 + y_3 + y_4)/4 \\
L_3 &= b_{12} = (y_1 - y_2 - y_3 + y_4)/4
\end{aligned} \tag{19}$$

2. Fractional Factorial Designs at Two Levels (2^{k-r})

A 2^k fractional factorial experimental design (7,9) is formed of k columns corresponding to the k variables, each of which can take two distinct levels, noted ($-$) and ($+$) and of $N = 2^{k-r}$ rows corresponding to the 2^{k-r} experiments. It represents the fraction $1/2^r$ of the full 2^k design: $2^{k-r} = 2^k/2^r$. It can be built from a 2^m full factorial design, with $\mathbf{m} = k - r$ independent variables, the structures of the columns of the k–m variables left being combinations of the m basic columns.

Example: 2^{6-3} Factorial Design

If we want to build a 2^{6-3} design ($k = 6$, $r = 3$, $m = 3$), this means that we want to study six factors whose corresponding coded variables can take each the two levels ($-$) and ($+$) in eight experiments. It is as if we had a full 2^3 factorial design. We actually build the 1/8 fraction ($= 2^{6-3}$) of the 2^6 full factorial design.

a. Construction, Independent Generators

The effect matrix built from the $\boldsymbol{m}$ variables is formed of $\boldsymbol{m}$ columns corresponding to the 2^m effects, which can be calculated independently. The $(m + 1)$th variable must be given the structure of one of the effects, the $(m + 2)$th variable the structure of another effect, and so forth, until the $\mathbf{r}$ variables left have been assigned. By going back to the example studied above, the 2^{6-3} design can be built from the effect matrix corresponding to the full 2^3 design, and by giving, for example, the variables X_4, X_5, and X_6 the structure of the X_1X_2, X_2X_3, and $X_1X_2X_3$ columns (see the following table), which leads to the experimental design of Table 12:

Table 12 Effect Matrix of a Full 2^3 Factorial Design

X_0	X_1	X_2	X_3	X_4 X_1X_2	X_1X_3	X_5 X_2X_3	X_6 $X_1X_2X_3$
+	−	−	−	+	+	+	−
+	+	−	−	−	−	+	+
+	−	+	−	−	+	−	+
+	+	+	−	+	−	−	−
+	−	−	+	+	−	−	+
+	+	−	+	−	+	−	−
+	−	+	+	−	−	+	−
+	+	+	+	+	+	+	+

matrix 2^3

Table 13 2^{6-3} Fractional Factorial Design

X_1	X_2	X_3	X_4	X_5	X_6
−	−	−	+	+	−
+	−	−	−	+	+
−	+	−	−	−	+
+	+	−	+	−	−
−	−	+	+	−	+
+	−	+	−	−	−
−	+	+	−	+	−
+	+	+	+	+	+

The column X_4 has the structure of the column X_1X_2. We write: $X_4 \equiv X_1X_2$ or **4** $\equiv$ **12**. The columns contain only (−) and (+). If we multiply the elements of the column X_4 by the corresponding elements of the column X_1X_2, we obtain a column that contains only (+). This column is represented by the letter I. We thus can write: **I** $\equiv$ **124**.

The column X_5 has the structure of the column X_2X_3. We can write $X_5 \equiv X_2X_3$ or **5** $\equiv$ **23**. The columns contain only (−) and (+). If we multiply the elements of the column X_5 by the corresponding elements of the column X_2X_3, we obtain a column that contains only (+). We represent this column by the letter I: **I** $\equiv$ **235**.

Similarly, if we combine the elements of the column X_6 and the corresponding elements of the column $X_1X_2X_3$, we also obtain a column that contains only (+), and we can also write: **I** $\equiv$ **1236**.

4 $\equiv$ **12, 5** $\equiv$ **23** and **6** $\equiv$ **123** are called generating defining relations (*r* relations)

124, 1236, and **235** are called independent generators (*r*-independent generators)

If we combine the independent generators 2×2, 3×3, . . . , $r \times r$, we also obtain columns that contain only (+); these are generators. We can obtain a *total* of $2^r - 1$ generators.

In the preceding example, $2^r - 1 = 2^3 - 1 = 7$ generators. The generators obtained by multiplication from the three independent generators are:

Defining relations for the 2^{6-3} fractional factorial design

$$\begin{aligned} 124 \circ 235 &\equiv 1345 \\ 124 \circ 1236 &\equiv 346 \\ 235 \circ 1236 &\equiv 156 \\ 124 \circ 235 \circ 123 &\equiv 2456 \end{aligned} \tag{20}$$

$$I \equiv 124 \equiv 235 \equiv 1236 \equiv 1345 \equiv 346 \equiv 156 \equiv 2456$$

The set of generators plus the column I represents the *defining relation*, which is composed of 2^r terms.

b. Calculation of Effects

Let us consider the full 2^k factorial design: we can build an effect matrix and from it calculate 2^k independent linear combinations, each combination allowing calculation of the estimation of an effect. If we consider only a fraction of the full factorial experimental design, we can still build an effect matrix comprising 2^k columns but only 2^m independent linear combinations. Each independent linear combination allows us to calculate the estimate of 2^r confounded (or aliased) effects. If we make the assumption that certain effects are negligible, we may be able to know the estimate of the effects that are of interest for us, provided that this assumption is acceptable. It is often considered that the interactions of an order superior to 1 are negligible and so we seek fractions such that in all of the independent linear combinations the effects that interest us are confounded with interaction effects of a higher order. Usually, the different fractions are classified using the notion of *resolution*. For the example studied, the experimental design being composed of $N = 2^3$ experiments ($N = 8$ experiments), only N independent pieces of information corresponding to the 8 independent linear combinations can be calculated. There are 2^k effects ($2^6 = 64$ effects) and each linear combination represents the estimate of the sum of 2^r confounded effects (2^3 effects). The defining relation being composed of 2^r terms (2^3 terms), each linear combination is the sum of N effects (8 effects).

The independent linear combinations for the 2^{6-3} fractional factorial design are:

$$\begin{aligned}
L_0 &= (+y_1 + y_2 + y_3 + y_4 + y_5 + y_6 + y_7 + y_8)/8 \\
L_1 &= (-y_1 + y_2 - y_3 + y_4 - y_5 + y_6 - y_7 + y_8)/8 \\
L_2 &= (-y_1 - y_2 + y_3 + y_4 - y_5 - y_6 + y_7 + y_8)/8 \\
L_3 &= (-y_1 - y_2 - y_3 - y_4 + y_5 + y_6 + y_7 + y_8)/8 \\
L_4 &= (+y_1 - y_2 - y_3 + y_4 + y_5 - y_6 - y_7 + y_8)/8 \\
L_5 &= (+y_1 - y_2 + y_3 - y_4 - y_5 + y_6 - y_7 + y_8)/8 \\
L_6 &= (+y_1 + y_2 - y_3 - y_4 - y_5 - y_6 + y_7 + y_8)/8 \\
L_7 &= (-y_1 + y_2 + y_3 - y_4 + y_5 - y_6 - y_7 + y_8)/8
\end{aligned} \tag{21}$$

that is,

$$\begin{aligned}
E(L_0) &= \beta_0 + \beta_{124} + \beta_{235} + \beta_{1236} + \beta_{1345} + \beta_{346} + \beta_{156} + \beta_{2456} \\
E(L_1) &= \beta_1 + \beta_{24} + \beta_{1235} + \beta_{236} + \beta_{345} + \beta_{134} + \beta_{56} + \beta_{12456} \\
E(L_2) &= \beta_2 + \beta_{14} + \beta_{35} + \beta_{136} + \beta_{12345} + \beta_{2346} + \beta_{1256} + \beta_{456} \\
E(L_3) &= \beta_{12} + \beta_4 + \beta_{135} + \beta_{36} + \beta_{2345} + \beta_{12346} + \beta_{256} + \beta_{1456} \\
E(L_4) &= \beta_3 + \beta_{1234} + \beta_{25} + \beta_{126} + \beta_{145} + \beta_{46} + \beta_{1356} + \beta_{23456}
\end{aligned}$$

$$E(L_5) = \beta_{13} + \beta_{234} + \beta_{125} + \beta_{26} + \beta_{45} + \beta_{146} + \beta_{356} + \beta_{123456}$$
$$E(L_6) = \beta_{23} + \beta_{134} + \beta_5 + \beta_{16} + \beta_{1245} + \beta_{246} + \beta_{12356} + \beta_{3456}$$
$$E(L_7) = \beta_{123} + \beta_{34} + \beta_{15} + \beta_6 + \beta_{245} + \beta_{1246} + \beta_{2356} + \beta_{13456}$$

Resolution of a Fractional Factorial Design

As we have just seen, in most studies it is generally considered as an initial hypothesis that the main effects are larger than the first-order interaction effects, themselves being larger than the second-order interaction levels, and so on. Thus it is in our interest to have:

Main effects confounded with higher order interaction effects
First-order interaction effects confounded with interaction effects of an order superior to 1

We now introduce a useful concept, associated with the 2^{k-r} two-level fractional factorial designs, which is that of *resolution **R***. The value of the resolution of a 2^{k-r} two-level fractional factorial design is defined as being equal to the length of the shortest generator of the defining relation.

1. A design is said of *resolution II* (noted $\boldsymbol{R}_{\mathrm{II}}$) if:
   - Some main effects are aliased with one another. Indeed, at least one of the generators is of the form *XY* and so the main effect *X* will be confounded with the main effect *Y*.
2. A design is said of *resolution III* (noted $\boldsymbol{R}_{\mathrm{III}}$) if:
   - No main effect is confounded with one or several other main effects.
   - Some main effects are confounded with first-order interaction effects. At least one generator is of the form *XYZ*, which means that the main effect *X* (or *Y* or *Z*) is confounded with the interaction effect *YZ* (or *XZ* or *XY*).
3. A design is said to be of *resolution IV* (noted $\boldsymbol{R}_{\mathrm{IV}}$) if:
   - The main effects are confounded with interaction effects of an order superior or equal to 2.
   - Some first-order interaction effects are confounded with one another. At least one generator is of the form *XYZT*; thus, for example, the interaction effect *XY* is confounded with the interaction effect *ZT*.
4. A design is said to be of *resolution V* (noted $\boldsymbol{R}_{\mathrm{V}}$) if:
   - The main effects are confounded with third-order interaction effects,
   - Some first-order interaction effects are confounded with the second-order ones.

In the example discussed above, the generator comprising the smallest number of symbols is **124** (or **235**), and thus the resolution of the experimental design is $\boldsymbol{R}_{\mathrm{III}}$.

Knowing the resolution of an experimental design provides an important piece of information. Actually, for a given number of factors and experiments, and unless otherwise specified, we usually seek the highest resolution. For example, if we want to build a 2^{4-1} design, and if, a priori, no interaction is to be studied, we can give the factor X_4 the structure of the columns **12, 13, 23**, or **123**. In the first three cases, the obtained design is of resolution III (generators: **124, 134**, or **234**) whereas in the last case it is of resolution IV (generator: **1234**).

In other words, in the absence of any particular knowledge, we would rather sacrifice a higher order interaction to have the possibility to calculate the lower order interactions. The definition shows that a *n*th order interaction effect is a variation of the $(n - 1)$th order effect. It is generally admitted that the higher order interaction effects are terms that correct the lower order interaction effects. However, this is not an absolute rule and there are exceptions.

The notion of resolution has been extended to experimental designs other than fractional factorial designs. For example, the Hadamard designs are said to be of resolution III, even when N is not equal to a power of 2.

Example: Study of the Chlofibric Acid Synthesis

In this study of the chlofibric acid synthesis, seven factors playing apart in the reaction were considered. The response measured was the reaction yield.

$$\underset{(A)}{CH_3COCH_3} + \underset{(B)}{CHCl_3} + \underset{(C)}{ClC_6H_4OH} + 4NaOH \rightarrow ClC_6H_4\text{-}O\text{-}C(CH_3)_2COONa + 3NaCl + 3H_2O \quad (22)$$

The seven factors, each of them fixed at two levels, are presented as follows:

Table 14 List of Factors Studied

Factors studied	Abbreviation
U_1: Solution C addition temperature	Sol. C add. T°
U_2: Solution B addition temperature	Sol. B add. T°
U_3: Stirring duration of the reactional medium before the addition of solution B	Initial stirring time
U_4: Solution B addition duration	Sol. B add.
U_5: Ratio NaOH/C	
U_6: Ratio B/C	
U_7: Soda nature	

The quantity of factor A is kept constant in the reaction.

It was surmised that some interactions between the seven factors might exist. The list of the two-factor interactions is presented in Table 15.

Table 15 List of Interactions Studied

24: Solution B addition temperature—solution B addition duration
35: Initial stirring duration—ratio NaOH/C
16: Solution C addition temperature—ratio B/C
56: Ratio NaOH/C—Ratio B/C
46: Solution B addition duration—ratio B/C
45: Solution B addition duration—ratio NaOH/C
15: Solution C addition temperature—ratio NaOH/C

a. Choice of the Experimental Design:

In order to determine the relative importance of the experimental factors and of first-order interaction effects, a fractional factorial design 2^{7-3} with three independent generators $G_1 \equiv 1237$, $G_2 \equiv 1345$, $G_3 \equiv 2346$ including 16 experiments was achieved.

The $2^{(7-3)}$ fractional design with coded variables is represented in Table 16. The corresponding plan of experimentation with the natural variables, containing the experimental results of yield, is shown in Table 17.

Table 16 $2^{(7-3)}$ Fractional Factorial Design

No. exp	X_1	X_2	X_3	X_4	X_5	X_6	X_7	Response
1	−1	−1	−1	−1	−1	−1	−1	Y1
2	1	−1	−1	−1	1	−1	1	Y2
3	−1	1	−1	−1	−1	1	1	Y3
4	1	1	−1	−1	1	1	−1	Y4
5	−1	−1	1	−1	1	1	1	Y5
6	1	−1	1	−1	−1	1	−1	Y6
7	−1	1	1	−1	1	−1	−1	Y7
8	1	1	1	−1	−1	−1	1	Y8
9	−1	−1	−1	1	1	1	−1	Y9
10	1	−1	−1	1	−1	1	1	Y10
11	−1	1	−1	1	1	−1	1	Y11
12	1	1	−1	1	−1	−1	−1	Y12
13	−1	−1	1	1	−1	−1	1	Y13
14	1	−1	1	1	1	−1	−1	Y14
15	−1	1	1	1	−1	1	−1	Y15
16	1	1	1	1	1	1	1	Y16

b. Calculation of the Effects:

Starting from the 16 experiments realized, we can calculate 15 independent

Table 17 Plan of Experimentation

No. exp	Sol. C addition T°	Sol. B addition T°	Initial stirring duration (min)	Sol. B addition duration (min)	Ratio NaOH/C	Ratio B/C	Soda nature	Response yield (%)
1	25°C	25°C	5	30	4	1	Pearl	32.6
2	45°C	25°C	5	30	5.6	1	Pastille	42.8
3	25°C	45°C	5	30	4	1.5	Pastille	14.5
4	45°C	45°C	5	30	5.6	1.5	Pearl	9.0
5	25°C	25°C	60	30	5.6	1.5	Pastille	25.5
6	45°C	25°C	60	30	4	1.5	Pearl	23.9
7	25°C	45°C	60	30	5.6	1	Pearl	57.8
8	45°C	45°C	60	30	4	1	Pastille	6.1
9	25°C	25°C	5	60	5.6	1.5	Pearl	45.0
10	45°C	25°C	5	60	4	1.5	Pastille	26.0
11	25°C	45°C	5	60	5.6	1	Pastille	58.3
12	45°C	45°C	5	60	4	1	Pearl	11.5
13	25°C	25°C	60	60	4	1	Pastille	24.0
14	45°C	25°C	60	60	5.6	1	Pearl	57.5
15	25°C	45°C	60	60	4	1.5	Pearl	9.4
16	45°C	45°C	60	60	5.6	1.5	Pastille	0.0

linear combination of effects (called *aliases*), which are determined from the following defining relation:

$$I = 1237 = 1345 = 2457 = 2346 = 1467 = 1256 = 3567 \tag{23}$$

Each independent linear combination represents the sum of eight effects.

Table 18 Aliases

L 0 = 0 + 1237 + 1345 + 2457 + 2346 + 1467 + 1256 + 3567
L 1 = 1 + 237 + 345 + 12457 + 12346 + 467 + 256 + 13567
L 2 = 2 + 137 + 12345 + 457 + 346 + 12467 + 156 + 23567
L 3 = 3 + 127 + 145 + 23457 + 246 + 13467 + 12356 + 567
L 4 = 4 + 12347 + 135 + 257 + 236 + 167 + 12456 + 34567
L 5 = 5 + 12357 + 134 + 247 + 23456 + 14567 + 126 + 367
L 6 = 6 + 12367 + 13456 + 24567 + 234 + 147 + 125 + 357
L 7 = 7 + 123 + 13457 + 245 + 23467 + 146 + 12567 + 356
L 8 = 24 + 1347 + 1235 + 57 + 36 + 1267 + 1456
L 9 = 15 + 2357 + 34 + 1247 + 4567 + 26 + 1367
L 10 = 35 + 1257 + 14 + 2347 + 2456 + 1236 + 67
L 11 = 45 + 13 + 27 + 2356 + 1567 + 1246 + 3467
L 12 = 16 + 2367 + 3456 + 1234 + 47 + 25 + 1357
L 13 = 46 + 1356 + 2567 + 23 + 17 + 1245 + 3457
L 14 = 56 + 1346 + 2467 + 2345 + 1457 + 12 + 37

Several representations of factor effects (effect graphic effects, normal plot, half-normal plot, etc.) permit us to determine the influence of the factors studied. We can see that factors X_1, X_2, X_5, X_6 are the only influent factors on the variation of the yield (Fig. 10–12), and the factors X_3 and X_4 have hardly any effect. An increase of the solution B addition temperature (X_2) from 25°C to 45°C entails a diminution of the yield.

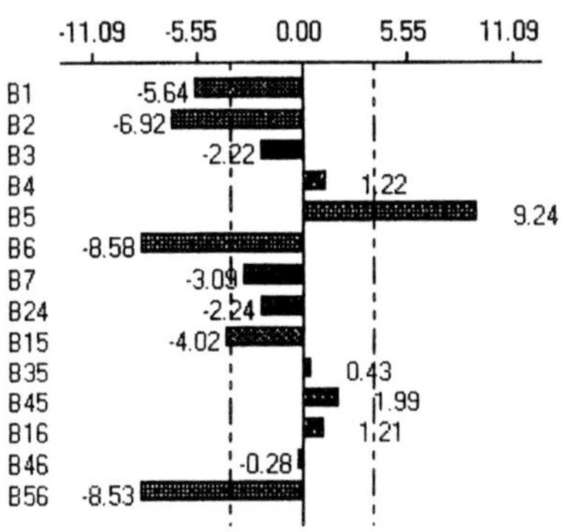

Figure 10 Effect graphic effects.

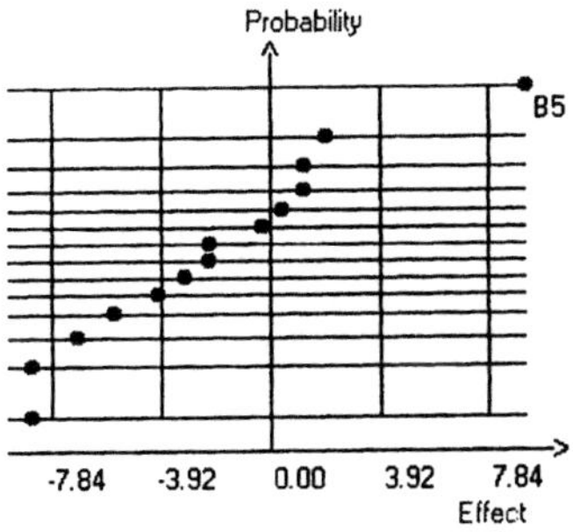

Figure 11 Normal plot.

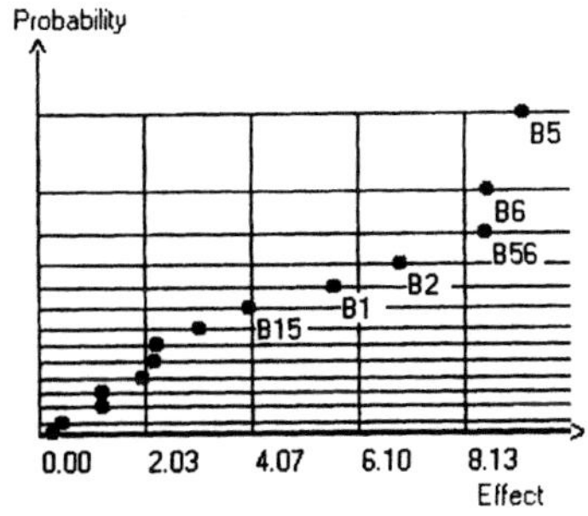

Figure 12 Half-normal plot.

There exist two interaction effects X_5X_6 and X_1X_5. Therefore, to interpret the effects of these three factors, the diagrams of these interactions (Fig. 13 and 14) will be made, in postulating every time that the aliased effects don't exist.

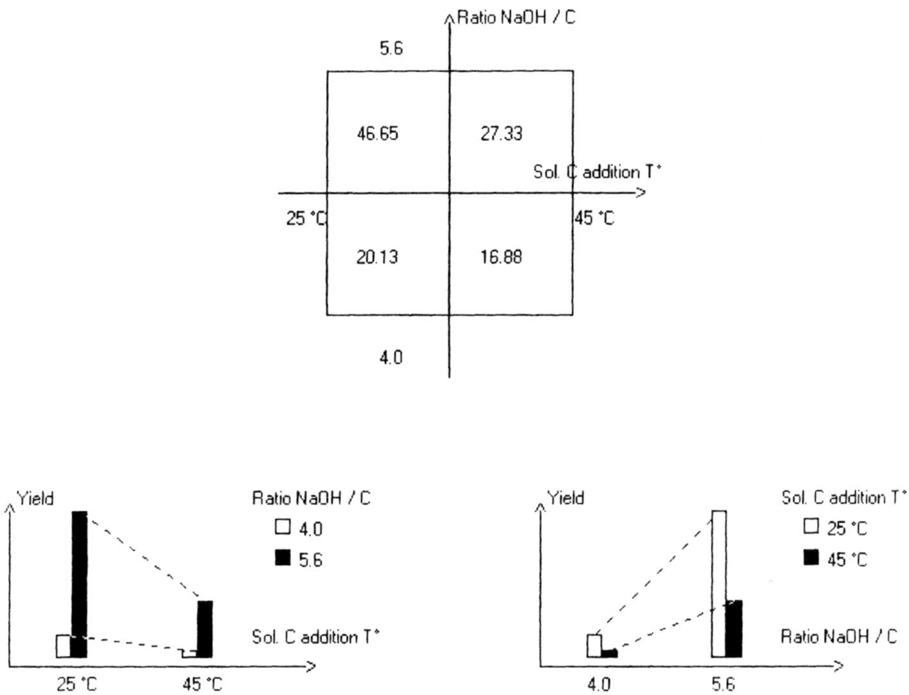

Figure 13 Representation of the interaction effect between the factors U_1 et U_5.

Interpretation of the Interaction Effect X_1X_5. For a low level of X_1, an increase of the ratio NaOH/C from 4 to 5.6 entails a strong augmentation of the yield, whereas for a high level of X_1, the effect of an increase of the ratio NaOH/C is smaller. For a low level of X_5, an increase of the solution C addition temperature from 25°C to 45°C entails a small diminution of the yield, whereas for a high level of X_5, an increase of X_1 entails a high diminution.

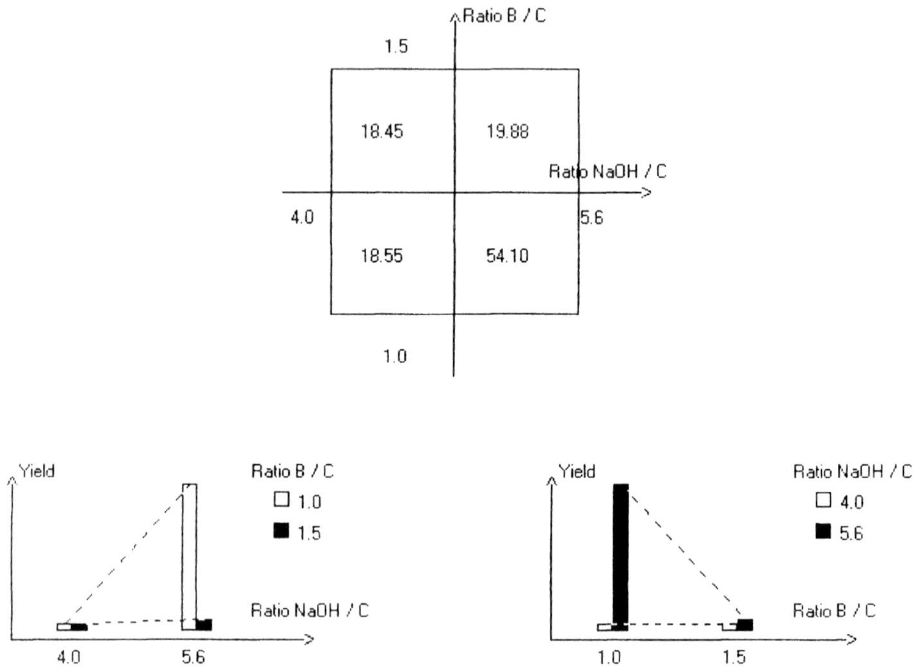

Figure 14 Representation of the interaction effect between the factors U_5 and U_6.

Interpretation of the Interaction Effect X_5X_6. For a low level of X_5, an increase of the ratio B/C from 1 to 1.5 has no effect on the yield, whereas for a high level of X_5, an increase of the ratio B/C entails a strong diminution of the yield. For a low level of X_6, an increase of the ratio NaOH/C from 4 to 5.6 entails a strong augmentation of the yield, whereas for a high level of X_6, an increase of X_5 has hardly any effect.

III. RESPONSE SURFACE METHODOLOGY

A. Introduction

In many applications, what interest is not in studying the effects of the factors or the importance of the interactions but in knowing how one or several measured characteristics behave in a well-defined experimental domain, which we call the *experimental domain of interest*. Actually, what interests us is *to know, in any point of the experimental domain of interest, the value of one or several experimental responses*. Indeed, the experimenter would like to be able to predict the value of the response *at any point in the experimental domain* (and possibly in the vicinity), without having to run the experiment. There are several circumstances in which the experimenter may need this possibility of prediction:

When he tries to replace arrays of results or charts by a mathematical model, such as to include it in a computer program
When he tries to determine the behavior of the phenomenon in the vicinity of a point, such as to anticipate the consequences of the variation of a factor
When he seeks an optimum, or a compromise of several responses simultaneously

For this, he has to find the relationships that exist between the factors and the responses. This part of the methodological tool is called *response surface methodology*.

Whatever the domain of application, the objectives are clear: we want to find a region in the experimental domain of interest where all of the properties studied meet the constraints, if possible with the greatest safety. We call this the *acceptable compromise area*.

In case there is no discontinuity and not many optima in the phenomenon studied in the experimental domain of interest, we try to link the factors and the responses. This is often done by modeling the phenomenon for a studied experimental response, i.e., by simplifying it in the form of a model:

$$\eta = f(x)$$

These models can be extremely varied and depend on the type of problem studied: linear or nonlinear models, differential equations, and so forth. It is obvious that if we knew these models it would be very easy for us to use them to make predictions.

1. What Qualities Must These Models Have?

These models are not required to account for the underlying mechanism of the phenomenon studied (which would be very interesting) since it would turn out either to be impossible or to require very long searching. They must simply have the following properties:

Be a good representation of the experimental response in the experimental domain of interest,

And, *providing that the above condition is respected*, allow one to obtain an estimate of the value of the response of acceptable quality. Indeed, a prediction is of little interest if its uncertainty is too high. The estimate of the value of the response studied, at any point of the experimental domain of interest, must therefore be known with an uncertainty that is not minimal but acceptable. We say that the quality of the estimated response at any point of this domain is acceptable if it can be compared with that obtained, for the same point, by running the experiment.

2. What Types of Models Shall We Use?

We may choose any type of model. It is not the structure or the shape of the model that is important but rather its ability to display the above-mentioned properties. We can thus use any type of model. They all belong to the class of *empirical models*, among which the most often used are the polynomial models, for two main reasons:

They are simple;
They allow a sequential approach.

We can also give a third reason, which may reassure the experimenters who do not really like to use empirical models. Indeed, if the experimental response studied does not display any discontinuity and has a certain regularity (continuously derivable as far as the order d), it can be represented by a polynomial of Taylor of degree d in the vicinity of the point considered (or center of interest). The precision of the development in series depends on the degree of the polynomial and on the volume of the experimental domain studied. The number p of coefficients to calculate for a model of degree d with k variables is equal to $(k + d)!/(k!\,d!)$. This number increases rapidly with k (number of variables) or d (degree of the polynomial). As the number of experiments should always be superior or equal to the number of coefficients to calculate, the degree of the model is rarely superior to two. Therefore, the number of experimental points to realize is reasonable, and very often the quality of the prediction is acceptable.

This can reassure the experimenters, but they should not forget that *all of the models are wrong, even if some are more useful than others*!

The most often used experimental designs are: the *composite designs* (10,11), the *Doehlert designs* (10,13), the *hybrid designs* (14), and the *Box-Benkhen designs* (15).

B. Experimental Domain

The experimental domain of interest can be of three different forms:

the *spherical domain*, chosen when the experimental domain of interest is defined as being the domain surrounding a central point that is the center of interest or center of the domain.

The *cubic domain*, chosen when the experimental domain of interest is considered as the domain defined by the domains of variations of the factors studied.

The *ordinary domain*, which is generally a spherical or cubic domain from which one or several parts have been excluded.

C. Validation of the Model

We have performed the experimentation, obtained the values of the experimental responses, and calculated the estimates of the model coefficients. We cannot use the model obtained yet, as we do not know if it represents the experimental response studied in the experimental domain of interest. We have to validate it, which can be done in several ways. There are two possibilities:

The model is validated, i.e., it is considered as being a good enough representation of the phenomenon in this domain. In this case, we have reached our objectives: we can use this model to make predictions at any point of the experimental domain.

The model is not accepted and thus we cannot use it. We have to return to the beginning, i.e., propose a (of course) different model. The interest of polynomial models is here obvious. A polynomial, as its name indicates, is a series of monomials. We can initially propose a polynomial model of degree d and, if this model is not accepted, add monomials to this model to obtain a polynomial model of degree $(d + 1)$. These models allow to use a sequential approach. We can thus reduce the domain of study, e.g., by dividing it into subdomains, or reconsider the factors that had been fixed theoretically during the experimentation in order to detect that or those which could have, while varying, distorted the values of the experimental responses.

If the empirical model is validated, we can then calculate the response studied at each point of the experimental domain. If the number of factors is big, it is not very easy to extract all of the existing information. There are tools that facilitate this interpretation (canonical analysis, study of optimal design, graphic representations, etc.) (9,12).

Let us take the same approach. We postulate a polynomial model:

$$\eta = \beta_0 + \sum_i \beta_i x_i + \sum_i \sum_{i<j} \beta_{ij} x_i x_j + \ldots \tag{24}$$

We have to know the values or at least the estimates of the model coefficients. For this, we are going to run experiments.

Can we choose them at random?
Which ones? How many?
How can we calculate the estimates of the coefficients of the postulated model?
How can we verify the validity of the model?
How can we control the quality of the prediction?

We should not forget that what interests us, in case the model is accepted, is to know the value of the experimental response with an acceptable quality at any point of the experimental domain of interest. We are going to try and find some answers to all these questions.

D. Choice of Experimental Design According to Model Proposed

1. First-Order Experimental Designs

The first-degree polynomial model can be written like this:

$$\eta = \beta_0 + \sum_i \beta_i x_i \tag{25}$$

The experimental designs that enable us to study this type of model are all established according to the same principles: They are composed of points that are regularly distributed on a circle ($k = 2$), a sphere ($k = 3$), or a hypersphere ($k > 3$).

Given:

k is the number of factors,
p is the number of coefficients of the model ($p = k + 1$)
N is the number of distinct points located on the sphere ($N \geq p$)
N_0 is the number of points located at the center of the domain

The principal experimental designs that allow us to study a first-degree model have been given different names corresponding to their ''discovery.'' We can cite:

* Hadamard or Plackett and Burman experimental designs (6)
* The full (2^k) or fractional (2^{k-r}) factorial experimental designs (7)
* The equiradial experimental designs (7)
* The simplex experimental designs (7,16)

However, a certain unity can be brought to these different types of designs and we can consider only two families:

The experimental designs containing only two level factors (± 1)
The experimental designs containing decimal levels

The first-order experimental design contains one of the experimental designs belonging to one of these two families, containing N points and N_0 points at the center of the domain. By definition, all of these experimental designs have these two important properties: orthogonality and isovariance by rotation.

2. Second-Order Experimental Designs

a. Composite Experimental Designs

The composite experimental designs (10–12) allow to study a second-degree polynomial model.

$$\eta = \beta_0 + \sum_{i} \beta_i x_i + \sum_{ii} \beta_{ii} x_i^2 + \sum_{i<j} \beta_{ij} x_i x_j \qquad (26)$$

The conventional second degree experimental designs can be separated in two categories: some of them allow to study a *cubic* experimental domain, while others allow to study a *spherical* domain. A cubic experimental domain is generally preferred when the area to explore is limited by certain requirements; if, conversely, the region of interest is limited to the vicinity of a point, it is advised to use the spherical form.

Construction. A composite experimental design contains:

A full (2^k) or fractional (2^{k-r}) two-level factorial experimental design, i.e., the vertices of the hypercube (-1, $+1$) that we shall note: [± 1, ± 1, . . . , ± 1], which will give us information on the coefficients (b_i) of the first-degree terms and the coefficients (b_{ij}) of the product terms. If the factorial experimental design is a fractional design, it should be of resolution V.

An axial (or crossed polytope) experimental matrix composed of points symmetrically spread on each axis, at a distance α of the center of the domain, and that we represent by [$\pm\alpha$, 0, . . . , 0], which will give us information on the coefficients (b_i) of the first degree terms and the coefficients (b_{ii}) of the squared terms.

One or several points spread at the center of the domain [0, 0, . . . , 0], which will give us information on the constant coefficient (b_0) and possibly an estimate of the experimental variance.

We represent by:

N_F is the number of points of the factorial experimental design: 2^k or 2^{k-r}
N_a is the number of points of the axial experimental design $= 2k$
N_0 is the number of points at the center of the domain
N is the total number of points

	X_1	X_2	....	X_k
N_F	±1	±1	....	±1
N_a	−α	0	....	0
	α	0	....	0
	0	−α	....	0
	0	α	....	0
	....	....	....	....
	....	....	....	....
	0	0	....	−α
	0	0	....	α
N_0	0	0	....	0
	....	....	....	....
	0	0	....	0

Figure 15 Composite design: general structure.

The value of N_0 is selected depending on the properties desired:

The property of quasi-orthogonality, by putting
The property of uniform precision, by putting

We recapitulate the different values in Table 19.

Table 19 Rules for the Construction of the Composite Experimental Designs Having the Property of Isovariance by Rotation and the Property of Orthogonality or the Property of Uniform Precision

k	2	3	4	5	5	6	6
Factorial design	2^2	2^3	2^4	2^{5-1}	2^5	2^{6-1}	2^6
N_F	4	8	16	16	32	32	64
α	1,414	1,682	2,000	2,000	2,378	2,378	2,828
N_a	4	6	8	10	10	12	12
$N_{0,\text{orthogonality}}$	8	9	12	10	17	15	24
$N_{0,\text{uniform precision}}$	5	6	7	6	10	9	15
$N = N_F + N_a + N_0$	16 13	23 20	36 31	36 32	59 52	59 53	100 91

Remarks. Several observations can be made:

1. A greater number of points is necessary at the center to respect the property of *near-orthogonality* than to respect the property of uniform precision. It is easy to understand why, since we have the choice, we prefer the property of uniform precision. It must be noted however that this property corresponds better to our objectives: "to know the value of the response studied at any point of the experimental domain." It is, however, preferable to know this estimated response with the same precision in all the experimental domain of interest.
2. Even if the property of uniform precision is respected, the number of points at the center is relatively great.

   What is the use of the points at the center?

   * To make certain properties respected, as we have just shown,
   * To obtain information on the variability of the phenomenon, which may allow the estimation of the experimental variance, if the number of points at the center is sufficient.
   * To test the validity of the model. For example, the presence of a curvature for a first-degree model, and the possible presence of non-negligible third-degree terms for a second-degree model.

   If the experimenter already has an estimate of the experimental variance, it will be very difficult to have him accept, say, 10 points at the center when 6 factors are under study. We consider that the values N given in this table are theoretical values that show the exactness of the mathematical development. We ask the experimenters to do at least three or four independent experiments at the center. We must say that this is not readily accepted.

3. The number of experiments of the composite experimental designs increases rapidly with the number of variables studied. This increase is mostly due to the factorial experimental design that must theoretically be a resolution *V* full (2^k) or fractional (2^{k-r}) factorial experimental design. Several authors (17,18) have tried to overcome this by reducing the size of the factorial part. Let us also cite Hartley (19), who proposed resolution III factorial experimental designs, and Westlaque (20), who used irregular fractions of the 2^k factorial experimental design. These proposals are not interesting because the general properties of these designs are less than excellent and other more economical experimental designs exist that have better qualities.

b. Doehlert Uniform Shell Designs

The experimental designs proposed here are known under the name of ''Doehlert uniform shell designs'' (10,15). We started using these designs in the 1970s and, although they have many properties, they are practically out of use in English-speaking countries (8). This can be seen upon examination of the conventional experimental designs proposed in software and specialized books.

Properties

Uniformly distributed experimental points. These experimental designs present a uniform distribution of the experimental points in the space of the coded variables. The uniform shell designs are particularly useful when one wants to cover an experimental domain, with an ordinary form, with a set of uniformly distributed points, in order to explore the whole of the domain (on and inside the boundaries) without proposing models that a priori represent the response studied. These experimental designs will also be useful to interpolate a response, this interpolation being always of constant quality since the shell design is uniform. The density of points will be determined by the form of the initial simplex.

Extension of the shell design in the experimental domain. In view of the obtained results, we may actually try to explore a neighboring domain, by reusing the points of the shell design adjacent to the new region to explore. Moreover, in a particular region, the shell design can be tightened, e.g., by dividing the size of the initial simplex by 2, in order to obtain a more accurate knowledge of the region.

Extension of the number of variables studied. It is always possible to add other variables to those under study without the design losing its quality. Let us suppose that our experimental domain is defined by 7 factors and that we can run only 3 experiments in the course of a week. If we postulate a second-degree polynomial model, the number of experiments is about 60, which will require about 20–25 weeks work. It is very difficult to remain isolated for 6 months to obtain the totality of the experimental responses in order to perform the interpretation. It is advisable to have the possibility of dividing the set of experiments in order to study first, say, 2 factors, perform the interpretation, then add a third factor, run the necessary additional experiments, perform the interpretation, add a fourth factor, and so on. This is possible if we use a Doehlert experimental design. This approach is highly flexible.

Let us cite the most common applications:

When it is necessary to periodically take into account the progress of the work,
When the priorities in the works of the firm are modified, which may force the experimenter to stop working
When experimental information appears in the course of the work to question all or part of the objectives,
When the number of experiments is small

The minimum number of distincts points N of a uniform shell design for a given number of factors k is:

$$\mathbf{N} = \mathbf{k}^2 + \mathbf{k} + \mathbf{1}$$

Differing number of levels of variables. The Doehlert design is an asymmetrical design: the first variable is at five levels, the last one at three levels, and all of the others (except for $k = 2$) at seven levels.

The Doehlert designs have none of the conventional properties of the response surface experimental designs (isovariance by rotation, orthogonality, uniform precision, etc.). It can indeed be shown by calculation that:

Certain odd moments of the information matrix $(\mathbf{X'X})$ are not null.
The even moments are not all constant.
The following relation is not verified:

$$[iiii]^4 = 3[ii, jj]^2$$

For example:

$$[ii]^2 = (k + 1)/N$$
$$[ii, jj]^2 = \{1 + (k + 1)/2j(j + 1)\}/N \qquad 1 < i < j < k$$
$$[iiii]^4 = \{i^3(k + 4) + 6i^2 + 3i + k + 1\}/N\{2i(i + 1)^2\}$$

In spite of this, the Doehlert experimental designs are of sufficient quality with regard to these properties, and they have other properties largely making up for it.

Construction. The Doehlert uniform shell designs are generated from a simplex.

Construction of the initial simplex. If we represent the number of variables by k, the regular simplex is a regular hyper polyhedral of dimension k, with $(k + 1)$ vertices. In table 20 we give the coordinates of the points of a simplex up to $k = 7$, that we call simplex I. We may take the desired submatrix for any k inferior to 7.

Table 20 Basic Simplex I for $k = 7$

No.	X_1	X_2	X_3	X_4	X_5	X_6	X_7
1	0,000	0,000	0,000	0,000	0,000	0,000	0,000
2	1,000	0,000	0,000	0,000	0,000	0,000	0,000
3	0,500	0,866	0,000	0,000	0,000	0,000	0,000
4	0,500	0,289	0,816	0,000	0,000	0,000	0,000
5	0,500	0,289	0,204	0,791	0,000	0,000	0,000
6	0,500	0,289	0,204	0,158	0,775	0,000	0,000
7	0,500	0,289	0,204	0,158	0,129	0,764	0,000
8	0,500	0,289	0,204	0,158	0,129	0,109	0,756

We recall that each simplex can be deduced from the simplex of immediately inferior size (k–1), by addition of a point of coordinates (Table 21).

For the k preceding points, the column k of the new variable introduced will be filled with zeros. Note that all the coordinates of the point $(k + 1)$ are identical to those of the point k, except the last two.

Table 21 Coordinates of the $(k + 1)$th Points of a Simplex II

Variable	$k - q(> 0)$	...	$k - 3$	$k - 2$	k
$(k + 1)$th point	$1/\sqrt{[2(k + 1 - q)(k - q)]}$	...	$1/\sqrt{[2(k - 1)(k - 2)]}$	$1/\sqrt{[2k(k - 1)]}$	$\sqrt{[(k + 1)/2k]}$

Obtention of the shell design. For each variable, we must subtract successively the coordinates of each point from all of the others. Of course, we do not need to subtract the coordinates of the point (0, . . . , 0) from the others since the points remain unchanged! We now illustrate this construction for the values of k equal to 2 and 3.

k = 2. The coordinates of the initial simplex are:

Table 22

No.	X_1	X_2
1	0.000	0.000
2	1.000	0.000
3	0.500	0.866

The experimental design of the uniform shell design allowing to calculate the coefficients of a second-degree response surface model is:

Table 23

No.	X_1	X_2	Obtained by
1	0.000	0.000	
2	1.000	0.000	
3	0.500	0.866	
4	−1.000	0.000	1–2
5	−0.500	−0.866	1–3
6	0.500	−0.866	2–3
7	−0.500	0.866	3–2

The geometrical figure generated by the six vertices (to the exclusion of the first point) is a regular hexagon.

k = 3. The coordinates of the initial simplex are:

Table 24

No.	X_1	X_2	X_3
1	0.000	0.000	0.000
2	1.000	0.000	0.000
3	0.500	0.866	0.000
4	0.500	0.289	0.816

The experimental design of the uniform shell design is:

Table 25

No.	X_1	X_2	X_3	Obtained by
1	0.000	0.000	0.000	
2	1.000	0.000	0.000	
3	0.500	0.866	0.000	
4	0.500	0.289	0.816	
5	−1.000	0.000	0.000	1–2
6	−0.500	−0.866	0.000	1–3
7	−0.500	−0.289	−0.816	1–4
8	0.500	−0.866	0.000	2–3
9	0.500	−0.289	−0.816	2–4
10	−0.500	0.866	0.000	3–2
11	0.000	0.577	−0.816	3–4
12	−0.500	0.289	0.816	4–2
13	0.000	−0.577	0.816	4–3

The geometrical figure generated by the 12 vertices (to the exclusion of the first point) is a cuboctahedral. This can be obtained by joining the midpoints of the cube edges.

Remarks. Let us take the above properties:

Extension of the shell design. One of the important characteristics of the Doehlert uniform shell designs is that they allow a sequential approach in the study of a second-degree response surface. Let us consider the following situation: in a first stage, we define the experimental domain of interest, considering that it contained the desired information (knowledge around an optimum, etc.). After a first series of experiments, we realize that the most interesting experimental domain is not in the previously chosen domain but in its vicinity. What shall we do? Shall we start all over again?

We define a new domain and we perform the same strategy. In contrast to the other known designs, it is easy to construct around one of the Doehlert's shell design points a new shell design that uses part of the points already done.

Let us recall the initial experimental design at two factors:

Table 26

No.	X_1	X_2
1	0.000	0.000
2	1.000	0.000
3	0.500	0.866
4	−1.000	0.000
5	−0.500	−0.866
6	0.500	−0.866
7	−0.500	0.866

Let us suppose that we want to construct a new matrix around point 7, considered now as the origin. The experimental design expressed in the first system of axes becomes:

Table 27

No.	X_1	X_2	
7	−0.500	0.866	*
3	0.500	0.866	*
8	0.000	1.732	
9	−1.500	0.866	
4	−1.000	0.000	*
1	0.000	0.000	*
10	−1.000	1.732	

* The points to which an asterisk has been assigned are common to the former and the new matrices; thus it is necessary to add only three new points (8, 9, 10) to have knowledge in the new domain.

Table 28

No.	X_1	X_2
8	0.000	1.732
9	−1.500	0.866
10	−1.000	1.732

Increase of the number of factors under study. Let us consider the Doehlert experimental design for two variables. The use of such a design supposes that the other variables (the corresponding factors) that can exist are kept to an average value that we can consider as equal to zero (for the corresponding reduced centered variable). After studying both variables, we may want to include the study of one of these variables kept constant. We can then consider two approaches:

We may construct a new experimental design containing 13 experiments.
We may add experiments to the previous experimental design so as to use at best the already obtained information.

Let us study the second approach. Let us examine the Doehlert uniform shell design for $k = 3$ (after rearrangement of the rows).

Table 29

No.	X_1	X_2	X_3
1	0.000	0.000	0.000
2	1.000	0.000	0.000
3	0.500	0.866	0.000
4	−1.000	0.000	0.000
5	−0.500	−0.866	0.000
6	0.500	−0.866	0.000
7	−0.500	0.866	0.000
8	0.500	0.289	0.816
9	−0.500	−0.289	−0.816
10	0.500	−0.289	−0.816
11	0.000	0.577	−0.816
12	−0.500	0.289	0.816
13	0.000	−0.577	0.816

We note that the first 7 points out of the 13 points are those of the experimental design for $k = 2$. There are thus only 6 points to add to pass from $k = 2$ to $k = 3$. In order to use this property, it is thus necessary that the nonstudied factors not be fixed to extreme values.

In Table 30 we give the various experimental designs up to $k = 6$. For any lesser value of k, we only have to isolate the corresponding subdesign enclosed in dots.

Table 30 Doehlert designs for $k = 2, \ldots, 6$

No.	X_1	X_2	X_3	X_4	X_5	X_6
1	0.000	0.000	0.000	0.000	0.000	0.000
2	1.000	0.000	0.000	0.000	0.000	0.000
3	−1.000	0.000	0.000	0.000	0.000	0.000
4	0.500	0.866	0.000	0.000	0.000	0.000
5	−0.500	−0.866	0.000	0.000	0.000	0.000
6	0.500	−0.866	0.000	0.000	0.000	0.000
7	−0.500	0.866	0.000	0.000	0.000	0.000
8	0.500	0.289	0.816	0.000	0.000	0.000
9	−0.500	−0.289	−0.816	0.000	0.000	0.000
10	0.500	−0.289	−0.816	0.000	0.000	0.000
11	0.000	0.577	−0.816	0.000	0.000	0.000
12	−0.500	0.289	0.816	0.000	0.000	0.000
13	0.000	−0.577	0.816	0.000	0.000	0.000
14	0.500	0.289	0.204	0.791	0.000	0.000
15	−0.500	−0.289	−0.204	−0.791	0.000	0.000
16	0.500	−0.289	−0.204	−0.791	0.000	0.000
17	0.000	0.577	−0.204	−0.791	0.000	0.000
18	0.000	0.000	0.612	−0.791	0.000	0.000
19	−0.500	0.289	0.204	0.791	0.000	0.000
20	0.000	−0.577	0.204	0.791	0.000	0.000
21	0.000	0.000	−0.612	0.791	0.000	0.000
22	0.500	0.289	0.204	0.158	0.775	0.000
23	−0.500	−0.289	−0.204	−0.158	−0.775	0.000
24	0.500	−0.289	−0.204	−0.158	−0.775	0.000
25	0.000	0.577	−0.204	−0.158	−0.775	0.000
26	0.000	0.000	0.612	−0.158	−0.775	0.000
27	0.000	0.000	0.000	0.632	−0.775	0.000
28	−0.500	0.289	0.204	0.158	0.775	0.000
29	0.000	−0.577	0.204	0.158	0.775	0.000
30	0.000	0.000	−0.612	0.158	0.775	0.000
31	0.000	0.000	0.000	−0.632	0.775	0.000
32	0.500	0.289	0.204	0.158	0.129	0.764
33	−0.500	−0.289	−0.204	−0.158	−0.129	−0.764
34	0.500	−0.289	−0.204	−0.158	−0.129	−0.764
35	0.000	0.577	−0.204	−0.158	−0.129	−0.764
36	0.000	0.000	0.612	−0.158	−0.129	−0.764
37	0.000	0.000	0.000	0.632	−0.129	−0.764
38	0.000	0.000	0.000	0.000	0.645	−0.764
39	−0.500	0.289	0.204	0.158	0.129	0.764
40	0.000	−0.577	0.204	0.158	0.129	0.764
41	0.000	0.000	−0.612	0.158	0.129	0.764
42	0.000	0.000	0.000	−0.632	0.129	0.764
43	0.000	0.000	0.000	0.000	−0.645	0.764

It is advisable to add two or three additional points at the center of the domain.

Remark. In cases where the number of distinct experiments is close to the number of coefficients, it will be necessary to add, besides the points at the center that allow us to have an estimate of the experimental variance, k or $(k + 1)$ points, in the experimental domain of interest, situated far from the points of the experimental design. This will enable us to test the validity of the model postulated.

E. Plan of Experimentation

After the experimental design has been chosen, it must be transformed into a design of experimentation. The experimenter must afterward perform a critical analysis of this design of experimentation by taking into account all of the information obtained on the problem under study. If the experimenter realizes that one or several experiments cannot be done, either because they are impossible to run or because they might cause incidents (explosion, clogging, etc.), it is pointless to perform this design of experimentation because the desired information would not be obtained. In this situation, the methodological approach must be considered again and the information previously neglected must be included. In the case of all experiments being possible, the experimenter must establish a schedule for the experiments and randomize them so that the information contained in the experimental results are not distorted.

F. Validation and Interpretation

Once the experimentation is finished, we calculate the estimates of the coefficients of the model postulated by the least-squares method. Although we have these estimates, we cannot use the model because we do not know if it represents the experimental response under study in the experimental domain of interest. We therefore have to validate it.

How can we validate the postulated model? We can consider two situations:

The number of distinct experiments is superior to the number of coefficients of the model ($N > p + 1$). We can thus use a statistical tool like the variance analysis which will enable us to disregard or not disregard the model with an acceptable risk.

The number of distinct experiments is equal or close to the number of coefficients of the model. In this case, no statistical tool will be of any help. We will have to accept the model and verify the interpretation that we do from it by running some experiments in the interesting areas.

G. Graphic Representation

We have validated the model that we postulated and we can at any point of the experimental domain of interest calculate the estimated response. We can materialize this knowledge by joining the points where the calculated response has the same value, and so we obtain isoresponse curves. It is very easy to represent these isoresponse curves graphically in a two-dimensional space. But as soon as the number of variables is higher, we must fix $(k - 2)$ variables and perform a projection of these curves onto the plane of the other two variables. As the number of variables increases, the number of projection planes increases too. This justifies the various interpretation helping tools that we have previously presented. From the information obtained, it is easy to find the most informative projection plane(s).

Even if we fix the $(k - 2)$ variables to values included in the variation domain of each of them, we will have to materialize on the projection plane the trace of the experimental domain. We must not forget that the model that we have validated is representative of the experimental response studied, only in this domain.

H. Example: Optimization of a Formulation [from data of a study done by the authors: Piccerelle Philippe, Eouani César, Joachim Joseph, Reynier Jean-Pierre (Laboratoire de Pharmacie Galénique Industrielle et de Cosmétologie, Faculté de Pharmacie, Marseille, France). Results not published.]

In the domains of cosmetology and dermatology, the sensorial qualities of the products used are of extreme importance. The choice of the consumer when comparing formulations for cosmetic use is determined by the esthetic of the packaging and mainly by their texture when applied to the skin.

Dermatologists that have been for a long time concerned only with the intrinsic efficiency of the product are now increasingly taking into account sensorial criterions. Indeed, a formulation that has got nearly cosmetic properties of texture can permit a better observation of the treatment and so better efficiency.

The aim of this study is to show the consequences of the variations of some factors of the fat phase of water/oil emulsions on the rheological and texture properties of the finished product.

The formulations have been achieved according to an experimental methodology by varying the volumetric fraction of the fat phase (oil), the value of the hydrophile-lipophile balance (HLB) in the W/O domain, and the quantity of the surfactant.

The responses taken into account in our study are cohesion and viscosity of the emulsion. These responses are obtained by rheological and texture analysis. The cohesion (nondimensional) represents the ratio between the penetration strength and the adhesion strength.

The domain of variation of the three factors is as follows:

Table 31 Experimental Domain

Factor	Unity	Center	Step of variation
Oil	%	50.0	14.1
Surfactant	%	3.00	2.83
HLB		5.05	4.17

To know the influence of the three factors on the two responses at the first time, and then to determine the optimal experimental conditions in the domain of interest, the surface response methodological approach is used. For this, the design realized is a Doehlert design (Table 32), where the change in the number of levels is minimized (each factor has only three levels of variation). The plan of experimentation is given in Table 33.

Table 32 Doehlert Experimental Design

No.	X_1	X_2	X_3
1	−0.7071	−0.7071	0.0000
2	0.7071	−0.7071	0.0000
3	−0.7071	0.7071	0.0000
4	0.7071	0.7071	0.0000
5	−0.7071	0.0000	−0.7071
6	0.7071	0.0000	−0.7071
7	−0.7071	0.0000	0.7071
8	0.7071	0.0000	0.7071
9	0.0000	−0.7071	−0.7071
10	0.0000	0.7071	−0.7071
11	0.0000	−0.7071	0.7071
12	0.0000	0.7071	0.7071
13	0.0000	0.0000	0.0000
14	0.0000	0.0000	0.0000
15	0.0000	0.0000	0.0000
16	0.0000	0.0000	0.0000

Table 33 Plan of Experimentation and Responses Measured

No.	Oil	Surfactant	HLB	Cohesion	Viscosity (mPa.s)
1	40.0	1.00	5.05	0.865	0.4280
2	60.0	1.00	5.05	0.651	0.4950
3	40.0	5.00	5.05	0.781	0.0980
4	60.0	5.00	5.05	0.842	1.3770
5	40.0	3.00	2.10	0.790	0.5970
6	60.0	3.00	2.10	0.742	2.7560
7	40.0	3.00	8.00	0.897	0.0900
8	60.0	3.00	8.00	0.715	0.6600
9	50.0	1.00	2.10	0.558	0.4480
10	50.0	5.00	2.10	0.618	1.5460
11	50.0	1.00	8.00	0.622	0.4740
12	50.0	5.00	8.00	0.752	0.6290
13	50.0	3.00	5.05	0.576	0.6150
14	50.0	3.00	5.05	0.699	0.4090
15	50.0	3.00	5.05	0.665	0.2490
16	50.0	3.00	5.05	0.649	0.3670

Remark: In Table 32, for the factor HLB, the coded level corresponding to the mean real value 5.05 should in fact be 0.1080.

The main interest of the surface response methodology is that we can know the value of each response studied, in every point of the experimental domain. If a mathematical approach is used, each response can be represented by a quadratic equation of the surface response:

$$\eta = \beta_0 + \beta_1 x_1 + \beta_2 x_2 + \beta_3 x_3 + \beta_{11} x_1^2 + \beta_{22} x_2^2 + \beta_{33} x_3^2 + \beta_{12} x_1 x_2 + \beta_{13} x_1 x_3 + \beta_{23} x_2 x_3 \tag{28}$$

Indeed, once the mathematical models are validated by classical statistical tools (e.g., ANOVA, or analysis of the variance; analysis of the lack of fit) (15), we can draw the response surfaces representing the evolution of the responses in the whole domain studied, when two factors are varying and the third one is fixed. From the different diagrams of isoresponse curves, we can determine the influence of the different factors considered on the responses.

1. Response Cohesion

The fitted polynomial model is:

$$y = 0.642 - 0.064x_1 + 0.051x_2 + 0.049x_3 + 0.286x_1^2 - 0.011x_2^2 + 0.002x_3^2 + 0.137x_1x_2 - 0.063x_1x_3 + 0.031x_2x_3 \tag{29}$$

From the isoresponse curves graphics below, we can see that the most influent factor is the quantity of oil. The effect of this factor on the cohesion depends on the percentage of oil, whatever the levels of the other factors may be. If we use a quantity of oil from 40% to 50%, the cohesion decreases, whereas percentages of oil from 50% up to 60% increase the cohesion.

The factor HLB has hardly any effect in the domain of variation studied.

The effect of the surface-active depends on the quantity of oil. In presence of large quantities of oil (more than 50%), the surface-active has a small effect: an increase of the percentage of surface-active from 1% to 5% entails a little increase of the cohesion. For percentages of oil less than 50, the surface-active has no effect.

2. Response Viscosity

The fitted polynomial model is:

$$y = 0.468 + 0.765x_1 + 0.345x_2 - 0.617x_3 + 0.441x_1^2 - 0.062x_2^2 + 0.672x_3^2 + 0.606x_1x_2 - 0.837x_1x_3 - 0.492x_2x_3 \quad (30)$$

The percentage of surface-active has only a small effect: the more important the percentage of surface-active, the higher the viscosity gets.

The most influent factors on the viscosity are the percentage of oil and HLB. Percentages of oil from 30% to 60% bring about an increase of cohesion, and this increase is more important in the presence of values of HLB less than 5.05 and percentages of surface-active more than 3.0%. The factor HLB has an effect on the viscosity that varies according to the values chosen. An increase of HLB from 2.1 to 5.05 entails an important decrease of the viscosity, and for values greater than 5.05, this factor has hardly any effect.

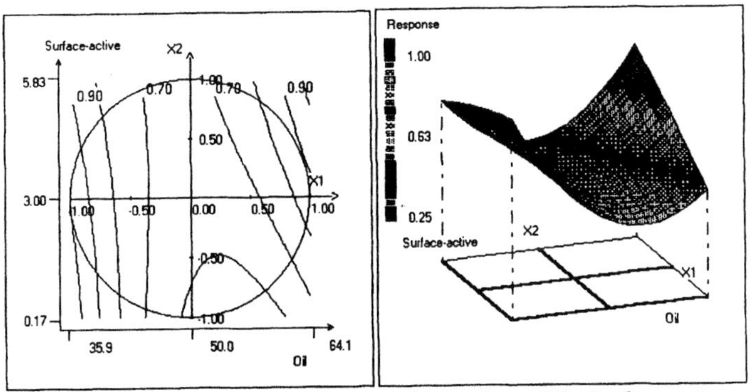

Variation of the response Cohesion - In the plan : Oil , Surface-active
FACTOR FIXED : - HLB = 5.05

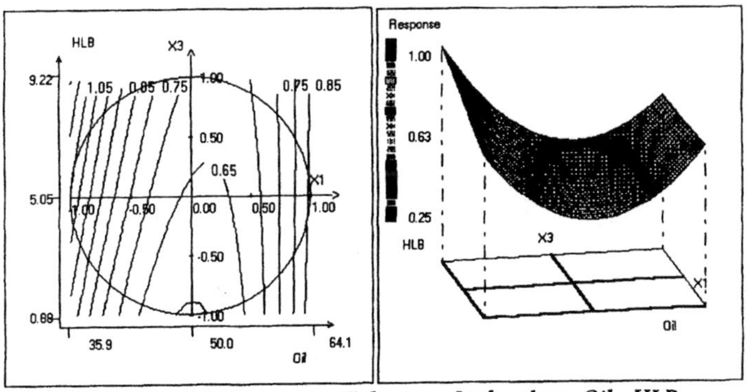

Variation of the response Cohesion - In the plan : Oil , HLB
FACTOR FIXED : Surface-active = 3.00%

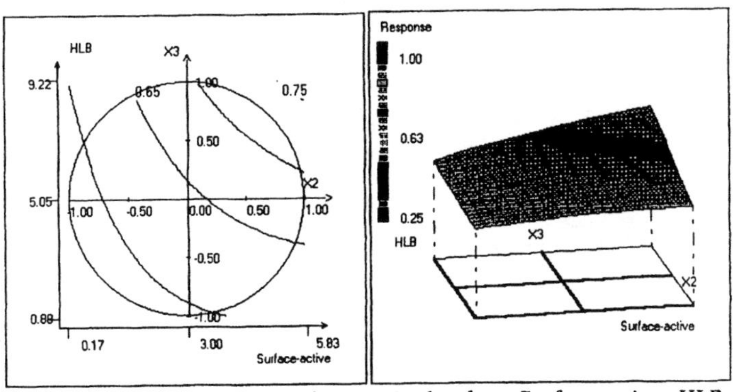

Variation of the response Cohesion - In the plan : Surface-active , HLB
FACTOR FIXED : Oil = 50.0%

Figure 16 Graphics of isoresponse curves for cohesion.

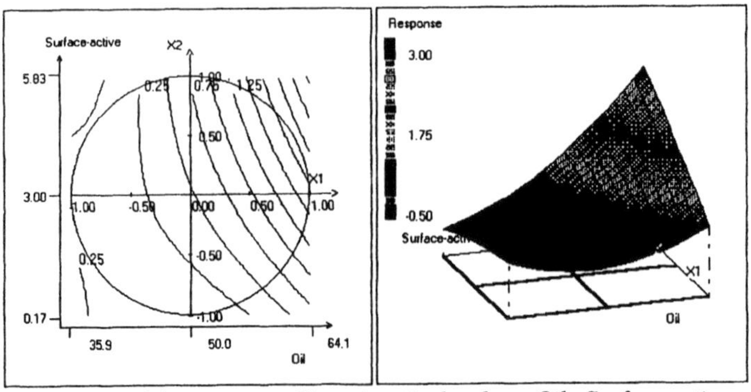

Variation of the response Viscosity - In the plan : Oil , Surface-active
FACTOR FIXED : HLB = 5.05

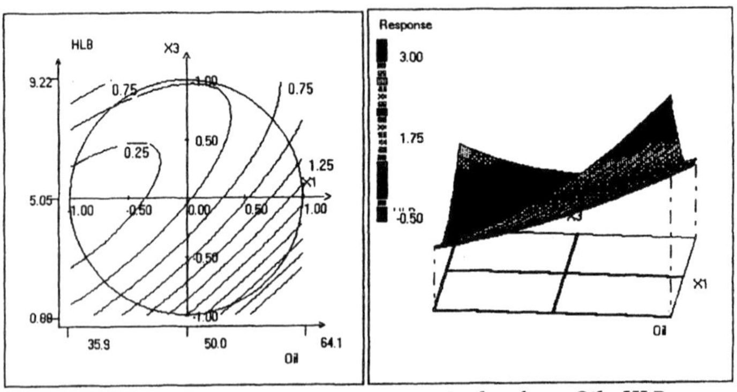

Variation of the response Viscosity - In the plan : Oil , HLB
FACTOR FIXED : Surface-active = 3.00%

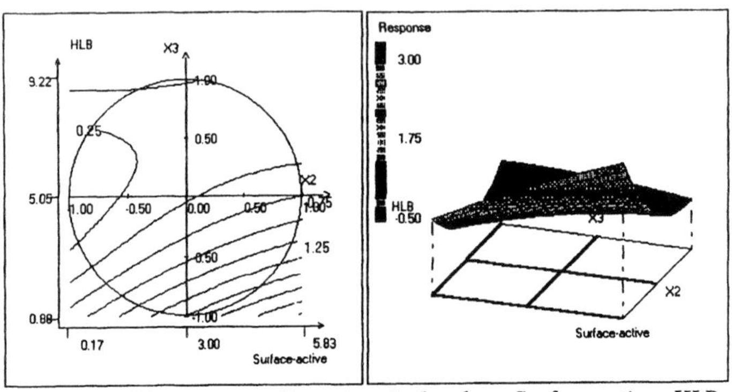

Variation of the response Viscosity - In the plan : Surface-active , HLB
FACTOR FIXED : Oil = 50.0%

Figure 17 Graphics of isoresponse curves for viscosity.

IV. MIXTURES

A. Introduction

In the case of mixtures, the factors are the proportions of the components. Therefore, they have two important characteristics:

Their sum is equal to the unity, and thus they are not independent.

Their values are dimensionless numbers, utterly comparable; a value of 0.1 has the same meaning, whether it represents the proportion of an antioxidizing agent in a custard or that of a wine variety in a wine. The number of components of a mixture is generally represented by q and the proportion of the component i by x_i.

We therefore have the following constraints:

$$x_i \geq 0 \qquad \sum_i x_i = 1 \qquad i = 1, 2, \ldots, q$$

These constraints on the values that x_i can take, account for the fact that the mixtures cannot be applied the same treatment as usual, where the values taken by the coded variables are not subject to constraints. Considering these constraints, any variation of the proportion of a component causes a variation of the proportions of the other components.

B. Experimental Domain

The possible experimental domain is a regular simplex of dimension $(q - 1)$. For $q = 3$, the experimental domain is an equilateral triangle (Fig. 18) for $q =$ 4, this domain is a regular tetrahedral. The coordinates representing the values of x_i, $i = 1, 2, \ldots, q$, are called *coordinates of the simplex*. The vertices of the simplex represent the pure components: ($x_i = 1$, $x_j = 0, j \neq i$). The inside points represent the mixtures in which all the components are present ($x_i > 0$, $i = 1$, $2, \ldots, q$).

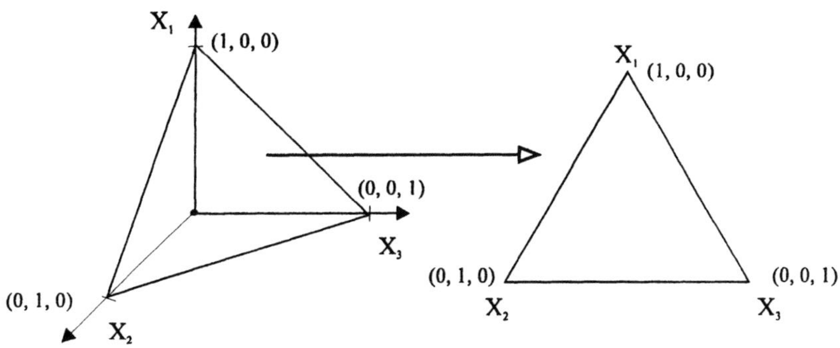

Figure 18 Experimental domain in the space of the three factors and simplex.

C. Choice of Experiments (Mixtures)

We want to know, for all possible mixtures situated in the experimental domain of interest, the value of the quality (or qualities) studied. In order to obtain the desired information, we have to perform measurements on mixtures.

1. Can We Choose These Mixtures at Random?

We must, after having established the mixture relation, be able to calculate the value of the studied quality for all of the possible mixtures. Let us consider a mixture M_i: if we know the composition of this mixture and the mixture relation, we can calculate the value of a quality Q_i. This calculated value, which we will call *forecast value*, should not be too different from the quality we would find if we actually performed the measurement on this mixture. We say that the prediction must be of good quality.

We demonstrate that the forecast quality depends on the form of the mixture relation and on the choice of the set of mixtures considered in the experimentation. It is independent of the results of the experiments. These mixtures cannot be chosen at random and among all the possible mixtures some yield more information than others. It will be our interest to choose these mixtures preferentially. They are represented by the term *informative mixtures*. The choice of these mixtures is done by applying the methodology of experimental research (8).

2. How Shall We Choose This Optimal Set?

For reasons of cost of experimentation and efficiency, the experimenter may try to minimize the number of necessary mixtures. The experimentation may be costly or even limited, and so the number of mixtures will have to be limited.

The property of sequentiality is very often utilized in mixture problems. That is to say, when the complexity of the problem increases, the experimentation is carried out in several steps. The choice of the *informative mixtures* for each step must take into account the mixtures already measured.

3. How Shall We Represent the Results?

When the mixture relation has been established, we can calculate, for any possible mixture belonging to the domain of interest, the value of the quality with a good precision. We have thus reached our objective: *to know, for all possible mixtures in the experimental domain of interest, the value of the quality studied.*

D. Isoresponse Curves

Knowing the mixture relation, we can, at any point of the experimental domain studied, calculate the value of the studied quality (or qualities). For a quality, this value will not be the same for all the mixtures and will vary depending on the composition of the mixture. If all the mixtures having the same value are brought together, a curve is obtained that we call the *isoresponse curve*. We can therefore represent under the form of isoresponse curves the set of measured qualities of all mixtures belonging to the experimental domain studied (Fig. 19). The mixtures situated on the same curve are of the same quality.

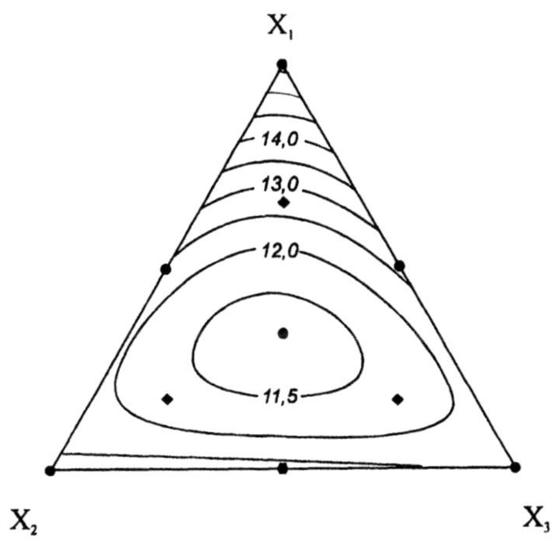

Figure 19 Isoresponse curves.

E. Response Surface Models

The first works dealing with this methodological approach started about 30 years ago. Claringbold (21) published the first ''reasoned'' experimental design in a work on the study of the effects of a mixtures of hormones. It is not until 1958 that Scheffé (22) published the bases of the optimal experimental strategy applied to mixtures. Scheffé's works can be considered as being those that established the bases in this field and have been at the origin of many applications (22,23).

1. Scheffé Simplex Lattice Designs

Scheffé (22) proposed that in the domain defined by the q components, the studied experimental response is represented by a polynomial of the form:

$$\eta = \alpha_0 + \sum_i \alpha_i x_i + \sum_i \sum_j \alpha_{ij} x_i x_j + \sum_i \sum_j \sum_n \alpha_{ijn} x_i x_j x_n \cdots \qquad (31)$$

The proportions of the components of the mixture must respect the constraints:

$$\sum_i x_i = 1 \qquad i = 1, 2, \ldots, q$$

In order to simplify this two-equation system, Scheffé proposed a simple transformation that makes it possible to have only one equation, called *canonical equation.*

To obtain the estimates of the model coefficients allowing the best forecast quality in the experimental domain studied, Scheffé proposes an experimental design that he calls *simplex lattice design.* In the case of q components and for a polynomial of degree m, the corresponding simplex lattice design is noted $\{q, m\}$. The coordinates of each point are multiples of $1/m$ and such that:

$$\sum_i x_i = 1 \qquad i = 1, 2, \ldots, q$$

Each component can thus take the (equally spaced) $m + 1$ values: $0, 1/m, 2/m, \ldots, m/m$. The Scheffé simplex lattice contains all of the points obtained by doing all possible combinations of these $(m + 1)$ values. The number of points of a simplex lattice design is given by the formula:

$$N = (m + q - 1)!/(q - 1)!\, m! \tag{32}$$

a. Examples

Canonical Equation for a Scheffé Model {3, 2}

$$\eta = \beta_1 x_1 + \beta_2 x_2 + \beta_3 x_3 + \beta_{12} x_1 x_2 + \beta_{13} x_1 x_3 + \beta_{23} x_2 x_3 \tag{33}$$

Simplex Lattice Design

Table 34 Simplex Lattice Design

No.	X_1	X_2	X_3
1	1.0	0.0	0.0
2	0.0	1.0	0.0
3	0.0	0.0	1.0
4	0.5	0.5	0.0
5	0.5	0.0	0.5
6	0.0	0.5	0.5

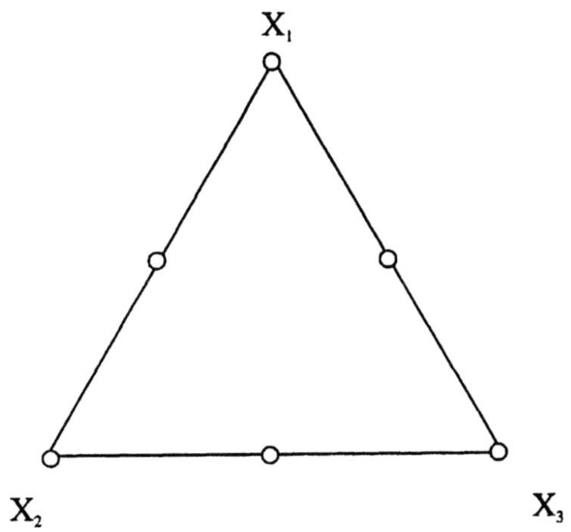

Figure 20 Simplex lattice design {3, 2}.

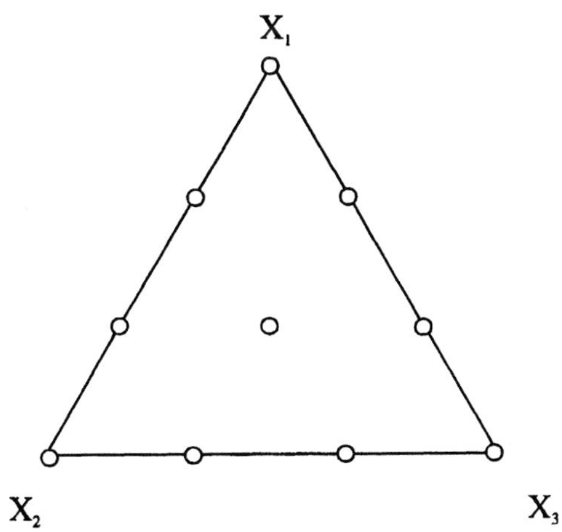

Figure 21 Simplex lattice design {3, 3}.

Canonical Equation for a Scheffé Model {4, 3}

$$\begin{aligned}\eta = {} & \beta_1 x_1 + \beta_2 x_2 + \beta_3 x_3 + \beta_4 x_4 + \beta_{12} x_1 x_2 + \beta_{13} x_1 x_3 \\ & + \beta_{23} x_2 x_3 + \beta_{14} x_1 x_4 + \beta_{24} x_2 x_4 + \beta_{34} x_3 x_4 \\ & + \gamma_{12} x_1 x_2 (x_1 - x_2) + \gamma_{13} x_1 x_3 (x_1 - x_3) + \gamma_{23} x_2 x_3 (x_2 - x_3) \\ & + \gamma_{14} x_1 x_4 (x_1 - x_4) + \gamma_{24} x_2 x_4 (x_2 - x_4) + \gamma_{34} x_3 x_4 (x_3 - x_4)\end{aligned} \tag{34}$$

Simplex Lattice Design

Table 35 Simplex Lattice Design

No.	X_1	X_2	X_3	X_4	No.	X_1	X_2	X_3	X_4
1	1.000	0.000	0.000	0.000	11.	0.666	0.000	0.000	0.333
2	0.000	1.000	0.000	0.000	12	0.333	0.666	0.000	0.000
3	0.000	0.000	1.000	0.000	13	0.000	0.666	0.333	0.000
4	0.000	0.000	0.000	1.000	14	0.000	0.666	0.000	0.333
5	0.333	0.333	0.333	0.000	15	0.333	0.000	0.666	0.000
6	0.333	0.333	0.000	0.333	16	0.000	0.333	0.666	0.000
7	0.333	0.000	0.333	0.333	17	0.000	0.000	0.666	0.333
8	0.000	0.333	0.333	0.333	18	0.333	0.000	0.000	0.666
9	0.666	0.333	0.000	0.000	19	0.000	0.333	0.000	0.666
10	0.666	0.000	0.333	0.000	20	0.000	0.000	0.333	0.666

b. Advantages and Disadvantages

Scheffé simplex lattice designs have many qualities:

They are easy to construct and require a minimal number of experiments. They contain as many points as there are coefficients in the corresponding polynomial, which allows direct resolution (not very interesting now that computers are used).

They allow a sequential approach, at least up to model of degree 3, since the approach is strictly sequential only up to the reduced cubic model included. When the models of a higher degree are studied, many already obtained points do not fit the new simplex lattice design. However, this disadvantage can be overcome by taking these points as test points.

The validity of the model can be tested by adding test points.

There is a forecast model that allows one to obtain a prediction of the response with a minimal variance.

On the other hand, there are some drawbacks:

Scheffé lattice designs require a great number of points (Table 36) when the degree increases. This is only due to the increase of the number of coefficients and not to the construction protocol. This partially explains why the study of polynomial models of degree superior to 4 is very rarely performed.

Table 36 Numbers of Points and Coefficients in a Simplex Lattice Design $\{q, m\}$

q m	2	3	4
3	6	10	15
4	10	20	35
5	15	35	70
6	21	56	126
8	36	120	330
10	55	220	715

The coefficients of a model, usually forecast, i.e., intended for the calculation of the response corresponding to a mixture in which all the components are present, are calculated from experiments in which are generally involved only simpler mixtures with q' components ($q' < q$) (pure substances, binary mixtures, etc.). This is particularly obvious for the first-degree model in which the experiments are performed only on pure substances.

It must also be noted that the appearance of a discontinuity or at least of strong variations at the boundaries of the domain can markedly distort the results and lead to aberrations.

Unlike simplex lattice designs of degree 1 or 2, which are optimal for the criteria of quality of the experimental designs, the higher degree simplex lattice designs are not optimal, not only concerning these former criteria but also for the other criteria usually considered (12) (trace $(\mathbf{X}'\mathbf{X})^{-1}$ minimal, isovariance by rotation, etc.).

Simplex lattice designs cannot be used in the form presented here if at least one of the components is subject to a constraint of the type:

$$0 \leq a_i \leq x_i \leq b_i < 1$$

To overcome certain drawbacks, several authors have proposed some modifications:

To reduce the number of points at the boundaries of the domain, Gammon (24) proposed replacing the experiments corresponding to each component by experiments performed at the gravity center of the opposite side. All other points of the simplex lattice design are preserved. But this modification is not possible in every case. In particular, when m is a multiple of $(q - 1)$ it is all the more useless that the degree is low and the number of components q is great, since the volume of the experimental domain decreases rapidly in this case.

With the same goal in view, Lambrakis (25) proposed to *interiorize* all of the points for which at least one coordinate is null (points situated at a vertex, on an edge, a surface, etc.). This leads to a reduction of the experimental domain different from that obtained with the modification of Gammon. This modification is useful only when q and m are large enough.

2. Scheffé Centered Simplex Lattice Designs

To overcome the problem of strict sequentiality, Scheffé (26) proposed a modification of his *simplex lattice designs*. He proposed a reduced empirical model containing only product terms. Like all of the other response surface models, this model is only an approximation and the product terms of increasing order can be added one after the other. The model proposed, called *centered simplex* model, is as follows:

$$\eta = \sum_i \beta_i x_i + \sum_i \sum_i \beta_{ij} x_i x_j + \cdots + \beta_{12\cdots q} x_1 x_2 \cdots x_q \tag{35}$$

It is of maximal degree equal to q and contains $2^q - 1$ terms at the maximum.

a. *Characteristics*

Scheffé also wanted to avoid the components that are not null in certain mixtures are in different proportions. In the experimental design called *centered simplex lattice design*, all components not null are in the same proportions. There is only one centered simplex lattice design for a number q of components. It is formed of the following points:

q points corresponding to the q possible permutations of the point $\{1, 0, \ldots, 0\}$

C_q^2 points corresponding to the C_q^2 permutations of the points $\{1/2, 1/2, 0, \ldots, 0\}$

C_q^3 points corresponding to the C_q^3 permutations of the points $\{1/3, 1/3, 1/3, 0, \ldots, 0\}$ etc.

1 point corresponding to the isobarycenter of the vertices of the hyperpolyhedral of $(q - 1)$ dimensions that compose the possible experimental domain: $\{ 1/q, 1/q, \ldots, 1/q \}$

There are as many points as there are coefficients in the centered simplex model. The centered simplex lattice design constructed for q components contains $2^q - 1$ points, which very quickly becomes prohibitive when q increases. We usually consider degrees 2 or 3. In the case of degree 3, the centered simplex model is often noted by the term *reduced cubic model.*

This type of experimental design still has one of the drawbacks of the initial simplex lattice design, since most of the responses used to determine the coefficients are obtained from mixtures formed of q' components ($q' < q$) and these experiments are situated at the limits of the domain (which is the major drawback).

b. Examples

Centered Simplex Model for $q = 3$

$$\eta = \beta_1 x_1 + \beta_2 x_2 + \beta_3 x_3 + \beta_{12} x_1 x_2 + \beta_{13} x_1 x_3 + \beta_{23} x_2 x_3 + \beta_{123} x_1 x_2 x_3 \quad (36)$$

Centered Simplex Lattice design

Table 37 Centered Simplex Lattice Design

No.	X_1	X_2	X_3
1	1.000	0.000	0.000
2	0.000	1.000	0.000
3	0.000	0.000	1.000
4	0.500	0.500	0.000
5	0.500	0.000	0.500
6	0.000	0.500	0.500
7	0.333	0.333	0.333

Centered Simplex Model for $q = 4$

$$\eta = \beta_1 x_1 + \beta_2 x_2 + \beta_3 x_3 + \beta_4 x_4 + \beta_{12} x_1 x_2 + \beta_{13} x_1 x_3 + \beta_{23} x_2 x_3 + \beta_{14} x_1 x_4 + \beta_{24} x_2 x_4 + \beta_{34} x_3 x_4 + \beta_{123} x_1 x_2 x_3 + \beta_{124} x_1 x_2 x_4 + \beta_{134} x_1 x_3 x_4 + \beta_{234} x_2 x_3 x_4 + \beta_{1234} x_1 x_2 x_3 x_4 \quad (37)$$

Centered Simplex Lattice Design

Table 38 Centered Simplex Lattice Design

No.	X_1	X_2	X_3	X_4	No.	X_1	X_2	X_3	X_4
1	1.000	0.000	0.000	0.000	9	0.000	0.500	0.000	0.500
2	0.000	1.000	0.000	0.000	10	0.000	0.000	0.500	0.500
3	0.000	0.000	1.000	0.000	11	0.333	0.333	0.333	0.000
4	0.000	0.000	0.000	1.000	12	0.333	0.333	0.000	0.333
5	0.500	0.500	0.000	0.000	13	0.333	0.000	0.333	0.333
6	0.500	0.000	0.500	0.000	14	0.000	0.333	0.333	0.333
7	0.500	0.000	0.000	0.500	15	0.250	0.250	0.250	0.250
8	0.000	0.500	0.500	0.000					

F. Addition of Test Points

It is advised to add test points to the simplex lattice designs. These points should help test the validity of the model postulated. We advise adding the points situated inside the experimental domain corresponding to mixtures containing all the components.

Table 39

No.	X_1	X_2	$\cdots$	X_q
1	$(q + 1)/2q$	$1/2q$	$\cdots$	$1/2q$
2	$1/2q$	$(q + 1)/2q$	$\cdots$	$1/2q$
$\cdots$	$1/2q$	$1/2q$	$\cdots$	$1/2q$
$\cdots$	$\cdots$	$\cdots$	$\cdots$	$\cdots$
q	$1/2q$	$1/2q$	$\cdots$	$(q + 1)/2q$

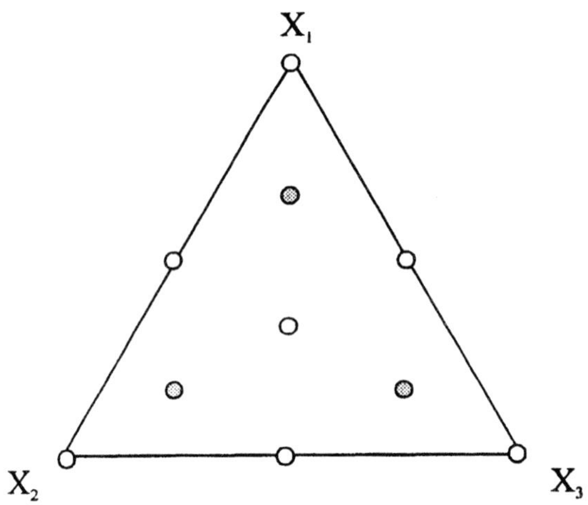

Figure 22 Centered simplex lattice design and test points for $q = 3$.

G. Models Using the Ratios

In some cases what interests us is not the study of the influence of the component proportions but the study of the ratios of the component proportions. Many ratios can be studied, e.g.:

q = 3

Table 40

Model	z_1	z_2
R_1	$x_1/(x_2 + x_3)$	x_2/x_3
R_2	$x_2/(x_3 + x_1)$	x_3/x_1
R_3	$x_3/(x_1 + x_2)$	x_1/x_2

z_1 and z_2 are independent variables.

q = 4

Table 41

Model	z_1	z_2	z_3
R_1	$x_1/(x_2 + x_3 + x_4)$	$x_2/(x_3 + x_4)$	x_3/x_4
R_2	$x_2/(x_3 + x_4 + x_1)$	$x_3/(x_4 + x_1)$	x_4/x_1
R_3	$x_3/(x_4 + x_1 + x_2)$	$x_4/(x_1 + x_2)$	x_1/x_2
R_4	$x_4/(x_1 + x_2 + x_3)$	$x_1/(x_2 + x_3)$	x_2/x_3

z_1, z_2, and z_3 are independent variables.

H. Mixtures with Constraints

Actually, in many problems we are not interested in the study of pure components. Therefore, the experimental domain of interest is not all of the possible experimental domain, and it must be clearly described. For this, the constraints on the component proportions must be well specified. There are two types of constraint:

Individual constraints of the form:

$$\begin{aligned} &0 < a_i \leq x_i \leq 1 \\ &0 \leq x_i \leq b_i < 1 \\ &0 < a_i \leq x_i \leq b_i < 1 \end{aligned} \tag{38}$$

which can affect one of several components,
Relational constraints:

$$C_i \leq a_1 x_1 + a_2 x_2 + \cdots + a_q x_q \leq D_i \tag{39}$$

Whatever their type, the constraints reduce the possible experimental domain, either by simple homothetical transformation or by deformation reduction. One or several modes of construction of the experimental designs corresponds to each type.

REFERENCES

1. Watson G.S. (1961). A study of the group screening method, *Technometrics*, 3, 371–388.
2. Kleijnen J.P.C. (1975). Screening designs for poly-factor experimentation, *Technometrics*, 18, 487–493.
3. Booth K.H.V., Cox D.R. (1962). Some systematic supersaturated designs, *Technometrics*, 4, 489–495.
4. Lin D.K. (1995). Generating systematic supersaturated designs, *Technometrics*, 37, 213–225.
5. Wu C.F.J., (1993). Construction of supersaturated design through partially aliased interactions, *Biometrika*, 80, 661–669.
6. Plackett R.L., Burman J.P. (1946). Design of optimum multifactorial experiments, *Biometrika*, 33, 305–325.
7. Box G.E.P., Hunter W.G., Hunter J.S. (1978). *Statistics for Experimenters: An Introduction to Design, Data Analysis and Model Building*, New York, Wiley.
8. Mathieu D., Phan-Tan-Luu R. (1995). Méthodologie de la Recherche Expérimentale, Stratégie Expérimentale et Procédés Biotechnologiques, *Lavoisier Technique et Document*, 9, 1–10.
9. Feneuille D., Mathieu D., Phan-Tan-Luu R. (1978). Méthodologie de la Récherche Expérimentale: Introduction; Outils mathématiques; Etudes des surfaces de réponses; Matrices de mélanges; Méthode du Simplex, Université d'Aix-Marseille III, Publications du LMRE (France).
10. Box G.E.P., Wilson K.B. (1951). On the experimental attainment of optimum conditions, *J. R. Stat. Soc. B*, 13, 1–45.
11. Myers R.H., Montgomery D.C. (1995). *Response Surface Methodology: Process and Product Optimization Using Designed Experiments*, New York, Wiley.
12. Droesbeke J.J., Fine J., Saporta G. (1997). Plans d'expériences: Applications à l'entreprise, Editions Technip.
13. Doehlert D.H. (1970). Uniform shell designs, *Appl. Stat.*, 19, 231–239.
14. Roquemore K.G. (1976). Hybrid designs for quadratic response surfaces, *Technometrics*, 18, 419–424.
15. Box G.E.P., Behnken D.W. (1960). Some new three level designs for the study of quantitative variables, *Technometrics*, 2, 455–475.
16. Spendley W., Heixt J.R., Himsworth F.M. (1962). Sequential Application of Simplex design in optimisation and evolutionary operation, *Technometrics*, Nov, 441–461.
17. Lucas J.M. (1976). Which response design is best, *Technometrics*, 18, 411–417.
18. Box M.J., Draper N.R. (1974). On minimum-point second order designs, *Technometrics*, 16, 613–616.
19. Hartley H.O. (1959). Smallest composite designs for quadratic response surfaces, *Biometrics*, 15, 611–624.
20. Westlake W.J. (1965). Composite designs based on irregular fractions of factorials, *Biometrics*, 21, 324–335.
21. Clarengbold P.J. (1955). Use of the simplex design in the study of the joint action of related hormones, *Biometrics*, 11, 174–185.

22. Scheffe H. (1958). Experiments with mixtures, *J. R. Stat. Soc. B*, 20, 344–360.
23. Cornell J.A. (1990). *Experiments with Mixtures: Designs, Models and the Analysis of Mixture Data*, Wiley, New York.
24. Gammon J. (1963). Contribution to the discussion on H. Scheffé's ''The simplex-centroid design for experiments with mixtures,'' *J. R. Stat. Soc. B*, 25, 255–256.
25. Lambrakis D.P. (1968). Experiments with mixtures: a generalization of the simplex lattice design, *J. R. Stat. Soc. B*, 30, 123–136.
26. Scheffe H. (1963). Simplex-centroid design for experiments with mixtures, *J. R. Stat. Soc. B*, 25, 235–263.

15

Application of Experimental Methodology to Emulsions and Suspensions

Françoise Nielloud, Gilberte Marti-Mestres, and Hélène Maillols
Université Montpellier I, Montpellier, France

I. INTRODUCTION

Emulsions and suspensions are thermodynamically unstable. We regard stability as being the preservation of dispersion for a long enough time for the preparation to perform as a pharmaceutical product. Viscosity, nature of surfactants, lipophilic phase, drug content, pH adjustment, concentration of each component, and methodology of formulation are factors of concern. It is unfeasible to test all factors at the various levels. Experimental design offers an excellent approach, reducing the charge of time and money by limiting the number of manipulations in the formulation of emulsions and suspensions. This methodology also retains a high quality of information. Emulsion instability appears in physical property changes, particularly, rheological properties and droplet size (1), and these characteristics are frequently utilized as responses in experimental designs. Particles of suspension systems will tend to settle under gravity, and the worst consequence of sedimentation is caking. Times of settlement and caking appearance are regularly investigated as responses.

To illustrate the previous chapter, practical examples are given here. Our purpose is to review the state of the art of experimental design in formulations with emphasis on pharmaceutical emulsions and suspensions. We do not intend to deal with all sorts of designs exhaustively but will give some representative examples used in the scope of emulsions and suspensions. This chapter is divided into two parts. Section II covers screening methodology and factor study used in the pharmaceutical field, and Sec. III deals with the most common optimizations, particularly central composite designs and mixture problems.

II. SCREENING APPROACH AND FACTOR STUDY

In screening experiments, the most common designs encountered in pharmacy are factorial and fractional factorial designs. Nevertheless, when the number of factors is too great, a screening approach based on Hadamard's matrix is used to pick out the factors displaying the most significant effects. These designs give first-order factor effects but they also suppose that they are no factor interactions. We must bear in mind that if factor interaction exists, the main effects could be pooled with the interaction effects.

Typically, one can find three methods to analyze factorial designs. The standard approach for calculating the standard error of effects and individual confidence intervals entails replication of experiments. However, when the design involves 2^4 or more design points, the design is rarely replicated. A point—frequently the center point—with a few replications can be added to the design. Unreplicated designs can be used with an associated graphic methodology such as Daniel's normal probability plot (2,3), Pareto's approach (4), and Lenth's technique (5). These procedures will be illustrated in the examples below.

A. Screening Approach, Hadamard Designs

These matrices, generally known as Plackett and Burman designs (6), are rarely described in the literature for dispersed systems where many interactions can occur. Wehrle et al. (7) studied formulations of oil/water (O/W) emulsions containing an antifungal agent in a sequential experimental design. In a first step, process and formulation parameters were investigated to obtain the smallest drop-

Table 1 Interval of Variation for the Controlled Variables

	X_1	X_2	X_3	X_4	X_5	X_6	X_7
Var. abbrev.	MCT	MICO	POLOX	PH	TPH	LS	TCH
Lower limit	10.0%	0.5%	0.5%	6.0	5 min	1.0%	3 min
Upper limit	20.0%	1.0%	1.0%	8.0	10 min	2.0%	7 min

Percentage of the oily phase (MCT), percentage of drug (MICO), percentage of poloxamer 188 (POLOX), pH adjustment before sterilization (pH), time of the high-pressure homogenization process in minutes (THP), percentage of the couple lipoid–stearylamine (LS), time of coarse homogenization process in minutes (TCH).
Source: Adapted from Ref. 7.

let size emulsions. The authors selected seven important parameters, four composition variables [drug content, amount of lipophilic phase (midchain triglyceride), poloxamer concentration, quantity of phospholipids–stearylamine couple], and three formulation variables (pH adjustment, time of coarse emulsification, and time of high-pressure homogenization). Eight experiments were needed to estimate the main linear effects of these seven parameters and the levels of each factor are reported in Table 1. A Pareto diagram represents the graphic classification of these effects (Fig. 1). Three factors greatly affected the droplet size. The oily phase volume increased the mean droplet size of the emulsion; on the contrary, poloxamer 188 and lipoid–stearylamine couple decreased the particle size. These matrices are also used in the development of nanoparticulate dosage forms

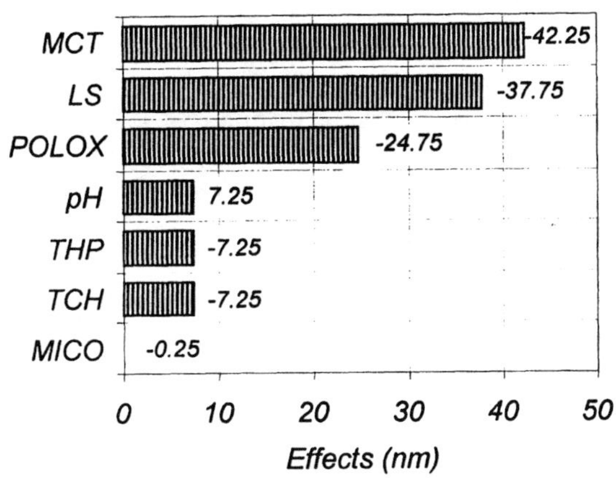

Figure 1 Pareto diagram for effects. (Adapted from Ref. 7.)

(8). This methodology may be suitable in all exploratory work in formulation where the aim is to quickly determine the influence of each factor.

B. Study of the Effects of Factors

Factorial designs are utilized in experiments where the effects of different factors on experimental responses are to be described. They are suitable designs for determination of the influence of several factors and for obtaining their interactions (9).

1. Full Factorial Designs

Factorial designs involve a certain number of levels for each factor. The simplest full factorial designs are those that require just two levels—low and high—for each factor (10,11). The aim of this methodology is to elucidate the effects of factors simultaneously, to presume their relative influence, and to establish if there are interactions between factors. Several examples of these designs can be found in the literature.

In the following example, parameters of formulation of O/W emulsions were performed with the aid of an unreplicated 2^5 design (12). For this full factorial design with five factors at two levels or variations, 32 trials were necessary. The validation of this model uses in particular Lenth's proposed method, with an alternative approach for standard error calculation. Five factors at two levels—mode of cooling, time of phase introduction, rate of homogenization, temperature of manufacture, and operators—are investigated. The study of the flow behavior of these emulsions corresponds to the quantitative response. The matrix is reported in Table 2, and the graphic treatment (Lenth's method) of data is given in Fig. 2. Lenth proposed an effective alternative method for formal analysis of unreplicated factorials. It is based on a simple formula for the standard error of the contrast estimates (see previous chapter). In the graphic (Fig. 2) the two calculated limits ME (margin error) and SME (simultaneous margin error) are drawn. Effect estimates that fall inside the inner limit could be described as inactive. Thus one can evaluate both the weight and ''significance'' of the effect by looking at just one graph. In this study only the main effect b_1 mode of cooling of emulsion preparation influences the response. The two levels of this formulation variable correspond to a progressive cooling (45 min of homogenization at room temperature) and a brutal cooling (30 min of homogenization at 15°C in a water bath). The results show a drastic difference in behavior between these two modes of cooling.

Recently, Prinderre et al. performed a similar study (13). They also studied the operating conditions in order to obtain stable O/W emulsions. They used a simple design at two levels and three factors: mixing rate (500 or 900 rev/min),

Table 2 Factorial design 2^5: variables, their levels, matrix, and data of the response K

Run	X_1	X_2	X_3	X_4	X_5	K	Factors	Levels −	Levels +
1	−	−	−	−	−	5.60	X_1 mode of cooling	BC	PC
2	+	−	−	−	−	19.68	X_2 time of phase introduction (s.)	3	30
3	−	+	−	−	−	6.11	X_3 rate of homogenization (rpm)	300	500
4	+	+	−	−	−	8.11	X_4 temp. of manufacture (°C)	70	80
5	−	−	+	−	−	2.36	X_5 operators	op1	op2
6	+	−	+	−	−	16.21			
7	−	+	+	−	−	4.17			
8	+	+	+	−	−	17.94			
9	−	−	−	+	−	14.17			
10	+	−	−	+	−	22.36			
11	−	+	−	+	−	9.48			
12	+	+	−	+	−	18.84			
13	−	−	+	+	−	5.10			
14	+	−	+	+	−	14.44			
15	−	+	+	+	−	3.83			
16	+	+	+	+	−	17.07			
17	−	−	−	−	+	7.60			
18	+	−	−	−	+	15.25			
19	−	+	−	−	+	6.74			
20	+	+	−	−	+	9.88			
21	−	−	+	−	+	4.39			
22	+	−	+	−	+	7.98			
23	−	+	+	−	+	3.86			
24	+	+	+	−	+	19.57			
25	−	−	−	+	+	9.49			
26	+	−	−	+	+	14.80			
27	−	+	−	+	+	9.10			
28	+	+	−	+	+	18.17			
29	−	−	+	+	+	5.85			
30	+	−	+	+	+	13.50			
31	−	+	+	+	+	6.90			
32	+	+	+	+	+	7.81			

Source: Adapted from Ref. 12.

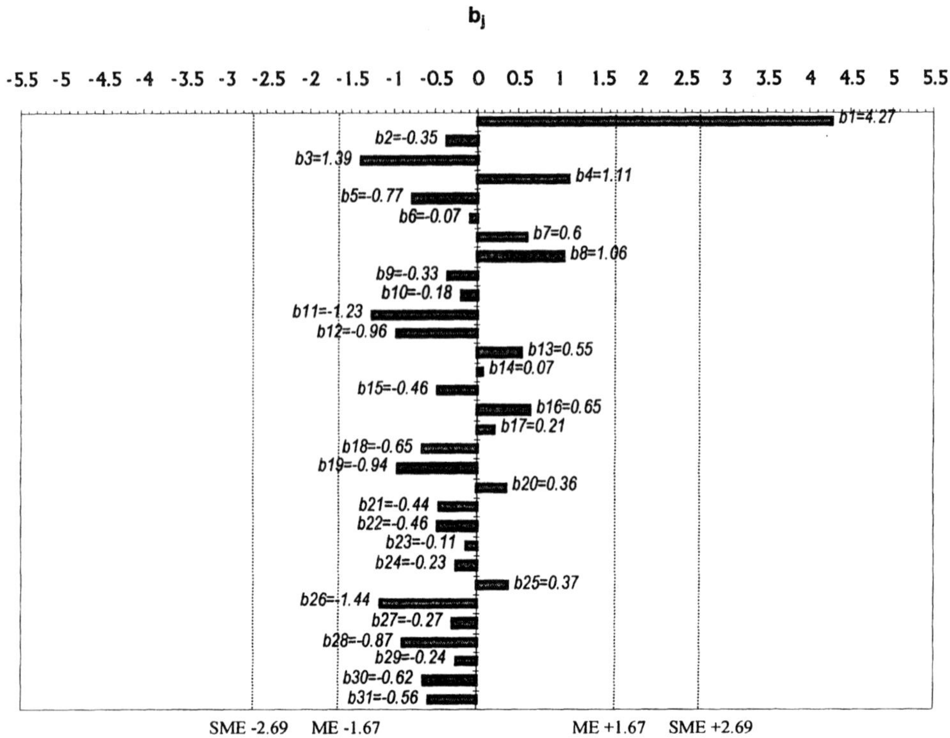

Figure 2 Experimental design 2^5. Graphic treatment of Lenth's proposed method. (Adapted from Ref. 12.)

homogenization at 15,000 rev/min (yes or no), mixing time (10 or 20 min). Two responses were evaluated: viscosity and particle diameter.

Two examples of 2^3 designs for the evaluation of topical formulations were given by Hwang et al. (14) and Giannakou et al. (15). The effect of different enhancers at various concentrations was studied in both cases. These vehicles are used for methotrexate and nitrendipine in order to find the most appropriate formulations for the transdermal application. In vitro skin permeation of drug represents an interesting response, and Giannakou classified several enhancers according to their efficiency.

Some investigators (16) recently proposed an experimental design for optimizing the solubility and the release of a new retinoid from a gel formulation. In this case a 2×3^2 factorial experiment was designed to investigate the effect of three formulation components. If one factor is to be tested at two levels and two others at three levels, we have a 2×3^2 in 18 units (17,18). The experiment involved discovering the concentration of a cosolvent (glycol) and two surfactants (poloxamer and polysorbate) that increased retinoid release and solubility in a topical formulation. Eighteen variations of a prototype gel were formulated. Two responses were investigated, i.e., the saturated solubility values and slope of the line obtained from the plot of percentage retinoid released vs. the square root of time. Finally, the formulation that provided optimum drug release and solubility contained twice the amount of polysorbate and the same concentration of glycol present in the prototype formulation, but did not contain poloxamer.

Factorial designs are also found in microsphere formulation studies. The factors influencing the preparation and release of salbutamol sulfate from poly(-lactide-co-glycolide) microspheres prepared by water-in-oil-in water (W/O/W) emulsion technique have been investigated by the same team in two studies with 2^3 factorial designs (19,20). The independent variables were drug loading, amount of gelatin, and concentration of polyvinyl alcohol. The responses were particle size and entrapment ratio. Another example is given with a 3^2 full factorial design to investigate the key variables affecting the preparation of diclofenac sodium microspheres (21). The amount of dichloromethane and the amount stabilizer were used as independent variables, while drug content and microsphere diameter were selected as dependent variables. Contour graphs were presented to visualize the impact of factors on the drug content in the microspheres.

To allow an independent estimation of the experimental error, an alternative methodology is given with replicates at the center of the design. Some investigators have recently proposed a 2^4 full factorial design to develop a reconstitutable suspension of rifampicin (22). The study illustrated the effect of the percentage of sucrose, avicel RC-591, hydrophilic aerosil, and aerosol-OT on the flowability and the bulk density of the dry mixture, as well as the viscosity, the sedimentation volume, and the redispersibility of the suspension. This work is particularly well conceived: five replicates at the center of the design make it possible to calculate the experimental error and to give the response surfaces. The authors added three supplementary formulas to check the validity of the model equation. The effects of each factor are detailed for all responses, and response surfaces are presented for significant effects.

2. Fractional Factorial Designs

With full factorial designs, the number of experiments increase expensively, even if the assays are not replicated. Thus, for example, a five- and six-factor two-level design requires 42 and 64 experiments, respectively. This methodology gives second- and third-order interactions, which are not always necessary. One approach for minimizing the large amount of experiments is to use fractional factorial designs (FFDs), in which all main effects and two-factor interactions can be estimated. In conclusion, when the number of variables is high, fractional factorial designs are used because they provide the needed information by executing only a part of the full factorial design.

To illustrate the foregoing statements, a first example of emulsion formulation is given by Carlotti (23). In the first part of his study, a 2^{5-2} factorial design was performed to determine the optimum conditions of formulation to achieve maximum stability. The following variables were investigated: amounts of oil, emulsifier, and coemulsifier; homogenization time; and speed of the homogenizer. The given response was the time (in minutes) necessary to break the emulsions by centrifugation at 6000 rpm for three different emulsifiers (the salts of hexadecyl hydrogen phosphate with lysine, arginine, and diethanolamine). The 2^{5-2} factorial design matrix and the response values are reported in Table 3. The effects of the variables appear to be quite different for the three surfactants, but as a general rule the percentages of emulsifier and coemulsifier have the largest effect on the stability of emulsions.

The same author, in another work, used a fractional factorial design to test the influence of lipopeptides, surfactants, oil, homogenization rate, homogenization time, and pH on the mean diameter of emulsions (24). In that case, a 2^{5-1} fractional factorial design was carried out.

A good example of orthogonal experimental designs for the formulation of chitosan microspheres containing cisplatin, an anticancer drug, was given by Min Wang et al. (25). Seven factors at three levels were selected and arranged in an $L_{27}3^{13}$ orthogonal experimental design (26). These chitosan microspheres were made by an emulsion–chemical crosslinking technique. The contribution

Table 3 Matrix of FFD N.1 and Responses

	Variables[a]					Responses[b]		
Run	I	II	III	IV	V	HPD	HPA	HPL
1	−	−	−	+	+	2	2	0
2	+	−	−	−	−	2	2	2
3	−	+	−	−	+	3	46	88
4	+	+	−	+	−	2	30	25
5	−	−	+	+	−	3	2	10
6	+	−	+	−	+	6	6	4
7	−	+	+	−	−	17	50	148
8	+	+	+	+	+	12	34	148

[a] Variables as for Table 1.

[b] Minutes necessary to break the emulsion by centrifugation at 6000 rpm. The three responses refer to the three emulsifiers, respectively: HPD = diethanolamine hexadecyl hydrogen phosphate; HPA = L-arginine hexadecyl hydrogen phosphate: HPL, L-lysine hexadecyl hydrogen phosphate.

Source: From Ref. 23.

of different factors (concentration of chitosan, volume ratio, stirring rate, amount of cisplatin/cisplatin + chitosan, oil phase type, chitosan type, and stabilization time) was investigated and possible interactions among the first three factors were also analyzed. Three responses were studied in replicate: trapping efficiency of cisplatin, drug content, and size distribution for each batch of microspheres. Then the authors introduced a desirability function. Using this procedure they obtained the optimum formulation of chitosan microsphere containing cisplatin.

In the field of emulsions, but concerning crude oils, Førdedal et al. (27) designed an interesting 2^{7-3} reduced factorial design to investigate the coalescence of water–in–crude oil emulsions by means of dielectric time domain spectroscopy at high electric fields. They showed in this work that a reduced factorial design can be used as a screening technique to investigate emulsion stability and to extract information about general trends of the important variables in crude oil systems.

III. OPTIMIZATION

A. Central Composite Designs

Factorial designs permit estimation of linear relation in particular and are not adequate to determine the optimal experimental conditions. More specific experiments are needed to have enough degrees of freedom for statistical analysis. One procedure is to use a central composite design (CCD), which is a full factorial design with added replicates positioned at the center of the design and a couple of experiments along each coordinate axis. Thus, central composite designs in two variables need 10–14 experiments and one in three variables 16–20 experiments. With a CCD it is possible to determine a response surface, which gives the optimal experimental conditions. The interest of a CCD is that it can be developed sequentially from a previous factorial design.

Several investigators have proposed this methodology for optimizing emulsion formulation but we have also found many examples concerning microsphere or nanosphere fields. The first examples that we propose were presented by Carlotti et al. in 1991 and 1992 (23,28). In their first work, Carlotti et al. present several CCDs to optimize emulsions. The response was the stability to centrifugation and, more precisely, the time of centrifugation at 6000 rpm required to break the emulsion. One of these designs concerned in optimization of the emulsion containing L-lysine hexadecyl hydrogen phosphate (HPL) as the emulsifier. The three variables investigated were a surfactant (HPL), a cosurfactant (hexadecan-1-ol), and an oily phase (isopropylmyristate). In this case the authors concluded that they obtained quite remarkable stability times. The maximum response in the explored domain corresponded to the following mixture: isopropylmyristate 26%, HPL 4%, and hexadecan-1-ol 3.5%. In the same article, the authors investigated another salt of hexadecyl hydrogen phosphate and showed that the counterion of this emulsifier affects the composition and the stability of an emulsion. In another effort to investigate the emulsion stability, Carlotti et al. propose an interesting application of chemometric procedures to the optimization of water-in-oil-silicon (W/O-S) emulsion. The central composite design used three variables and three replicates at the central point. The following variables were investigated to determine their effects on the stability of the emulsions: amounts of dimethylsiloxane copolyol in cyclodimethicone as surfactant, octamethyltetracyclosiloxane and 2-ethylhexylpalmitate as oils. This step, performed by the experimental design and the response surface method, led to an emulsion of very high stability and thus cleared the way for the second aim of this work, which was to compare the influence of a series of esters.

More recently, Wehrle performed a very good example on a positively charged O/W emulsion containing an antifungal agent developed for ophthalmic use (7). The sequential statistical procedure is well illustrated by a first screening approach based on Hadamard's matrix (see Sec. II of this chapter) in order to select the parameters displaying the most significant effects on response parameters. A second step is a 2^2 experimental design with the two main factors selected previously in order to describe both their effects in detail and the first-order interaction. Here the main factors chosen were the concentration of poloxamer 188 and the quantity of the phospholipids–stearylamine couple. The responses were the droplet size of the emulsions, their standard deviation, and their polydispersity. As the statistical test showed that the first-order model with interaction was not adequate to describe the response, the design was completed with experiments according to Box and Wilson's method. This design was determined by the 1.414 value and five replicates at the center. The complete design and the results are reported in Table 4; the experiments are represented in Fig. 3. The statistical validations of the quadratic models are given in Table 5 and they show that the second-order polynomial was suitable to describe the particle size as well as its standard deviation. The model for the polydispersity could not be validated ($R^2 =$ 0.627). For the first two responses, the amount of poloxamer and of lipoid–stearylamine showed a significant effect, but the interaction between these two variables was not significant. The response surface and the contour plots help to clarify this phenomenon. The superimposition of contour plots (Fig. 4) gives the optimal formulation conditions leading to the smallest droplet size and to the smallest standard deviation. This optimal area can be confirmed by the realization of an experiment inside this zone. A very close correlation between the expected and the observed values was obtained and validated the model.

The results from our laboratory deal with the optimization of a pharmaceutical suspension (29,30). The approach used an antimicrobial agent (sulfadiazine) and three factors were investigated: a surfactant (polysorbate 80), an electrolyte (sodium chloride), and a thickener (methylcellulose). The levels of these variables

Table 4 Experiments and Results of the Composite Design

Lipoid–stearylamine		Poloxamer 188		Response variables		
Real value (%)	Coded level	Real value (%)	Coded level	Particle size (nm)	SD (nm)	Polydispersity
2.25	1	1.25	1	136	35	0.09
2.25	1	0.25	−1	161	44	0.11
0.75	−1	1.25	1	180	44	0.08
0.75	−1	0.25	−1	212	57	0.10
1.50	0	0.75	0	153	43	0.12
0.44	−1.414	0.75	0	201	56	0.11
2.56	1.414	0.75	0	139	35	0.09
1.50	0	0.04	−1.414	185	50	0.10
1.50	0	1.46	1.414	137	30	0.06
1.50	0	0.75	0	157	34	0.06
1.50	0	0.75	0	160	38	0.08
1.50	0	0.75	0	166	39	0.07
1.50	0	0.75	0	157	36	0.07

Source: Ref. 7.

are listed in Table 6. Two responses were analyzed; the volume of dispersed sediment and the caking level. The authors considered that the caking level should be equal to zero and that the volume of dispersed sediment is acceptable between 90% and 100%. The major factor affecting the two responses was the thickener and the use of graphic illustration such as contour plot responses allowed delimitation of the optimal zone of formulation.

In the field of micro- and nanoparticles, many examples of manufacturing processes have been described by several authors with generally three-factor CCD (31–35). Julienne et al. (36) reported results from a study on the use of a solvent evaporation method to prepare acid- and glycolic acid–based copolymer nanoparticles. Three formulation variables—phase volume ratio of the emulsion formed, polymer concentration in organic phase, and homogenization pressure—were investigated. The particle size distribution was analyzed as the response with precise mean diameter, coefficient of variation, and polydispersity. In this work, a significant influence of volume phase ratio and polymer concentration on mean particle size and coefficient of variation was shown. It was noted that to obtain a small mean particle size with a narrow unimodal distribution it is necessary to prepare particles operating with a low phase volume ratio and a relatively low concentration of polymer. The opposite conditions could also lead to the same results.

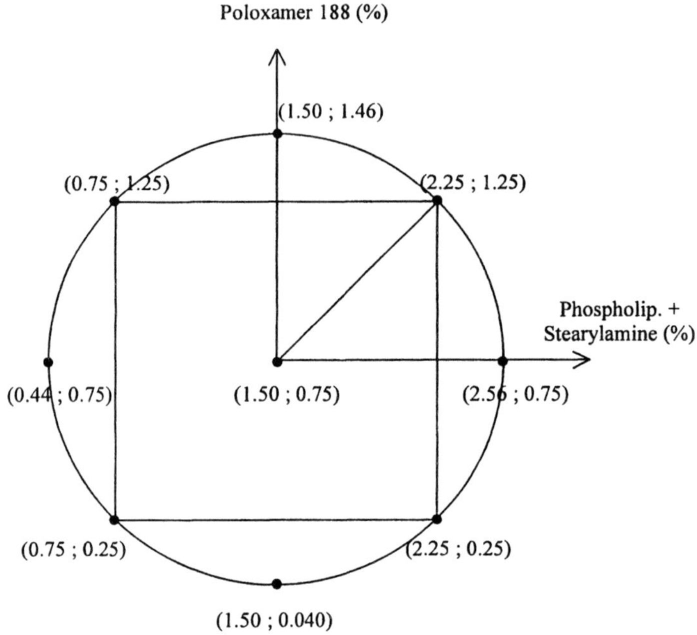

Figure 3 Representation of the composite design. (Adapted from Ref. 7.)

Later authors (37) have studied extensively the preparation of polycaprolactone nanoparticles by solvent displacement. Five factors of formulation affecting the particle size characteristics and the percentage of drug entrapment were studied: aqueous medium temperature, diameter of the needle employed to inject the organic into the aqueous phase, organic phase volume, amounts of both polymer and surfactant. In this CCD, the authors specified that the results were the mean

Table 5 Quadratic Models and Their Statistical Validation

Variable	ANOVA, *F* ratio	Sig. level	R^2
Particle size	3.46	0.372	0.912
SD	25.62	0.144	0.988
Polydispersity	0.33	0.817	0.501

Source: Ref. 7.

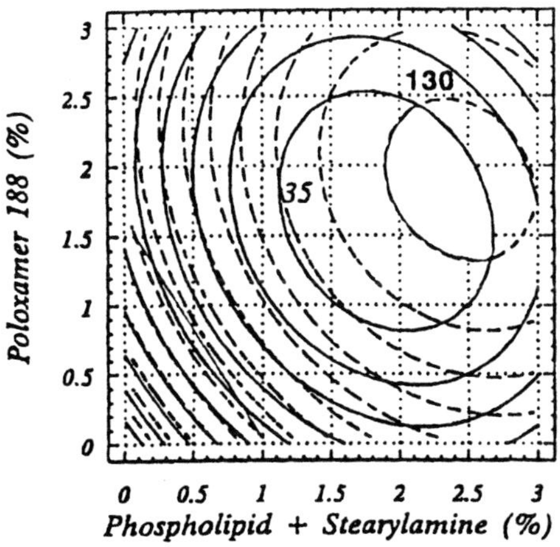

Figure 4 Superimposed contour plots for the particle size (------) and the standard deviation (———). (Reprinted from Ref. 7.)

of three assays each performed twice. They concluded that the amounts of polymer and surfactant and the diameter of the needle used in the preparation of nanoparticles significantly affected the percentage of entrapped drug. The mean particle size was significantly affected by all of the formulation variables except for the amount of surfactant. As stated above, the particle size is the most prominent response used in this field and it can be controlled by the injection rate of the needle gauge and the polymer concentration. The aim of this work was to prepare nanoparticulates easily and reproducibly using the nanoprecipitation method.

B. Experiments with Mixtures

In the mixture problem, the studied response is supposed to depend only on the proportions of the ingredients present in the mixture and not on the quantity of the mixture (38). This final part of the chapter corresponds to a narrow field of application. Examples of mixture experiments are developed. Recently, Elkheshen used a simplex lattice design for the optimization of the microencapsulation of a water-soluble drug using poly(lactic acid) and poly(lactide-co-glycolide) copolymer (39). Surface response methodology optimizes the characteristics of nicotinic acid microcapsules prepared by the W/O emulsion/solvent technique. This design changes the ratio of poly(lactic acid) of molecular weight 2000, poly(lactic acid) of molecular weight 100,000, and poly(lactide-co-glycolide) of molecular weight 18,000 while keeping the total amount of these polymers constant

Table 6 Experimental Domain and Coding of the Three Factors[a]

Exp	X_1	X_2	X_3	Responses Caking ml	Responses Sedim. %	Factors	Levels −1.682	−1	0	+1	+1.682
1	−1	−1	−1	17	10.11	X_1 % electrolyte	0	0.18	0.45	0.72	0.90
2	+1	−1	−1	16	10.79	X_2 % surfactant	0	0.41	1	1.60	2
3	−1	+1	−1	18	13.21	X_3 % thickener	0	0.304	0.750	1.196	1.50
4	+1	+1	−1	20	13.37						
5	−1	−1	+1	0	96.47						
6	+1	−1	+1	0	97.14						
7	−1	+1	+1	0	94.14						
8	+1	+1	+1	4	14.33						
9	−1.682	0	0	10	6.47						
10	+1.682	0	0	16	14.11						
11	0	−1.682	0	7	7.55						
12	0	+1.682	0	14	8.0						
13	0	0	−1.682	18	11.3						
14	0	0	+1.682	0	100.0						
15	0	0	0	12	5.74						
16	0	0	0	12	7.47						
17	0	0	0	10	8.52						
18	0	0	0	10	9.49						
19	0	0	0	10	7.86						
20	0	0	0	10	9.09						

[a] Results for the two responses: sediment volume (%) and caking level (mL).

Source: Ref. 29.

in all experiments. Once more the studied characteristics included particle size distribution, yield, and percentage drug loading. The author gives a quadratic model for all responses and analyzes separately the effect of different polymers on each response with the aid of the response surfaces.

Pharmaceutical and cosmetic gels also illustrate this methodology (40,41). For example, Forestier et al. studied the formulation of a gel with a ternary diagram at three components (X_1: C12–C15 alcohol benzoate; X_2: cyclomethicone; X_3: ethylene vinyl acetate copolymer in water). The authors carried out their study using a systematic sequential walk strategy. Ternary diagrams for formulations can exhibit discontinuities. To avoid this problem, in a first step the authors determined the incompatibility zone. The initial region of interest was chosen inside the simplex to study the form of the surface that is defined by fixing of lower boundaries X_1, X_2, and $X_3 > 0.1$ on the constituent proportions. Three first points were performed and one of them was heterogeneous (0.1: C12–C15 alcohol benzoate; 0.8: cyclomethicone; and 0.1: ethylene vinyl acetate copolymer in water). A new matrix was considered where the incompatibility zone was eliminated. The new proportions of the vertices were:

1. 0.8: X_1; 0.1: X_2; 0.1: X_3
2. 0.36: X_1; 0.54: X_2; 0.1: X_3
3. 0.1: X_1; 0.1: X_2; 0.8: X_3

In a second step, a simplex centroid design was realized with seven design points and six check points. Two responses were optimized consistency and whiteness. Using the second-order equations from this matrix, it is possible to obtain isoresponse graphs. The contour plots are given in Figs. 5 and 6. The optimal gel must correspond to a gel similar to petrolatum. This optimal zone is reported in Fig. 7 where contour plots of the two responses are superimposed.

This work shows that the use of matrices based on the Scheffé algorithm is particularly useful in the study of these types of mixture. Isoresponse graphs provide an exceptional tool for the interpretation of results and for decision-making.

Discontinuity response (stable or unstable emulsion) or binary response such as appearance of a gel, an emulsion, or a microemulsion in a ternary diagram remain a problem in mixture formulations. Conventional methods such as dilution or lattice methods for identification of different zones in a ternary mixture diagram require many experiments that do not provide information. Mathieu et al. have proposed a new application of the simplex method to resolve these difficulties (42). This study did not use a classical concept in the design and analysis of experiments with mixtures. The authors used a method for exploring the ternary diagram based on the optimization technique called the simplex method (43). The simplex is an equilateral triangle whose size and orientation are chosen by the experimenter (generally a submultiple of that of the ternary diagram). The

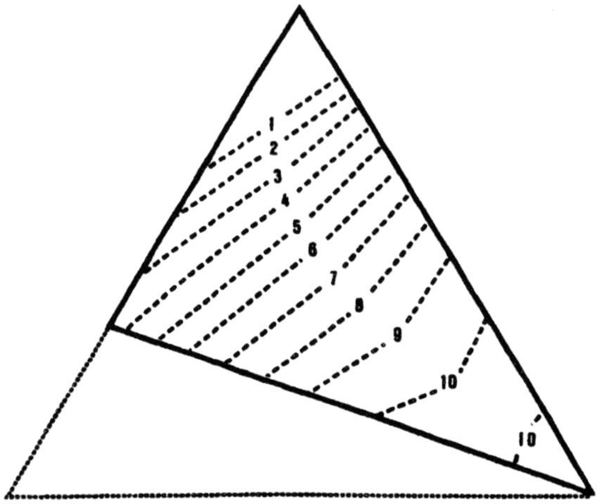

Figure 5 Estimated consistency surface optioned with second-degree model. (Reprinted from Ref. 41 with permission of Academic Press, Inc.)

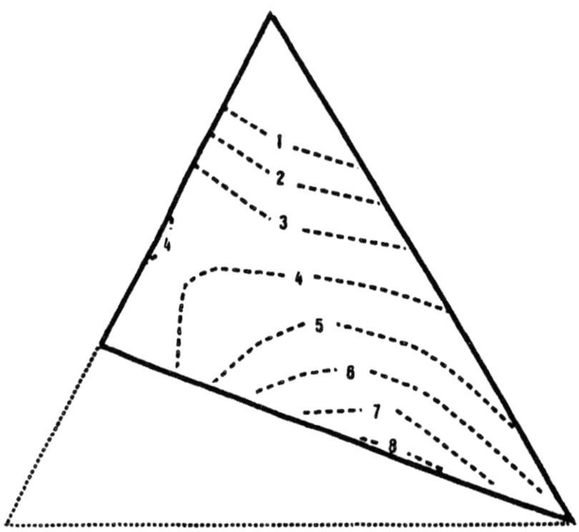

Figure 6 Estimated whiteness surface optioned with second-degree model. (Reprinted from Ref. 41 with permission of Academic Press, Inc.)

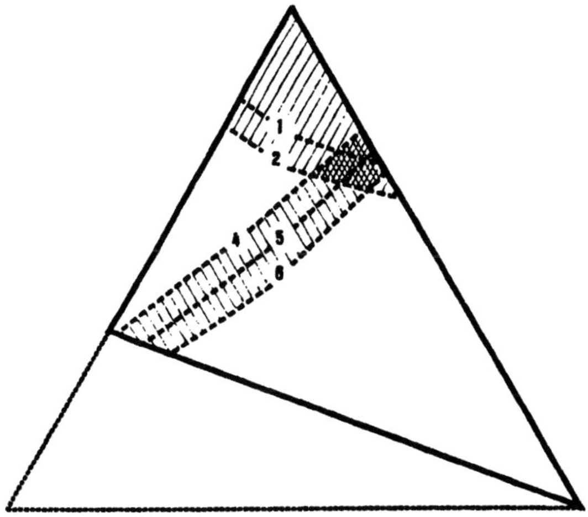

Figure 7 Superimposed contour plots for the consistency and for the whiteness. (Reprinted from Ref. 41 with permission of Academic Press, Inc.)

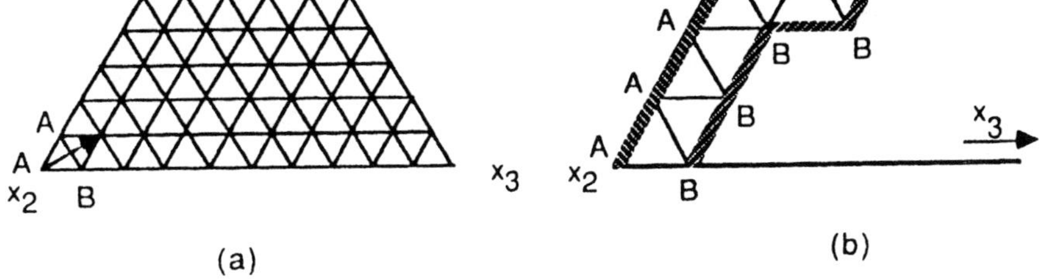

Figure 8 Examples of simplex progression and meeting an intersection of three phases. (A) initial simplex AAB, and first reflection; (B) progression and meeting with phase C. (Reprinted from Ref. 42, with permission of Elsevier Science-NL, The Netherlands.)

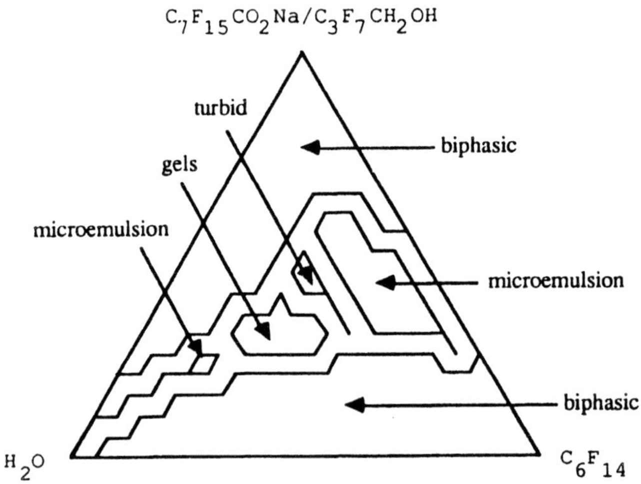

Figure 9 Envelopes of the various zones obtained in a simulation of a microemulsion, which consists of a mixture of four components. (Reprinted from Ref. 42, with permission of Elsevier Science-NL, The Netherlands.)

next experiment is the symmetrical point of the vertex giving the worst response. The goal is to attain the desired limit zone as fast as possible. If a third phase appears on one of the reflections, two directions must be explored. An example of simplex procedure and intersecting of three phases is given in Fig. 8. The authors have chosen in this article an example from a microemulsion study with water, a mixture of surfactant (sodium perfluorooctanoate: $C_7F_{15}CO_2Na$) and co-surfactant (heptafluoro-1-butanol: $C_3F_7CH_2OH$), and oil (perfluorihexane: C_6F_{14}). Each step of the simplex represents 5% of the whole and the first simplex is located at the water vertex. The exploration of the experimental domain leads to four zones: microemulsion zones, a gel zone, a turbid zone, and a biphasic zone illustrated in Fig. 9. With this methodology, even if the number of experiments required is quite high, no experiment is unnecessary.

IV. CONCLUSION

Experimental designs are useful tools for reducing the number of experiments and may also contribute to a better understanding of formulation parameters. This methodology is broadly used in the cosmetics (44), agricultural (45), food (46–48), and chemical fields (49). From present knowledge, experimental designs are sparsely used in pharmaceutical emulsions and suspensions, and little is available in the literature concerning their application. Nevertheless a variety of strategies have been explored to facilitate these formulations. Moreover, several software packages specifically intended for experimental design and optimization purposes are now available. These programs provide facilities for the calculation and analysis of designs and promising applications should be developed with exciting future prospects.

REFERENCES

1. D. Attwood, A. T. Florence. Surfactant Systems. London: Chapman and Hall, 1983, pp 469–566.
2. C. Daniel. Use half-normal plots in interpreting factorial two-level experiments. Technometrics 1: 311–341, 1959.
3. R. Sundberg. Interpretation of unreplicated two-level factorial experiments, by examples. Chemometr. Intell. Lab. 24: 1–17, 1994.
4. D. Mathieu. R. Phan-Tan-Luu, M. Sergent. Criblage et étude des facteurs. Marseille: LPRAI; 1996.
5. R. V. Lenth. Quick and easy analysis of unreplicated factorials. Technometrics 31: 469–473, 1989.
6. R. L. Plackett, J. P. Burman. The design of optimum multifactorial experiments. Biometrica 33: 305–332, 1946.
7. P. Wehrle, D. Korner, S. Benita. Sequential statistical optimization of a positively-charged submicron emulsion of miconazole. Pharm. Dev. Tech. 1: 97–111, 1996.
8. S. Benita. Microencapsulation. Methods and Industrial Applications. New York: Marcel Dekker, 1996, pp 93–131.
9. S. Bolton. Pharmaceutical Statistics. Practical and Clinical Applications. New York: Marcel Dekker, 1990, pp 308–337.
10. D. L. Massart, B. G. M. Vandeginste, S. N. Deming, Y. Michotte, L. Kaufman. Chemometrics: A Textbook. New York: Elsevier, 1988, pp 271–282.
11. N. A. Armstrong, K. C. James. Pharmaceutical Experimental Design and Interpretation. London: Taylor and Francis Ltd, 1996, pp 131–167.
12. F. Nielloud, G. Marti-Mestres, J. P. Laget, C. Fernandez, H. Maillols. Emulsion formulations: study of the influence of parameters with experimental designs. Drug Dev. Ind. Pharm. 22: 159–166, 1996.
13. P. Prinderre, Ph. Piccerelle, E. , Cauture, G. Kalantzis, J. P. Reynier, J. Joachim. Formulation and evaluation of o/w emulsions using experiment design. Int. J. Pharm. 163: 73–79, 1998.
14. G. C. Hwang, A. Y. Lin, W. Chen, R. J. Sharpe. Development and optimization of a methotrexate topical formulation. Drug Dev. Ind. Pharm. 21(17): 1941–1952, 1995.
15. S. A. Giannakou, P. P. Dallas, D. M. Rekkas, N. H. Choulis. Development and in vitro evaluation of nitrendipine transdermal formulations using experimental design techniques. Int. J. Pharm. 125: 7–15, 1995.

16. E. Keyhani-Morrison, S. Nadkarni, A. Sakr. Optimization of gel formulation of a novel retinoid based on experimental design. Pharm. Ind. 57(4): 333–336, 1995.
17. W. G. Cochran, G. M. Cox. Experimental Designs. New York: Wiley, 1992, pp 244–275.
18. M. G. Vigier. Pratique des plans d'expériences. Paris, Les Editions d'Organisation, 1988, pp 49–65.
19. N. Erden, N. Celebi. Factors influencing release of salbutamol sulphate from poly(lactide-co-glycolide) microspheres prepared by water-in-oil-in-water emulsion technique. Int. J. Pharm. 137: 57–66, 1996.
20. N. Celebi, N. Erden, A. tûrkyilmaz. The preparation and evaluation of salbutamol sulphate containing poly(lactic acid-co-glycolic acid) microspheres with factorial design-based studies. Int. J. Pharm. 136: 89–100, 1996.
21. G. K. Jani, M. C. Gohel. Effects of selected formulation parameters on the entrapment of diclofenac sodium in ethyl cellulose microspheres. J. Controlled Release. 43: 245–250, 1997.
22. S. A. Elkheshen, S. S. Badawi, A. A. Badawi. Optimization of a reconstitutable rifanpicin using 2^4 factorial design. Drug Dev. Ind. Pharm. 22(7): 623–630, 1996.
23. M. E. Carlotti, R. Carpignano, M. R. Gasco, M. Trotta. Optimization of emulsions. Int. J. Cosmet. Sci. 13: 209–219, 1991.
24. M. E. Carlotti, F. Pattarino, M. R. Gasco, P. Brusasca. Optimization of parameters in the emulsification process by two different methods. Int. J. Cosmet. Sci. 15: 245–259, 1993.
25. Y. Min Wang, H. Sato, I. Adachi, I. Horikoshi. Optimization of the formulation design microspheres containing cisplatin. J. Pharm. Sci. 85(11): 1204–1210, 1996.
26. Y. Benoist, Y. Tourbier, S. Germain-Tourbier. Plans d'expériences. Construction et analyse. Paris, Tech & Doc. Lavoisier, 1994, pp 60–145.
27. H. Førdedal, Ø. Midttun, J. Sjöblom, O. M. Kvlheim, Y. Schildberg, J. L. Vollet. A multivariate screening analysis of W/O emulsions in high external electric fields as studied by means of dielectric time domain spectroscopy. II. Model emulsions stabilized by interfacially active fractions from crude oils. J. Collord Interface Sci. 182: 117–125, 1996.
28. M. E. Carlotti, R. Carpignano, M. R. Gasco, F. Pattarino. Optimization of W/O-S emulsions and study of the quantitative relationships between ester structure and emulsion properties. J. Disp. Sci. Tech. 13, 3: 315–336, 1992.
29. F. Nielloud, G. Marti-Mestres, R. Fortuné, H. Maillols. Utilisation d'un plan central composite en vue de la formulation d'une suspension pharmaceutique. Colloque Chimiometrie 97, expériences et modèles. P24, 1–4. Lyon 1997.
30. F. Nielloud, G. Marti-Mestres, R. Fortuné, J. P. Mestres, H. Maillols. Formulation of insoluble drugs. Study of pharmaceutical suspensions by response surfaces methodology. Analusis 26: 277–281, 1998.
31. M. R. Gasco, S. Morel, R. Carpignano. Optimization of incorporation of deoxycorticosterone acetate in lipospheres. Eur. J. Pharm. Biopharm. 38(1): 7–10, 1992.
32. E. E. Hassan, R. C. Parish, J. M. Gallo. Optimized formulation of magnetic chitosan microspheres containing the anticancer agent, oxantrazole. Pharm. Res. 9(3): 390–397, 1992.
33. E. E. Hassan, A. G. Eshra, A. H. Nada. Formulation of prolonged release lipid micro-

pellets by emulsion congealing: optimization of ketoprofen entrapment and release. Int. J. Pharm. Pharmacol. 121: 149–155, 1995.

34. K. A. Abu-Izza, L. Garcia-Contreras, D. R. Lu. Preparation and evaluation of sustained release AZT-loaded microspheres: optimization of the release characteristics using response surface methodology. J. Pharm. Sci. 82(2): 144–149, 1996.
35. B. G. Müller, H. Leuenberger, T. Kissel. Albumin nanospheres as carriers for passive drug targeting: an optimized manufacturing technique. Pharm. Res. 13(1): 32–37, 1996.
36. M. C. Julienne, M. J. Alonso, J. L. Gomez Amoza, J. P. Benoit. Preparation of poly(D,L-lactide/glycolide) nanoparticles of controlled particle size distribution: application of experimental designs. Drug Dev. Ind. Pharm. 18(10): 1063–1077, 1992.
37. J. Molpeceres, M. Guzman, M. R. Aberturas, M. Chacon, L. Berges. Application of central composite designs to the preparation of polycaprolactone nanoparticles by solvent displacement. J. Pharm. Sci. 85(2): 206–213, 1996.
38. J. A. Cornell. Experiments with Mixtures. New designs, Models, and the Analysis of Mixture Data. New York: Wiley, 1990, pp 1–20.
39. S. Elkheshen. Simplex lattice design for the optimization of the microencapsulation of a water soluble drug using poly(lactic acid) and poly(lactide co-glycolide) copolymer. J. Microencaps. 13(4): 447–462, 1996.
40. J. S. Chu, G. L. Amidon, N. D. Weiner, A. H. Goldberg. Pharm. Res. 8(11): 1401–1407, 1991.
41. J. P. Forestier, E. Puech, J. L. Tichadou. Application d'une méthodologie expérimentale à l'étude du diagramme ternaire d'un gel. Int. J. Cosmet. Sci. 7: 219–233, 1985.
42. D. Mathieu, E. Puech-Costes, M. T. Maurette. The simplex method applied to the detection and following of a discontinuity. Chemometrics Intell. Lab. Syst. 20: 25–34, 1993.
43. W. Spendley, G. R. Hext, F. R. Himsworth. Sequential application of simplex designs in optimisation and EVOP. Technometrics 4: 441–461, 1962.
44. G. Marti-Mestres, F. Nielloud, R. Marti, H. Maillols. Optimization with experimental design of non-ionic, anionic, and amphoteric surfactants in mixed system. Drug Dev. Ind. Pharm. 23(10): 993–998, 1997.
45. E. Schrevens, J. Cornell. Design and analysis of mixture systems: applications in hydroponic, plant nutrition research. Plant Soil 154: 45–52, 1993.
46. E. Dumay, N. Souchet, J. C. Cheftel. Multifactorial study of gelled oil/water emulsions from sardine surimi. Sci. Aliment. 13: 119–140, 1993.
47. K. K. S. Nair, D. N. Rao, N. Chand, R. B. Nair, P. Puttarajappa, B. L. Amla. Tandoori chicken: processing optimized by response surface methodology. J. Food Qual. 18: 103–117, 1995.
48. S. Y. Hsu. Optimization of surimi processing system with a central composite design method. J. Food Eng. 24: 101–111, 1995.
49. L. Davies. Efficiency in Research, Development, and Production. Cambridge: The Royal Society of Chemistry, 1993.

16
Rheology of Suspensions and Emulsions

María Isabel Briceño
University of Los Andes, Merida, Venezuela

I. INTRODUCTION

In this era of constant innovation, development of new products is one of the most outstanding trends. New products are continuously created or transformed to meet the ever-changing customer taste and expanding requirements. The pharmaceutical, cosmetic, and food industries do not escape this innovation urge.

Many new products are composites of two, three, or more well-differentiated phases, such as gels and pastes. These are nonhomogeneous (from a chemical point of view) dispersed systems.

We will define a dispersed system as a fluid consisting of one or more internal phases, suspended in a liquid (aqueous or organic), external phase. In this sense, we may distinguish suspensions, which consist of one or more solid phases suspended in a liquid, and emulsions, which are one or more liquid phases, dispersed in an immiscible liquid. Mixed solids and liquid suspended phases may also be present, as is the case of many food and cosmetic products.

Although rheology has been often associated with the domain of fluid mechanics, it is increasingly recognized as an important aspect in the development of many new products. Some customers requirements, such as consistency, rubbing, and creaminess, which are linked to touch and taste, may be related to the rheological behavior of the product. Furthermore, the efficiency of drug delivery of many pharmaceutical preparations may be enhanced by means of an appropriate rheological assessment. In addition, handling of dispersions involving operations such as pumping, draining, and pouring require previous knowledge of rheology (or, at least, viscosity), in order to determine process energy and equipment requirements in an efficient manner.

Section II contains a review of the basic concepts of viscosity, rheology, rheological behavior, and well-known and proven evaluation techniques. The discussion will be focused exclusively on shear flow, excluding elongational flow and nonlinear viscoelasticity which, despite their importance, are not often used in product formulation and for this reason will not be discussed. In fact, elongational effects are more often negligible in suspensions and emulsions.

The main physical and chemical properties of dispersed systems, which influence viscosity and give rise to the so-called non-Newtonian behavior, are discussed in Sec. III. It will be seen that the viscosity and rheological behavior of dispersed systems are the result of the interplay of many variables, and at least a qualitative knowledge of these effects is required to control, to some extent, their flow properties.

Finally, Sec. IV begins with a brief discussion of how to carry out correct rheological evaluations, as well as which are the most common experimental errors and how to avoid them. Many desirable characteristics (or undesirable, depending on the application) such as formation of aggregates, stability, and drug delivery efficiency, among others, that are related to the microstructure of the fluid may be fine-tuned by means of the appropriate rheological techniques. Specific cases will be discussed dealing with pharmaceutical, food, and cosmetic applications. It is the hope of the author that this discussion will convince the reader of the importance and pertinence of rheology, despite the somewhat intimidating mathematics involved.

II. BASIC CONCEPTS

A. Viscosity and Rheology

Viscosity is a transport property and denotes the conductivity of momentum through homogeneous medium. It can also be interpreted as the opposition to flow that all fluids exhibit under shear. To understand the concept of viscosity, let us describe the following experiment (1). Imagine a layer of fluid contained by two solid plates, as shown in Fig. 1. The plates have a contact area with the fluid denoted by A. At the beginning of the experiment, a force F is applied to the upper plate. Once steady conditions have been reached, the plate moves with a constant velocity, v_p. Simultaneously, momentum (mv_x, where m is mass, v is velocity, and subscript x refers to the flow direction) is transferred to the fluid under the plate inducing the former to flow. The layer of fluid next to the moving plate moves with a velocity equal to v_p, while the fluid next to the stationary plate is stationary (this is called the nonslip condition). The intermediate layers move as if they were sliding on each other, and velocity decays from the upper to the lower plate. Because of this flow configuration, a velocity gradient develops in the direction of the momentum transfer. The velocity gradient, or shear rate or shear strain, $\dot{\gamma}$, is expressed as

$$-\frac{dv_x}{dy} = \dot{\gamma} \tag{1}$$

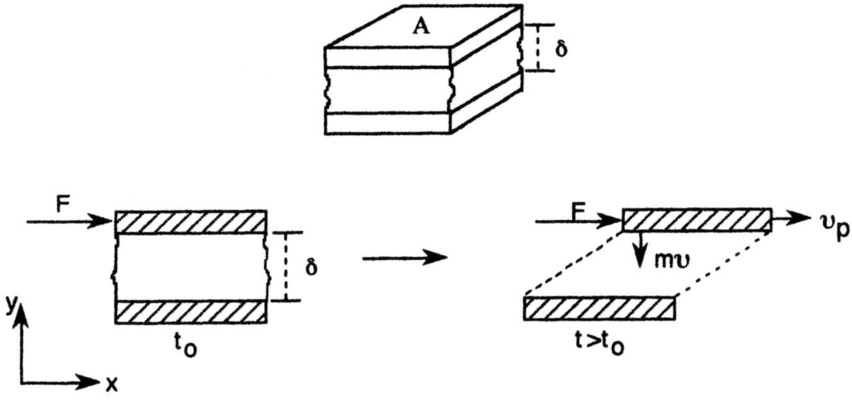

Figure 1 Transference of momentum between two parallel plates.

The ''amount'' of momentum transferred by unit area and unit time, or shear stress τ, may be written as

$$\frac{mv_x}{At} = \tau \tag{2}$$

Both quantities, shear stress and shear rate, are related in the following manner:

$$\tau = \mu\dot{\gamma} \tag{3}$$

where μ is viscosity. The quantities defined above, i.e., shear rate, shear stress, and viscosity, have units of s^{-1}, Pa, and Pa.s (SI units), respectively. Viscosity is also expressed in Poise, or P, and 10^{-2}P, or cP (1 cP is equal to 1000 mPa.s). Equation [3] is better known as Newton's law. Fluids that behave according to Eq. [3], or Newtonian fluids, exhibit a linear relationship between τ and $\dot{\gamma}$, as shown in Fig. 2. The latter is called flow curve.

The flow behavior of homogeneous, low molecular weight liquids (water, organic and inorganic oil, aqueous solutions, and diluted polymer solutions) is accurately described by Newton's law. Fluids that behave as described in Eq. [3] are called Newtonian. The viscosity of Newtonian fluids is dependent only on temperature. Nevertheless, most fluids that are not homogeneous (e.g., suspensions and emulsions) or have a high molecular weight (melted polymers and polymer solutions) do not behave as established by Newton law, and are called non-Newtonian. In fact, were it not for non-Newtonian fluids, the scientific discipline of rheology would not exist!.

Rheology is the study of flow and deformation of matter and this includes everything from elastic solids to Newtonian fluids (2). The former types of fluids

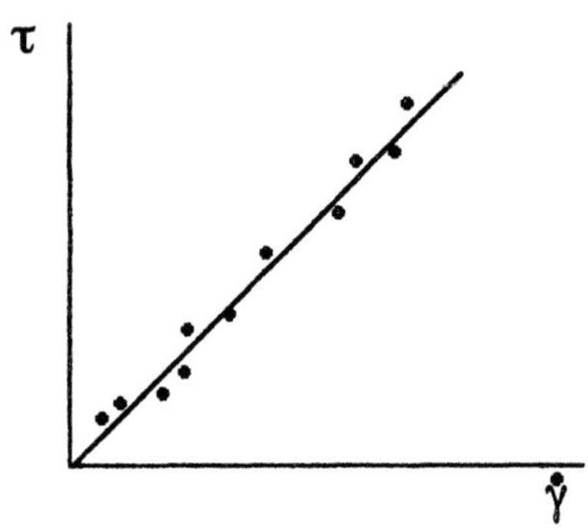

Figure 2 Typical flow curve for a Newtonian fluid.

are the two extremes of the spectrum of behavior that can be encountered in nature. At one extreme are elastic solids that obey Hook's law:

$$\tau = G\gamma \tag{4}$$

where G is the elastic modulus and γ is the relative length change (ratio of change in length to unchanged length) or strain. At the other extreme are viscous or Newtonian liquids. Many in-between materials may display both solid-like and fluid-like properties, depending on the deformation conditions and the time scales involved.

As an example, let us consider the case of mayonnaise which, when placed in a jar, looks like a solid. If the jar is tilted, it does not flow. If the jar is hit softly, the mayonnaise hardly wobbles. Yet anyone can dig a hole with a spoon and spreading with a knife is even easier. Another example of non-Newtonian fluid is beach sand. Hardening of the sand may be clearly perceived under the pressure of a foot, which first sinks a few millimeters with relative ease.

The next sections deal with the description of the most typical non-Newtonian behaviors and the mathematical models, or constitutive equations, that describe them. For complementary information on these topics, see Refs. 2 and 3.

B. Non-Newtonian Fluids

Suspensions and emulsions are non-Newtonian fluids. From a practical point of view, having to deal with non-Newtonian fluids means that viscosity is not enough to describe their response to deformation. This property no longer depends on temperature alone but on many other factors such as shear rate, shear stress, strain, time, and measuring geometry. Under shear conditions, the flow response is now called apparent viscosity and it is defined as

$$\eta = \frac{\tau}{\dot{\gamma}} \tag{5}$$

where η is apparent viscosity. Equation [5] does not imply a linear relationship between shear stress and shear rate. It only asserts that a given shear rate will bring about a given shear stress (or the other way around) and the viscosity for this given shear rate is calculated as in Eq. [5]. This idea is illustrated in Fig. 3, which is a flow curve corresponding to a shear-thinning fluid. It can be seen that the slope of the curve flow decreases as shear rate progresses, hence η diminishes. Apparent viscosity can also be calculated as $\eta = t_g\alpha$.

Only diluted dispersions behave according to Newton's law, provided the suspended particles do not interact. Particle interactions are the main basis of non-Newtonian bahavior and the extent of these interactions will dictate how complex the flow behavior is. The flow characteristics derived may be classified

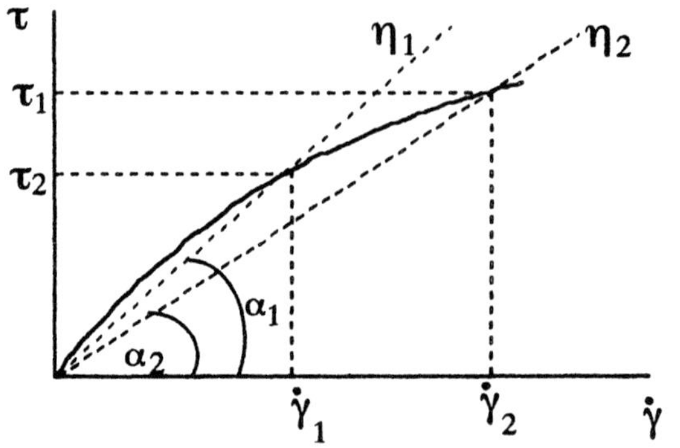

Figure 3 Schematic flow curve of a non-Newtonian fluid exhibiting a shear thinning behavior.

as shear thinning or shear thickening; viscoplastic (yield stress), thixotropic, or viscoelastic. Many materials display one, two, or all of the attributes listed above, depending again on the degree of particle interactions, the extent of deformation, and the time scales involved. In the following sections each type of behavior is described. In-depth discussion on the basis of non-Newtonian behavior will be left to Sec. III.

1. Shear-Thinning and Shear-Thickening Fluids

As indicated by their names, shear-thinning and shear-thickening fluids exhibit a decrease or increase of viscosity under shear. Most suspensions and emulsions display a shear-thinning viscosity and this is the most common behavior of non-Newtonian fluids.

Figure 4 depicts the typical flow curves that can be obtained for dispersions. A shear-thinning fluid displays a decreasing opposition to flow (Fig. 4a) and therefore its viscosity also tends to diminish (Fig. 4b). The opposite can be said of a shear-thickening suspension or emulsion.

An alternative classification for shear thinning and thickening is pseudoplasticity and dilatancy; however, these flow categories are rather limited to the so-called power law fluids. The flow behavior of a power law fluid may be described by the expression

$$\tau = k\dot{\gamma}^n \tag{6}$$

where n and k are constant parameters. If τ and $\dot{\gamma}$ are plotted in a log-log scale, a straight line is obtained. The slope of the log-log curve is the parameter n. If

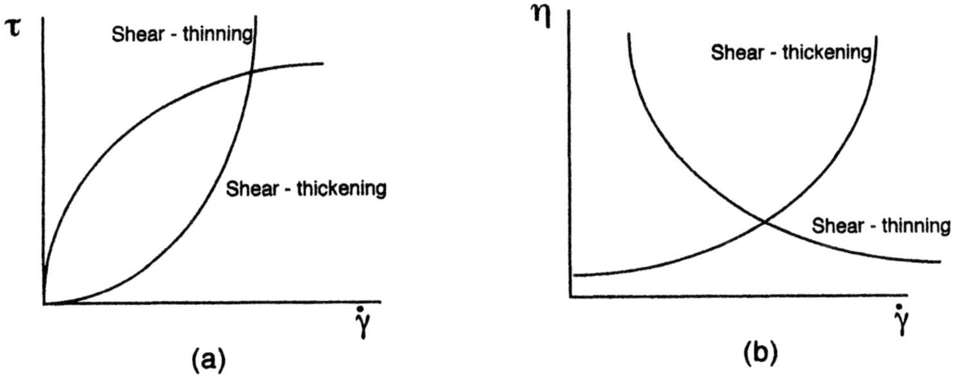

Figure 4 Typical curves of shear stress (a) and apparent viscosity (b) as a function of shear rate, for shear-thinning and shear-thickening fluids.

$n > 1$, the fluid is dilatant; if $n < 1$, the fluid is pseudoplastic. Newtonian fluids have an n value of unity.

Equation [6] can be written in terms of apparent viscosity (dividing by $\dot{\gamma}$ both sides of the equation), and

$$\eta = k\dot{\gamma}^{n-1} \tag{7}$$

It is worth mentioning that truly pseudoplastic and dilatant flow is very rare. Most fluids exhibit a linear relationship between log τ and log $\dot{\gamma}$ for only two or three decades of τ or $\dot{\gamma}$. Over a wider measuring range it is more often verified that n changes, especially for the case of concentrated suspensions and emulsions. These fluids are said to exhibit a structural viscosity.

Typically, a structural fluid displays shear-thinning behavior at intermediate shear rates, while being Newtonian at very low and very high shear rates. Fluids such as blood, polyacrylamide solutions, and aqueous latex show a behavior as depicted in Fig. 5.

An equation that represents this behavior is the Carreau model:

$$\eta(\dot{\gamma}) = \eta_\infty + \frac{\eta_0 - \eta_\infty}{[1 + (\lambda\dot{\gamma})^2]^p} \tag{8}$$

This model includes four parameters: the low-shear limiting viscosity, η_0; the high-shear limiting viscosity, η_∞; a shear index, p; and a time constant, λ. An interesting aspect of this model is that, after convenient simplifications, different typical flow models can be obtained. As an example, let us consider the case for intermediate shear rates, in which $\eta_\infty \ll \eta \ll \eta_0$ (η_∞ can be neglected) and the

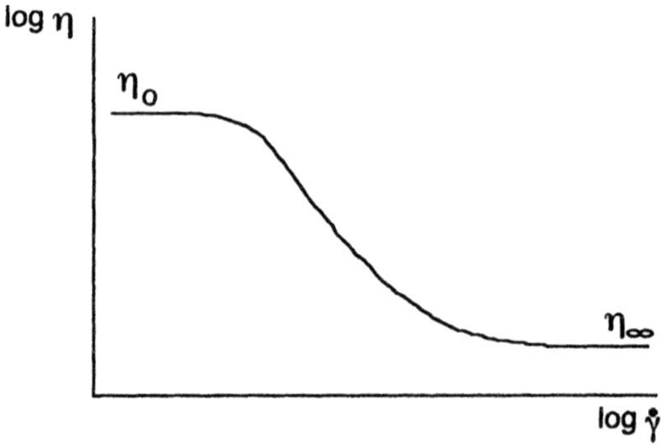

Figure 5 Behavior of fluids that exhibit a structural viscosity.

term $(\lambda\dot{\gamma})^2 \gg 1$. Setting $k = \eta_0/\lambda^{2p}$ and $n = (1 - 2p)$, and multiplying both sides of Eq. [7] by $\dot{\gamma}$, the power law model (Eq. [6]) is obtained.

The Bingham model for viscoplastic fluids (Eq. [10], see next section) is obtained when $\eta \ll \eta_0$ (intermediate to high shear rates), the term $(\lambda\dot{\gamma})^2 \gg 1$ and p is equal to 1/2. As a result,

$$\tau_0 = \frac{\eta_0}{\lambda} \quad \text{and} \quad \eta_\infty = \mu_\infty \tag{9}$$

Finally, if $\eta \ll \eta_0$, the term $(\lambda\dot{\gamma})^2 \gg 1$ and setting $n = (1 - 2p)$ and $k = \eta_0/\lambda^{2p}$, the Herschel-Bulkley model for viscoplastic fluids (Eq. [11], see next section) is attained.

2. Viscoplastic Fluids

The flow behavior of a viscoplastic fluid is identified by the appearance of a yield stress, i.e., the fluid flows in a viscous manner only after a threshold has been exceeded. Below this threshold, or yield stress, the behavior of the fluid is similar to an elastic solid and should obey Eq. [4] when subjected to a strain or stress sweep. The simplest type of viscoplastic fluid is the so-called Bingham plastic, and its behavior can be expressed by means of the following mathematical model:

$$\tau = \tau_0 + \mu_\infty\dot{\gamma} \tag{10}$$

Where τ_0 is the yield stress and μ_∞ is the shear limiting or plastic viscosity. A Bingham plastic displays a behavior as depicted in Fig. 6.

A more complex viscoplastic behavior may be modeled by means of the Herschel-Bulkley equation:

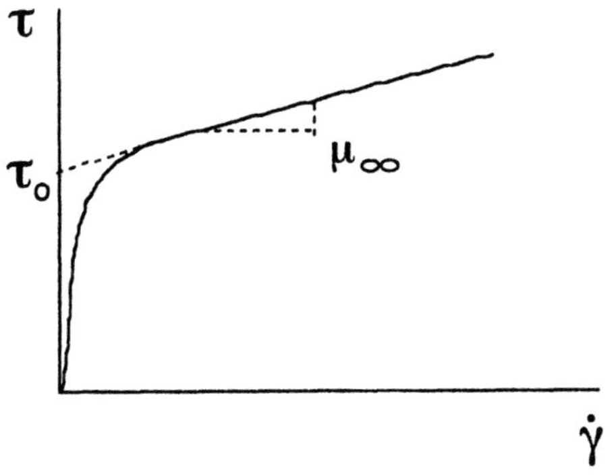

Figure 6 Flow curve of a viscoplastic, Bingham-type fluid.

$$\tau = \tau_0 + k\dot{\gamma}^n \tag{11}$$

The latter combines the first elastic response with a power law behavior (generally shear thinning) after $\tau > \tau_0$. In general, it is found that if the shear rate measuring range is rather low (but large enough to exceed τ_0), or if the study of the low shear range is required, Eq. [11] is usually most appropriate. For a wide measuring range, Eq. [10] is more convenient.

An additional well-known mathematical expression for viscoplastic fluids is the Casson model:

$$\sqrt{\tau} = \sqrt{\tau_0} + \sqrt{\mu_\infty \dot{\gamma}} \tag{12}$$

The most adequate model for a viscoplastic fluid will depend on the fluid response to deformation and how well the experimental data fit the model. Well-behaved viscoplastic fluids will most likely adjust to any of the former three expressions. However, concentrated or gelled suspensions and emulsions often exhibit a more complex behavior, as depicted in Fig. 7. It can be observed that there is not a well-defined yield, since a maximum may be obtained for low shear rates. Three different yields have been identified for this type of behavior (14): a static yield stress or spur value τ_s; a dynamic yield τ_d; and an engineering (or practical) yield τ_i (see Fig. 7). The static yield stress or spur value generally appears when the sample is subjected to shear for the first time; subsequent deformations will not show the static yield. In fact, the static yield is found usually after prolonged storage and disappears afterward. The rheological behavior of fluids exhibiting one or several yields are very likely thixotropic and viscoelastic (see the next sections), and their behavior depends on the history of previous deformation, i.e., on how the sample is manipulated prior to the rheological measurement.

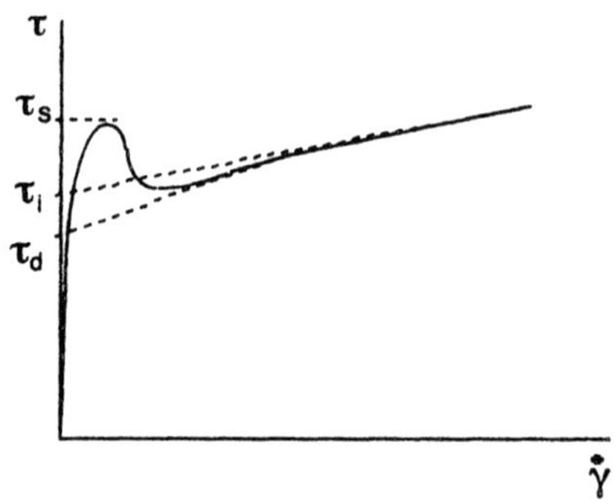

Figure 7 Schematic flow curve of a complex viscoplastic fluid displaying a static and a dynamic yield stress.

3. Thixotropic Fluids

The rheological behavior of thixotropic fluids exhibits a pattern that is time-dependent, i.e., the flow response and viscosity vary depending on how long or on what conditions the fluid is sheared. A typical thixotropic curve is shown in Fig. 8. This figure depicts the flow behavior of a fluid subjected to subsequent cycles of shearing. Hysteresis loops are obtained indicating that the flow resistance changes as a function of the flow history. If a hysteresis loop is obtained in which the flow resistance diminishes as the shearing loop progresses, the fluid is said to exhibit ''positive'' thixotropy. Many shear-thinning fluids are also thixotropic. The opposite behavior is called ''negative'' thixotropy or antithixotropy. Shear

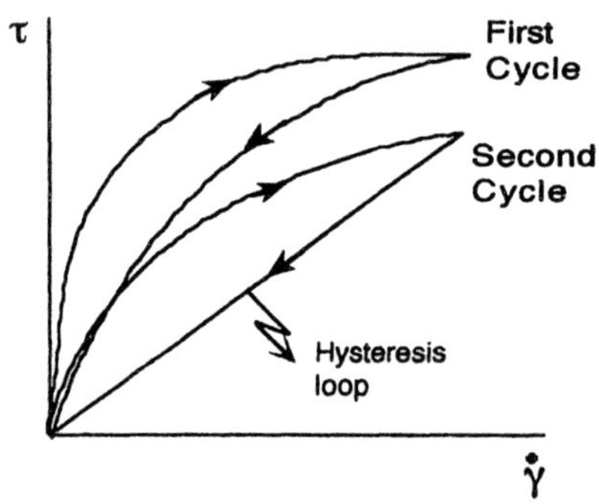

Figure 8 Typical hysteresis loops for a thixotropic fluid during two consecutive shear rate runs.

thickening is often associated with negative thixotropy. Another category of thixotropic fluids includes rheopexy. A rheopectic fluid exhibits an increase of viscosity under *low* shear rates.

Many thixotropic fluids are not so ''well behaved'' as shown in Fig. 8 and may display complex behaviors, most often not reproducible, as shown in Fig. 9. The latter figure suggests that complex thixotropic fluids may exhibit a yield stress, or several yields as depicted in Figs. 7 and 9. Yield stress of thixotropic fluids often vanishes after consecutive shearing loops.

The thixotropic behavior of suspensions and emulsions is a rather difficult property to measure, and this is further complicated by the contribution of viscoelastic behavior (see next section). The main reason for this difficulty lies in the nature of the phenomenon involved. Thixotropy arises because of the existence of interparticle forces that produce three-dimensional microstructures called *flocculates* or *aggregates* (see Sec. III). Depending on the magnitude of these forces, these microstructures are more or less prone to destruction by shearing. In consequence, any manipulation of a thixotropic sample may induce structural breakdown, thereby changing, most often irreversibly, the viscosity and yield stress of the sample.

The response of a thixotropic fluid to stepwise changes of shear rate is illustrated in Fig. 10. At low shear rates, the rate of viscosity reduction is much more gradual than at higher rates. This phenomenon is again associated with the degree of structural breakdown induced by shear. High shear rates are more likely to provide enough energy to overcome interparticle forces and, in consequence, are more efficient in microstructure breaking.

Despite the difficulties, one may be interested in evaluating thixotropy as a measure of the degree of microstructure that has built up. The latter is proportional to the area of the hysteresis curve shown in Fig. 8. A good account on thixotropy and modeling of thixotropic behavior is found in Ref. 5.

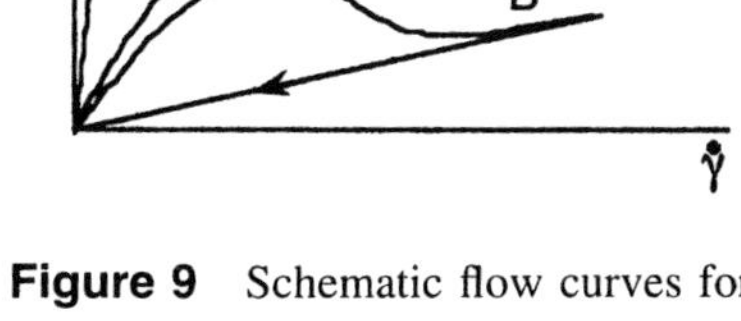

Figure 9 Schematic flow curves for complex thixotropic fluids having a yield stress.

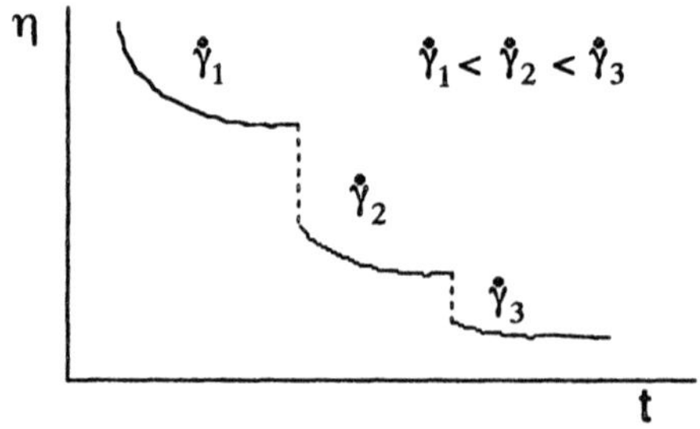

Figure 10 Time-dependent viscosity of thixotropic fluids undergoing a stepwise increase of shear rate.

4. Viscoleastic Fluids

The word *viscoelastic* encompasses many fluids that exhibit both elasticity (solid-like behavior) and flow (liquid-like behavior) when sheared. Most concentrated pastes, emulsions, and gels are viscoelastic. Under small deformations, viscoelastic fluids literally behave as elastic solids; under higher deformations they flow as liquids.

An example of this behavior is jelly. The gravitational pull does not deform a jelly and it behaves as a solid. If the jelly's container is gently stroked, the jelly wobbles, just like an elastic solid. The jelly has an internal structure capable of storing energy that is used to recover its original configuration when the perturbation is small. At last, if jelly is vigorously mixed it flows exhibiting a viscosity during and after destruction of the gel structure.

The simplest description of ideal viscoleastic behavior (or linear viscoelastic behavior) of viscoelastic liquids is the Maxwell model. It consists of an elastic spring and a dashpot in series, as shown in Fig. 11. This configuration is called a Maxwell element. The spring represents the elastic response of the fluid and the dashpot the viscous one. If a strain is applied, the first to respond is the spring, followed by the movement of the dashpot. If the strain ceases, the spring, which has stored energy, recoils to its original size. The dash pot, which has dissipated the energy, cannot recover.

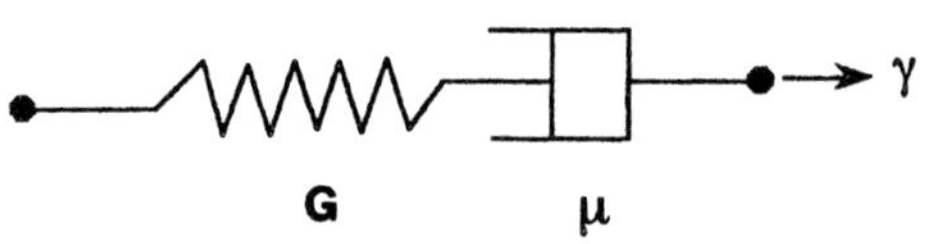

Figure 11 Maxwell model for a viscoelastic fluid.

For a simple viscoelastic fluid, the latter behavior may be described mathematically in terms of the stress and the shear rate:

$$\tau + \lambda \frac{d\tau}{dt} = \eta_0 \dot{\gamma} \tag{13}$$

where λ is a rate constant called the relaxation time. The term η_0 is also equal to the product $G_0\lambda$, where G_0 is the elastic modulus. More complex viscoelastic liquids may be described using several Maxwell elements in parallel.

The viscoelastic behavior is evaluated by means of two types of methods: static tests and dynamic tests. In the first category a step change of stress or strain is applied and the stress or strain response is recorded as a function of time. Stress relaxation, creep compliance, and creep recovery are static methods. The dynamic tests involve the imposition of an oscillatory strain or stress. Every technique is described in the following sections.

5. Stress Relaxation

The stress relaxation experiment consists of applying a constant relative strain, γ_0, and recording the shear stress response, as shown in Fig. 12. Typical relaxation curves for elastic solids, viscoelastic solids or liquids, and Newtonian liquids are shown in the latter figure. The elastic solid is able to store energy and, in consequence, can maintain deformation and does not relax under the applied strain. The other extreme is the Newtonian flow that relaxes completely and flows.

The in-between behavior corresponds to the viscoelastic fluid, which relaxes exponentially according to

$$\tau = \eta\gamma_0[1 - \exp(-t/\lambda)] \tag{14}$$

Viscoelastic liquids relax to a value of zero while viscoelastic solids relax to a constant value, τ_e, or equilibrium stress.

The relaxation modulus, $G(t)$, may be written as

$$G(t) = \frac{\tau(t)}{\gamma} \tag{15}$$

When the relaxation modulus is only a function of time, the behavior is called linear viscoelastic. Generally, the latter is true below a critical value of strain, γ_c. The former equation is only valid for strains below γ_c. Beyond this threshold the behavior is called nonlinear viscoelastic.

The exponential decay of $G(t)$ is expressed as a function of the G_0 and λ:

$$G(t) = G_0 e^{-t/\lambda} \tag{16}$$

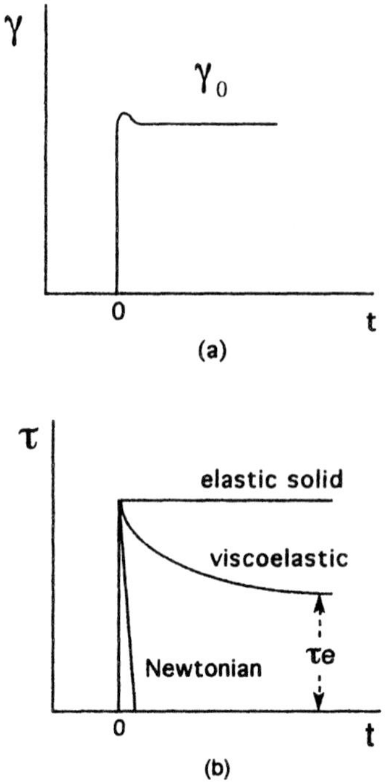

Figure 12 Graphic description of a stress relaxation experiment: (a) applied strain γ_0; (b) stress response according to material type.

Simple viscoelastic fluids show a single relation time but more complex fluids may present various relaxation times, i.e.,

$$G(t) = \sum_{k=1}^{n} G_k e^{-t/\lambda_k} \tag{17}$$

If the flow behavior can be described by means of several Maxwell elements in parallel, G_k and λ_k correspond to the elastic modulus and relaxation time of the kth Maxwell element.

6. Creep Compliance

In this type of experiment, a constant stress τ_0 is applied for a period of time t_0, as shown in Fig. 13a, and the subsequent deformation or strain is recorded as a function of time. The response of elastic solids and viscoelastic fluids is displayed in Fig. 13b. The elastic solid responds with a short-term deformation and, once

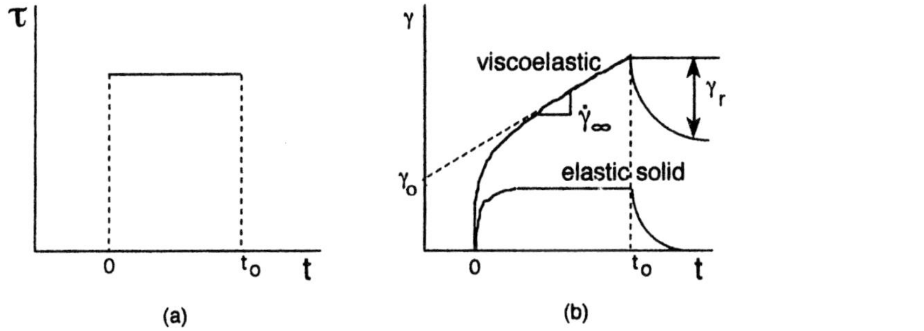

Figure 13 Graphic description of creep compliance experiment followed by the creep recovery response: (a) applied shear stress τ_0; (b) material deformation response.

the stress is removed, the elastic solid recovers completely its original configuration. The viscoelastic fluid deforms continuously until it reaches a steady rate of strain, $\dot{\gamma}_\infty$, and, after removal of the stress, exhibits only a small recovery, γ_r. Most energy is lost through viscous dissipation or flow.

Usually, the data in Fig. 13 are represented in terms of the creep compliance parameter, $J(t)$, which is

$$J(t) = \frac{\gamma(t)}{\tau_0} \tag{18}$$

$J(t)$ has units of $1/G$, although they are not equivalent for most cases. With this change of variables, Fig. 14 may be drawn. For steady creeping regime, i.e., when the slope of the curve becomes constant,

$$J(t) = \frac{\gamma_0}{\tau_0} + \frac{t\dot{\gamma}_\infty}{\tau_0} \tag{19}$$

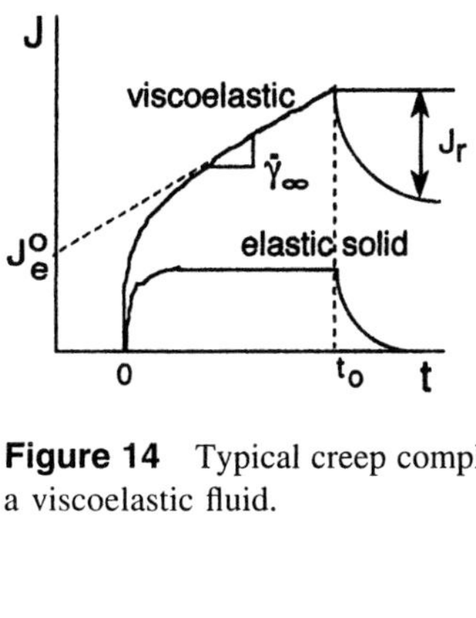

Figure 14 Typical creep compliance and creep recovery curve for an elastic solid and a viscoelastic fluid.

The term γ_0/τ_0 is denoted as J_e^0, the steady state creep compliance, and $\dot{\gamma}_\infty/\tau_0$ is the inverse of η_0, the viscosity in the low-shear rate limit. Note that an elastic solid, after the initial deformation, has a slope equal to zero ($\dot{\gamma}_\infty \to 0$; $\eta_0 \to \infty$), meaning that the solid does not flow.

Creep behavior of simple viscoelastic fluids showing a single relaxation time may be expressed as

$$J(t) = \frac{1}{G_0} + \frac{t}{\eta_0} = \frac{1}{G_0} + \frac{t}{\lambda G_0} \tag{20}$$

However, a single λ may not account for the material behavior; therefore, $J(t)$ can be written as

$$J(t) = \sum_k J_k(1 - e^{-t/\lambda_k}) + \frac{t}{\eta_0} \tag{21}$$

This expression allows for several relaxation times (λ_k) and can model the behavior of complex materials.

7. Creep Recovery

Creep recovery experiments measure the capacity of the fluid to recover its original configuration, i.e., how more elastic than viscous a fluid is. In this sense, the portion of the curve after stress removal (see Fig. 14) is analyzed. An ideal elastic solid reverts the deformation completely (recoils), while a Newtonian fluid does not recover at all. For a viscoelastic fluid, the creep recovery function can be defined as

$$J_r(t) = \frac{\gamma_r(t)}{\tau_0} \tag{22}$$

where γ_r is the strain recovered after stress removal.

Creep recovery experiments can only be performed if a steady-state creep compliance, i.e., $\dot{\gamma}(t) = \dot{\gamma}_\infty$ (constant slope), has been reached. The last is true when it is verified that

$$\lim_{t \to \infty} J_r(t) = J_e^0 \tag{23}$$

This is valid when equilibrium conditions are attained during the creep recovery experiment.

8. Oscillation Experiments

An oscillation experiment consists of submitting a fluid to a sinusoidal strain or stress. Mathematically, a sinusoidal strain can be written as

$$\gamma = \gamma_0 + \sin(\omega t) \quad (24)$$

where γ_0 is the amplitude of the strain wave and ω is the frequency of oscillation. The corresponding stress response is

$$\tau = \tau_0 \sin(\omega t + \delta) \quad (25)$$

where δ is the phase angle. The behavior described by Eqs. [24] and [25] is depicted in Fig. 15. The phase angle δ is zero for elastic solids and 90° for Newtonian fluids. A viscoelastic fluid exhibits a phase angle between zero and 90°.

This type of behavior is well described by means of complex numbers, and the stress response can be decomposed in two components, an in-phase component and an out of-phase component. In this sense, a complex elasticity modulus may be defined as

$$G^* = G' + iG'' \quad (26)$$

The in-phase signal, G', is called the storage modulus, or energy stored per cycle, and it is calculated as

$$G' = \frac{\tau_0}{\gamma_0 \cos \delta} \quad (27)$$

The out of-phase signal, G'', is called the loss modulus, or energy dissipated per cycle, and it is expressed as

$$G'' = \frac{\tau_0}{\gamma_0 \sin \delta} \quad (28)$$

Note that $\tan \delta = G''/G'$

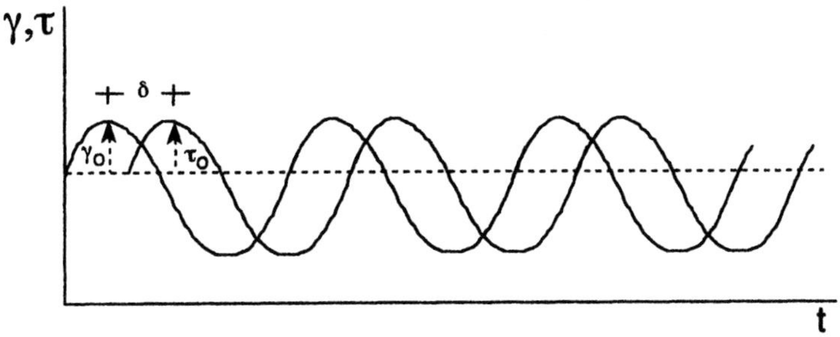

Figure 15 Graphic description of an oscillatory experiment showing the applied strain signal and the shifted stress response.

A complex apparent viscosity, η^*, can be defined as

$$\eta^* = \frac{G''}{\omega} - i\frac{G'}{\omega} = \eta' - i\eta'' \tag{29}$$

where η' is called the dynamic viscosity and η'' is the elastic component of η^*. It is worth noting that the yield stress can be estimated as $G^*\gamma_C$. G^* is the complex modulus at the limit strain γ_C, or critical strain in the linear viscoelastic region.

C. Characterization of Rheological Behavior

The following section deals with the description of instruments, or rheometers, that are most commonly used to evaluate the rheological behavior. A rheometer measures the stress developed by the material under a given deformation or the deformation history of a material under a given stress. In a rheometer, the relationship between τ and γ, or τ and $\dot{\gamma}$, is evaluated, often as a function of time, frequency (ω); and temperature. Then the experimental data are fitted to a convenient model or constitutive equation, e.g., Eqs. [5]–[29], and the corresponding material functions (η, τ_0, G', G^*, etc.) and associated parameters (e.g., n, k, λ, δ) are derived as a result.

There are different kinds of rheometers and the simplest is a viscometer, which permits measurement of the shear viscosity of a fluid, $\eta(\dot{\gamma})$. In principle, the rheometric measurement is valid only when there exists a precise expression for the relationship between shear stress and shear rate. If this relationship is not known precisely for a given geometry, the results can only be used for comparison purposes. These types of instruments are called *indexers*.

Rheometers may be classified depending on the sensor geometry and the flow configuration. In this sense, they can be divided in drag rheometers and pressure-driven rheometers. In the first class of instruments, the material is deformed by the drag induced by the movement of a solid boundary; the most common geometries that use this principle are concentric cylinders, cone-and-plate, and parallel plate systems. Falling-ball and rolling-ball, which operate under this principle, do not classify as rheometers (they are indexers), yet they allow for a good assessment of viscosity. In the second class of instruments (pressure-driven), the material is forced to flow by means of a pressure difference across a conduit, capillary, or pipe.

The choice of the most adequate rheometer or indexer depends on the application. This is discussed in Sec. IV. The following sections are devoted to description of the most established instruments for rheological evaluations.

1. Concentric Cylinders

Concentric cylinders are a preferred configuration for the rheological evaluation of suspension and emulsions. They are relatively easy to use and less expensive than most sensor tools. They consist of two cylinders placed one inside the other, an inner cylinder of radius R_i and an external or outer cylinder of radius R_e, as shown in Fig. 16. Either the internal or external cylinder, or both, rotates.

The sample is contained in the annulus or gap between the cylinders, $R_e - R_i$. When, say, the inner cylinder rotates at an angular velocity of Ω, the fluid opposes deformation and produces a torque T, against the direction of flow. The shear stress (if measured on the inner cylinder) is calculated as

$$\tau = \frac{T}{2\pi R_i^2 L} \tag{30}$$

where L is the length of the cylinder in contact with the sample. If the torque is measured on the outer cylinder, the term R_i in Eq. [30] is replaced by R_e. The exact expression for the shear rate is

$$\dot{\gamma} = r\frac{d\Omega}{dr} \tag{31}$$

The derivative in Eq. [31] has to be evaluated to obtain a precise relationship between τ and $\dot{\gamma}$. If the gap is narrow, i.e., $R_i/R_e > 0.97$, the shear rate reduces to

$$\dot{\gamma} = \frac{\Omega(R_e + R_i)}{2(R_e - R_i)} \tag{32}$$

Some commercial sensors meet this criterion, $R_i/R_e > 0.99$. However, they are only adequate to homogeneous or colloidal fluids, which have particles below

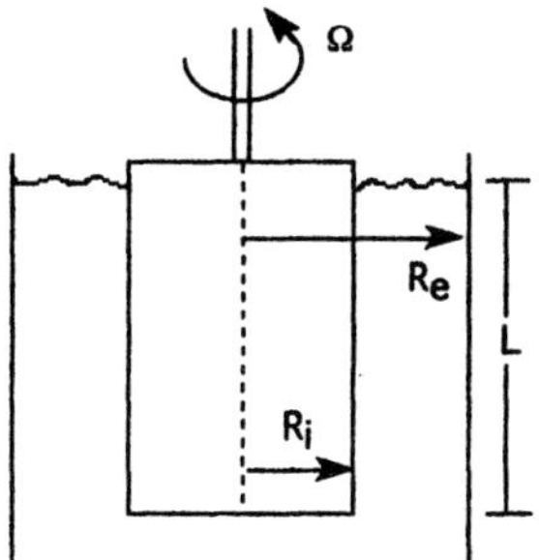

Figure 16 Concentric cylinders sensor tool.

the submicrometer range. Most commercial sensor present a $R_i/R_e < 0.99$; in this case, the shear rate is no longer homogeneous and for $0.5 < R_i/R_e < 0.99$,

$$\dot{\gamma}(R_i) = \frac{2\Omega}{n\left(1 - \dfrac{R_i^{2/n}}{R_e}\right)} \tag{33}$$

when the internal cylinder rotates. The latter is true if the shear rate in the gap can be modeled by the power law (see Eq. [6]). If the external cylinder rotates,

$$\dot{\gamma}(R_e) = \frac{-2\Omega}{n\left(1 - \dfrac{R_i^{2/n}}{R_e}\right)} \tag{34}$$

The term n in Eqs. [33] and [34] is

$$n = \frac{\mathrm{d} \ln T}{\mathrm{d} \ln \Omega} \tag{35}$$

For power law fluids n is a constant, and for Newtonian fluids n is equal to 1. The shear strain across the annulus for narrow gaps is

$$\gamma = \frac{\theta(R_e + R_i)}{2(R_e - R_i)} \tag{36}$$

where θ is the angular displacement. The more general case (wide gaps and more complex materials besides power law fluids) requires Eq. [31] to be solved for different flow configurations. The reader is recommended to consult specific literature on the topic (2,3).

2. Cone and Plate

A schematic representation of the cone-and-plate sensor tool is presented in Fig. 17. It consists of a lower stationary plate and an upper rotating cone, with a small angle α. The sample is contained in the space between the cone and the plate.

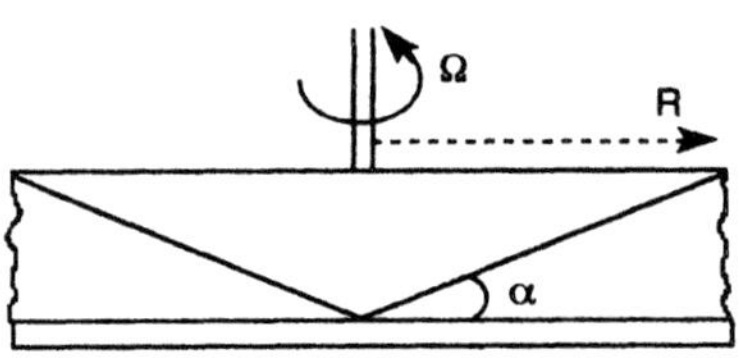

Figure 17 Cone-and-plate sensor tool.

The measuring principle is the same as for cylindrical cylinders and the shear stress is

$$\tau = \frac{3T}{2\pi R^3} \tag{37}$$

where T is the measured or imposed torque and R is the cone radius. If $\alpha < 4°$, the shear rate across the gap may be considered constant and equal to

$$\dot{\gamma} = \frac{\Omega}{\alpha} \tag{38}$$

and the strain is

$$\gamma = \frac{\theta}{\alpha} \tag{39}$$

where θ is the cone angular displacement.

Given that the shear rate is constant across the gap, the use of cone and plate is very convenient since the shear rate calculation is simple to perform. Apparent viscosity can be calculated straight from Eqs. [37] and [38] (as $\tau/\dot{\gamma}$), no corrections being needed. However, this tool is not convenient for suspensions and emulsions containing particles or droplets larger than the submicrometer size (see Sec. IV), being more adequate for polymers and colloid dispersions.

3. Parallel Plates

The parallel plate system is depicted in Fig. 18. It consists of two parallel disks separated by an adjustable distance h. The sample is contained between the plates, while the upper plate rotates. For Newtonian fluids, the measured torque is related to the Newtonian shear stress, τ_N, by means of the following expression,

$$\tau_N = \frac{2T}{\pi R^3} \tag{40}$$

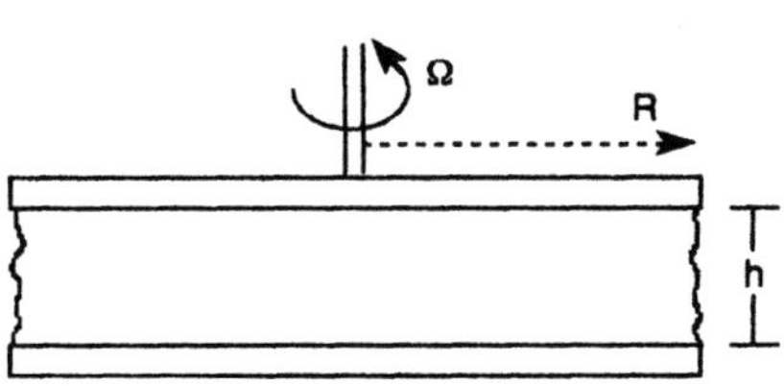

Figure 18 Parallel plates sensor tool.

which is the shear stress for $r = R$, the plate radius. The shear stress for the more general case, or non-Newtonian fluid, is

$$\tau = \tau_N \frac{1}{4} \left| 3 + \frac{d \ln T}{d \ln \dot{\gamma}_R} \right| \tag{41}$$

where

$$\dot{\gamma}_R = \frac{\Omega R}{h} \tag{42}$$

is the shear rate corresponding to the radius R.

The shear rate is not homogeneous across h and varies with the radius, attaining its maximum value in $r = R$. The corresponding strain is calculated as

$$\gamma = \frac{\theta r}{h} \tag{43}$$

Despite the fact that shear rate is not homogeneous across the distance h, requiring longer calculations, the parallel plate configuration is very adequate to perform dynamic measurements (creep and oscillation) on suspensions and emulsions. The distance h can be adjusted according to the dispersion particle size.

4. Capillary or Tube Viscometer

The capillary viscometer is one of the simplest and most accurate tools for measuring viscosity. It is also the most adequate geometry when it is required to simulate a process involving pumping, draining, or extruding. The measuring section of a tube or capillary consists of a cylindrical conduit of radius R and length L. The sample is driven through the tube by a pressure difference or by gravity, depending on the viscometer type. The working equations for a capillary or tube viscometer are rather simple. The shear stress is calculated at $r = R$ (wall shear stress or τ_w):

$$\tau_w = \frac{R\Delta P}{2L} \tag{44}$$

where ΔP is the pressure drop along the capillary of radius R and length L. If the flow is solely driven by gravity,

$$\Delta P = \rho g L \sin \beta \tag{45}$$

where ρ is the fluid density, g is the acceleration of gravity, and β is the tilt angle of the capillary. The shear rate for a Newtonian fluid, $\dot{\gamma}_{wN}$, calculated on the wall, is

$$\dot{\gamma}_{wN} = \frac{4Q}{\pi R^3} \tag{46}$$

where Q is the volumetric flow rate or volume flowing per unit time. For a non-Newtonian fluid,

$$\dot{\gamma}_{w} = \frac{\dot{\gamma}_{wN}}{4}\left[3 + \frac{\mathrm{d}\ln Q}{\mathrm{d}\ln \Delta P}\right] \tag{47}$$

The most used capillary viscometers are gravity-driven and correspond to the Canon-Fenske, Ostwald, and Ubbelohde types; Fig. 19 shows a scheme of an Ostwald capillary. In order to obtain viscosity, it is only necessary to measure the time t required for a given volume V of fluid to pass between the two marks, L_1 and L_2. Thus, the Newtonian viscosity is

$$\mu = \frac{\pi R^4 \Delta P t}{8LV} \tag{48}$$

It is worth noting that the former type of viscometers (Ostwald and similar) are not rheometers and qualify as indexers. In this sense, they are only precise for Newtonian fluids.

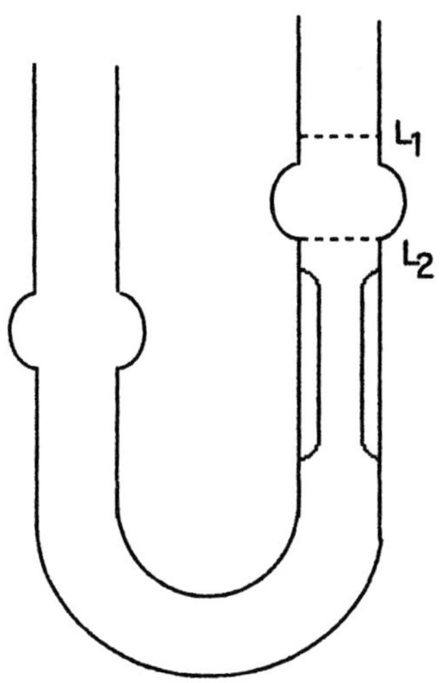

Figure 19 Capillary viscometer of the Ostwald type.

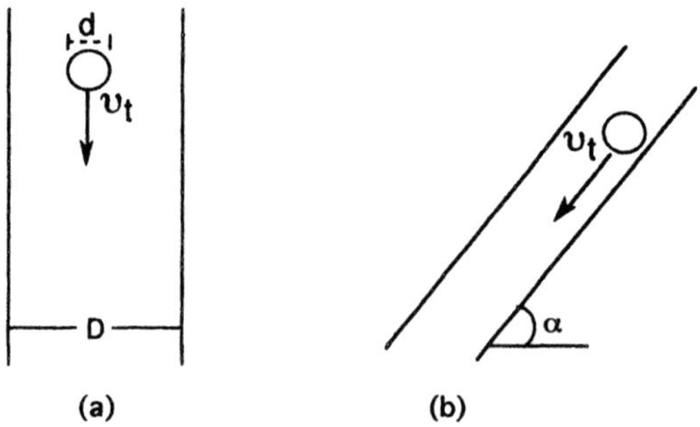

Figure 20 Schematic representation of (a) a falling-ball viscometer and (b) a rolling-ball viscometer.

5. Falling- and Rolling-Ball Viscometers

The falling-ball and the rolling-ball viscometers, shown in Fig. 20, are also simple devices for viscosity measurement. However, they do not qualify as rheometers but as indexers.

The measuring procedure is simple. The time t that a ball of diameter d falls a distance L in a fluid, contained in a cylinder of diameter D, is measured. Apparent viscosity is calculated as

$$\eta = \frac{d^2(\rho_b - \rho_f)g \sin \beta}{18 \frac{L}{t}} \tag{49}$$

where ρ_b and ρ_f are the ball and the fluid density, respectively; β is the tilt angle (90° for the falling-ball device) and g is the acceleration of gravity. The term L/t is the terminal or steady velocity, v_t, of the ball.

Given the sample density, Eq. [49] produces a good estimation of viscosity for Newtonian fluids, provided $D/d < 10$ for the falling-ball configuration, and for $D/d \ll 10$ for the rolling-ball. However, the results are not accurate therefore should only be used for comparison purposes. Note that the sample density, ρ, is necessary for the viscosity calculation. Otherwise, it is the kinematic viscosity, $\upsilon = \mu/\rho$, which is directly measured.

II. RHEOLOGY OF SUSPENSIONS AND EMULSIONS

The concept of viscosity is based on the assumption that the fluid is homogeneous. Therefore, it can be treated as a continuum. However, suspensions and emulsions exhibit a behavior that may strongly deviate from the one observed on a homogeneous fluid, due to the presence of different phases and interfaces. The rheological behavior may evolve from Newtonian for dilute systems of noninteracting particles, to complex non-Newtonian (combined yield stress, thixotropy, and viscoelasticity) for dilute systems of interacting particles or concentrated dispersions.

The properties of the solid–liquid or liquid–liquid interfaces are also important, often so important that their influence overshadows any other factor. The contribution of hydrodynamic driven phenomena such as slip flow, secondary flow, edge and end effects, viscous heating, and inertia may also play an important role (see Sec. IV). Good experimental and calculation procedures should ensure either that these factors are absent or that the data are corrected to eliminate their contribution. These will be discussed in Sec. IV. In the following sections, the main physicochemical factors that influence rheological behavior and viscosity are discussed. For the sake of clarity, a distinction will be made between the factors that are related to physical properties such as composition and particle size, and physicochemical aspects, especially interfacial properties.

A. Physical Factors

The physical properties that influence rheological behavior are internal phase content; size, shape, and particle size distribution; viscosity and rheological behavior of the continuous phase; and temperature. For the case of emulsions two additional parameters, droplet deformability and viscosity of the dispersed phase, should also be considered.

1. Internal Phase Content

The internal phase content, usually denoted as ϕ, is the most important factor among the parameters listed above. It is generally defined as the ratio of volume of the internal phase to total volume of dispersion, expressed as fraction or percentage. Parameter ϕ may also be written in units of g/dL or weight fractions or percentages, but from a theoretical point of view, it is more correct to use volume fractions.

In the absence of physicochemical effects, the following rule of thumb may apply. The more concentrated the system, the more viscous the suspension or emulsion, and the more complex the rheological behavior. Figure 21 shows typical curves of viscosity as a function of shear rate for various ϕ values. Most suspensions and emulsions exhibit a shear-thinning behavior that becomes more pronounced as ϕ increases. This behavior is displayed by many types of dispersions such as aerosol suspensions, human blood, food oil-in-water emulsions, and yogurt (6–9), to mention a few examples. Both increasing viscosity and shear-thinning behavior as ϕ increases are illustrated in Fig. 22. The latter (Fig. 22) corresponds to apparent viscosity as a function of ϕ and shear rate for sunflower oil-in-water emulsions (8).

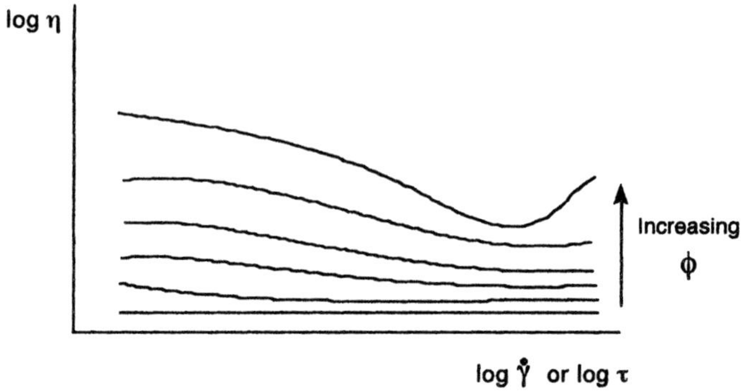

Figure 21 Typical curves of apparent viscosity as a function of the internal volume fraction, φ, for suspensions and emulsions.

Shear thickening is often observed in suspensions of solid particles after a shear thinning phase, especially in dispersions of nonspherical particles (10). It seems that at relatively low shear rates particles arrange into two-dimensional layers oriented in the direction of flow. These structures are often associated with shear-thinning behavior. After a critical shear rate, they become disrupted and

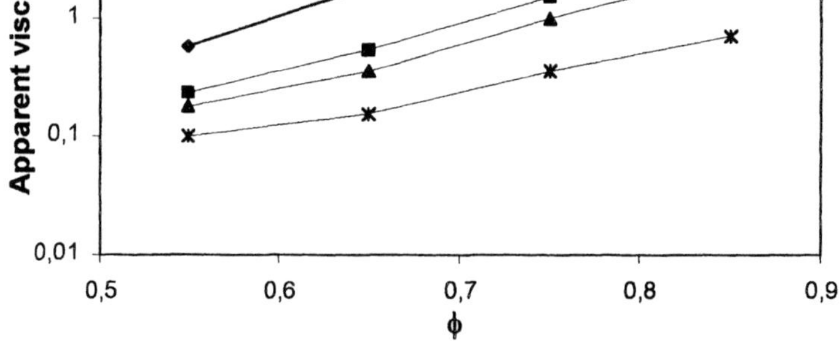

Figure 22 Apparent viscosity as a function of internal volume fraction and shear rate for sunflower oil-in-water emulsions. (Adapted from Ref. 8.)

the energy required to produce a new arrangement is translated into a viscosity increase (11). Nevertheless, shear thickening is still a poorly understood phenomenon.

Increasing ϕ may also increase viscoelastic behavior and the yield stress value. Figure 23 displays values of storage modulus G' as a function of clay concentration and pH, for sodium montmorillonite suspensions aged for 200 min (12). It can be observed that the higher the clay content, the higher G'.

There exists an equation based on theory, which relates the viscosity to the internal phase content. This is the well-known Einstein equation:

$$\eta_s = \eta_0(1 + 2.5\phi) \tag{50}$$

where η_s is the viscosity of the suspension and η_0 is the viscosity of the external or continuous phase. Equation (50) is exact for very dilute suspensions ($\phi <$ 0.02) of hard, noninteracting spheres and establishes that the viscosity of the suspending medium is increased by the hydrodynamic perturbation introduced by the presence of particles. In a dilute dispersion, particles are too far apart from each other to interact.

However, most practical cases involve concentrated systems, in which multiple particle interactions occur. Rotating doublets or triplets of particles may form temporarily, and more energy is dissipated as compared to the same number of single particles. The formation of permanent or temporary aggregates has another important effect and this is the entrapping of continuous phase, which amounts to an apparent increase of internal phase content. This apparently augmented internal phase content is defined as the effective volume fraction, ϕ_e. It is "effective" from the rheological point of view since "$1 - \phi_e$" refers to the continuous phase that is available to reduce particle interactions. The reader is

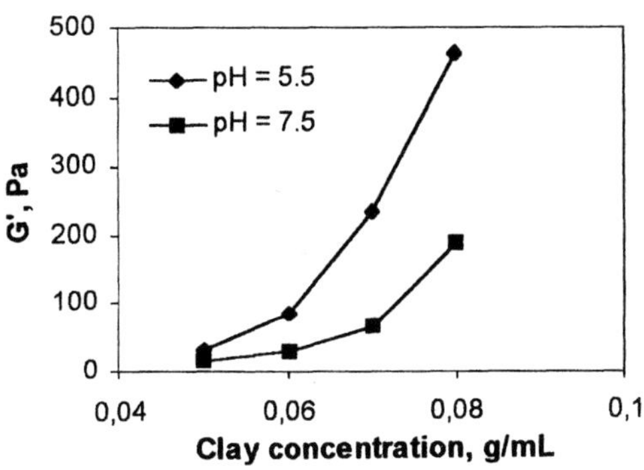

Figure 23 Elastic modulus, G', as a function of solid concentration and pH, for suspensions of clay. (Adapted from Ref. 12)

recommended to bear this concept in mind since most of phenomena that will be discussed in the next sections will involve the idea of an effective volume fraction.

Interactions of the first and second order may be described as (13)

$$\eta_s = \eta_0(1 + 2.5\phi + a\phi^2) \tag{51}$$

where a is the two-body interaction parameter. Note that the third term in Eq. [51] indicates an increase of viscosity due to the formation of doublets. Particle interactions of a higher order have resulted in expressions such as the semiempirical Thomas equation (13):

$$\eta_s = \eta_0[1 + 2.5\phi + 10.05\phi^2 + 0.00273\exp(16.6\phi)] \tag{52}$$

which should allow prediction of η_s up to ϕ of 0.6.

For small ϕ values, the third and fourth terms in parentheses are of little importance and Eq. [52] tends to Einstein's equation, which predicts a linear relationship for ϕ sufficiently low. As ϕ increases, the last term becomes more important, giving an exponential increase of viscosity. This behavior has been verified experimentally in many types of suspensions and emulsions (5,13–16).

A second expression that is often cited is the Krieger-Dougherty equation (14):

$$\eta_s = \eta_0\left(1 - \frac{\phi}{\phi_m}\right)^{-p[\eta]} \tag{53}$$

where p and $[\eta]$ are phenomenological parameters and ϕ_m is the so-called maximum packing fraction. The latter parameter (ϕ_m) denotes the maximum volume of particles that a dispersion may contain without deformation of the particles. For an ordered, cubical, or hexagonal close packing of monosized spherical particles, ϕ_m is 0.75 (15). Random packing of fairly monodisperse particles would yield a value of ϕ_m between 0.6 to 0.64 approximately (17).

It is worth noting that neither Eq. [52] nor Eq. [53] takes into account the dependency of viscosity with shear rate, which may be encountered for ϕ values from about 0.2 (in the absence of other type of interactions). In addition, the aforementioned equations do not take into account the influence of particle size and shape, or particle size distribution (see following sections). To date, the complexity of the problem has made it very difficult to obtain a good, all-encompassing theoretical or empirical model.

2. Particle Size

The viscosity of dilute dispersions is independent of particle size up to a value of ϕ about 0.4, as long as particle interactions do not intervene. In concentrated systems, particle size is important depending on the counterbalance of hydrodynamic and Brownian forces and on how rigid or soft the particles are.

A way to relate particle size of hard spheres (solid or colloidal particles, or very viscous drops) to shear rate and thermal energy (Brownian motion) is by means of a dimensionless parameter, the Peclet number (*Pe*), which is

$$Pe = \frac{\dot{\gamma}\eta_c d^3}{kT} \tag{54}$$

where η_c is the viscosity of the continuous phase, d is particle diameter, k is the Boltzman constant, and T is absolute temperature. If Pe is smaller than unity (small d or $\dot{\gamma}$ and large T), Brownian forces dominate the flow behavior and viscosity is independent of particle size. If Pe is large (large d or $\dot{\gamma}$ and small T), hydrodynamic forces dominate motion and again viscosity is independent of particle size. At intermediate Pe, viscosity increases with decreasing particle size.

Deformable or soft particles and droplets show the same behavior as rigid particles. However, the high shear rate viscosity of deformable particles is lower than that observed for rigid particles of the same internal phase content.

On the other hand, concentrated systems ($\phi > 0.7$) can be expected to develop higher yield stresses and viscoelasticity as droplet size decreases (18,19). Figure 24 shows how the storage modulus G' augments when increasing ϕ and decreasing droplet size, for water-in-oil emulsions of two different droplet sizes,

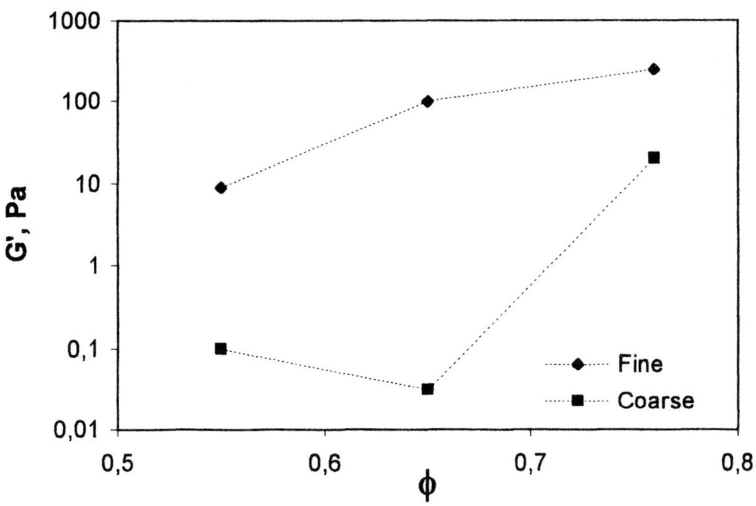

Figure 24 Elastic modulus, G', as a function of internal phase content for water-in-oil emulsions of two different droplet sizes, fine and coarse. (Adapted from Ref. 19.)

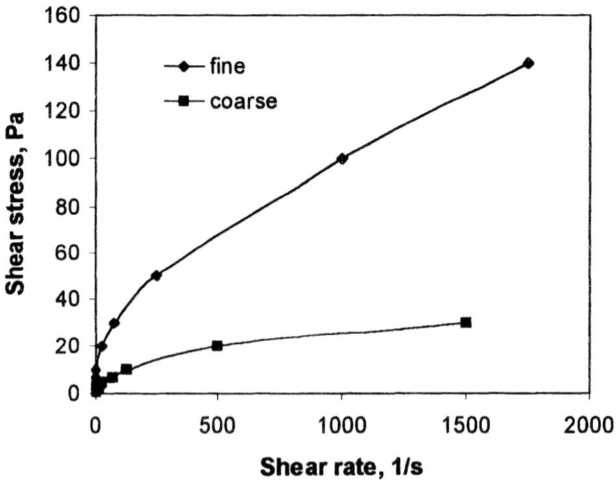

Figure 25 Flow curves of oil-in-water emulsions of equal internal phase volume (ϕ = 0.745) but different droplet size, fine and coarse. (Adapted from Ref. 19)

fine and coarse (20). Further, Fig. 25 displays curves of shear stress as a function of shear rate for oil-in-water emulsions exhibiting the same oil content (ϕ = 0.745) but two different droplet sizes, fine and coarse (20). It can be observed in the latter figure that the fine droplet emulsion presents a yield stress of about 35 Pa while the yield stress of the coarse emulsion is hardly perceivable. In addition, the fine emulsion produces higher shear stresses; hence, it is more viscous.

Particle size also affects the thixotropic behavior. Thixotropy is associated to the degree of microstructures building driven by Brownian motion and interfacial forces. As a consequence, dispersions of large particles aggregate at a smaller rate than dispersions of small particles.

3. Particle Shape

It seems that deviations from the spherical shape yield higher viscosity, as shown in Fig. 26 (10). In fact, ϕ_m is greatly decreased as shape departs from sphericity. Glass plates (100 × 400 μm) show a ϕ_m of 0,382 while for glass rods (30 × 700 μm) it is 0,268 (10). It can be seen in Fig. 26 that the behavior becomes increasingly shear-thickening, as particles diverge from sphericity. Non-spherical particles present a higher degree of contacts under the influence of a shear field, amounting to an increase of the effective volume fraction and, hence, of viscosity.

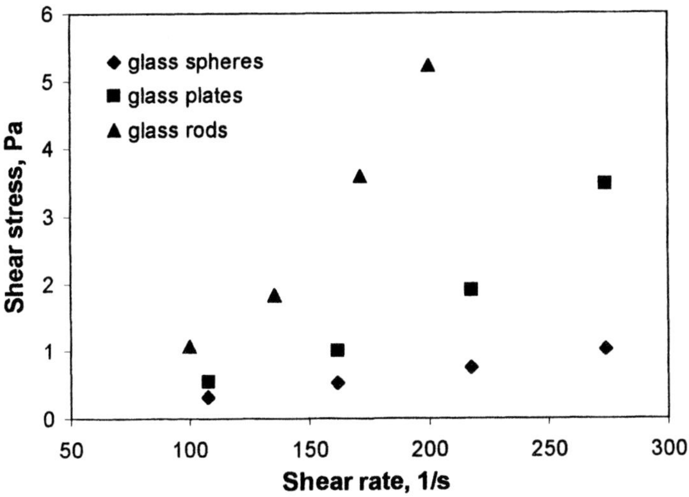

Figure 26 Flow curves of suspensions of glass particles of similar size but different shapes. (Adapted from Ref. 10.)

Strong shear-thinning behavior can also be verified for suspensions of non-spherical particles under shear. A typical case is fibers suspension. Non-sheared dispersed fibers may exhibit a disordered array that is hardly disturbed under low shear rates. However, as shear increases, fibers adopt a configuration (as shown in Fig. 27) that minimizes interactions, thus decreasing viscosity.

4. Particle Size Distribution

Viscosity is also a function of width of distribution. It appears that viscosity is reduced as polydispersity increases. Figure 28 depicts two particle size distributions that exhibit the same mean size; however, the viscosity of the wider distribu-

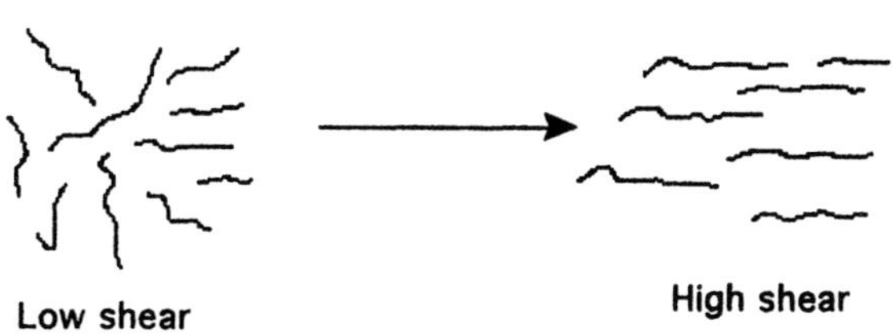

Figure 27 Fibers in a suspension that undergo a rearrangement from a random to a shear-oriented configuration.

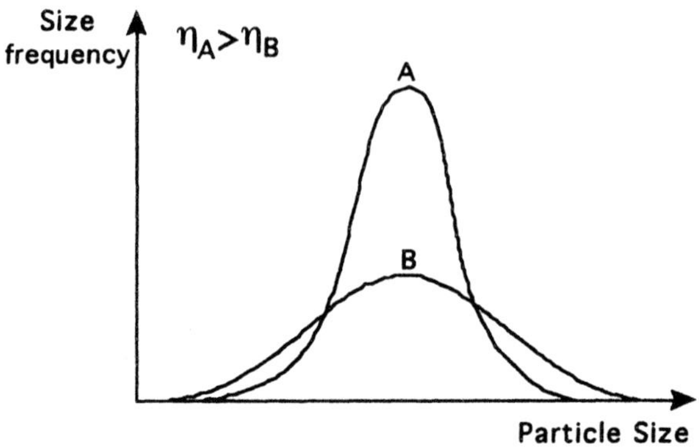

Figure 28 Size frequency diagrams for typical dispersions. Distributions A and B have similar mean and mode values but different widths.

tion is lower. Shape of distribution also affects viscosity in a significant way. Figure 29 shows drop size distributions of two oil-in-water emulsions that have nearly the same width (21). Both are bimodal and differ in the degree of overlapping of the two modes. The emulsion exhibiting the least degree of overlapping (bimodal 2, viscosity 35 mPa.s) is also less viscous than the more overlapping one (bimodal 1, viscosity 75 mPa.s).

All of the effects mentioned above are linked to how efficient is the packing of particles in a suspension or emulsion. Very fine particles or droplets may fit between the voids of large particles, reducing the interactions between the latter. It is as if the small particles and the continuous phase became a *pseudocontinuous* phase to the large particles, thereby reducing the effective internal volume phase content.

The size ratio and the relative amount of fines are also important. On the one hand, as the size ratio (size of large particles to small) is increased, viscosity diminishes. Theoretically, it has been determined that the greater reduction of viscosity occurs when the size ratio is about 10, for mixtures of unimodal particles (22). On the other hand, viscosity is strongly dependent on the fraction of fine particle. As the latter increases (for a constant overall ϕ), viscosity decreases until it reaches a minimum. If fine fraction is further increased, colloidal interactions among fine particles become important, thereby overshadowing the effective volume fraction reduction due to the formation of the pseudocontinuous medium (23). In consequence, viscosity increases.

This effect is not exclusive of monodispersed particles and has also been found in polydisperse, polymodal suspensions and emulsions (16,24). Figure 30 shows the viscosity of oil-in-water emulsions having the same overall internal

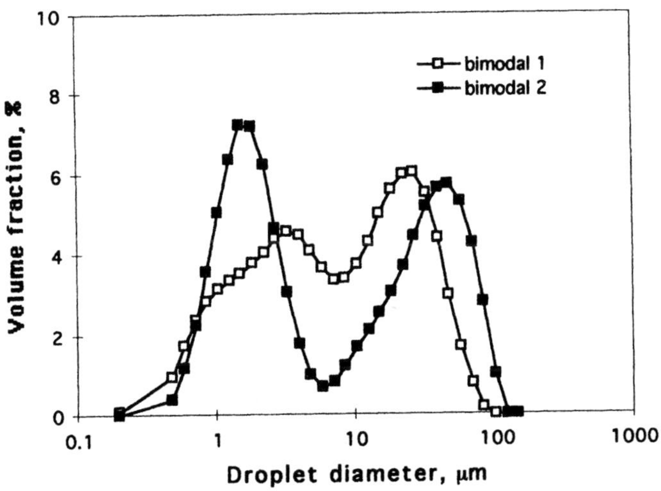

Figure 29 Droplet size distribution of bimodal lubricant oil-in-water emulsions of similar width but different shape. Viscosity of bimodal 1: 75 mPa.s. Viscosity of bimodal 2: 35 mPa.s. (Adapted from Ref. 20.)

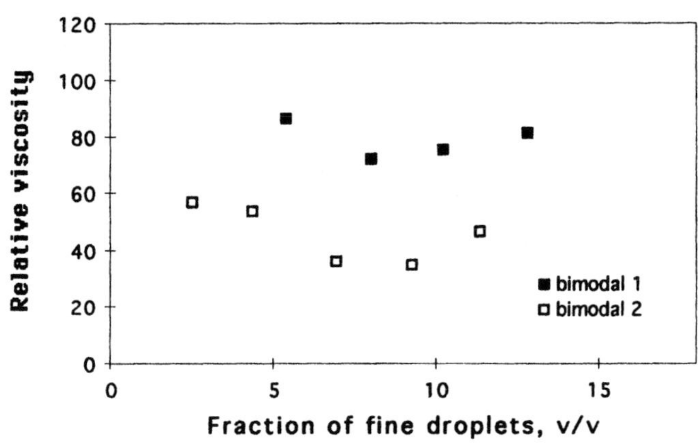

Figure 30 Relative viscosity as a function of fines volume fraction (size ratio $\simeq 10$) for lubricant oil-in-water emulsions of an internal phase volume of 0.65. (Adapted from Ref. 20.)

phase content ($\phi = 0.65$), as a function of the fraction of fine droplets (size ratio of about 10). The emulsion viscosity exhibits a minimum for a fine fraction of about 8% (21).

Particle size distribution also affects viscoelastic behavior. Monodispersed, noninteracting particles exhibit a measurable elastic modulus for volume fractions of 0.64, near the critical packing fraction (25). Polydispersed dispersions may reach higher volume fractions before G' is detectable.

5. Properties of the Continuous Phase

The viscosity and rheological behavior of the continuous phase also modifies the behavior of emulsions and suspensions. In fact, Eqs. [50]–[53] establish that the larger the viscosity of the continuous phase (η_0), the larger the viscosity of the suspension (η_s). Furthermore, if the continuous phase is strongly viscoelastic, die swelling and rod climbing may appear (26). Shear thickening has also been observed in concentrated emulsions of very viscous internal phase (27).

6. Temperature

It is generally accepted that rising temperature produces a reduction of viscosity and, in some cases, the rheological behavior may become less non-Newtonian (16). The latter cannot be explained solely by considering the reduction of the continuous phase viscosity produced by the augmentation of temperature. Temperature also modifies the interfacial properties and may induce or reduce flocculation (see next section). For the case of colloidal size particles, temperature increases Brownian dispersion forces. If the internal phase is a liquid, increasing temperature will make the droplets more deformable due to the simultaneous reduction of its viscosity and interfacial tension. An additional effect for both emulsions and suspensions is the different volume expansion of the phases that is bound to occur when changing temperature. This implies that the volumetric ratio, ϕ, is also changed. The combined effects of the variables mentioned above usually produce a decrease of viscosity with increasing temperature. However, viscosity may increase or remain constant if increasing temperature promotes extensive flocculation, aggregation, and swelling as for the case of clay particles. An increase of viscosity with increase of temperature has been observed on organophilic clay particles dispersed in a vegetable oil (28).

B. Chemical Factors

In the previous discussion, it was stated that internal phase content is the most important parameter in terms of viscosity and rheological behavior. Anything that slightly modifies the effective internal phase fraction produces a steep change of viscosity (remember the exponential dependency of η_s with ϕ). We shall see that most of the particle interactions, which are chemical in origin, also modify the effective internal phase content and, hence, viscosity and rheological behavior. These are interparticle forces, electroviscous effects, solvation layers, invivition adsorption, and osmotic diffusion in multiple emulsions.

1. Interparticle Forces

Interparticle forces are a determinant factor for most properties of dispersions, including rheological behavior. They are produced by the molecular forces on the surfaces of the particles, due to their nature or to adsorbed molecules, that modify the interface. These are electrical forces arising from charges on the particles and London–van der Waals attraction forces. The role of these forces on suspension stability has been extensively study and is known as the DLVO theory. In addition, sterical forces encountered on dispersions stabilized with nonionic species also exert an important influence on rheological behavior. The nature of these forces will not be considered since they are matters of discussion in Chapters 1–4. However, from a rheological point of view it is important to understand how these factors modify the flow characteristics of dispersions.

The repulsion forces (Ψ_R), either electrical or sterical, and the attraction ones (Ψ_A) sum up to give the total interaction potential (Ψ_T), which could have a shape as shown in Fig. 31, for electrically stabilized suspensions. Depending on the extent of the interactions, two attraction "wells" may be present: a primary minimum and a secondary minimum. The primary minimum is associated with

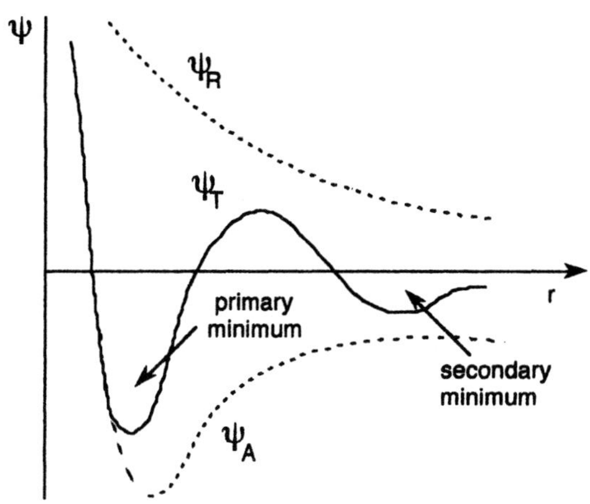

Figure 31 The interplay of repulsion and attraction forces between particles interphases according to DLVO theory.

the irreversible formation of aggregates (and coalescence for the case of emulsions) and the secondary minimum is associated with the formation of reversible aggregates or flocculates.

Flocculation is of primary importance from a rheological point of view. First, continuous phase becomes entrapped in the structure of flocculates, and hence $\phi_e > \phi$. Second, the attraction forces between these flocculates are relatively weak and can be overcome by shear, which induces floc destruction. Flocculation between particles and added interfacial agents may occur by means of two mechanisms: bridging flocculation and depletion flocculation. Flocculates produced by bridging are stronger to shear fields than those brought about by depletion (3).

Let us consider a dispersed system that has undergone flocculation. If p is the number of particles in a floc, the effective internal phase content is, for a flocculated dispersion,

$$\phi_e = \phi p^{(3/D)-1} \tag{55}$$

where p is the number of particles in a floc and D is the radius of a sphere enclosing the floc (29). Since p is a decreasing function of shear, ϕ_e also diminishes under an applied shear. In consequence, viscosity is reduced as shear progresses.

This phenomenon is illustrated in Figs. 32–34. In Fig. 33, flocculates are depicted as small spheres (particles) attached by springs. This is an imaginary representation of the forces that maintain the flocs. If a strong enough shear stress or shear rate is applied to the floc for a sufficient time, or shear is gradually increased to a sufficiently high value, the flocculate may be destroyed. An example of this phenomenon is shown in Fig. 33, which is average size of aggregates of salbutamol sulfate particles suspended in a propellant, as a function of particles

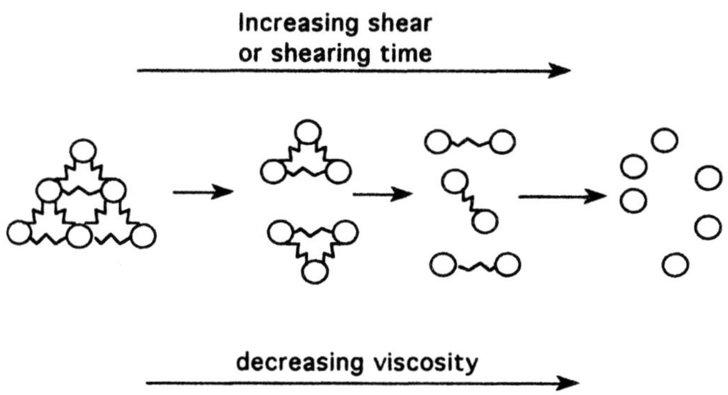

Figure 32 Ideal representation of the effect of increasing shear or increasing shearing time on the size of aggregates and viscosity for a flocculated dispersion.

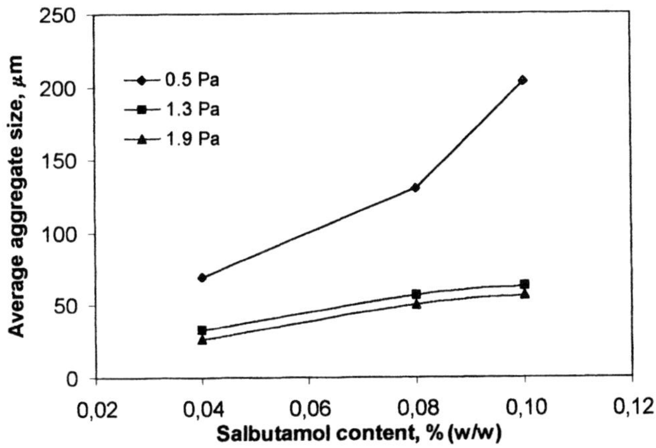

Figure 33 Average aggregate size as a function of salbutamol sulfate content and shear stress for propellants suspensions. (Adapted from Ref. 28.)

content and applied shear stress (30). It can be observed that as shear stress augments, the average size of aggregates diminishes.

The rheological behavior that goes along with floc destruction is illustrated in Fig. 34, which depicts shear stress as a function of shear rate for a strongly flocculated dispersion. A simple explanation of the events occurring on a microscopic level can be presumed. When shear is started, flocculates are first distorted without breaking, producing an elastic response. However, once shear overcomes the attraction forces, flocculates start to come apart and a yield stress appears.

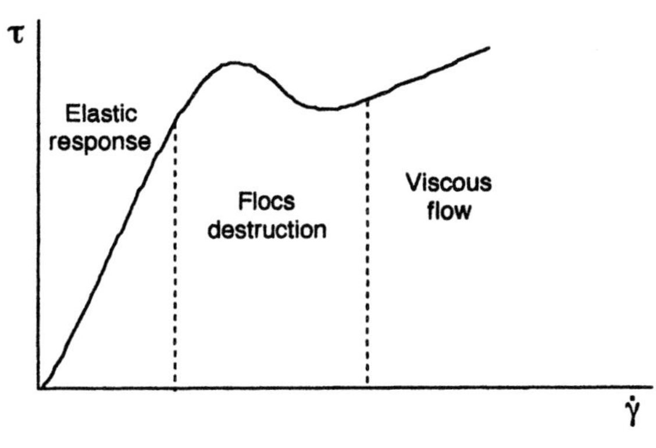

Figure 34 Typical flow curve of a strongly flocculated or very concentrated dispersion.

Then the gradual floc destruction produces a shear-thinning behavior, until no flocs remain but single particles, and the flow becomes Newtonian. In fact, the motion of the droplets under oscillatory strains has been observed by means of diffusing wave spectroscopy (31). Near the yield point, only a small fraction of drops flow while the rest remain as solid blocks.

The phenomena described above are the basis of the structural viscosity and viscoelastic behavior mentioned in Sec. II. The region of elastic response and the region of viscous flow depicted in Fig. 33 correspond to the first and to the second Newtonian plateau of Fig. 5, respectively. Further, the floc destruction region shown in Fig. 34 corresponds to the shear-thinning portion between both plateaus of Fig. 5.

Thixotropy, viscoelasticity, and yield stress appear when the strength of interparticle forces is large as in the case of concentrated ($\phi > 0.6$) emulsions. In fact, the higher the internal phase content, the closer the particles and droplets are; therefore, flocculation is more likely to occur. Many suspensions may show all of the behaviors mentioned above at very low concentration of solids, as in the case of suspended clay particles (12,32,33). Depending on factors such as electrolyte concentration, pH, and adsorbed macromolecules, a weight fraction below 0.1 is enough to produce a highly non-Newtonian behavior. An example of the tight relation between composition (ϕ) and formulation can be found in Fig. 23, which shows the effect of both pH and ϕ on the degree of viscoelasticity of suspensions of clay particles. Another example of how increasing internal phase content promotes flocculation is shown in Fig. 35, which is apparent viscosity as a function of shear rate for suspensions of lactose particles in chloroform

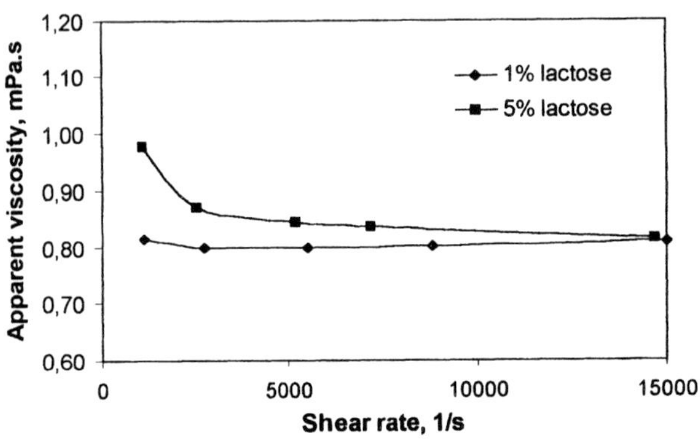

Figure 35 Apparent viscosity of suspensions of lactose particles in chloroform as a function of shear rate, for two lactose contents, 1% and 5%. (Adapted from Ref. 5.)

(6). The decrease of viscosity produced by flocs disruption under shear can also be observed in the latter figure.

2. Electroviscous Effect

Electroviscous effects are present in particles in which an electric double layer takes place. Three types of effects have been encountered:

The first electroviscous effect, which is described as the increase of frictional interparticle forces when the electric double layer is distorted by a shear field. An uneven distribution of charges is produced and, as a consequence, viscosity increases.

The second electroviscous effect, which occurs when particles collide. Their electric double layers overlap and strong repulsive forces arise. This also results in a viscosity increase.

The third electroviscous effect, in varying pH or ionic strength, produces a change in the electric double layer leading to strong flocculation; hence, viscosity increases.

The first two effects are only important in colloidal size particles, whereas the third and last effect may occur in particles stabilized with polyelectrolytes.

3. Solvation

Great emphasis has been placed on the fact that viscosity and rheological behavior are a function of the effective volume fraction. Solvation is another mechanism that alters ϕ_e and it occurs when a thick adsorbed layer of continuous phase or surfactant develops on the surface of particles. In this sense, ϕ_e may be estimated in two ways. In the first,

$$\phi_e = \phi\left[1 + \frac{k\delta}{a}\right] \tag{56}$$

where δ is the thickness of the adsorbed layer and k is a constant equal to 2 or 6 (20). This model predicts that the importance of the adsorbed layer decreases with increasing particle size, provided δ is independent of particle size. The second model does not take into account particle size (34) and predicts that

$$\phi_e = \phi\left[1 + \frac{m_{bc}}{m_p}\left(\frac{\rho_p}{\rho_c}\right)\right] \tag{57}$$

where m_{bc} is the mass of bound continuous phase; m_p is the mass of particles, and ρ_p and ρ_c are the densities of particles and continuous phase, respectively.

If m_{bc} is small respective to m_p, which means that δ is small, the contribution of the adsorbed layer is negligible. Solvation can be an important component of the rheological behavior for particles below 100 nm.

4. Invivition Adsorption and Osmotic Diffusion

Invivition adsorption is encountered in porous solid particles that are wet by the continuous phase. The latter diffuses into the pores, thereby increasing ϕ_e. Osmotic diffusion occurs in multiple emulsions, in which there is an electrolytic unbalance between the innermost internal phase and the continuous phase. This condition induces migration of liquid from the region of high osmotic pressure to low osmotic pressure. Transference of continuous phase into the droplets produces an increase of ϕ_e, whereas transference from the droplets to the continuous phase produces a decrease of ϕ_e. Both effects account for an increase or decrease of viscosity, respectively. Invivition and osmotic diffusion are of great importance in concentrated systems in which a relatively small increase of ϕ_e leads to large increases of viscosity and on the complexity of the rheological behavior.

IV. APPLICATION OF RHEOLOGY TO PRODUCT FORMULATION

A. Procedures for a Correct Rheological Evaluation

In this section, a series of guidelines regarding the selection of a correct rheological procedure are presented. This discussion is by no means exhaustive, its purpose being to apprise the reader of practical aspects of experimental rheology that are rarely mentioned in the literature. In this sense, useful references that may help the practitioner rheologist are given.

1. Errors To Be Avoided

As with any experimental techniques, it is important to ensure that the rheological evaluation is carried out correctly and the results are adequately interpreted. Experimental errors such as deficient instruments calibration, for those that require a calibration, are to be avoided. Other type of errors that have to be prevented are viscous heating, end effects, and secondary flow.

Viscous heating. This may occur when a viscous fluid is sheared for long periods and the heat generated by friction is not adequately withdrawn. A Newtonian fluid under this condition would exhibit a false shear-thinning behavior.

End effects. There occur at the boundaries of the sensor system, such as the entrance and exit of a capillary, and the bottom and upper part of concentric cylinders. When these effects are important, extra stresses are recorded making the sample appear more viscous than real. Regarding concentric cylinders, end effects can be eliminated if the ratio of the depth of liquid to the gap between the cylinders is larger than 100.

Secondary flow. This may rise when shearing low-viscosity fluids under high shear rates in concentric cylinders. The inconvenience of having secondary flow are two fold, i.e., the equations presented in Sec. II.C do not describe this type of flow behavior and a Newtonian fluid would show a false shear-thickening behavior. Strategies to avoid or minimize these experimental errors are found in Ref. 2.

A second category of experimental errors is related to the preparation and physical properties of the sample. A careful experimental procedure involves the following stratagems:

The sample must be thoroughly homogeneous before introduction in the sensor tool, which should not be over- or underfilled.

Sample temperature has to be always verified before running a test, and a constant-temperature bath or jacket should always be used.

Some samples may contain volatile components that evaporate during a run, changing the composition of the sample. In this case, measurements have to be carried out at low enough temperatures or the sensor tool has to be covered to minimize solvent loss. When using concentric cylinders (which are more convenient than cone-and-plate and parallel plates to reduce solvent loss), a thin film of a nonvolatile liquid may be poured on top of the sample.

Samples prone to particle sedimentation during a run should be reformulated to avoid this condition, or measured with special devices such as anchors or spindles.

Some samples require a routine thermal or mechanical treatment before the rheological test. These kind of treatments may be necessary in order to erase the sample's "prior memory," which often leads to nonreproducible results.

The presence of dispersed particles or droplets in suspensions and emulsions may introduce a further complication, when particles are too "large" for a measuring system. A criterion to determine if particles are too large is by means of the size ratio, δ/d_p. In the former expression, δ is the characteristic length of the sensor tool (e.g., gap size in concentric cylinders, plate–plate distance, or the capillary diameter) and d_p is the droplet or particle size. If this ratio is smaller than 50, the particles may interact with the solid boundaries causing the so-called slip-stick effect (35). The latter is induced by the shear field that produces the local formation of particles lumps (stick) that are later dispersed (slip). During the measurement of the flow curve, frequent peak values are registered that correspond to the jamming–unjamming of particles across the solid boundaries.

An unfortunate consequence of the latter is that cone-and-plate sensor systems are not convenient for suspensions and emulsions unless truncated cones are used. Another problem that arises is that large gap sizes or plate–plate separations have to be used, which requires more involved calculations of the shear rate across the gap (see Sec. II.C).

A second type of problem, common in very concentrated or jellified suspensions and emulsions, is the occurrence of edge failures and shear rate discontinuities. When this occurs, a portion of the sample may be expelled out of the sensor or rotate rigidly with the moving boundary. Figure 36a displays a rotating parallel plate sensor in which a portion of sample has already been expelled and another portion is rotating at the same speed of the solid boundaries (a, shaded area). No velocity gradient as shown in Fig. 1 is present in the rigid portion of sample.

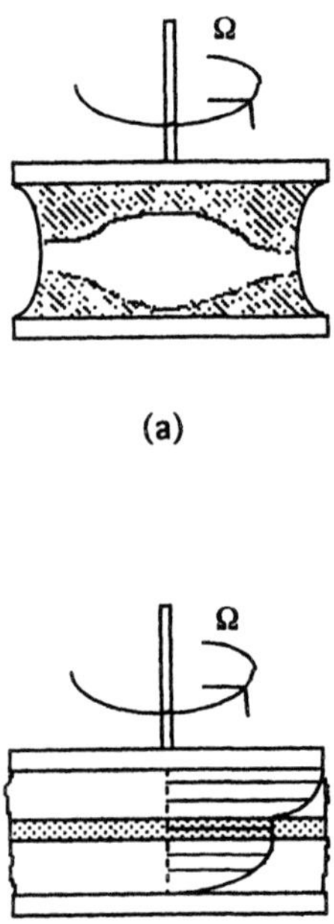

(a)

(b)

Figure 36 Schematic of rotating parallel plates in which (a) edge failure and (b) shear rate discontinuities (shaded areas) have occurred.

Both edge failure and shear rate discontinuity is easy to detect since the stress or strain readings become erratic or nondetectable. However, some types of shear rate discontinuities, characterized by the slip of nonsheared layers of sample between layers of sheared ones, is not so easily detected although it may produce some flattening of the flow curve. This case is illustrated in Fig. 36b, which shows the velocity vectors of the sheared layers and the nonsheared ones (shaded area). Again, these conditions invalidate the equations shown in Sec. II.C.

Another category of troublesome phenomenon is slip flow, a common problem of suspensions and emulsions. It is characterized by the apparent reduction of viscosity when the characteristic length δ of the sensor tool diminishes. Slip is mainly caused by the migration of particles and droplets out of the region of high shear stress, usually on the solid boundaries, leaving a depleted layer of continuous phase (for a complete review of slip, see Ref. 36). This condition reduces the stress required for a given shear rate, which means that less energy is required for the same flow rate as δ decreases. A direct consequence is that apparent viscosity diminishes.

Slip flow is detected when running flow curves for different δ values. If the curves separate as displayed in Fig. 37, slip flow has developed. There are some strategies to overcome this problem and a typical one is to use roughened or grooved tool surfaces in order to reduce slip. A second strategy is to perform a series of experiments as depicted in Fig. 37 and to correct the data as explained in Refs. 37 and 38. The first reference (37) presents a procedure for the correction

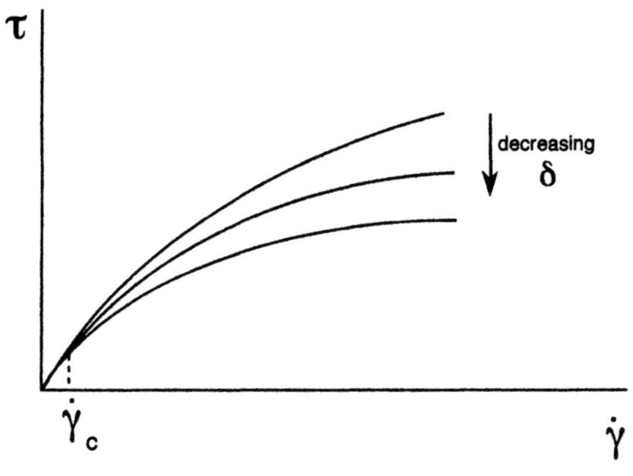

Figure 37 Flow curves of fluids as a function of the characteristic length of the sensor tool, δ, when slip flow occurs.

of data generated in concentric cylinders and parallel plates, while the second reference (38) proposes a corrective method for capillary viscometers.

All experiments that undergo edge failure, shear rate discontinuities, and slip should be discarded. However, it is often difficult to avoid slip flow altogether. In this case, it is recommended to perform rheological evaluations below the critical shear rate or stress that trigger the slip phenomenon (see Fig. 37). If data are needed only for comparison purposes, all samples to be compared should be measured with exactly the same sensor tool under the same measuring conditions.

2. Selection of Adequate Geometry and Measuring Conditions

The process of choosing the adequate sensor geometry is made after answering two key questions: (a) Why is a rheological assessment required? (b) How much money are we able or willing to spend? A rheological assessment may be required in any of the following cases: calculation of process energy requirements, process control, product quality control and research, and development of new products.

Calculation of process energy requirements. This is the field of a process engineer. However, simple rheological tests may be used with this purpose in mind. Capillary viscometers are a good choice to measure viscosity, which is always needed to predict pipe pressure drop or flow rates (see Eq. [48]). Mixer torque rheometers and extruders are adequate for evaluating mixing and extrusion of solids or wet masses. An example of a rheological evaluation carried out by means of an extruder may be found in Ref. 38. The latter describes how formulation is adjusted to obtain a satisfactory extrudate. A simple rule of thumb may be drawn from the previous discussion: always use the measuring geometry that most resembles the process geometry.

Measuring range. This is the interval of shear in which a sample is evaluated. The measuring range also has to agree with the process shear rate. In this sense, if the rheological evaluation is needed to assess the behavior of a sample when it undergoes processes such as pumping or mixing, tests should be run between 10 and 1000 s^{-1}. Extrusion and draining should be evaluated in a range of 0.1 to 100 s^{-1}, chewing and coating between 10 and 100 s^{-1}, while rubbing and spraying between 10^3 and 10^5 (5). Useful information about the measuring ranges of many commercial rheometers can be found in Ref. 40.

Process control and product quality control. These procedures are generally carried out using simple devices when only an assessment of viscosity is required. Capillary, falling- or rolling-ball viscometers, infinite sea viscometers (a rotating cylinder immersed in the sample), and free-fall penetrometers (the latter for very viscous materials) may be sufficient. These types of instruments are inexpensive and easy to use. However, these devices present a serious inconvenience as far as non-Newtonian fluid are concerned. The shear rate is not well known and although good estimates of viscosity for Newtonian fluids are obtained, experimental data for non-Newtonian fluids may be used only as a reference or for comparison purposes.

Some cases may require monitoring of complex material functions such as yield stress and G' as a measure of a sample internal microstructure. Yield stress may be readily assessed by means of shear viscometers, preferring concentric cylinders or parallel plates geometry to cone-and-plate, if the sample has particles larger than the colloidal size. Shear rate interval should be as low as possible (10^{-6}–10^{-4}). Many commercial instruments are computer-controlled and have routines that permit the calculation of τ_0. Yield stress may also be estimated using a simple graphic technique described in Ref. 40. The vane method (40–42) is also a recommended procedure for measuring yield stress since the sample is hardly disturbed by the insertion of the vane, whose results are more reliable. This is especially important in the case of samples whose microstructure is greatly disturbed by small, previous perturbations. Useful guidelines for measuring this type of sample may be found in Ref. 4. Yield stress is also evaluated by means of creep compliance experiments, such as that illustrated in Fig. 13b. Below the yield stress, a liquid fluid exhibits a behavior of an elastic solid, while above the yield stress, the fluid flows (slope different than zero). Performing several creep experiments at increasing τ values permits determination of the yield stress as the τ value for which viscous flow begins.

Measurement of G' requires more sophisticated and expensive rheometers and more involved experimental procedures. It must be remembered that experiments have to be carried out below the critical strain value (see Sec. II), or in the region of linear viscoelastic behavior. This region is determined by measuring the complex modulus G^* as a function of the applied strain at a constant oscillation frequency (usually 1 Hz). Up to γ_c, G^* does not vary with the strain; above γ_c, G^* tends to drop. The evaluation of oscillatory parameters is more often restricted to product formulation studies and research. However, a controlled-fall penetrometer may be used to compare the degree of elasticity between different samples. Creep compliance and creep relaxation experiments may be obtained by means of this type of device. In fact, a penetrometer may be the only way to assess viscoelasticity when the sample does not adhere to solid surfaces, or adheres too well, or cures to become a solid or semisolid. This is the case of many dental products such as fillings, impression putties, sealants, and cements.

Research and product development are the last cases in which rheology may be, in many cases, the most accurate way to formulate a product or to investigate its properties. This is a matter of further discussion in the next section.

B. Product Formulation

The next sections present a discussion of how the rheological evaluations described in Sec. II can be used to develop a pharmaceutical product. The underlying principle is to relate a specific rheological parameter to a required property or flow behavior. The applications of rheology to product development are many, but only three of the most important applications are discussed, i.e., product stability, drug delivery, and flow behavior. The discussion is accompanied with a few examples that illustrate the topic.

1. Stability

Stability encompasses various compound properties and the first to be discussed is the stability to flocculation or aggregation. Lack of stability alludes to the sedimentation or flocculation of particles or droplets. This condition is either a required or an undesired aspect, depending on the application.

When a dispersion flocculates or sedimentation occurs, the latter being very often a consequence of the former, an internal network builds up that can be monitored by means of rheological tests. In other words, rheological testing allows one to infer the microstructure of a fluid. The simplest procedure is to obtain a flow curve from which useful parameters such as viscosity (at given shear rates), η_0, and yield stress may be calculated. All of these parameters tend to increase as microstructure builds up and, depending on the strength of the aggregates, shearing a sample may result in their partial or total destruction.

Let us consider the case of aerosol suspensions. The suspended particles must be well dispersed so as to penetrate the lung. In this sense, an ideal aerosol suspension does not sediment and form aggregates. However, particles eventually do aggregate, so it is convenient to formulate them in such a way as to promote weak flocculates that can be easily redispersed by shear (e.g., agitation of the aerosol container). The formulation can be adjusted measuring viscosity as a function of shear rate or shear stress, as depicted in Refs. 6 and 30.

Another case is the formulation of jellifying agents such as clay particles and polymers. Both type of compounds promote the formation of a strong microstructure whose main function is to entrap solid particles to avoid sedimentation or to confer adequate flow properties, such as the right consistency for topical application.

Shear viscosity and yield stress can also be related to the gel strength (43,44) and used as comparison tools between formulations. Structure breakdown and buildup parameters can be deduced from the aforementioned rheological variables and from the evaluation of hysteresis curves of thixotropic samples (4,45–47). Reference 48 contains an interesting account on the use of shear rheometry to assess the influence of particle morphology (size and shape) on viscosity and rheological behavior for the formulation of stable suspensions of rbSt oil.

However, shear rheometry presents an important drawback, i.e., the sample microstructure may be destroyed during the test run. This is especially true for weak gels or pastes, even at low shear rates or shear stresses. Much information about the fluid microstructure is lost in this manner. In this case, creep and oscillatory rheometry are more convenient. The creep and viscoelastic parameters may be related to the strength of flocculation or aggregation, and a better knowledge of the microstructure is obtained. Useful information about the application of these methods for the study of suspensions of clay particles, creams, and concentrated suspensions of latex particles can be found in Refs. 12, 49, and 50, respectively.

The chemical degradation of a compound is also another form of lack of stability and many compounds may undergo undesirable changes during storage, thermal treatments, or intense shear. Measuring viscosity as a function of time is also a way of monitoring chemical degradation; modifications on a molecular level also produce a change of viscosity. As an example, the chemical stability of glyceride bases, which are essentially Newtonian fluids, has been evaluated by means of viscosity measurements (51). However, great care has to be taken when interpreting viscosity data from non-Newtonian fluids. Changes in viscosity during storage may also be produced by changes in the microstructure (e.g., flocculation, coalescence, and sedimentation) and not only by molecular modification of the compound.

2. Controlled Drug Delivery

It has been found that the release of a drug may be controlled if the active compound is suspended in a carrier matrix, which is often a polymeric solution that confers viscoelastic properties to the fluid (52–54). Rates of release have been correlated either with creep parameters such as G_0 (52) or with viscoelastic parameters, G' and G'' (53,54). It seems that increasing G' (i.e., the more the carrier is gel-like or viscoelastic) retards the rate of release of active compounds. Once more, rheology may be used to assess either the microstructure of the carrier or the rate of release by monitoring the rheological properties.

3. Flow Properties

Adjusting the flow properties of a compound is the more obvious use of rheology. From a pharmaceutical point of view, an adequate flow behavior is often required to ensure a good therapeutic or cosmetic function. For example, a cream must exhibit the proper consistency for application to the skin. The consistency parameter in fact encompasses various rheological properties. On the one hand, a cream should be strongly shear thinning such that it can be poured into and from the container and to facilitate rubbing on the skin. On the other hand, the cream, once spread, has to form a strong film that does not run and that allows the delivery of the active or cosmetic compound. A behavior such as the one just described requires of a fluid to form structures under rest that can be disaggregated under shear. A cream consistency may be evaluated by means of standard procedures or by measuring shear viscosity, τ_0, creep behavior, and viscoelastic parameters (45,46,55).

Another example of function associated with flow behavior is ease of swallowing of a medicament. Some people have some difficulty swallowing capsules or pills; oral dosages have to be formulated as jelly-like preparations (56). Therefore, rheological techniques that evaluate microstructure should be used to adjust formulation. A third example consists of the infusion of a drug in the body. Many drug carriers may exhibit a non-Newtonian behavior and, in consequence, their viscosity is bound to change as they infuse through an ever-changing flow path. Shear rheometry can be used to adjust the viscosity of a drug in such a way that it performs adequately (57).

REFERENCES

1. Darby, R. Chemical Engineering Fluid Mechanics. Marcel Dekker, New York, 1996, chapters 1 and 2.
2. Macosko, C. Rheology: Principles and Measurements. VCH, New York, 1994.
3. Barnes, H.A.; Hutton, J.F.; Walters, K. An Introduction to Rheology. Elsevier, Amsterdam, 1989.
4. Cheng, D.C-H. Yield stress: a time-dependent property and how to measure it. Rheologica Acta, 25: 542–554, 1986.
5. Barnes, H.A. Thixotropy—a review. J. Non-Newton Fluid Mech., 70: 1–33, 1997.
6. Sidhu, B.K.; Washington, C.; Davies, S.S.; Purewal, T.S. Rheology of model aerosol suspensions. J. Pharm. Pharmacol., 45: 597–600, 1993.
7. Mills, P.; Rubi, J.M.; Quemada, D. In: Rheology, Vol. 2. Editors G. Astarita, G. Marrucci; L. Nicolais. Plenum Press, New York, 1980.
8. Partal, P.; Guerrero, A.; Berjano, M., Muñoz, J., Gallego, G. Flow behavior and stability of oil-in-water emulsions stabilized by a sucrose palmitate. J. of Texture Stud., 25: 331–348, 1994.
9. de Kee, D.; Turcotte, G.; Code, R.K. Rheology, Vol. 3. Editors G. Astarita, G. Marrucci; L. Nicolais. Plenum Press, New York, 1980.
10. Clarke, B. Rheology of coarse settling suspensions. Trans. Inst. Chem. Eng., 45: 1251, 1967.
11. van Egmond, J.W. Shear-thickening in suspensions, associating polymers, worm-

like micelles, and poor polymer solutions. Curr. Opin. Colloid Interface Sci., 3: 385–390, 1998.
12. Khandal, R.K.; Tadros, T.F. Applications of viscoelastic measurements to the investigation of the swelling of sodium montmorillonite suspensions. J. Colloid Interface Sci., 125(1): 122–128, 1988.
13. Thomas, D.G. Transport properties of suspension: VII. A note on the viscosity of Newtonian suspensions of uniform spherical particles. J. Colloid Sci., 20: 267–277, 1965.
14. Krieger, I.; Dougherty, T. Trans. Soc. Rheol., 3: 137–152, 1959.
15. Sherman, P. Rheological properties of emulsions. In: Encyclopedia of Emulsions, Editors P. Becher. Marcel Dekker. New York, 1983, chapter 7.
16. Briceño, M. Study of the main variables that influence the rheology of bitumen-in-water emulsions (Spanish). MSc thesis, University Simon Bolivar, Caracas, 1994.
17. Polinski, A.; Ryan, M.; Gupta, R.; Seshadi, S.; Frechette, F. J. Rheol., 32(7): 703–735, 1988.
18. Princen, H.H.; Kiss, A.D. Rheology of foams and highly concentrated emulsions. IV. An experimental study of the shear viscosity and yield stress of concentrated emulsions. J. Colloid Interface Sci., 128(1): 176–187, 1989.
19. Princen, H.H.; Kiss, A.D. Rheology of foams and highly concentrated emulsions. III. Static shear modulus. J. Colloid Interface Sci., 112: 427, 1986.
20. Pal, R. Effect of droplet size on the rheology of emulsions. AlChE J., 42(11): 3181–3190, 1996.
21. Briceño, M.; Ramirez, M.; Bullon, J.; Salager, J.-L. Customizing drop size distribution to change emulsion viscosity. Proceedings of the Second World Congress on Emulsion, Paper No. 2-1-094/01, Bordeaux, September 23–26, 1997.
22. Farris, R.J. Prediction of the viscosity of multimodal suspensions from unimodal viscosity data. Trans. Soc. Rheol., 12(2): 281–301, 1968.
23. Hoffman, R. Factors affecting the viscosity of unimodal and multimodal colloidal dispersions. J. Rheol., 36(5): 947–965, 1992.
24. Sengun, M.Z.; Probstein, R.F. Bimodal model of slurry viscosity with application to coal-slurries. 1. Theory and experiment. Rheologica Acta, 28: 382–393, 1989.
25. Lequeux, F. Emulsion rheology. Curr. Opin. Colloid Interface Sci., 3: 408–411, 1998.
26. Barnes, H.A. Rheology of emulsions—a review. Colloid Surfaces A: Physicochem. Eng. Aspects, 91: 89–95, 1994.
27. Otsub, Y.; Prud'homme, R.K. Rheologica Acta, 33(1): 29, 1994.
28. Urbina, R. Biodegradable emulsions as a base of drilling fluids (Spanish). FIRP Technical Report No. 9704, University of Los Andes, Mérida, 1997.
29. Wessel, R.; Ball, R.C. Phys. Rev. A, 46(6):R3009, 1992.
30. Bower, C.; Washington, C., Purewal, T.S. The effect of surfactant and solid phase concentration on drug aggregates in model aerosol propellents. J. Pharm. Pharmacol., 48: 342–346, 1996.
31. Hebraud, P.; Lequeux, F.; Munch, J.P.; Pine, D.J. Yielding and rearrangements in disordered emulsions. Phys. Rev. Lett., 78: 4657–4660, 1997.
32. Ormsby, W.C.; Diamond, S. Effect of $Ca(OH)_2$ on the flow properties of montmorillonite-water systems. Trans. Br. Cer. Soc., 65(9): 553–558, 1966.

33. Miano, F.; Rabaioli, M.R. Rheological scaling of montmorillonite suspensions: the effect of electrolytes and polyelectrolytes. Colloid Surfaces A: Physicochem. Eng. Aspects, 84: 229–237, 1994.
34. Hiemenz, P.C. Principles of Colloid and Surface Chemistry. Marcel Dekker Inc., New York, 1977, pp. 42–83.
35. Cheng, D.C-H. Flow Properties of Suspensions of Inert Spheres. Ed. Warren Spring Laboratory, Stevenage, 1984.
36. Barnes, H.A. A review of the slip (wall depletion) of polymer solutions and particle suspensions in viscometers: its cause, characteristics and cure. J. Non-Newton. Fluid Mech., 56: 221–251, 1995.
37. Yoshimura, A., Prud'homme, R.K. Wall slip corrections for Couette and parallel disk viscometers. J. Rheol., 32(1): 53–67, 1988.
38. Mooney, M. Explicit formulas for slip and fluidity. J. Rheol. 2(2): 210–222, 1931.
39. Dixon, F.K.; Rowley, G.; Hurst, J.; Sharkey, I.M.; Coulthard, M.G. Investigation of the rheological behavior of aqueous granulations containing the unpalatable hygroscopic drug cysteamine hydrochloride using the ram extrusion technique. J. Pharm. Pharmacol., 46(2): 1039, 1994.
40. Ma, L.; Barboza-Canovas, G.V. Review: Instrumentation for the rheological characterization of foods. Food Sci. Technol. Int., 1: 3–17, 1995.
41. Yoshimura, A., Prud'homme, R.K.; Princen, H.M.; Kiss, A.D. A comparison of techniques for measuring yield stress. J. Rheol., 31(8): 699–710, 1987.
42. Nguyen, Q.D.; Boger, D.V. Direct yield stress measurement with the Vane method. J. Rheol., 29(3): 335–347, 1984.
43. Rao, M.A.; Steffe, J.F. Measuring yield stress of fluid foods. Food Technol., 51(2): 50–52, 1997.
44. Mohamed, Y.; Traisnel, M.; Gayot, A. Rhéologie et Viscosimétrie: etude Rheologique de quelques Gélifiants et Epaississants utilisés en Pharmacie. STP Pharma Pratiques, 4(2): 117–121, 1994.
45. Eros, I.; Thaleb, A. Rheological studies of creams. I. Rheological functions and structure of creams. Acta Pharm. Hung., 64(3): 101–103, 1994.
46. Thaleb, A.; Eros, I. Rheological studies of creams. III. Effect of lipophilic phase on consistency. Acta Pharm. Hung., 66(2): 77–81, 1996.
47. Moore, F. Trans. Br. Cer. Soc., 58: 470–494, 1959.
48. Pena, L.E.; Possert, P.L.; Stearns, J.F.; Lee, B.L.; Hageman, M.J. Rheological characterization of rbSt oil suspensions. Int. J. Pharm., 113: 89–96, 1995.
49. Tamburic, S.; Craig, D.Q.M.; Vuleta, G.; Milic, J. Int. J. Pharm., 137: 243–248, 1996.
50. Tadros, T.; Hopkinson, A. Use of viscoelastic measurements for investigating the stability/flocculation of concentrated dispersions. Faraday Disc. Chem. Soc., 90/3, Paper 0/02267E, 1990.
51. Sutanata, W.; Craig, D.Q.M.; Newton, M. An Investigation into the effects of preparation conditions and storage on the rate of drug release from pharmaceutical glyceride bases. J. Pharm. Pharmacol., 47: 355–359, 1995.
52. Walters, P.A.; Rowley, G.; Pearson, J.T.; Taylor, C.J. Low shear rheological characterization and stability of triglyceride gels for liquid filling into hard gelatin capsules. J. Pharm. Pharmacol., 47(12B), 1995.

53. Talukdar, M.M.; Vinckier, I.; Moldenaers, P.; Kinget, R. Rheological characterization of xanthan gum and hydroxypropylmethyl cellulose with respect to controlled-release drug delivery. J. Pharm. Sci., 85(5): 537–540, 1996.
54. Mortazavi, S.A.; Smart, J.D. Factors Influencing Gel-strengthening at the Mucoadhesive-mucus Interface. Journal of Pharmaceutical and Pharmacology, 46: 86–90, 1994.
55. Jiménez, M.M.; Fresco, M.J.; Sellas, E. Proposal and Pharmacotechnical Study of a Modern Dermo-pharmaceutical Formulation for Cold Cream. Boll. Chim. Farm., 135(6): 364–373, 1996.
56. Hanawa, T.; Watanabe, A.; Tsuchiya, T.; Ikowa, R.; Hidaka, M.; Sugihara, M. New Oral Dosage Form for Elderly Patients: Preparation and Characterization of Silk Fibroin Gel. Chemical Pharm. Bulletin, 43(2): 284–288, 1995.
57. Natsume, H.; Imamizu, M.; Sugibayashi, K.; Shibata, T.; Miyagawa, A.; Morimoto, Y. Oily Drug Carriers in Cancer Chemotherapy. II. Preparation of Viscous Ethyl Oleate for Intra-arterial Infusion Therapy and its Disposition in Rats and Hamsters. Chemical Pharm. Bulletin, 40(6): 1554–1552, 1992.

17

Determination of Size Distributions of Submicrometer Particle Dispersions by Photon Correlation Spectroscopy

E. R. Pike
King's College, London, United Kingdom

I. INTRODUCTION

In this chapter we will review the light scattering method of photon correlation spectroscopy (PCS) for the sizing of dispersed suspensions of submicrometer particles of interest in pharmacology. These particles may be macromolecules of varying types including liposomes and other laminar vesicles, micelles, and droplets in microemulsions.

Microemulsions have attracted much interest in recent years in terms of their drug delivery potential. Both water-in-oil and oil-in-water microemulsions can enhance the oral bioavailability of drugs; surfactants in the formulation may give rise to enhanced absorption. Submicrometer-sized, oil-in-water emulsions are also used for the intravenous administration of physiological water-insoluble triglycerides. The mean size and polydispersity of the particle distributions are important parameters which influence their biopharmaceutical fate and knowledge of these parameters may be used, for example, to limit possible severe side effects of intravenous emulsions (1). Particle charge is also an important physical parameter and measurements of ζ-potential can also be made with an adaptation of the conventional PCS apparatus.

From the point of view of instrumentation and technique, we should perhaps say immediately that the sizing of microscopic particles of interest in pharmacology by this method is, in essence, the same as particle sizing of other types of macromolecular suspension, e.g., proteins, enzymes, viruses, and polymers. Given careful sample preparation, in particular scrupulous cleanliness to avoid the introduction of dust or other contaminant particles, sufficient dilution to avoid particle–particle interactions, and complete dispersion to break up aggregations, the technique is easy, fast, accurate, absolute, and nonperturbative. In the author's experience, lack of attention to these factors is the major cause of difficulties found by users in practice. Prealigned or carefully aligned optics and standard detection electronics, used within their known limits, should not give trouble and, as provided by a number of commercial manufacturers, can be found in routine use in many laboratories worldwide.

After a discussion of the origins of the technique and the basic physics, including charge measurement, we will consider experimental methods and their evolution. The routine determination of the mean value and variance of the translational diffusion coefficient of macromolecules in dilute suspension will be explained and related to the mean particle diameter and polydispersity. We will then deal with the difficult problem of obtaining further information about the particle size distribution from the correlation function. A discussion of methods for dealing with more concentrated suspensions follows, where interactions between particles may be studied and information gained even from light scattering in turbid media. Finally, we illustrate the use of the various techniques in topics of interest in pharmacology with a review of some selected papers in the literature.

II. BASIC PHYSICS

Although a fluctuating pattern of diffraction from small water droplets formed by breathing on a glass plate, which we now call "speckle," the random granular pattern seen in coherent light that has been scattered from a rough surface, was described as long ago as 1877 (2), it seems that the realization that the thermal diffusive Brownian movement of microscopic particles in suspension could be studied by coherent light scattering first occurred to Sir C. V. Raman. In 1919 (3) he had also briefly reported the observation of a speckle effect. In 1943, Ramachandran (4) correctly formulated and verified the basics of speckle pattern formation. With a layer of lycopodium powder on a glass slide, in both plane and spherical wavefronts, by painstaking visual classification of the scattered light patterns he demonstrated the exponential law of Rayleigh for the distribution of intensity from a fixed number of interfering, randomly phased oscillators (Rayleigh's exponential law for scattered intensities arises from a Gaussian law for the scattered amplitudes). In the conclusion to his paper he says that Raman had suggested that Brownian motion might be studied by observing the fluctuations of these speckle patterns and that he was proposing to study this possibility experimentally. Unfortunately, there seems to be no record of such further work from him. There is, however, a description of an experiment by Raman himself in his 1959 *Lectures in Physical Optics* (5). He viewed a thin film of milk illuminated by a mercury lamp behind a very small aperture and said that "bright points of illumination continually appear in the field and others disappear. These changes become less rapid and ultimately stop when the film is dry."

In prelaser days, in addition to the group of Raman, the fact that optical sources should theoretically show similar beating phenomena to those well known in the radiowave region was pointed out by Gorelik (6) and independently by Forrester et al. (7), both in 1947, and such light beating experiments had been performed by Forrester et al. (8) and by Hanbury Brown and Twiss (9) in the mid 1950s. Forrester et al. illuminated a square-law detector with the light from a Zeeman-split Hg 546.1-nm emission line and observed a component in the photocurrent at the magnetic field splitting of around 1010 Hz, while Hanbury Brown and Twiss correlated photocurrent fluctuations from two detectors measuring light from a single star focused onto them by two separate concave mirrors. The stellar diameter could be deduced by the dependence of the correlations as a function of the mirror separation. It is interesting to note that the interpretation of these experiments caused considerable controversy at the time in spite of the theoretical predictions of Gorelik and Forrester published some years before.

The invention of the laser made available a bright coherent source that allowed light scattering experiments to be performed much more effectively than before. That such light scattering experiments could give *quantitative* information about the Brownian motion, and thus the sizes of suspended dispersed scatterers, was first shown in pioneering experiments by Cummins and his colleagues at Johns Hopkins University in 1964 (10). They scattered a red 633-nm beam of a helium-neon laser, discovered 2 years previously by White and Rigden (11), from a suspension of polystyrene spheres. They heterodyned the scattered signal with a ''reference'' signal split off from the same laser beam which was shifted in frequency by passing through an ultrasonic ''Bragg cell.'' This is the principle used to detect radio and television signals by beating them against a frequency-shifted ''local oscillator'' in a nonlinear detector. They measured the frequency spectrum of their photomultiplier detector output current (in the kHz range) with a wave analyzer. As we shall see later, for a monodisperse suspension of spherical macromolecular particles the amplitude spectrum of the fluctuating optical speckle pattern is predicted to be of Lorentzian profile and in this experiment this spectrum is seen directly in the fluctuations of the photocurrent. The smaller the particles, the faster is the Brownian motion, the faster the intensity fluctuates, and the broader the linewidth becomes. The spectral linewidth thus gives a measure of the particle size.

At the time of the first experiments of Cummins et al. on particle scattering, the coherence of the laser was being used for other light beating studies in the laboratory of Benedek at the Massachusetts Institute of Technology (MIT) where, for example, very beautiful experiments were performed on the critical opalescent fluctuations of pure SF_6 (12), experiments being refined even to this day to study critical fluctuations in xenon in the microgravity environment of the U.S. space shuttle (13). As was done by Raman, in these experiments Ford and Benedek observed the fluctuations of the scattered light field without the addition of the reference beam. Fluctuations which appear in the output current can be considered to be due to the interference or ''beating together'' in the detector of the different spectral components of the electromagnetic field. For a dilute particle suspension, where the particles are spaced at separations of many optical wavelengths, each particle can be considered to be one of Rayleigh's randomly phased model oscillator sources; in the common situation where there are a large number of particles in the observed scattering volume Rayleigh's exponential distribution of detected intensities is accurately observed.

Such a ''self-beating'' experiment on biologically interesting molecules was first published by Dubin et al. (14), who made measurements of bovine serum albumin, ovalbumin, lysozyme, and polystyrene latex spheres. They also studied tobacco mosaic virus and DNA and found that the nonsphericity of these molecules leads to deviations from the Lorentzian spectrum. A theoretical analysis by Pecora (15), which predicts angular-dependent, multi-Lorentzian spectra, can be used to obtain shape information of such molecules.

The field is thus sometimes called light beating spectroscopy or intensity fluctuation spectroscopy. It also goes by the name of dynamic light scattering. In the absence of a reference beam it is called ''self-beating'' spectroscopy and also still quite commonly (although more problematic since it conflicts with heterodyning with a reference beam at base band in radio and microwave terminology) ''homodyne'' spectroscopy. True homodyne spectroscopy in the radiofrequency sense, i.e., heterodyning with an unshifted reference beam, is also perfectly feasible in laser scattering (16) and (as also does heterodyning) gives an improved signal-to-noise ratio of almost an order of magnitude over the self-beating method since only the signal beam out of the beating pair of signals fluctuates. However, the difficulties in the optical region of aligning wavefronts to the required wavelength scale interferometric accuracy at varying scattering angles has precluded the widespread utilization of this advantage and the self-beating method has become standard. This may change soon with optical fibre implementation. With the work to be described shortly by the author's group in Malvern in the late 1960s, (17), in which detection and analysis was performed by the now universally used sensitive digital photon counting and correlation techniques, the description ''photon correlation spectroscopy'' (PCS) came into widespread use and we shall use this terminology here. Photon correlation turned out to be a fortunate ''marriage'' of the one-bit nature of optical detection, which turns one optical photon into one photoelectron (or in a semiconductor into one electron-hole pair) and the digital revolution in silicon integrated electronics which made available, even at that time, 50-ns clock speeds for digital processing. By light beating spectroscopy we now tend to classify the detection and analysis of analog photocurrents which predated the digital photon correlation methods.

When the particles are charged and an electric field is applied, they undergo a bulk mean velocity in addition to their Brownian motion. That this mean velocity could also be measured by photon correlation was first shown by the author (18) in 1972. Either a strong reference beam is added to give a heterodyne signal, or two equal-strength incident beams are used and the difference of the two Doppler shifts in the scattered light is measured. This latter method obviates the need for the interferometric alignment accuracy mentioned above and is generally adopted. The correlation function has an oscillatory component at the Doppler or Doppler-difference frequency. The ζ-potential measurements are now routinely obtained in this way.

III. EXPERIMENTAL METHODS

The measurement of particle size from a few nanometers to a few micrometers in diameter by PCS is now a mature field with a number of commercial companies offering equipment and software packages. Nevertheless, experimental methods continue to evolve. The first instruments were marketed in 1971 by the Malvern (UK) company Precision Devices and Systems (now Malvern Instruments), who licensed patents taken out by the UK Ministry of Defense at the Royal Radar

Establishment (RRE), Malvern. At the center of the setup was a goniometer, which carried a photomultiplier tube assembly and collection optics on a rotating arm, which could collect light at angles between about 20° and 160°. The photomultipliers were originally so-called star-tracker FW 130 tubes, designed by Eberhard at ITT Fort Worth, USA. Due to an electron focusing arrangement at their front end, we found them to be the only ones available at that time suitable for counting individual photon detections.

The sample was contained in a 1-cm-diameter cylindrical cell in an index-matching bath on the axis of rotation. A thermostatted housing kept the sample and bath at a constant temperature. The laser, mounted on a fixed second arm, was usually helium-neon of about 3–5 mW, but for very dilute samples of low molecular weight an argon-ion laser was used which could supply typically a few hundred milliwatts. The laser beam was focused down into the cell to a few hundred micrometers diameter with a simple lens.

Later goniometers were to have motor-driven angle drives coupled to the system software and could be adapted, simply by changing pinholes and software package, to record the scattered intensity vs. angle for use in the classical molecular weight determination technique known as the Zimm plot. Modern versions of this technique use sophisticated data reduction procedures and are a valuable complement to the PCS facility since, with supplementary measurements of the dependence of refractive index with concentration, they give absolute values of molecular weights. Molecular weight can also be determined by combining PCS data with independently determined values of sedimentation coefficient and partial specific volume in the Svedberg equation, also well known in the classical literature of the physical chemistry of macromolecules.

In some PCS instruments a fixed scattering angle of 45°, 60°, 90°, or 135° is used; this is sufficient for spherical particles in the Rayleigh scattering region ($[2\pi/\lambda]r \ll 1$), where λ is the laser wavelength and r the particle radius, since the angular dependence of the scattered intensity is then negligible.

In the "Malvern correlator," as it came to be called, spectral analysis of the photomultiplier tube currents used in the first experiments at Johns Hopkins and at MIT described above was superseded by the digital photon correlation technique developed at RRE. For the type of (stationary) optical signals encountered in light scattering from macromolecular suspensions the spectrum of the optical field fluctuations is given by the Fourier cosine transformation of its autocorrelation function (the Wiener-Khinchine theorem). The "analog" anode photocurrent is, in fact, the superposition of the amplified individual photoelectron emissions from the tube cathode plus lower level noise from the tube base and its associated circuitry. By using a discrimination threshold the latter components can be rejected and, with sufficiently fast circuits so that the pulses do not overlap, a replica of the digital pulse train of photon detections at the cathode can be output in standardized pulses, each typically of the order of a few volts in amplitude and some tens of nanoseconds in width. These then went forward to a 50-ns digital correlator. The autocorrelation function of the detected pulse train is called the digital photon correlation function. It is defined by

$$G^{(2)}(\tau_i) = \sum_j n(t_j)n(t_j + \tau_i) \qquad [1]$$

where $n(t_j)$ is the number of photon detections registered in the interval t_j to t_j + T, T is a sampling time chosen to be sufficiently short compared to the time scale of the fluctuations that it does not appear explicitly in the theory, and τ_i is the delay time of the ith sample. The superscript (2) refers to the fact that the correlation function is describing the fluctuations of the intensity $I(t)$ (photon arrival rate), which by quantum detection theory can be shown to be given by the modulus of the *square* of the electromagnetic field. This function is a good approximation to the continuous intensity correlation function $\int < I(0)I(\tau) > d\tau$. $G^{(2)}(\tau)$ is more conveniently written in the normalized form $g^{(2)}(\tau)$, where

$$g^{(2)}(\tau) = \text{Const}\left\{\frac{G^{(2)}(\tau)}{[G^{(2)}(\infty)]} - 1\right\} + 1 \qquad [2]$$

The constant, which will be greater than 1, depends on optical aperture sizes, in particular the size of the detector pinhole; it decreases as the detector pinhole is increased to collect more light. The detection optics should be arranged so that the constant is as close to 1 as possible, consistent with having, say, 10 or more photon detections in the mean decay time of the correlation function. In practice, $G^{(2)}(\infty)$ is estimated at a long delay time where the fluctuations are independent; it measures the square of the mean intensity. Since the zero-delay channel is not normally measured, it is difficult to normalize $g^{(2)}$ accurately to its theoretical value of 2 at the origin and the constant is therefore carried along but does not affect the time dependence of the functions, which is used to find the particle sizes.

As we have seen, our optical signals are not only stationary but have Gaussian amplitude statistics; this leads to a relation between $g^{(2)}$ and the autocorrelation function, $g^{(1)}$, of the optical field (19):

$$g^{(2)}(\tau) = 1 + |[g^{(1)}(\tau)]|^2 \qquad [3]$$

The first commercial electronic circuits used Schottky TTL LSI and the high speeds required for forming the photon correlation functions in real time were achieved by the technique of "single clipping" proposed by Jakeman and Pike (20). This was as an equivalent, for a digital train of photon detection pulses, of Van Vleck and Middleton's microwave technique (21) of one-bit "hard limiting" of signals, i.e., using only their zero crossings. In the digital case the signal $n(t_j)$ is given the value zero if the number of counts in the sample time falls below the "clipping level," taken to be an integer near the mean value, and to be unity otherwise, while the signal $n(t_j + \tau_i)$ is not changed. In this way the multiplications required in Eq. [1] are reduced to additions which can be done quickly using a simple "and" gate. The information loss in constructing the spectrum in this way is surprisingly small and the clipping technique has been used successfully for many years. However, the increase in the speed of semiconductor circuitry in recent times means that autocorrelation of at least four bits of the full digital signal, and in some cases the full signal, is now the norm for the particle sizing problem (although single-clipped correlators are still being designed and built for high-speed applications). We will therefore only consider the theory of full multibit correlation functions here.

Miniaturization of electronic components has made possible spectacular changes in the size and performance of these instruments. Most photon correlators are now available in the form of plug-in boards for personal computers. One of these designed by the author and his group at King's College, London, even runs very quickly as a software application on a general-purpose CPU. This gives more flexibility than does dedicated hardware. EMI later entered the photon-counting photomultiplier market and more recently other miniature tubes and solid state avalanche photodiodes (APDs) have been used (22). The same is true on the optical side, following work by R. G. W. Brown at RSRE Malvern in the late 1980s (23,24). Modular systems using semiconductor lasers, solid state avalanche photodiodes, and single-mode optical fibers are in use today in space shuttle experiments (25) and are also available in some recent commercial instruments, where there are specific requirements for very high sensitivity.

IV. MEAN PARTICLE DIAMETER AND POLYDISPERSITY

The data presented in the photon correlation function is a set of values of $G^{(2)}(\tau_i)$ at delay times, which may be spaced linearly:

$$\tau_i = T, 2T, 3T, 4T, \ldots,$$

or, more soundly, as we shall explain later, geometrically:

$$\tau_i = T, \delta T, \delta^2 T, \delta^3 T, \ldots,$$

where δ is called the "dilation factor." The sample time, T, is chosen so that, on a logarithmic time scale, $G^{(2)}(\tau_i)$ varies little at both low and high values of delay time so that the "action," as it were, takes place well within the delay times covered. To determine the mean particle diameter and the polydispersity of a distribution of spherical or near-spherical particles we must first realize that the raw data give information about particle sizes only by relating the first-order correlation function $g^{(1)}(\tau)$ to their translational diffusion coefficients, D_T. We firstly utilize the Seigert relation of Eqs. [2] and [3] to deduce $g^{(1)}$ from the measured $G^{(2)}$ (to within a constant factor). The primary relations of the particle sizing technique are, then, that for scattering from dilute monodisperse particle suspensions at an angle θ from the forward direction we have the single exponential relation:

$$g^{(1)}(\tau_i) = \exp(-K^2 D_T \tau_i) \qquad [4]$$

where K is the magnitude, $4\pi/\lambda \sin^2(\theta/2)$, of the scattering vector, and that for dilute polydisperse suspensions with distributions $P(D_T)$ of translational diffusion coefficient we have the linear superposition of exponentials for each D_T:

$$g^{(1)}(\tau_i) = \int_{D_T(\min)}^{D_T(\max)} P(D_T) \exp(-K^2 D_T \tau_i) dD_T \qquad [5]$$

where we have cut off the distribution at suitable upper and lower limits according to a priori knowledge available to the experimenter. It is convenient to denote $K^2 D_T$ by Γ so that this equation becomes

$$g^{(1)}(\tau_i) = \int_{\Gamma(\min)}^{\Gamma(\max)} P(\Gamma) \exp(-\Gamma \tau_i) d\Gamma \qquad [6]$$

Equation [4] can be derived in several ways. Cummins and Pusey (26) show that

$$|g^{(1)}(\tau)| = \langle \exp(-i\mathbf{K} \cdot [\mathbf{r}(0) - \mathbf{r}(\tau)]) \rangle \qquad [7]$$

where $\mathbf{r}(t)$ denotes particle position and $\langle\ \rangle$ denotes a long-time average. By definition we have

$$\mathbf{r}(0) - \mathbf{r}(\tau) = \int_0^\tau \mathbf{v}(t) dt \qquad [8]$$

where $\mathbf{v}(t)$ denotes velocity at time t. Using an expansion of the exponential and keeping only even terms in which the pairs of times lie sufficiently close to each other, the required result can be obtained with the identification (Kubo formula):

$$D_T = \int_0^\infty \langle \upsilon(0)\upsilon(t) \rangle dt \qquad [9]$$

where υ is the linear velocity in a fixed direction. The Wiener-Khintchine theorem gives the Lorentzian spectrum mentioned in Sec. II above, since this is the Fourier transform of the (symmetrical) exponential.

To determine the mean value $\langle\Gamma\rangle$ (first moment or cumulant) and the Γ polydispersity (second central moment, second cumulant or variance),

$$\mu_2 = \int P(\Gamma)(\Gamma - \langle\Gamma\rangle)^2 d\Gamma \qquad [10]$$

it suffices to fit the normalized correlation function to an expansion in these first two moments of the form:

$$\frac{1}{2}\ln\left\{\frac{G^{(2)}(\tau)}{[G^{(2)}(\infty)]}-1\right\} = \text{constant} - \langle\Gamma\rangle\,\tau + \frac{1}{2!}\,\mu_2\tau^2 \qquad [11]$$

This is a simple and accurate (and widely used) method for recovering these parameters of the distribution. The scattering cross-section of a macromolecule in the Rayleigh region goes as the square of its molecular weight, M_i, so that for a polydisperse suspension the mean value of $\langle\Gamma\rangle/K^2$ gives a weighted average of diffusion coefficients, $\sum c_iM_i^2\, D_i/\sum c_iM_i^2$, where c_i is the concentration of molecules of molecular weight M_i. This is called the "*z*-average" molecular weight, M_z, and is, happily, just the average needed in the Svedberg equation (assuming constant partial specific volumes) to give the weight-average molecular weight, M_w. For narrow distributions the value of $\mu_2/\langle\Gamma\rangle^2$ can be related to polydispersity indices such as M_z/M_w by assuming model forms, such as random coils, for the scatterers (27).

The link with particle size is then made by using the Stokes-Einstein formula:

$$D_T = \frac{kT}{6\pi\eta r} \qquad [12]$$

where k is Boltzmann's constant, T is the absolute temperature (not the sample time here), η is the solvent viscosity, and r is an "equivalent spherical radius," i.e., it is the actual radius if the particle is a sphere but an estimated measure of equivalent size if not. Note that for polydisperse suspensions the polydispersity index in r will again be weighted in a way that could be difficult to interpret for broad distributions.

V. HIGHER ORDER MOMENTS

Formula [11] gives the first two terms of an expansion of the correlation function in a series of cumulants first derived by Koppel (28):

$$\ln|g^{(1)}(\tau)| = \sum_{m=1}^{\infty}\frac{(-\tau)^m}{m!}K_m \qquad [13]$$

where K_m is the mth cumulant (semiinvariant) of $P(\Gamma)$. In principle, therefore, the higher order cumulants can be found by fitting $\ln|g^{(1)}(\tau)|$ to a polynomial function. The first three cumulants of a distribution are identical to its first three central moments but the fourth and higher cumulants are certain combinations of higher order moments. Since, in practice, these cannot be found accurately from experimental data, for reasons that we shall explain in the next section, we will not discuss this method further here.

VI. POLYDISPERSITY ANALYSIS

The reason for the lack of success of the cumulant expansion to investigate properties of the distribution beyond its variance are very subtle and have only been understood more recently in terms of a theory of information (29) that generalizes that of Shannon (30). In simple terms, it turns out that there is a unique expansion for the distribution in which each successive order term is protected as best possible against the inevitable noise present. This is called a ''singular function'' expansion. (Mathematically speaking, any expansion or inversion method would do if there were no noise.) The cumulant expansion is thus not the best way to recover higher order polydispersity information and one can say the same of any other ad hoc recipes for this inversion problem. The singular function expansion is determined by analyzing Eq. [6], which is a Fredholm equation of the first kind with compact kernel, and which suffers, as do all such equations, from a problem known as ill posedness or, in its digitized form, ill conditioning. This can be understood by considering the exponential function in the integrand, the kernel, as a smoothing function for the distribution $P(\Gamma)$. If $P(\Gamma)$ has any fine oscillatory structure it will not contribute strongly to the integral since multiplication by the smoother exponential function and integration will iron it out. The mathematical expression of this phenomenon is the Riemann-Lebesgue theorem:

$$\mathrm{Lim}_{n\to\infty} \int_0^\infty \sin(n\Gamma) K(\Gamma, \tau) d\Gamma = 0 \qquad [14]$$

where $K(\Gamma, \tau)$ is any integrable function of Γ. Each Fredholm equation of the same type as Eq. [6] has its own unique ''best'' expansion series, determined by its particular kernel function. These series are analogous to Fourier series expansions in that successive terms have finer and finer structure and their coefficients are easily found by projection since each function is orthogonal to the others, i.e., the functions in which the solution is to be expanded form a complete orthonormal set. However, Fourier analysis gives only one of many such decompositions and is not particularly ''tuned'' to any given problem in general. It is a sort of Jack of all trades but master of none (or at best only one, i.e., that of temporal or spatial translationally invariant problems). The other generalization from Fourier analysis is that in general two such ''basis'' sets of functions are required, one, $\upsilon_n(\tau)$, in which to expand the data, $g^{(1)}(\tau)$, and one, $u_n(\Gamma)$, in which to expand the solution, $P(\Gamma)$; $n = 0, 1, 2, 3, \ldots$. The integral equation transforms each orthogonal component of $P(\Gamma)$ into the corresponding component of $g^{(1)}(\tau)$.

$$\alpha_n \upsilon_n(\tau) = \int u_n(\Gamma) \exp(-\Gamma\tau) d\Gamma \tag{15}$$

where α_n is a real constant called the nth singular value and which determines the strength at which the nth component is "transmitted" by the experiment into the data values. Thus if the diffusion-coefficient distribution were to be given by the function $u_n(\Gamma)$ the correlation function would be $\alpha_n \upsilon_n(\tau)$. The values of α_n decrease to zero as $n \to \infty$. Of course, all of the u_n except u_0 have negative regions, so that a positive-valued combination of u_n is needed to be a real distribution but the theory is the same.

The Fredholm equation of our problem, Eq. [6], is a finite Laplace transform and with a suitable digitization in Γ, which can be fine enough to avoid any need for interpolation, its "singular system" comprising its singular functions, $u_n(\Gamma)$ and $\upsilon_n(\tau)$, and singular values, α_n, can be found using standard numerical routines to be found in any comprehensive mathematical software library. The singular system of the full Laplace transform was first found by McWhirter and Pike in 1978 (31) in closed analytical form. The singular functions, where now u and υ are the same "eigenfunctions," have the form of cos(log) functions and the eigenvalues were found to be given by

$$\alpha_\omega = \sqrt{\frac{\pi}{\cosh(\pi\omega)}} \tag{16}$$

where the integer order index n is replaced by the continuous index ω (as happens when going from Fourier series to Fourier integrals).

The eigenvalues show a rapid exponential decrease in the strength at which higher order functions contribute to the correlation function and the same effect occurs for the singular values, of course, in the finite transform case. Given the singular system of the equation, the mathematical solution to the inverse problem posed by Eq. [6] is given in an analogous way to Fourier analysis by

$$P(\Gamma) = \sum_{n=0}^{\infty} \frac{\int g^{(1)}(\tau)\upsilon_n(\tau) d\tau}{\alpha_n} u_n(\Gamma) \tag{17}$$

Division of the expansion coefficients by smaller and smaller values of α_n must be stopped and the series truncated or tapered off to not include amplified noise in the reconstruction. In practice, with the order of noise achievable in real experiments, it turns out that only three or four independent components can be reliably recovered in the data analysis process.

Using the analogy with the sampling theory of Fourier analysis and the geometrically spaced zeros of the singular functions found by McWhirter and

Pike, Ostrowsky et al. (32) proposed an "exponential sampling" method of data reduction in which the solution is reconstructed at geometrically spaced Γ values within the range chosen. Due to the ill conditioning, only a small number of sample values can be used and an interpolation, equivalent to the $\sin(x)/x$ of Fourier sampling theory, but now stretched out exponentially, is used between the points. Calculation of the first two moments, as described above, is first performed to choose a suitable range of Γ for sampling. The sampling and interpolation theory of the Laplace transform inversion problem can, in fact, be formally mapped onto the Whittaker-Shannon theory of the Fourier transform inversion problem by geometrical dilatation of the variables and this is the reason for the use of geometrical spacings for both the data and the solution in this problem. Just as in the Fourier transform case, there is no further information to be found within the selected bandwidth between such sample points. The use of such geometrical sample spacings reduces both hardware requirements and data reduction times, and has been perhaps one of the most significant developments in PCS particle sizing since the method was first introduced. More recent developments in data processing have included improved methods (33), initiated by Ross et al. (34), for handling the reduction of the measured data $g^{(2)}$, which avoid taking the explicit square root in the Siegert relation, and more effective ways of including the a priori knowledge of the positivity of the particle size distribution (35) using recent advances in mathematical programming theory due to Borwein and Lewis (36).

VII. INTERACTING PARTICLES

There are two areas of interest when the particles in suspension interact. The first is more easily dealt with and concerns the case where, although the suspension remains dilute so that the dynamics of most of the individual particles is controlled by the unperturbed Brownian motion, aggregation by the formation of dimers and higher polymers occurs. This results in one form of the polydispersity dicussed above and the data reduction by singular system or other form of analysis will quantify the amount and degree of aggregation according to Eq. [5] above. All we will say here is that, since the light scattered by a particle increases very quickly with its radius (to the sixth power in the case of forward scattering in the Rayleigh regime), the technique is exquisitely sensitive to aggregation and has been used to study both fast (diffusion-limited) and slow (reaction-limited) aggregation mechanisms. The reverse side of this coin is that poor specimen preparation, which can result in unknown or unwanted aggregates being present, will give poor results.

The second case is that in which the concentration of the particle suspension is sufficiently high for either direct or hydrodynamic interactions to occur between the particles. This usually happens in circumstances where the high concentration gives rise to multiple scattering of the laser beam, which we will discuss in the next section. However, such interactions can be studied effectively by light scattering in two ways. The first is by deliberate refractive index matching or near-matching of the particles and solvent; this method has been particularly exploited by Pusey and his colleagues who have used it to make fundamental studies of colloidal suspensions of high density (37). The second is by using a "two-color" technique (38), which makes use of the fact that fluctuations due to single scattering in each laser beam color can be arranged, by looking at the same "speckle," to be correlated and therefore observed by cross-correlation, while multiple scattering is not and therefore does not contribute to the time dependence of the cross-correlation function. The second technique is much more difficult than the first. A combination of the two methods can also be effective (39).

VIII. TURBID "WHITE" MEDIA: MULTIPLE SCATTERING

Two general regimes of turbid media need to be considered, namely, those of low and high absorption. In the case of dense solutions of particles with low absorption, multiple scattering occurs but considerable progress has been made in recent years. We shall discuss the high-absorption case in the next section.

There is a considerable literature, beginning with the work of Ivanov and Kostko (40) in 1983 and, in the West, by Maret and Wolf (41) in 1987, on multiple scattering. Following Pine et al. (42), this field has been named "diffusing-wave" spectroscopy. In the first experiments the conventional cylindrical geometry of the PCS method was adopted save that the laser beam and detector pinhole were each focused at the near-wall of the cell. The scattering angle θ may now more appropriately be called the angle of observation. Following work on the transport theory of light in such media by Ishimaru (43,44) in which the photons are assumed to follow random walk trajectories, Sorensen et al. (45) arrived, for monodisperse particles in the Rayleigh region, at an equation for the first cumulant for multiple scattering:

$$\Gamma_m = N\Gamma(90°) \qquad [18]$$

where N is a "mean multiplicity" of scattering derived by averaging over all random walks and $\Gamma(90°)$ is the dilute solution value of Γ at a scattering angle of 90°. The Russian group formulated a diffusion approach for the calculation of N. They proposed that the method could thus be used to determine the concentration of scatterers and their size.

Maret and Wolf used a new "slab geometry" for which the diffusion equation has an exact solution in terms of elementary functions and not only the first cumulant but also the shape of the photon correlation could be calculated. They looked at backscattered light and found that for a thick slab:

$$N = \frac{L^2}{2ll^*}, \qquad [19]$$

where L is the slab thickness, l is the photon mean free path, and l^* is the transport mean free path. For large particles l^* may be an order of magnitude greater than l. Pine et al. give the formula for the first-order correlation function:

$$g^{(1)}(\tau) = \frac{L}{\gamma l^*} \frac{\sinh[\gamma(6\tau/\tau_0)^{1/2}]}{\sinh[L/l^*(6\tau/\tau_0)^{1/2}]} \qquad [20]$$

where γ is a constant of the order of unity. In diffusing-wave spectroscopy experiments Γ_m can be several orders of magnitude greater than $\Gamma(90°)$, so that the correlation function falls off much more quickly. Thus, the requirements on signal-to-noise and correlator speed are more severe than for conventional PCS. Reviews by Kostko (46) and Weitz et al. (47) may be consulted.

IX. TURBID "BLACK" MEDIA: ABSORPTION

In this section we will mention the case of turbid media of absorbing particles for completeness, although as yet this may not have great relevance for pharmacology. The best solution here is to confine the scattering to a very small volume so that multiple scattering is ruled out as much as possible by geometry. One successful approach developed for the Institut Française de Pétrole has been to use backscattering at either 45° or 135° from a film whose thickness can be controlled by a micrometer-driven mechanism while monitoring beam transmission through it. When the film is sufficiently thin for a significant level of transmission to occur, then single scattering is the main process and conventional PCS analysis of the scattered light gives good results for particle size. Other techniques include using corners or edges of the sample cell so that the incident and scattered light does not have to travel far within the specimen and the use of fiberoptic probes where the scattering takes place in a small volume near the tip of the probe.

X. APPLICATIONS

Finally we give some examples of applications relevant to pharmacology chosen from university and industrial research. These have not been selected in any particular manner but provide a glimpse of various possibilities that have been explored to date.

We start by describing very early experiments of PCS measurements from solutions of ionic micelles near the critical micelle concentration (48) that date back to 1977. They illustrate typical problems that could arise with liposomes or any emulsions. It is well known that amphiphilic molecules in water may self-aggregate to shield their hydrophobic portions from the solvent. These aggregates are called micelles. They form above a given concentration of amphiphile called the critical micelle concentration (CMC). The transition has been studied both by classical light scattering and by PCS. This paper was one of the first to use the PCS technique for such studies but also used the concentration dependence above the CMC of the scattered intensity, calibrated by the known Rayleigh ratio of the solvent, together with a measurement of the refractive index increment, dn/dc, to calculate molecular weights. Highly purified sodium dodecyl sulfate in aqueous solution at three different NaCl concentrations was used.

One experimental difficulty was the formation of large aggregates just below the CMC due to the presence of dodecanol as an impurity; ultrafine filtration with 25-nm Millipore filters was used to remove them. Another, which is common to all light scattering measurements on surfactant solutions, was the presence of small bubbles. Checks at different scattering angles were made and careful degassing of the solvent cured this problem. A third was the weak scattering of the micelles by comparison with the solvent near the CMC. An argon-ion laser was needed for this work to provide sufficient scattered light intensity.

At 0.10 M NaCl at 25°C the CMC was found to be 0.5 mg/cm^3, and the molecular weight was found to be 26,000, corresponding to an aggregation number of 90. The electronic charge was deduced to be $14e^-$ per micelle from both the dependence of the intensity and of the diffusion coefficient on concentration. Good single-exponential photon correlation functions were obtained. Results showed a linear dependence of the diffusion coefficient on concentration above the CMC which varied with salt concentration; analysis showed that this was due to the charge interactions and that the size of the micelles remained constant. At the CMC a value of 0.98×10^{-6} cm^2/s was found for the translational diffusion coefficient, which from the Stokes-Einstein relation, Eq. [12] above, gave a hydrodynamic radius of about 2.4 nm for the micelle.

We next describe a study of small unilamellar liposomes containing fluorinated steroids (49). The aim of this work was to investigate the value of the PCS method, together with ^{19}F NMR spectroscopy and high-performance liquid chromatography for physicochemical characterization for various purposes, namely, the management of liposomal formulation, for pharmaceutical control on the industrial production line and for control preceding injection at the clinical site.

Flumethasone or dextramethasone was encapsulated in small unilamellar liposomes of dipalmitoylphosphatidylcholine. Ultracentrifugation eliminated unwanted particulates and multilamellar vesicles. Dialysis against physiological saline and suspension in a 0.9% NaCl sterile solution at 20°C completed the preparation procedures. Correlograms were taken at a number of angles between 10° and 90° to facilitate the study of a minor population of large liposomes that scatter relatively more strongly at low angles. The calculated mean value of the small vesicles came out to be the same at all angles in an analysis of the particle size distribution which "binned" populations of increasing size with a dilatation factor of approximately 2. This is a reasonable resolution under typical PCS conditions. The proportions of the two populations showed a variation with scattering angle, which is to be expected when the large fraction is of a size to be near or out of the Rayleigh scattering regime, since the scattering strength is a function of angle, but unexplained variations in the size of the minor population also occurred. A scattering angle of 90° was selected for routine use and the mean particle size, calculated from the first cumulant, was between 55 and 60 nm in all batches. The major population showed mean diameters of 30 $\pm$ 12 nm depending on formulation and preparation procedures, while a minor population, $\sim$3%, contained structures of the order of 70 nm. The much greater scattering of the larger particles distorts the apparent mean and needs to be corrected out. After 30 h of storage the mean diameter showed a small increase. The high sensitivity of the method to aggregation allowed the authors to estimate that 2–7% of the small liposomes were converted to larger structures. Repeated PCS analysis on seven different batches of liposomes convinced the authors of the reproducibility of the data obtained. They concluded that the technique was of interest for the rapid nondestructive screening of vesicle size and morphology, for the determination of physicochemical stability, and for determination of the reproducibility of preparation in the context of the development of liposomal pharmaceutical preparations for clinical use.

Phospholipid-stabilized intravenous emulsions have been widely used for parenteral nutrition and have also been introduced as drug carrier systems, especially for lipophilic compounds. The aim of the authors in the next papers we review here (50,51) was to consider in detail various methods, e.g., PCS, nuclear magnetic resonance (NMR), transmission electron microscopy (TEM), and small-angle x-ray diffraction studies (SAXS), to determine parameters related to the internal structure of the particles in a model intravenous emulsion stabilized by phospholipids. An emulsion with an extremely high fat load and a classical emulsifier was chosen. PCS measurements were used to derive a particle size distribution and this was then used to calculate the total oil droplet surface area. The result indicated that there should be an excess of surfactants of 150%. Such an excess was not confirmed by either NMR or SAXS measurements and the discrepancy was blamed on the fact that PCS underestimated the total surface area since it did not detect particles smaller than 150 nm; these were present in large numbers as could be shown by TEM micrographs. These small particles were also seen by the PCS system when fractionated off by centrifugation. The conclusions of the study were that the model intravenous emulsion had a complex inner structure. It consisted of particles with different structures, namely, oil droplets covered by an emulsifier monolayer, oil droplets covered by emulsifier oligolayers, double-emulsion droplets, and possibly small unilamellar vesicles. In addition, the emulsion contained measurable amounts of water-soluble degradation products of the emulsifier.

The PCS equipment used, which is typical of semiindustrial installations in that it may possibly compromise on dynamic range and ''hide'' the difficulties of software data reduction (but gives an answer anyway), was clearly not able to cope with such a mélange. The particle distribution given in fact displayed fractions far in excess of a resolution that could be feasibly achieved, even in ideal circumstances, and thus would give a false sense of accuracy for the above surface area calculation. These papers are thus instructive in that they reinforce the fact that, given good technique, although mean particle size and, to some extent, polydispersity are well-conditioned parameters, the determination of more information about the shape of the particle size distribution is *not* a ''point-and-shoot'' operation, for the reasons that we have discussed in great detail in Sec. VI above. In fact, as explained there, PCS is not a suitable technique when fine details of such distributions are required and it may indeed be impossible to tackle such problems as are attempted here, where the total amount of scattering from the small-particle fraction may be essentially invisible when compared to that from the larger particles, even if research grade equipment with more adaptive methods of analysis were to be used.

The next paper we discuss (52) is concerned with water-in-oil emulsions that are diluted with water until phase reversal to an oil-in-water or water-in-oil-in-water emulsion occurs. The effects of temperature, pH, ionic strength, storage, and addition of a peptide were investigated. A commercial PCS system was used with an argon ion laser and no difficulties in measuring mean diameter or polydispersity were reported. It was found that the water-in-oil microemulsions, freshly prepared or stored, and with or without peptide, when diluted to induce phase inversion produced fine microemulsions with mean diameters from 5 to 200 nm and polydispersities between 0.1 and 0.4, respectively. Only under extreme conditions of storage (temperatures above 60°C or with cloud point microemulsions) were large increases in particle size and polydispersity detected.

The penultimate paper in this selection is an academic-industrial collaboration (53) on phase studies and particle size analysis of oil-in-water phospholipid microemulsions. Commercial samples of egg and soya lecithin with ethanol, iso-propyl myristate (IPM), and water at different ethanol/water ratios were investigated at 37°C. Stable oil-in-water microemulsion regions were identified. Both PCS at 90° and classical light scattering were used to determine droplet size. An argon ion laser was again required. Data reduction was by a single exponential fit. It was found that the droplets had radii between approximately 10 and 60 nm, which increased with increasing volume fraction of IPM. A series of experiments were performed on oil-in-water emulsions in which the lecithin content and the percentage of ethanol in the aqueous phase was kept constant and the IPM content was varied across the emulsion region. The scattering data from both techniques had to be corrected for nonideality arising from interparticle interaction in these concentrated systems. For the PCS data this was done by expressing the variation of the apparent diffusion coefficient D_{app} with ϕ, the IPM volume fraction, as

$$D_{\text{app}} = D(1 + k_{\text{d}}\phi) \qquad [21]$$

where the calculation of k_{d} involves consideration of both the interactions due to interparticle forces and the concentration dependence of the frictional forces arising from the motion of the droplets. D is then used in the Stokes-Einstein relation to find the hydrodynamic radius. The agreement between the two techniques was, unfortunately, rather poor with PCS giving somewhat smaller particle sizes, probably due to the approximate nature of the corrections for the interactions.

We end these selected reviews of the literature with another academic-industrial collaboration (54). This concerns ''biovectors.'' A biovector is an association between a charged maltodextrin gel fragment and a phospholipid, and can be used as a drug delivery system or in vaccine formulations as an alternative to phospholipid vesicles. The aim of the work was to characterize the shape and size of the maltodextrin fragments both with and without the associated phospholipid. A standard multiangle PCS equipment using a 35-mW He-Ne laser was used with a temperature-controlled bath at 21°C. The material used for sample preparation was carefully cleaned and the sample was filtered through 0.22-μm Millipore disposable units directly into the scattering cell. Static light scattering using the Zimm plot technique and TEM were also used. Both Koppel's cumulant fit and an iterative nonlinear data reduction scheme for inversion were tried and gave similar results. Moderately disperse unimodal distributions were found for both positively and negatively charged polysaccharide gel fragments. The hydrodynamic radii varied between 48.6 nm for a positively charged polyelectrolyte in a PBS buffer to 100.3 nm for a negatively charged fragment in water. Further work is required on these systems to elucidate the molecular organization and the location of the phospholipids. Implantation of antigenic molecules on the surface might lead to new vaccine formulations.

XI. CONCLUSIONS

We have reviewed a number of ways of using photon correlation spectroscopy, under various conditions of absorption and turbidity, of dispersed macromolecular particles in liquid suspension or solution. One may gain valuable information including mean size, distribution of size, concentration, molecular weight and polydispersity, shape information, hydrodynamic interactions, and fast and slow aggregation mechanisms. We have emphasized that many of these methods are widely and routinely used across disciplines in a number of industries but that the knowledge base is still growing at a formidable rate. We have discussed selected papers with relevance to pharmacology from the early days of the field through to the present.

By application of these methods and by innovative research into its own particular problems, the field of pharmacology is in a position at the present time both to benefit from and to contribute to exciting and important developments in photon correlation spectroscopy.

REFERENCES

1. Singh, M. and Ravin, L. J., 1986, Parenter. Sci. Technol., *40*, 34.
2. Exner, K., 1877, *Sitzb. kais. Akad. Wien. Mathem. Naturw.-Klasse Abt II, 76*, 522–552.
3. Ramaseshan, S. (ed), 1988, *Scientific Papers of C. V. Raman III—Optics*, Indian Academy of Sciences, Bangalore, p. xi.
4. Ramachandran, G. N., 1943, *Proc. Ind. Acad. Sci., A18*, 190.
5. Raman, C. V. 1959, *Lectures in Physical Optics, Part I*, Indian Academy of Sciences, Bangalore, p. 160.
6. Gorelik, G., 1947, *Dokl. Acad. Nauk, 58*, 45.
7. Forrester, A. T., Parkins, W. E. and Gerjuoy, E., 1947, *Phys. Rev., 72*, 728.
8. Forrester, A. T., Gudmunsen, R. A. and Johnson, P. O., 1955, *Phys. Rev., 99*, 1691.
9. Hanbury Brown, R. and Twiss, R. Q., 1956, *Nature, 177*, 27; *178*, 1046, 1447.
10. Cummins, H. Z., Knable, N. and Yeh, Y., 1963, *Phys. Rev. Lett., 12*, 150.
11. White, A. D. and Rigden, J. D., 1962, *Proc. IRE, 50*, 1697.
12. Ford, N. C. and Benedek, G. B., 1965, *Phy. Rev. Lett., 15*, 649.
13. Gammon, R. W., Shaumeyer, J. N., Briggs, M. E., Boukari, H. and Gent, D. A., 1997, in *Light Scattering and Photon Correlation Spectroscopy*, NATO ASI Series, High Technology, Vol 40, E. R. Pike and J. B. Abbiss (eds), Kluwer, Dordrecht, p. 389.
14. Dubin, S. B., Luncek, J. H. and Benedek, G. B., 1967, *Proc. Natl. Acad. Sci. USA, 57*, 1164.
15. Pecora, R., 1964, *J. Chem. Phys., 40*, 1604.

16. Jakeman, E., Oliver, C. J. and Pike, E. R., 1975, *Adv. Phys. 24*, 349.
17. Foord, R., Jakeman, E., Oliver, C. J., Pike, E. R., Blagrove, R. J., Wood, E. and Peacock, A. R., 1970, *Nature, 227*, 242.
18. Pike, E. R., 1972, *J. Phys. D: Appl. Phys., 5*, L23.
19. Seigert, A. J. F., 1943, MIT Radiation Laboratory Report, No. 465.
20. Jakeman, E. and Pike, E. R., 1968, *J. Phys. A: Mathematical and General, 1*, 627.
21. van Vleck, J. H. and Middleton, D., 1966, *Proc. IEEE, 54*, 2.
22. Pike, E. R., Gran, M. and Pailharey, E., 1997, in *Light Scattering and Photon Correlation Spectroscopy*, NATO ASI Series, High Technology, Vol 40, E. R. Pike and J. B. Abbiss (eds), Kluwer, Dordrecht, p. 117.
23. Brown, R. G. W., Ridley, K. D. and Rarity, J. G., 1986, *Appl. Optics, 25*, 4122.
24. Brown, R. G. W., Jones, R., Rarity, J. G. and Ridley, K. D., 1987, *Appl. Optics, 26*, 2383.
25. Smart, A. E., 1997, in *Light Scattering and Photon Correlation Spectroscopy*, NATO ASI Series, High Technology, Vol 40, E. R. Pike and J. B. Abbiss (eds), Kluwer, Dordrecht, p. 373.
26. Cummins, H. Z. and Pusey, P., 1977, in *Photon Correlation Spectroscopy and Velocimetry*, H. Z. Cummins and E. R. Pike (eds), NATO ASI Series B23, Plenum Press, New York, p. 164.
27. Pusey, P. N., 1973, in *Industrial Polymers: Characterisation by Molecular Weight*, J. H. S. Green and R. Dietz (eds), Transcripta Books, London, p. 26.
28. Koppel, D. E., 1972, *J. Chem. Phys., 57*, 4814.
29. Pike, E. R., McWhirter, J. G., Bertero, M. and De Mol, C., 1984, *Proc. IEE, 131*, 660.
30. Shannon, C. F. and Weaver, W., 1949, *The Mathematical Theory of Communication*, University of Illinois Press, Chicago.
31. McWhirter, J. G. and Pike, E. R., 1978, *J. Phys. A, 11*, 1729.
32. Ostrowsky, N., Sornette, D., Parker, P. and Pike, E. R., 1981, *Opt. Acta, 28*, 1059.
33. McNally, B., Patin, P. and Pike, E. R., 1997, *Appl. Optics*, 1997, *36*, 7531.
34. Ross, D. A., Dhadwall, H. S. and Suh, K., April 4–8, 1994, *SPIE Proceedings, Orlando*, Orlando, Florida.
35. Pike, E. R., and McNally, B. M., 1997, in *Scientific Computing*, F. T. Luk and R. J. Plemmons (eds), Springer, Berlin, p. 101.
36. Borwein, J. M. and Lewis, A. S., 1992, *Mathematical Programming, 57*, 49.
37. de Schepper, I. M., Cohen, E. G. D., Pusey, P. N. and Lekkerkerker, H. N. W., 1990, *Physica A, 164*, 12.
38. Segré, P. N., van Megen, W., Pusey, P. N., Schätzel, K. and Peters, W., 115, *J. Mod. Opt., 42*, 1929.
39. Pusey, P. N., Segré, P. N., Behrend, O. P., Meeker, S. P. and Poon, W. K. C., 1997, in *Light Scattering and Photon Correlation Spectroscopy*, NATO ASI Series, High Technology, Vol 40, E. R. Pike and J. B. Abbiss (eds), Kluwer, Dordrecht, p. 1.
40. Ivanov, D. Yu. and Kostko, A. F., 1983, *Opt. Spektrosk., 55*, 950 [English translation: *Opt. Spectrosc., 55*, 573].
41. Maret, G. and Wolf, P. E., 1987, *Z. Phys. B, 65*, 409.
42. Pine, D. J., Weitz, D. A., Chaikin, P. M. and Herbolzheimer, E., 1988, *Phys. Rev. Lett., 60*, 1134.

43. Ishimaru, A., 1975, *Radio Sci., 10*, 45.
44. Ishimaru, A., 1978, *Wave Propagation and Scattering in Random Media*, Academic Press, New York.
45. Sorenson, C. M., Mockler, R. C. and O'Sullivan, W. J., 1977, *Phys. Rev., A16*, 365.
46. Kostko, A. F., 1997, in *Light Scattering and Photon Correlation Spectroscopy*, NATO ASI Series, High Technology, Vol 40, E. R. Pike and J. B. Abbiss (eds), Kluwer, Dordrecht, p. 325.
47. Weitz, D. A., Zhu, J. X., Durian, D. J., Gang, H. and Pine, D. J., 1993, *Physica Scripta, T49*, 610.
48. Corti, M. and Degiorgio, V., 1977, in *Photon Correlation Spectroscopy and Velocimetry*, H. Z. Cummins and E. R. Pike (eds), NATO ASI Series *B23*, Plenum Press, New York, p. 450.
49. Devoisselle, J-M., Vion-Dury, J., Confort-Gouny, S. Coustaut, D. and Cozzone, P. J., 1992, *J. Pharm. Sci., 81*, 249.
50. Westerson, K., and Wehler, T., 1992, *J. Pharm. Sci., 81*, 777.
51. Westerson, K., and Wehler, T., 1993, *Colloids and Surfaces A: Physico-chemical and Engineering Aspects, 78*, 125.
52. Constantinides, P. P. and Yiv, S. H., 1995, *Int. J. Pharm., 115*, 225.
53. Saint Ruth, H., Attwood, D., Ktistis, G. and Taylor, C. J., 1995, *Int. J. Pharm., 116*, 253.
54. Castanho, M. A. R. B., Prieto, M. J. E., Betbeder, D. and Santos, N. C., 1997, in *Light Scattering and Photon Correlation Spectroscopy*, NATO ASI Series, High Technology, Vol 40, E. R. Pike and J. B. Abbiss (eds), Kluwer, Dordrecht, p. 173.

Index